WAVELETS
AND THEIR
APPLICATIONS

Baum, R. J., *Philosophy and Mathematics*
ISBN 0-86720-514-2

Eisenbud, D., and Huneke, C., *Free Resolutions in Commutative Algebra and Algebraic Geometry*
ISBN 0-86720-285-8

Epstein, D.B.A., *Word Processing in Groups*
ISBN 0-86720-241-6

Epstein, D.B.A., and Gunn, C., *Not Knot Supplement*
ISBN 0-86720-297-1

Geometry Center, University of Minnesota, *Not Knot*
ISBN 0-86720-240-8

Gleason, A., *Fundamentals of Abstract Analysis*
ISBN 0-86720-238-6

Loomis, L.H., and Sternberg, S., *Advanced Calculus*
ISBN 0-86720-122-2

Protter, M.H., and Protter, P.E., *Calculus, Fourth Edition*
ISBN 0-86720-093-6

Redheffer, R., *Differential Equations: Theory and Applications*
ISBN 0-86720-200-9

Ruskai, M.B. *et al.*, *Wavelets and Their Applications*
ISBN 0-86720-225-4

Serre, J.-P., *Topics in Galois Theory*
ISBN 0-86720-210-6

WAVELETS AND THEIR APPLICATIONS

Edited by

Mary Beth Ruskai
Department of Mathematics
University of Massachusetts • Lowell
Lowell, Massachusetts

Gregory Beylkin
Program of Applied Mathematics
University of Colorado at Boulder
Boulder, Colorado

Ronald Coifman
Department of Mathematics
Yale University
New Haven, Connecticut

Ingrid Daubechies
AT&T Bell Laboratories
Murray Hill, New Jersey

Stephane Mallat
Courant Institute of
Mathematical Sciences
New York University
New York, New York

Yves Meyer
CEREMADE
University of Paris–Dauphine
Paris, France

Louise Raphael
Department of Mathematics
Howard University
Washington, D.C.

JONES AND BARTLETT PUBLISHERS
Boston   London

Editorial, Sales, and Customer Service Offices
Jones and Bartlett Publishers
20 Park Plaza
Boston, MA 02116

Library of Congress Cataloging-in-Publication Data

Wavelets and their applications / edited by Mary Beth Ruskai. . . [et al.].

    p. cm.

    Includes bibliographical references and index.

    ISBN 0-86720-225-4

    1. Integrals, Singular. 2. Maximal functions.

QA403.5.W38 1992

515'.2433—dc20   91-42742

                                            CIP

ISBN 0-86720-225-4

Typesetting: *Superscript Typography*
Printing and Binding: *Hamilton Printing Company*

Printed in the United States of America
96 95 94 93 92   10 9 8 7 6 5 4 3 2 1

CONTENTS

PREFACE

This book had its genesis in the NSF/CBMS conference on Wavelets held
at the University of Lowell in June, 1990. In accordance with the usual
CBMS (Conference Board of the Mathematical Sciences) procedures for
this conference series, Ingrid Daubechies' main series of ten lectures are
being published by SIAM. Although there was sufficient interest in wavelets to merit publishing many of the other talks as well, formal conference
proceedings did not seem quite appropriate. A few months before the conference, I was approached by Klaus Peters about the possibility of editing a
book of essays on wavelets, using the invited lectures from the NSF/CBMS
conference as a core, but including other contributors as well. Gregory
Beylkin, who organized a special session on Wavelets at the July, 1990
meeting of SIAM, faced a similar dilemma; however, he graciously agreed
to also forego conference proceedings and join our project.

Nine of the ten invited lecturers at Lowell agreed to contribute to this
book; in addition, we asked several speakers from the SIAM special session as well as several "at-large" experts to contribute. Although we had
very much hoped to include an essay on music and audio signals by Alex
Grossmann and Richard Kronland-Martinet, unforeseen delays have forced
postponement of their chapter to another book. Because of the extremely
rapid development of wavelet theory, the topics in these chapters by no

means exhaust the applications of wavelets, but merely give some indication of the state of this art.

Special thanks are due to Gerald Kaiser for his help in organizing the NSF/CBMS conference at Lowell and for reviewing several manuscripts, and to Wayne Lawton for reviewing and proofreading most of the manuscripts during the final hectic month.

Mary Beth Ruskai
Lowell, June, 1991

CONTRIBUTORS

Bradley Alpert, NIST, MC 881, 325 Broadway, Boulder, Colorado 80303

A. Arneodo, Centre de Recherche Paul Pascal, Avenue Schweitzer, 33600 Pessac, France

F. Argoul, Centre de Recherche Paul Pascal, Avenue Schweitzer, 33600 Pessac, France

P. Auscher, Université de Rennes I, Départment de Mathématiques et Informatique, Campus de Beaulieu, 35042 Rennes Cedex, France

Guy Battle, Mathematics Department, Texas A&M University, College Station, Texas 77843

G. Beylkin, Program of Applied Mathematics, University of Colorado at Boulder, Box 526, Boulder, Colorado 80309

Charles K. Chui, Center for Approximation Theory, Texas A&M University, College Station, Texas 77843

Albert Cohen, CEREMADE, University of Paris–Dauphine, 75775 Paris, Cedex 16, France

Ronald R. Coifman, Department of Mathematics, P.O. Box 2155, Yale Station, Yale University, New Haven, Connecticut 06520

Ingrid Daubechies, Room 2C-371, AT&T Bell Labs, Murray Hill, New Jersey 07974

Marie Farge, Laboratoire de Météorologie Dynamique du CNRS, Ecole Normale Supérieure, 24 rue Lhomond, F-75231 Paris, France

Jean-Christophe Feauveau, Matra MS2i, Laboratoire de Traitement de l'Image et du Signal, 38 Blvd Paul Cézanne, 78052 Saint Quentin, Enyvelines, France

Hans Feichtinger, Institut für Mathematik, Universität Wien, Strudlhofg 4, A-1090, Wien, Austria

Michael Frazier, Department of Mathematics, Michigan State University, East Lansing, Michigan 48824

E. Freysz, Centre de Physique Moléculaire Optique et Hertzienne, Université de Bordeaux I, 351 Avenue de la Libération, 33405 Talence Cedex, France

K. Gröchenig, Department of Mathematics U9, University of Connecticut, Storrs, Connecticut 06269

B. Jawerth, Department of Mathematics, University of South Carolina, Columbia, South Carolina 29208

J. Liandrat, IMST, 12 Avenue Général Leclerc, F-13003 Marseille, France

Stephane Mallat, Courant Institute of Mathematical Sciences, NYU, 251 Mercer St., New York, New York 10012

J. F. Muzy, Centre de Recherche Paul Pascal, Avenue Schweitzer, 33600 Pessac, France

Yves Meyer, CEREMADE, Université de Paris–Dauphine, 75775 Paris, Cedex 16, France

T. Paul, CEREMADE, Université de Paris–Dauphine, Place de Lattre de Tassigny, 75775 Paris, Cedex 16, France

V. Perrier, Laboratoire d'Analyse Numérique, Centre d'Orsay, Bat 425, 91405 Orsay, France

B. Pouligny, Centre de Recherche Paul Pascal, Avenue Schweitzer, 33600 Pessac, France

Louise Raphael, Department of Mathematics, Howard University, Washington D.C. 200059

V. Rohklin, Department of Computer Science, P.O. Box 2158, Yale Station, Yale University, New Haven, Connecticut 06520

Mary Beth Ruskai, Department of Mathematics, University of Massachusetts • Lowell, Lowell, Massachusetts 01854

Kristian Seip, Division of Mathematical Sciences, University of Trondheim, N-7034 Trondheim, NTH, Norway

Ph. Tchamitchian, Centre de Physique Théorique, CNRS-Luminy, Case 907, F-13288 Marseille, Cedex 09, France

Bruno Torrésani, Centre de Physique Théorique, CNRS-Luminy, Case 907, F-13288 Marseille, Cedex 09, France

Martin Vetterli, Department of Electrical Engineering, Room 1342, SW Mudd Bldg., Columbia University, New York, New York 10027

Victor Wickerhauser, Department of Mathematics, Washington University, 1 Brookings Drive, St. Louis, Missouri 63130

Sifen Zhong, University of California, Department of Mathematics, 405 Hilgard Avenue, Los Angeles, California 90024

I
INTRODUCTION

INTRODUCTION

MARY BETH RUSKAI

Department of Mathematics
University of Lowell
Lowell, Massachusetts

The term *wavelets* refers to sets of functions of the form $\psi_{ab}(x) = |a|^{-1/2}\psi\left(\frac{x-b}{a}\right)$, i.e., sets of functions formed by dilations and translations of a single function $\psi(x)$ sometimes called, variously, the "mother wavelet", "basic wavelet" or "analyzing wavelet". A few authors even reserve the term "wavelet" for analyzing wavelets with specified properties. The dilation and translation parameters a, b may range over either a continuous or a discrete set $\mathcal{S}_{DT}$. In general, the term wavelets is reserved for situations in which the set $\mathcal{S}_{DT}$ corresponds to some subgroup of an affine group, and for which the corresponding set of functions $\{\psi_{ab}\}$ has sufficient members to allow any function f in L^2 to be reconstructed from its wavelet coefficients $\langle f, \psi_{ab}\rangle$ where $\langle \, , \, \rangle$ denotes the standard L^2 inner product. For simplicity, only the one-dimensional situation, in which $f \in L^2(\mathbf{R})$ and $\mathcal{S}_{DT}$ is a suitable subset of $\mathbf{R}^+ \times \mathbf{R}$, will be discussed here. (Some aspects of two-dimensional wavelets are considered in Mallat's chapter and of multi-dimensional wavelets in Vetterli's chapter in this volume, as well as in Daubechies [D1]. A very general discussion of multi-dimensional wavelets, including the possibility that the dilation parameter a is replaced by a matrix, is given in the monograph by Meyer [Me1]. For more recent developments involving the construction of multi-dimensional orthonormal wavelet

3

bases, see Lawton and Resnikoff [LR], Gröchenig and Madych [GM], Kovacevic and Vetterli [KV], and Cohen and Daubechies [CD2].)

Because a very thorough and readable discussion of wavelets is now available in Daubechies' CBMS lecture notes [D1] (hereafter referred to as D-CBMS), to which this volume should be considered as a companion, this introduction will summarize only the most basic definitions. In addition, the reader can consult Meyer's book [Me1] and Daubechies' earlier reviews and articles, especially [D2], [D3], and [D4]. Additional references on special topics will be mentioned later. As is well-known (and thoroughly discussed in D-CBMS and elsewhere, e.g. [D1-3]), wavelets can be chosen to have extremely desirable time and frequency localization properties, particularly when compared to other sets of functions, such as the Gaussian coherent states or windowed Fourier transform. These properties make wavelets a very useful and promising tool for analyzing such mathematical entities as signals, data, singular operators, etc. which arise in many different applications. The primary purpose of this volume is to present reviews of some of these applications, as well as some recent theoretical developments which were not treated in D-CBMS. More detailed bibliographies on these topics are given with the chapters; for additional recent developments, one can also consult the forthcoming special issue on "Wavelet Transforms and Multiresolution Signal Analysis" of the *IEEE Transactions on Information Theory* [DMW].

If $\mathcal{S}_{DT} = \mathbf{R}^+ \times \mathbf{R}$ so that (a,b) varies over all of $\mathbf{R}^+ \times \mathbf{R}$, the map $f \to \langle f, \psi_{ab} \rangle$ is called the continuous wavelet transform. The function f can then be recovered using the identity

$$f = \frac{1}{C_\psi} \int_{-\infty}^{\infty} \int_{-\infty}^{\infty} \frac{da\,db}{a^2} \langle f, \psi_{ab} \rangle \, \psi_{ab} \tag{1}$$

where $C_\psi = \int_{-\infty}^{\infty} \frac{1}{|\xi|} |\hat{\psi}(\xi)|^2 \, d\xi$ and $\hat{\psi}$ denotes the Fourier transform of ψ. The identity (1) is very reminiscent of the "resolution of the identity" formula for other sets of functions, such as the Gaussian coherent states. The condition $C_\psi < \infty$ implies $\hat{\psi}(0) = \int_{-\infty}^{\infty} \psi(x)\,dx = 0$. Applications of the continuous wavelet transform are considered in the chapters in this volume by Cohen, by Farge, by Mallat and Zhong, and by Tchamitchian and Torresani.

Although there are many ways of restricting (a,b) to a discrete subset of $\mathbf{R}^+ \times \mathbf{R}$, the most common choice is $a = 2^{-k}$, $b = an$ where $k, n \in \mathbf{Z} \times \mathbf{Z}$, i.e., dilation by (both positive and negative) powers of 2 and translation by integers. The corresponding discrete wavelets $\psi_{kn}(x) = 2^{-k/2}\psi(2^{-k}x - n)$ can then be parameterized by a pair of integers k, n rather than a, b. In order to insure that one can retrieve f from a set of discrete wavelet coefficients $\langle f, \psi_{kn} \rangle$ it is sufficient that the set of functions $\{\psi_{kn}\}$ constitute a frame, i.e., for some $A, B > 0$ they satisfy the condition

$$A\|f\|^2 \leq |\langle f, \psi_{kn} \rangle|^2 \leq B\|f\|^2 \quad \forall f \in L^2(\mathbf{R}) \ . \tag{2}$$

If one defines the operator $S = \sum_{k,n} P_{kn}$ where P_{kn} denotes the (one-dimensional) projection onto the subspace spanned by the single function ψ_{kn} (and if $\|\psi\| = 1$), then (2) is equivalent to

$$AI \leq S \leq BI \tag{3}$$

and the requirement $A > 0$ implies that S is invertible. It is then straight-forward to verify that

$$f = \sum_{k,n} \langle f, \psi_{kn} \rangle \, \tilde{\psi}_{kn} \tag{4}$$

where $\tilde{\psi}_{kn} = S^{-1}\psi_{kn}$. Although (4) is reminiscent of a biorthogonal expansion, the sets of functions $\{\psi_{kn}\}$ and $\{\tilde{\psi}_{kn}\}$ are not biorthogonal in general. On the contrary, the set $\{\psi_{kn}\}$ may be linearly dependent; in that case, (4) is simply one of many possible ways of writing f as a linear combination of $\tilde{\psi}_{kn}$. However, as discussed elsewhere [D1-3], expansions of the form (4) have certain advantages. Moreover, S^{-1} can easily be computed as a power series in $I - \frac{2S}{A+B}$ whose rate of convergence is at least as fast as $\frac{B-A}{B+A}$ so that when $A \approx B$ the set of dual functions $\{\tilde{\psi}_{kn}\}$ can efficiently be computed from $\{\psi_{kn}\}$; for further details see [D1-3].

Although attention has been restricted here to frames composed of wave-lets, other sets of functions can also form frames. One advantage to wavelet frames is that, unlike frames of windowed Fourier transforms which must satisfy the Balian-Low theorem (proved and extended to such frames by Coifman and Semmes, using the Zak transform, as described in [D3] and D-CBMS), it is possible to construct wavelet frames for which both $x\psi(x)$ and $\xi\hat{\psi}(\xi)$ are in $L^2(\mathbf{R})$. In the chapter by Feichtinger and Gröchenig, a number of different classes of functions which form frames are discussed together with their connection with group representations. Some of these topics were also reviewed by Heil and Walnut [HW].

The following special classes of wavelet frames are of particular interest.

a) If $A = B = 1$, then $S = S^{-1} = I$ and the set of functions $\{\psi_{kn}\}$ forms an orthonormal basis.

b) If the functions $\{\psi_{kn}\}$ are linearly independent, then the sets $\{\psi_{kn}\}$ and $\{\tilde{\psi}_{kn}\}$ are biorthogonal.

c) If $A = B$, then $S^{-1} = \frac{I}{A}$ and the set of functions $\{\psi_{kn}\}$ is called a tight frame.

d) If the set of functions $\{\tilde{\psi}_{kn}\}$ are also wavelets, i.e. there is a function $\tilde{\psi}$ such that $\tilde{\psi}_{kn}(x) = 2^{-k/2}\tilde{\psi}(2^{-k}x - n)$, then the map $f \rightarrow \langle f, \psi_{kn} \rangle$ is referred to as a ϕ-transform.

The study of ϕ-transforms originated with Frazier and Jawerth [FJ] independent of the development of wavelet frames. The theory of ϕ-transforms and related topics in harmonic analysis is discussed in their chapter in this book. All tight frames, and some biorthogonal wavelets, define ϕ-transforms and every ϕ-transform defines a frame; however, there are examples of frames which are not ϕ-transforms. While a great deal of attention has been given to the discovery that it is, in fact, possible to form orthonormal bases of wavelets — and that such bases can even be compactly supported — the other cases, both redundant and biorthogonal wavelets, remain of interest as is evident from several of the chapters in this volume.

Because the construction of orthonormal bases of compactly supported wavelets is thoroughly discussed in D-CBMS and her original paper [D4] is very readable, this topic is not treated in detail in this monograph* (although related constructions are discussed in the chapters by Battle and by Chui). However, both the construction and many applications of orthonormal wavelets use the important concept of multi-resolution analysis developed by Mallat [Ma] and Meyer [Me2]. Therefore, a brief review of this concept is included here. A multi-resolution analysis consists of a family of subspaces V_k satisfying

a) $\quad \ldots \subset V_2 \subset V_1 \subset V_0 \subset V_{-1} \subset V_{-2} \ldots$ $\qquad$ (5)

b) $\quad \cap_{k \in \mathbf{Z}} V_k = \{0\}, \quad \text{and} \quad \overline{\cup_{k \in \mathbf{Z}} V_k} = L^2(\mathbf{R})$ $\qquad$ (6)

c) $\quad f(x) \in V_k \Leftrightarrow f(2x) \in V_{k-1}$ $\qquad$ (7)

d) $\quad \exists \, \phi \in V_0$ such that $\{\phi_{0n}\}$ is a Riesz basis for V_0 .

A Riesz basis is a basis which is also a frame. Note that (c) and (d) imply that $\{\phi_{kn}\}_{n \in \mathbf{Z}}$ is a Riesz basis for V_k. Moreover, one can choose ϕ so that the set of translates $\{\phi_{0n}\} = \{\phi(x - n)\}$ is actually an orthonormal basis. If the original ϕ does not generate an orthonormal set, then $\tilde{\phi}$ will do so,

---

*Note Added in Proof: For a very readable introduction to the construction of orthogonal wavelets, the reader can also consult a forthcoming article by Strichartz [Sz].

where $\tilde{\phi}$ satisfies

$$\hat{\tilde{\phi}}(\xi) = \frac{\hat{\phi}(\xi)}{\sqrt{\sum_{k \in \mathbf{Z}} |\hat{\phi}(\xi + 2k\pi)|^2}} . \tag{8}$$

Unfortunately, the modified generator $\tilde{\phi}$ may not have all the desirable properties (such as compact support), that the original ϕ does.

Once one has a set of multi-resolution subspaces, one can generate an orthonormal basis of wavelets as follows. Define W_k as $V_k^\perp$ where the orthogonal complement is taken in V_{k-1} so that

$$V_{k-1} = V_k \oplus W_k \quad \text{and} \quad V_k \perp W_k . \tag{9}$$

It then follows that the set of subspaces $\{W_k\}$ are mutually orthogonal and $\oplus_{k \in \mathbf{Z}} W_k = L^2(\mathbf{R})$. It can be shown that one can find a function ψ such that $\{\psi_{0n}\} = \{\psi(x - n)\}$ is an orthonormal basis for W_0; it then follows that $\{\psi_{kn}\}_{n \in \mathbf{Z}}$ is an orthonormal basis for W_k and $\{\psi_{kn}\}$ is an orthonormal basis for $L^2(\mathbf{R})$. In fact, (assuming that integer translates of ϕ generate an orthonormal basis for V_0) ψ can be constructed as follows. By (5) and (7) there exist c_n such that

$$\phi(x) = \sum_{n \in \mathbf{Z}} c_n \phi(2x - n) . \tag{10}$$

Then $\psi(x)$ is given by

$$\psi(x) = \sum_{n \in \mathbf{Z}} (-1)^n c_{n+1} \ \phi(2x + n) . \tag{11}$$

It should be noted that the convention of decreasing subspaces used here, while followed by many authors including Daubechies and Mallat, is not universal. Meyer [Me1] uses exactly the opposite convention, in which case (7) and (9) are replaced by

$$f(x) \in V_k \Leftrightarrow f(2x) \in V_{k+1} \tag{7'}$$

$$V_{k+1} = V_k \oplus W_k \quad \text{and} \quad V_k \perp W_k . \tag{9'}$$

Roughly speaking, in the Daubechies-Mallat convention the functions in V_k scale like 2^k, whereas in the Meyer convention they scale like 2^{-k}. *In this volume, many authors follow the convention of Meyer.*

The idea of analyzing at successively more refined scales, as one does in a multi-resolution analysis, has many similarities with the renormalization group analysis which has had considerable success in mathematical physics. A renormalization group approach to the construction of orthonormal wavelet bases is discussed in the chapter by Battle. In general, a renormalization group analysis is initially defined on a lattice; applications to quantum field theory then require a continuum limit, i.e. the lattice spacing → 0. Proceeding from this viewpoint, Battle describes a construction which generates smooth inter-scale orthogonal wavelets in the continuum limit. Other authors, namely Federbush [F] and Balaban, O'Carroll and Schor [BOS] have described an approach in which the scale decomposition of the renormalization group persists after the continuum limit so that a renormalization group type of analysis can be carried out directly on $\mathbf{R}^d$, rather than on a lattice. Although these authors do not construct wavelets explicitly, inter-scale orthogonal wavelets can be obtained from their analysis together with a suitable averaging procedure. However, the averaging procedures used in [F] and [BOS] do not generate very smooth wavelets; for example, a "block spin" averaging procedure, together with the identity operator in the [BOS] formalism yields the Haar basis. By constrast, Battle uses the continuum limit of a more complex averaging procedure on a lattice to construct smooth wavelets, which can be chosen to have arbitrary regularity.

It should be emphasized that this renormalization group approach leads very naturally to wavelets which are not orthonormal, but only orthogonal at different scales, i.e.

$$\langle \psi_{jm}, \psi_{kn} \rangle = 0 \text{ if } j \neq k \text{ but, } \langle \psi_{km}, \psi_{kn} \rangle \text{ may be non-zero} . \quad (12)$$

Battle [Ba] called such sets of functions "pre-ondelettes" because orthogonal wavelets can be obtained from them by a subsequent orthogonalization procedure (as contrasted with the pre-orthogonalization procedure given by (8) above). However, as before, the orthogonalization process may destroy some properties, and for some purposes it may be preferable to sacrifice intra-scale orthogonality and retain compact support. In particular, Chui and Wang [CW], proceeding from the viewpoint of approximation theory, recently constructed compactly supported wavelets formed from cardinal B-splines which are orthogonal only on different scales; their construction is discussed in the chapter by Chui in this volume. Since such partially orthogonal bases are also frames, by remark (b) above, they automatically generate biorthogonal bases. Biorthogonal bases also arise naturally from subband coding schemes in signal analysis. In fact, orthonormal bases of wavelets correspond to subband coding schemes in which the analysis and synthesis filters coincide, while biorthogonal bases are associated with schemes in which the analysis and synthesis filters differ. The resulting bases typically lack inter-scale as well as intra-scale orthogonality, but they have the advantage that both ψ and its dual $\tilde{\psi}$ can be symmetric as well as compactly supported. (Symmetry is impossible for smooth, real-valued orthonormal wavelets [D4], [D5]. Although the inter-scale orthogonal wavelets in [CW] are symmetric, only ψ, but not $\tilde{\psi}$, is compactly supported.) Such biorthogonal bases have recently been constructed and studied independently by several groups — from the mathematical perspective by Cohen, Daubechies and Feauveau [CDF], from a signal analysis perspective by Vetterli and Herley [VH], and from an approximation theory perspective by DeVore, Jawerth, and Popov [DJP]. The question of when a subband coding scheme defines an orthonormal basis of wavelets was considered slightly earlier by Lawton [La]; a review and extension of Lawton's results is contained in [CDF]. For further discussion of the relation between wavelets and coding schemes, one can also consult the chapter by Vetterli in this volume.

In a different direction, the question of when one can obtain orthonormal bases of wavelets using dilation factors other than 2 is also of interest. Auscher studied the case of rational dilation factors a in the region $1 < a < 2$; his results are summarized in his chapter in this volume. Meyer [Me1] has considered orthonormal bases obtained using integer dilation factors $a > 2$; such bases require two or more sets of wavelets. It should be mentioned that the multi-resolution scheme described above requires some modification if dilation factors other than 2 are used. Kaiser [K] has shown that a complex structure can be associated with the standard $(a = 2)$ multi-resolution analysis; however, the existence of a similar algebraic structure associated with other dilation factors remains an open question. Recently, Cohen and Daubechies [CD1] observed that the splitting of the spaces W_k which occurs naturally in the modified multi-resolution analysis for dilation factor $a = 4$ can also be used with the usual dilation factor $a = 2$ to obtain orthonormal bases with better frequency resolution. Their construction is a special case of the general concept of bases constructed from several families of wavelets, which has been termed "wavelet packets" by Coifman, Meyer and Wickerhauser; the theory of such bases is discussed in their chapter entitled "Size Properties of Wavelet Packets" and some applications are considered in their chapter on "Wavelet Analysis and Signal Processing".

The development of wavelets during the past decade has been closely associated with advances in signal analysis; both the mathematical theory and the applications seem to have benefitted from this synergy. Vetterli has given a concise, yet thorough, introduction to the discrete filter banks used in signal analysis and their relation to discrete wavelets; his chapter in this volume also includes a description of the associated pyramid algorithms for signal analysis and synthesis. A more technical discussion of one class of pyramid algorithms then follows in the chapter by Feauveau. In contrast to these discrete algorithms, Mallat and Zhong's chapter demonstrates the utility of the continuous wavelet transform for edge detection in vision analysis; their approach yields an algorithm for signal compression and reconstruction with impressive results. Additional applications to signal analysis are considered in the chapters by Cohen; by Coifman, Meyer and Wickerhauser; and by Tchamitchian and Torresani. The skeleton extraction techniques developed by Tchamitchian and Torresani were exploited by M. Farge to study turbulence as is discussed in her chapter.

The localization properties of wavelets also make them extremely useful for numerical analysis of systems with singular behavior. In many cases, not only are fewer basis functions required with wavelets than with such traditional bases as Fourier series, but such annoying anomalies as the "Gibbs' phenomenon" are minimized. As wavelet theory has advanced, numerical algorithms have simultaneously been developed which exploit these advantages. In particular, Daubechies' compactly supported orthonormal wavelets can be used to develop a "Fast Wavelet Transform" which appears to be superior to the the fast Fourier transform for many purposes. The Fast Wavelet Transform was first proposed by Mallat in his original paper on multi-resolution analysis [Ma], using truncated versions of infinitely supported wavelets. A numerical algorithm using the compactly supported wavelets of [D4], thereby avoiding the error due to truncation, was subsequenlty implemented by Beylkin, Coifman and Rokhlin [BCR]. Another class of numerical algorithms arise from the fact that many operators, including integral operators of Calderón-Zygmund type and pseudo-differential operators, have wavelet representations involving sparse matrices, leading to a variety of efficient algorithms. (In contrast to traditional orthonormal bases which diagonalize Sturm-Liouville operators, wavelets can not be obtained as eigenfunctions of any reasonable operator.) A survey of these aspects of numerical analysis is given in the review chapter by Beylkin, Coifman and Rokhlin; some additional developments involving sparse matrix representations are then discussed in the chapter by Alpert. In a different direction, two groups have used wavelets to study non-linear differential equations. As discussed in their chapter, Liandrat, Perrier, and Tchamitchian have used periodic spline wavelets to analyze Burgers equation; a group at AWARE [GLRT] has reported success in using Daubechies' orthonormal wavelets in a Galerkin treatment of non-linear equations, including Burgers equation.

While wavelets have obviously made a major impact on signal processing and numerical analysis, wavelets are useful in other areas as well. The two-scale difference equation (10) used in multi-resolution analysis is reminiscent of the self-similarity equations for fractals. The ability of wavelets to "zoom in" and study details at an appropriately fine scale, is another interesting feature, important for many applications. Arneodo and collaborators have exploited this latter aspect of wavelets to perform physical experiments for the optical analysis of fractals; their work is described in a chapter in this volume. In a different direction, M. Farge has advocated searching for the hidden coherent structure in turbulence; in her chapter, she summarizes this approach to turbulence and describes how wavelets can be used as a tool for the analysis of the underlying coherent structure.

Finally, it is worth recalling that wavelets are generated by the affine group, while the traditional Gaussian coherent states of mathematical physics are generated by the Weyl-Heisenberg group and have provided the basis for a semi-classical analysis of quantum theory. In their chapter, Paul and Seip demonstrate that affine coherent states (which are actually Fourier transforms of a simple class of wavelets) can also be used to provide a semi-classical description of some quantum systems. In particular, Paul shows that the Bohr rule for the energy levels of hydrogen arises naturally from this type of semi-classical analysis — a recent development that Schrödinger [S] seemed to have anticipated in his original paper on coherent states.

ACKNOWLEDGEMENT

This paper was written while the author was visiting the Institute for Theoretical Atomic and Molecular Physics at the Harvard-Smithsonian Center for Astrophysics, and thereby supported by a grant from the National Science Foundation to that institute; her work is also supported by NSF grant DMS-89-13319.

REFERENCES

[Ba]   G. Battle, "A Block Spin Construction of Ondelettes:," Part I "Lemarié Functions" *Commun. Math. Phys.* **110**, 601–615 (1987); Part III "A Note on Pre-Ondelettes," preprint (1990).

[BCR]   G. Beylkin, R. Coifman and V. Rokhlin, "Fast Wavelet Transforms and Numerical Algorithms," *Commun. Pure Appl. Math.*, to appear.

[BOS]   T. Balaban, M. O'Carroll, and R. Schor "Block Averaging Renormalization Group for Lattice and Continuum Euclidean Fermions: Expected and Unexpected Results," *Lett. Math.*

Phys. **17**, 209–214 (1989); "Block Renormalization Group for Euclidean Fermions," *Commun. Math. Phys.* **122**, 233–247 (1989).

[CW]   C.K. Chui and J.Z. Wang, "A Cardinal Spline Approach to Wavelets," *Proc. AMS*, to appear.

[CD1]   A. Cohen and I. Daubechies, "Orthonormal Bases of Compactly Supported Wavelets, III. Better Frequency Resolutions," AT&T Bell Laboratories preprint.

[CD2]   A. Cohen and I. Daubechies, "Nonseparable Bidimensional Wavelet Bases," AT&T Bell Laboratories preprint.

[CDF]   A. Cohen, I. Daubechies, and J.C. Feauveau "Biorthogonal Bases of Compactly Supported Wavelets," *Commun. Pure Appl. Math.*, to appear.

[D1]   I. Daubechies, *Ten Lectures on Wavelets*, CBMS-NSF Series in Applied Mathematics, in press (SIAM, 1991).

[D2]   I. Daubechies, "The Wavelet Transform: A Method for Time-Frequency Localization," pp. 366–417 in *Advances in Spectrum Analysis and Array Processing*, Vol. 1, ed. S. Haykin (Prentice Hall, 1990).

[D3]   I. Daubechies, "The Wavelet Transform, Time-Frequency Localization and Signal Analysis," *IEEE Trans. Info. Theory* **41**, 961–1005 (1990).

[D4]   I. Daubechies, "Orthonormal Bases of Compactly Supported Wavelets," *Commun. Pure Appl. Math.* **41**, 909–996 (1988).

[D5]   I. Daubechies, "Orthonormal Bases of Compactly Supported Wavelets, II. Variations on a Theme," AT&T Bell Laboratories preprint.

[DMW]   I. Daubechies, S. Mallat, and A. Willsky, editors, "Wavelet Transforms and Multiresolution Signal Analyis" *IEEE Trans. Info. Theory* to appear (1992).

[DJP]   R.A. DeVore, B. Jawerth, and V. Popov, "Compression of Wavelet Decompositions," preprint.

[F]   P. Federbush, "A Phase Cell Approach to Yang-Mills Theory" *Commun. Math. Phys.* **107**, 319- 329 (1986).

[GLRT]   R. Glowinski, W. Lawton, M. Ravachol, and E. Tennebaum, "Wavelet Solution of Linear And Nonlinear Elliptic, Parabolic, and Hyperbolic Problems in One Space Dimension," pp. 55–120 in *Proc. of Ninth Conference on Computing Methods in Applied Science and Engineering* (SIAM, Philadelphia, 1990).

[GM]   K. Gröchenig and W. Madych, "Multiresolution Analysis, Haar Bases, and Self-Similar Tilings of $\mathbf{R}^n$," preprint.

[HW]   C.E. Heil and D.F. Walnut, "Continuous and Discrete Wavelet Transforms," *SIAM Review* **31**, 628–666 (1989).

[K]   G. Kaiser, "An Algebraic Theory of Wavelets, I. Operational Calculus and Complex Structure," *SIAM J. Math. Anal.*, to appear.

[KV]   J. Kovacevic and M. Vetterli, "Multi-dimensional Nonseparable Filter Banks and Wavelets" *IEEE. Trans Info. Theory*, to appear.

[L]   W.M. Lawton, "Tight Frames of Compactly Supported Affine Wavelets," *J. Math. Phys.* **31**, 1898–1901; "Necessary and Sufficient Conditions for Constructing Orthonormal Wavelet Bases," *J. Math. Phys.* **32**, 57–61 (1991); Multi-resolution Properties of the Wavelet Galerkin Operator *J. Math. Phys.* **32**, 1440–1443 (1991).

[LR]   W.M. Lawton and H.L. Resnikoff, "Multi-dimensional Wavelet Bases," AWARE preprint.

[Ma]   S. Mallat, "Multi-resolution Approximations and Wavelet Orthonormal Bases of $L^2(\mathbf{R})$," *Trans. AMS* **315**, 69–88 (1989).

[Me1]   Y. Meyer, *Ondelettes* (Hermann, 1990).

[Me2]   Y. Meyer, "Ondelettes, Fonctions Splines and Analyses Graduées," lectures given at the University of Torino, Italy (1986).

[S]   E. Schrödinger, "Der Stetige Übergang von der Mikro- zur Makromechanik" *Naturwissenschaften* **14**, 664–666 (1926).

[Sz]   R.S. Strichartz, "How to Make Wavelets" *MAA Monthly*, to appear (1992).

[VH]   M. Vetterli and C. Herley, "Linear Phase Wavelets," preprint.

II
SIGNAL ANALYSIS

WAVELETS AND FILTER BANKS FOR DISCRETE-TIME SIGNAL PROCESSING

MARTIN VETTERLI[1]

*Department of Electrical Engineering
and Center for
Telecommunications Research
Columbia University, New York, NY*

1. INTRODUCTION

Digital filter banks, that is the filtering of a signal by several filters in parallel followed by subsampling, have been studied over the last fifteen years because of their widespread use in speech and image coding systems [12, 11, 49, 41, 51, 42, 45, 52] ("filter" stands for a convolution operator in the engineering literature). Such coding or compression systems aim at reducing the bitrate needed to represent a signal for transmission and are called subband coding systems because the signal's frequency spectrum is split into bands. Filter banks also find application in telecommunication systems as transmultiplexers [4, 52]. Pyramid coding schemes, which are "oversampled" subband coding schemes (because more samples are actually used than is necessary), have been used in vision and coding [7].

The derivation of the first orthogonal two-channel filter bank [39, 35, 41] and of general perfect reconstruction filter banks [50, 40, 51, 42, 44, 52] (biorthogonal and multichannel cases) lead to a well understood theory

[1]Work supported in part by the National Science Foundation under grants ECD-88-11111 and MIP-90-14189.

of splitting discrete sequences into subsequences by multirate filtering allowing perfect synthesis of the original. The effort concentrated on filters having rational transfer functions for obvious implementation reasons. Limiting cases, like infinite iterations, were not considered mainly because applications did not require it.

Independently of this work, the theory of wavelets [22] emerged both as a signal analysis method [20] and a mathematical tool for constructing original bases for many function spaces [22, 13, 30, 34, 16]. While mainly concerned with continuous functions, the wavelet theory lead also to an interesting connection with filter banks, with the multiresolution work of S. Mallat [29, 30] and the work of I. Daubechies [13]. In particular, it is shown in [13] that filter banks can be used, under certain conditions, to generate orthonormal bases of compactly supported wavelets. This method can be extended to generate biorthogonal bases as well [8, 54].

The purpose of the present paper is first to review this connection between filter banks and wavelets, and then to explore it and indicate a number of results from filter banks which can be useful for the design of wavelets. Conversely, results from wavelets that influence the design of filter banks will be discussed.

Thus, we start with a brief review of the two-channel filter bank case, showing the relation to unitary matrices and indicating factorization results, which are central to both design and implementation of such filter banks. We then recall the connection between filter banks and wavelets through the regularity result of Daubechies [13] and the two scale equation property. We show the design of orthogonal and biorthogonal wavelets, and how they can be improved using the algebraic structure of filter banks.

Multichannel filter banks are considered next, including tree structures, and it is shown that they lead to wavelet bases as well. Filter banks with rational sampling rate changes are described. The discrete short-time Fourier or Gabor transform is also discussed, as well as cosine-modulated filter banks. Then, we consider the multidimensional case, especially with two channels and non-separable filters, and show wavelets in that case. Finally, numerical behavior, computational complexity and applications are discussed.

2.   TWO CHANNEL DISCRETE-TIME FILTER BANK

Consider real or complex sequences in $l^2(Z)$, $x(n)$, $n \in Z$. The inner product is defined as $< a(n), b(n) > = \sum a^*(n)b(n)$ (where $*$ denotes the complex conjugate). It will be convenient to use z-transforms of signals and filters (note that z^{-1} is the delay or right shift operator):

$$X(z) = \sum_{n=-\infty}^{\infty} x(n)z^{-n} \quad . \tag{1}$$

Note that the Fourier transform is obtained by evaluating the z-transform on the unit circle, $z = e^{j\omega}$ $(j = \sqrt{-1})$.

Given a filter with impulse response $h(n)$, we refer to its z-transform $H(z)$ as its *transfer function*. This is because the z-transform $Y(z)$ of the output of the filter is related to its input $X(z)$ by $Y(z) = H(z)X(z)$, that is, convolution becomes multiplication in the z-transform domain. By abuse of language, we will often say "filtering with $H(z)$" and mean "convolving with a filter having impulse response $h(n)$".

Consider a subband analysis/synthesis system as shown in Figure 1. Analysis stands for splitting a signal into frequency bands, while synthesis stands for reconstructing a signal from various frequency components.

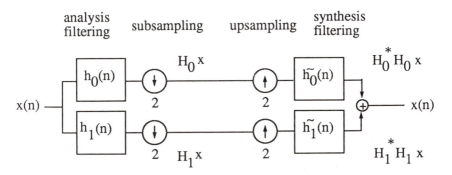

FIGURE 1. Orthogonal two channel discrete-time filter bank with analysis filters $h_0(n)$ and $h_1(n)$ and synthesis filters $\tilde{h}_0(n)$ and $\tilde{h}_1(n)$ (in the general non-orthogonal case, the synthesis filters would be $g_0(n)$ and $g_1(n)$).The $\downarrow$ and $\uparrow$ stand for sub- and upsampling by 2 respectively, while the $\oplus$ stands for summation.

Specifically, from the original sequence, we want to derive two subsampled sequences by filtering with $H_0(z)$ and $H_1(z)$ and dropping all samples having odd indices. Reconstruction is achieved by resampling at the original sampling rate (replacing the dropped samples by zero), filtering with $G_0(z)$ and $G_1(z)$ and summing up. Such a system is shown in Figure 1 for the orthogonal case (where $G_0(z) = \tilde{H}_0(z)$ and $G_1(z) = \tilde{H}_1(z)$ as we shall see). Using the fact that subsampling followed by upsampling by 2 (that is, replacing odd-indexed samples by 0) can be written, for example for $X(z)$, as $1/2[X(z) + X(-z)]$, we find that the output of the system is equal to:

$$\hat{X}(z) = \frac{1}{2}[G_0(z)\ G_1(z)] \begin{bmatrix} H_0(z) & H_0(-z) \\ H_1(z) & H_1(-z) \end{bmatrix} \begin{bmatrix} X(z) \\ X(-z) \end{bmatrix} . \tag{2}$$

The matrix in (2) is often called $\mathbf{H}_m(z)$, where the subscript m stands for modulation [52].

2.1. THE UNITARY CASE

We will show that unitary filter banks (corresponding to unitary bases) have transfer function matrices which are unitary as well, and are thus related to classic structures in filtering and network theory called lossless systems [3, 48]. A filter with a rational transfer function is called *allpass* if it satisfies:

$$\tilde{H}(z) \cdot H(z) = 1 \tag{3}$$

where $\tilde{H}(z) = H_*(z^{-1})$ and subscript $*$ means conjugation of coefficients but not of z [45]. On the unit circle, $\tilde{H}(e^{j\omega}) = [H(e^{j\omega})]^*$, and thus, an allpass filter is unitary on the unit circle. Note that $z \to z^{-1}$ because the adjoint operator of delay (or right shift) is the advance operator (or left shift). Also, the only finite impulse response (FIR) filters that are allpass are pure delays (z^{-l}) [36].

A square matrix with rational entries is called *paraunitary* if it satisfies [45]:

$$\tilde{\mathbf{H}}(z) \cdot \mathbf{H}(z) = \mathbf{H}(z) \cdot \tilde{\mathbf{H}}(z) = c \cdot \mathbf{I} \tag{4}$$

where $\tilde{\mathbf{H}}(z) = \mathbf{H}_*^T(z^{-1})$. On the unit circle, $\tilde{\mathbf{H}}(e^{j\omega}) = [\mathbf{H}(e^{j\omega})]^*$ (superscript denotes hermitian transpose), and thus, a paraunitary matrix is unitary on the unit circle (within a scale factor). In the rational case, paraunitary matrices which are *stable* [36] are called *lossless* matrices [48]. For the matrix in (2) to be paraunitary, the $H_i(z)$'s have to satisfy:

$$H_i(z)\tilde{H}_i(z) + H_i(-z)\tilde{H}_i(-z) = c \tag{5}$$

$$H_0(z)\tilde{H}_1(z) + H_0(-z)\tilde{H}_1(-z) = 0 . \tag{6}$$

Since $H_i(z)\tilde{H}_i(z)$ is the z-transform of the autocorrelation sequence of $h_i(n)$ (that is the convolution of $h_i(n)$ with $\tilde{h}_i(-n)$) and (5) has only even terms, we have (choosing $c = 2$):

$$< h_i(n), h_i(n+2m) >= \delta(m) \tag{7}$$

where $\delta(.)$ stands for the discrete delta function. Thus, the filter impulse responses are orthogonal to their even translates. Because $[H_1(z),\ H_1(-z)]^T$

has to be orthogonal to $[\tilde{H}_0(z), \tilde{H}_0(-z)]^T$, $H_1(z)$ has to be of the form:

$$H_1(z) = -z^{-1} \cdot A(z^2) \cdot \tilde{H}_0(-z) \tag{8}$$

where $A(z)$ has to be an allpass filter. In particular, if we desire an FIR solution, then $H_1(z) = -z^{-2l-1} H_*(-z^{-1})$.

Because of (5,8), it follows that on the unit circle:

$$\mid H_0(e^{j\omega}) \mid^2 + \mid H_0(e^{j(\omega+\pi)}) \mid^2 = \mid H_0(e^{j\omega}) \mid^2 + \mid H_1(e^{j\omega}) \mid^2 = 1 \tag{9}$$

that is, the filters are power complementary [41, 45]. Because $H_0(z)\tilde{H}_1(z)$ is the z-transform of the crosscorrelation of $h_0(n)$ and $h_1(n)$ (that is the convolution of $h_0(n)$ with $h_1(-n)$), it follows from (6) that:

$$< h_0(n + 2l), h_1(n + 2m) >= 0 \tag{10}$$

that is, the two filters and their even translates are orthogonal to each other.

Finally in (2), choose $G_0(z) = \tilde{H}_0(z)$ and $G_1(z) = \tilde{H}_1(z)$ (that is, from (8), $G_1(z) = z \cdot \tilde{A}(z^2) \cdot H_0(-z)$), and assume that $H_0(z)$ satisfies (5) with $c = 2$. Then, it is easy to check that $\hat{X}(z) = X(z)$, that is, we have perfect reconstruction with the adjoint synthesis filters.

Thus, we have shown that when the matrix in (2) is paraunitary (with $c = 2$), then the set:

$$\{h_0(n + 2l), h_1(n + 2m)\} \tag{11}$$

is an orthonormal basis for $l^2(Z)$.

Alternatively, call $\mathbf{H}_i$ the operator corresponding to filtering with $h_i(n)$ followed by subsampling by 2. Each line contains the impulse response of $h_i(n)$ from right to left because convolution involves time-reversal. Each line is shifted by 2 to the right from the line above it (due to the subsampling by 2). Then, (7) is equivalent to:

$$\mathbf{H}_i \cdot \mathbf{H}_i^* = \mathbf{I} \ . \tag{12}$$

Upsampling followed by filtering with $h_i(-n)$ corresponds to the operator $\mathbf{H}_i^*$. Because of (10), the crossproduct:

$$\mathbf{H}_0 \cdot \mathbf{H}_1^* = \mathbf{0} \ . \tag{13}$$

Therefore, in the unitary case, because the filters and their even translates form an orthonormal set, the two channels correspond to orthogonal projections onto spaces spanned by their respective impulse responses, and we have the perfect reconstruction as the direct sum of the projections:

$$\mathbf{H}_0^* \cdot \mathbf{H}_0 + \mathbf{H}_1^* \cdot \mathbf{H}_1 = \mathbf{I} \ . \tag{14}$$

If one gathers the filter impulse responses $\{h_0(n), h_1(n)\}$ into a single transform matrix $\mathbf{T}$ by simply interleaving the operators $\mathbf{H}_0$ and $\mathbf{H}_1$, then $\mathbf{T}$ is unitary. This matrix is block Toeplitz (with blocks of size 2 by 2) and is banded if the filters are FIR [53].

Note that the matrix in (2) is redundant since every filter coefficient appears twice. A more compact form separates even and odd coefficients of the filters (called *polyphase components* [4, 51]). The filters can be written as:

$$H_i(z) = H_{i0}(z^2) + z^{-1}H_{i1}(z^2) \tag{15}$$

where $H_{i0}(z)$ and $H_{i1}(z)$ are the first and second polyphase component of $H_i(z)$, respectively. The polyphase matrix $\mathbf{H}_p(z)$ corresponding to the filter bank is obtained from the matrix $\mathbf{H}_m(z)$ in (2) as:

$$\mathbf{H}_p(z^2) = \begin{bmatrix} H_{00}(z^2) & H_{01}(z^2) \\ H_{10}(z^2) & H_{11}(z^2) \end{bmatrix} = 1/2 \cdot \mathbf{H}_m(z) \begin{bmatrix} 1 & 1 \\ 1 & -1 \end{bmatrix} \begin{bmatrix} 1 & 0 \\ 0 & z \end{bmatrix} . \tag{16}$$

Now, the matrix $\mathbf{H}_p(z)$ will be paraunitary if and only if $\mathbf{H}_m(z)$ is, since they are related by unitary operations. Because of the relation between the filters $H_0(z)$ and $H_1(z)$ indicated by (8), the form of a paraunitary $\mathbf{H}_p(z)$ is:

$$\mathbf{H}_p(z) = \begin{bmatrix} 1 & 0 \\ 0 & A(z) \end{bmatrix} \cdot \begin{bmatrix} H_{00}(z) & H_{01}(z) \\ -H_{01*}(z^{-1}) & H_{00*}(z^{-1}) \end{bmatrix} . \tag{17}$$

Note that it follows from (5) and (8) that the determinant of $\mathbf{H}_m(z)$ is equal to $2z^{-1} \cdot A(z^2)$, that is an allpass function, when the matrix is paraunitary. From (16), the same holds for $\mathbf{H}_p(z)$. In particular, it follows that:

$$H_{00}(z)H_{00*}(z^{-1}) + H_{01}(z)H_{01*}(z^{-1}) = 1 \tag{18}$$

which can also be seen directly from (5). That is, on the unit circle, the polyphase components of the filter are also power complementary.

How can we design paraunitary filter banks, or equivalently, sets of filters which are orthogonal to their even translates? First, only one filter has to be designed, since the other is closely related following (8). We will concentrate on the real FIR case in the following. Two approaches are possible.

The first consists in finding an autocorrelation sequence that satisfies (5) [39, 41]. In particular, it has to be a positive function on the unit circle, that is, zeros on the unit circle have to appear in pairs. Then, one factors the autocorrelation into its roots, and takes one out of each pair at $(\alpha, 1/\alpha^*)$ into $H_0(z)$. Taking zeros from inside or outside the unit circle will change the phase behavior (but linear phase, that is a phase of the form

$\phi(\omega) = \phi_0\omega$, is excluded except in the trivial Haar case). This procedure is like taking the "square root" of the positive function $H_0(e^{j\omega})H_0(e^{-j\omega})$. However, factorization becomes numerically ill conditioned as the filter size grows, and thus, the resulting filters are usually only approximately orthogonal.

An alternative and numerically well conditioned procedure relies on the fact that paraunitary, just like unitary matrices, possess canonical factorizations into elementary paraunitary matrices [45, 46]. Thus, all paraunitary filter banks with FIR filters of length $L = 2N$ can be reached by the following lattice structure (here $H_1(z) = z^{-L+1}H_0(z^{-1})$):

$$\begin{bmatrix} H_{00}(z) & H_{01}(z) \\ H_{10}(z) & H_{11}(z) \end{bmatrix} = \mathbf{R}_0 \cdot \prod_{i=1}^{N-1} \begin{bmatrix} 1 & 0 \\ 0 & z^{-1} \end{bmatrix} \cdot \mathbf{R}_i \qquad (19)$$

where $\mathbf{R}_i$ is a 2 by 2 rotation matrix:

$$\mathbf{R}_i = \begin{bmatrix} cos(\alpha_i) & -sin(\alpha_i) \\ sin(\alpha_i) & cos(\alpha_i) \end{bmatrix} . \qquad (20)$$

That the resulting structure is paraunitary is easy to check (it is the product of paraunitary elementary blocks). What is much more interesting is that all paraunitary matrices of a given degree can be written in this form [46].

For example, choosing $\alpha_0 = 2\pi/24$ and $\alpha_1 = -2\pi/6$ leads to a lowpass filter with coefficients $1/(4\sqrt{2})[1 + \sqrt{3}, 3 + \sqrt{3}, 3 - \sqrt{3}, 1 - \sqrt{3}]$. This filter has a double zero at $z = -1$, and is known as D_4, since it was derived in direct form by I. Daubechies in [13] and is regular (see next section).

2.1.1. REMARKS

(i) IIR solutions

If one replaces z^{-1} by a general allpass filter in (19), one still obtains a paraunitary filter bank, corresponding now to infinite impulse response (IIR) filters. In particular, if one uses a first order allpass:

$$A_i(z) = \frac{1 + a_i z^{-1}}{a_i + z^{-1}}, \quad | a_i | < 1 \qquad (21)$$

then one obtains again a complete characterization of paraunitary filter banks of a given degree [18]. One problem is that if the analysis is causal (the impulse response goes from 0 to $+\infty$), the synthesis will be anticausal (the impulse response goes from $-\infty$ to 0).

(ii) State-space realizations

Consider the state-space realization of a stable allpass filter or a lossless (paraunitary and stable) transfer function matrix $\mathbf{H}(z)$. Such a state-space realization is characterized by a quadruple $\{\mathbf{A}, \mathbf{B}, \mathbf{C}, \mathbf{D}\}$:

$$\mathbf{v}(n+1) \;=\; \mathbf{A}\mathbf{v}(n) + \mathbf{B}\mathbf{x}(n) \tag{22}$$

$$\mathbf{y}(n) \;=\; \mathbf{C}\mathbf{v}(n) + \mathbf{D}\mathbf{x}(n) \tag{23}$$

where $\mathbf{v}(n)$ is the state vector, $\mathbf{x}(n)$ and $\mathbf{y}(n)$ are vectors of the size of the input and output respectively, and the matrices $\mathbf{A}$, $\mathbf{B}$, $\mathbf{C}$ and $\mathbf{D}$ have appropriate sizes. A fundamental result is that $\mathbf{H}(z)$ is lossless if and only if there exists a minimal realization such that the matrix:

$$\mathbf{R} = \begin{pmatrix} \mathbf{A} & \mathbf{B} \\ \mathbf{C} & \mathbf{D} \end{pmatrix} \tag{24}$$

is unitary [48]. This gives an alternative way to parametrize lossless systems in state-space domain by factorizing $\mathbf{R}$ into elementary unitary matrices (e.g. Givens rotations). If finite impulse response (FIR) systems are desired, $\mathbf{A}$ has to be forced to have all zero eigenvalues (modes), which can be achieved by parametrizing it to be lower triangular for example.

(iii) Numerical properties

A very attractive feature of the cascade or lattice structure in (19) is that quantization of the coefficients of the matrices $\mathbf{R}_i$ will not destroy the paraunitary property of the filter bank. This is easy to see: if instead of $cos(\alpha_i)$ and $sin(\alpha_i)$ we have $cos(\alpha_i) + \delta_i$ and $sin(\alpha_i) + \epsilon_i$ in $\mathbf{R}_i$, then:

$$\mathbf{R}_i \cdot \mathbf{R}_i^T = \mathbf{R}_i^T \cdot \mathbf{R}_i = (1 + \delta_i^2 + \epsilon_i^2 + \delta_i cos(\alpha_i) + \epsilon_i sin(\alpha_i)) \cdot \mathbf{I} \tag{25}$$

that is, $\mathbf{R}_i$ remains unitary, up to a scale factor. It follows that $\mathbf{H}(z)$ remains paraunitary (also up to a scale factor, which might have to be corrected), the filter responses $\{h_0(n+2l), h_1(m+2k)\}$ form an orthogonal set, and the two filters have the same norm (possibly not equal to 1). This robustness in the presence of quantization is powerful, and quite unlike the case where filters are obtained by factoring an autocorrelation function. In that case (besides the problem of finding accurate roots), any quantization of the coefficients will actually destroy orthogonality.

(iv) Orthogonal solutions with symmetries

A rational filter having linear phase $(\mathrm{Arg}[H(e^{j\omega})] = \phi_0\omega)$ has either a symmetric or an antisymmetric impulse response [36]. It is well known that there are no paraunitary linear phase (symmetric/antisymmetric filters) solutions in the real FIR case (except for the trivial Haar case). But if we allow IIR filters (which will not be causal), such solutions exist [23]. For example, the filters with z-transforms:

$$H_0(z) = A(z^2) + z^{-1}A(z^2), \quad H_1(z) = -A(z^2) + z^{-1}A(z^2) \qquad (26)$$

with $A(z)\tilde{A}(z) = 1$ form an orthonormal set with respect to even translates and have symmetric and antisymmetric impulse responses (around $n = 1/2$).

2.2. GENERAL TWO CHANNEL SOLUTIONS

Orthogonal filter banks have many nice features (conservation of energy, identical analysis and synthesis) but also some restrictions (no real FIR linear phase solutions). Therefore, it will be of interest to consider general perfect reconstruction filter banks. We will concentrate on the finite length (FIR) case. It can be verified that perfect reconstruction FIR solutions are possible if and only if the matrix $\mathbf{H}_p(z)$ (or equivalently $\mathbf{H}_m(z)$) has a determinant equal to a delay, that is [55]:

$$H_{00}(z)H_{11}(z) - H_{01}(z)H_{10}(z) = z^{-l} \qquad (27)$$

$$H_0(z)H_1(-z) - H_0(-z)H_1(z) = 2z^{-2l-1} \quad . \qquad (28)$$

The synthesis filters are then equal to (up to shifts):

$$G_0(z) = z^{-k}H_1(-z), \quad G_1(z) = -z^{-k}H_0(-z) \quad . \qquad (29)$$

In operator notation, in a similar fashion to (12-14) we have:

$$\mathbf{H}_i \cdot \mathbf{G}_i^* = \mathbf{I} \qquad (30)$$

$$\mathbf{H}_0 \cdot \mathbf{G}_1^* = \mathbf{H}_1 \cdot \mathbf{G}_0^* = \mathbf{0} \qquad (31)$$

$$\mathbf{G}_0^* \cdot \mathbf{H}_0 + \mathbf{G}_1^* \cdot \mathbf{H}_1 = \mathbf{I} \qquad (32)$$

which is referred to as biorthogonal relations [8, 55].

It follows immediately from (27,28) that perfect FIR reconstruction is possible only if the polyphase components of the filters are coprime (that is, they have no common factors except possibly at z^{-i}). This is equivalent to the filters not having zeros in pairs at $(\alpha_i, -\alpha_i)$. Because the polyphase components have to satisfy the Bezout Identity given by (27), it is easy to find a perfect reconstruction complementary filter $H_1(z)$ given a filter $H_0(z)$

having coprime polyphase components. One simply runs Euclid's algorithm to find 2 polynomials such that:

$$a(z)H_{00} + b(z)H_{01}(z) = 1 \qquad (33)$$

and thus $H_1(z) = a(z^2) - z^{-1}b(z^2)$ will yield a perfect reconstruction filter bank with synthesis filters given by (29). To find all other solutions, we can use the theory of Diophantine equations [55]. In particular, solutions $[a'(z), b'(z)]$ to the equation:

$$a'(z)H_{00} + b'(z)H_{01}(z) = 0 \qquad (34)$$

can be added to $[a(z), b(z)]$ without affecting (33), and thus all other solutions can be reached.

Of particular interest is the case when both $H_0(z)$ and $H_1(z)$ are linear phase (symmetric or antisymmetric) filters. Then, like in the paraunitary case, there are certain restrictions on the possible filters [51, 55]. In particular, linear phase perfect reconstruction FIR solutions have either of the following forms:

(a) Both filters are symmetric and of odd lengths, differing by an odd multiple of 2.

(b) One filter is symmetric and the other is antisymmetric; both lengths are even, and are equal or differ by an even multiple of 2.

(c) Filters are of even and odd lengths, but all their zeroes are on the unit circle at roots of ± 1.

Perfect reconstruction filter banks with linear phase filters can be designed by the factorization method that was discussed in the paraunitary case, except that now, the factors have to be kept symmetric. It is also possible to solve a system of linear equations to find a linear phase complementary filter to a given linear phase filter [51]. Finally, linear phase solutions can be extended by using the Diophantine equation method discussed above [55].

2.2.1. REMARKS

(i) Binomial filter

This filter is given by $B_N(z) = (1 + z^{-1})^N$. Its impulse response coefficients are therefore the binomial coefficients. Obviously, it has no pairs of zeros at $(\alpha_i, -\alpha_i)$, and thus, there is a complementary filter which can be found for example using Euclid's algorithm. In particular, one can find a minimal length complementary filter so that the product $B_N(z)H_1(-z)$ is an autocorrelation sequence (positive on the unit circle). This will be the solution leading to regular filters as shown in [13].

(ii) Lattice structure for linear phase filters

Unlike in the paraunitary case, there are no *complete* canonical factorizations for general matrices of polynomials. But there are lattice structures that will produce, for example, linear phase perfect reconstruction
filters [53]. The lattice

$$\mathbf{H}_p(z) = \begin{pmatrix} 1 & 1 \\ 1 & -1 \end{pmatrix} \prod_{i=1}^{N-1} \begin{pmatrix} 1 & 0 \\ 0 & z^{-1} \end{pmatrix} \cdot \begin{pmatrix} 1 & \alpha_i \\ \alpha_i & 1 \end{pmatrix} \tag{35}$$

produces length $L = 2N$ symmetric (lowpass) and antisymmetric (highpass) filters leading to perfect reconstruction banks. Note that the structure is incomplete [53]. Particular choices of the α_i's produce the binomial filter as a lowpass, and its complementary filter as a highpass [55].
Again, just as in the paraunitary lattice (19), perfect reconstruction is
structurally guaranteed within a scale factor (in the synthesis, replace simply α_i by $-\alpha_i$).

3.   WAVELETS DERIVED FROM FILTER BANKS

It is clear that an orthogonal filter bank computes a discrete-time wavelet
transform when iterated on the lower band (a so-called octave-band filter
bank), as shown in Figure 2. A deeper relationship exists and links, under
certain conditions, discrete-time filter banks with continuous-time wavelet
transforms. This was explored both in the multiresolution approach of
S. Mallat [29, 30] and Y. Meyer [34] and in I. Daubechies's construction
of compactly supported wavelets [13]. We will briefly review the latter
construction.

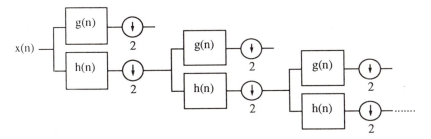

FIGURE 2.   Discrete wavelet transform implemented with filter banks.

3.1. THE REGULARITY CONDITION

We will assume FIR filters in what follows. The idea is to associate a continuous function with the iterated lowpass filter in Figure 2. In order to do this, we first derive a filter equivalent to i-steps of lowpass filtering and subsampling. Notice that subsampling by 2 followed by filtering with $H(z)$ is equivalent to filtering by $H(z^2)$ followed by subsampling by 2. Thus, the equivalent filter at step i is:

$$H^{(i)}(z) = \prod_{l=0}^{i-1} H(z^{2^l}) \qquad i = 1, 2 \cdots \tag{36}$$

followed by subsampling by 2^i. Its impulse response is given by $h^{(i)}(n)$ and we define $h^{(0)}(n) = \delta(n)$. If $h(n)$ has L coefficients, then $h^{(i)}(n)$ has $L^{(i)} = (2^i - 1)(L - 1) + 1$ coefficients. Now define the following equivalent piecewise constant function:

$$f^{(i)}(x) = 2^{i/2} \cdot h^{(i)}(n) \qquad\qquad n/2^i \leq x < (n+1)/2^i \tag{37}$$

which is constant over intervals of length $1/2^i$. The support of $f^{(i)}(x)$ is contained in $[0, L - 1]$ (from $L^{(i)}$ divided by 2^i). The normalization is such that $\| h^{(i)}(n) \|_2 = \| f^{(i)}(x) \|_2$. That different filters can lead to vastly different behavior is shown in Figure 3, where $h(n) = [1, \alpha, \alpha, 1]$ is iterated for $\alpha \in [-3, 3]$.

For $\alpha = 3$, the iterated function tends to a continuous and derivable function. As α decreases, first non-derivable points and then discontinuous points appear, and at $\alpha = -3$, the iterated function tends to a fractal function (discontinuous at every dyadic point). Also, at $\alpha = 0$ there is not even pointwise or L_2 convergence [13].

I. Daubechies gave a sufficient condition on the filter $h(n)$ for convergence to a continuous function [13]. Assume $H(z)$ has at least one zero at $z = -1$, and that its impulse response is orthogonal to its even translates. Then, because of (7) and $H(-1) = 0$, it follows that $H^2(1) = 2$ or $\sum h(n) = \sqrt{2}$. Define $m_0(z) = 1/\sqrt{2}H(z)$ so that $m_0(1) = 1$. Factor $m_0(z)$ as:

$$m_0(z) = [1/2(1 + z^{-1})]^N F(z) \tag{38}$$

where by assumption $N \geq 1$ and $F(1) = 1$. Define:

$$B = \sup_{\omega \in [0, 2\pi]} |F(e^{j\omega})| . \tag{39}$$

Then:

Proposition 3.1. (Daubechies 1988) *If $B < 2^{N-1}$, then the piece-wise constant function $f^{(i)}(x)$ defined in (37) converges pointwise to a continuous function $f^{(\infty)}(x)$.*

Additional zeros at $z = -1$ will lead to higher regularity (the limit function will possess continuous derivatives as well), and a detailed discussion of regularity can be found in [14]. In particular, better regularity estimates exist, but the one indicated above is an easy one to use as a filter design criterion.

Assume in the following that Proposition 3.1. is satisfied. Because of (36), the filter impulse response $h^{(i)}(n)$ satisfies the recursion:

$$h^{(i)}(n) = \sum_{k=0}^{L-1} h(k)h^{(i-1)}(n - 2^{i-1}k) \ . \tag{40}$$

Expressing $f^{(i)}(x)$ in terms of $h^{(i)}(n)$, and then replacing $h^{(i-1)}(n)$ by $f^{(i-1)}(2x)$ (which is constant over intervals of length $1/2^i$, consistent with $f^{(i)}(x)$), we get:

$$f^{(i)}(x) = 2^{1/2} \sum_{k=0}^{L-1} h(k)f^{(i-1)}(2x - k) \ . \tag{41}$$

Now, calling $\phi(x) = Lim_{i\to\infty} f^{(i)}(x)$ (which is well defined and continuous following Proposition 3.1.), we see that

$$\phi(x) = 2^{1/2} \sum_{k=0}^{L-1} h(k)\phi(2x - k) \ . \tag{42}$$

Thus, $\phi(x)$, which is called the *scaling function* associated with the discrete filter $h(n)$, satisfies a two-scale equation [13, 14, 43].

Recall (from (7)) that the filter $h(n)$ is orthogonal to its even translates, $< h(n), h(n + 2l) >= \delta(l)$. Then it is easy to show recursively that

$$< f^{(i)}(x), f^{(i)}(x - l) >= \delta(l) \tag{43}$$

holds, assuming that it holds for $f^{(i-1)}(.)$ and recalling that $f^{(0)}(x)$ is simply the indicator function of $[0, 1[$. Taking again the limit we find:

$$< \phi(x), \phi(x - l) >= \delta(l) \ . \tag{44}$$

So far we have dealt with the lowpass filter $h(n)$ from the filter bank. Recall that the highpass filter $g(n)$ equals $(-1)^{n+1}h(L - n - 1)$ in an FIR orthogonal filter bank. Now, define the wavelet in a similar fashion to (42) but with $g(k)$ instead of $h(k)$:

$$\psi(x) = 2^{1/2} \sum_{k=0}^{L-1} g(k)\phi(2x - k) \quad . \tag{45}$$

Conceptually, it corresponds to the iteration of lowpass filters (which converge to $\phi(x)$) followed by one highpass filter. This leads to a bandpass filter. From (44,45), it is easy to verify that:

$$< \psi(x), \psi(x - l) >= \delta(l) \quad . \tag{46}$$

Using $< h(n), g(n + 2l) >= 0$, it also follows from (42,43):

$$< \phi(x), \psi(x - l) >= 0 \quad . \tag{47}$$

It remains to show that the wavelet $\psi(x)$ is orthogonal across scales. Consider two adjacent scales first, and use (42):

$$< \psi(x), \psi(2x - l) > \quad = \quad < \sum g(k)\phi(2x - k), \psi(2x - l) > \tag{48}$$

$$= \quad \sum g(k) < \phi(2x - k), \psi(2x - l) >= 0 \tag{49}$$

following (47). The same idea is applied across several scales by decomposing the "longer" wavelet into scaling functions until (47) can be used. Defining:

$$\psi_{mn} = 2^{m/2}\psi(2^m x - n) \tag{50}$$

we have thus verified that:

$$< \psi_{mn}, \psi_{lk} >= \delta(m - l)\delta(n - k) \quad . \tag{51}$$

That the orthonormal set $\{\psi_{mn}\}$, $m, n \in Z$ constitutes a basis for $L^2(R)$ can be shown by verifying that it is a tight frame with framebound equal to 1 [13].

The biorthogonal case (see section 2.2.) leads to similar relations, except that there is now an analysis and a synthesis scaling function ($\phi_a(x)$ and $\phi_s(x)$) as well as an analysis and a synthesis wavelet ($\psi_a(x)$ and $\psi_s(x)$). The difficulty in the design of such wavelets from filter banks is that now both $h_0(n)$ and $g_0(n)$ (the analysis and synthesis lowpass filter) have to be regular. In the example of Figure 3, the length-4 dual filter to $h_0(n) = [1, \alpha, \alpha, 1]$ is $g_0(n) = [1, -\alpha, -\alpha, 1]$, and it is clear from the figure that there is no regular analysis and synthesis with length-4 filters (note that using longer synthesis filters, regularity could be achieved).

3.2. DESIGN OF WAVELETS BASED ON FILTER BANKS

The design goal is to derive perfect reconstruction filter banks (orthogonal or general ones) so that the lowpass filter ($h_0(n)$ in the orthogonal case, both $h_0(n)$ and $g_0(n)$ in the biorthogonal case) are maximally regular. Following Proposition 3.1., this amounts maximizing the number of zeros at $z = -1$, minimizing the supremum of the Fourier transform magnitude of the remainder $F(z)$ (39) while maintaining perfect reconstruction (possibly also orthogonality). Obviously, this is a very constrained problem with conflicting requirements.

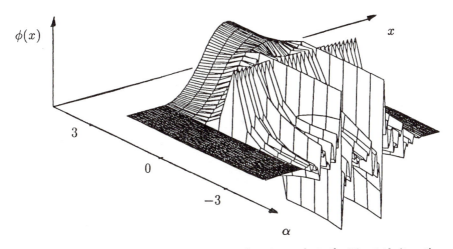

FIGURE 3. Iterated filter $h(n) = [1, \alpha, \alpha, 1]$ and $\alpha \in [-3, 3]$. The 6-th iteration is shown (at $\alpha = 0$, there is actually no convergence).

Let us consider the orthogonal case first, which is simpler because only a single filter has to be designed to be regular. The problem is to design an autocorrelation function meeting (5) and with $h_0(n)$ satisfying regularity. In particular, it will have an even number of zeros at $z = -1$. The design procedure amounts to finding a symmetric polynomial $R(z)$ of even degree such that $(1 + z^{-1})^{2k} R(z)$ is a positive function of $z = e^{j\omega}$ and has a single non-zero odd coefficient (at the center of symmetry). This can be solved as a system of linear equations, or a closed form for $R(z)$ is possible when it is of minimal degree $2(k-1)$ [13]. We can also compute the linear phase lattice (35), which leads to $H_0(z) = (1 + z^{-1})^{2k-1}$ and has $(1 - z^{-1})R(-z)$ as its complementary filter [55]. Then, one extracts the "square root" of $H_0(z)H_1(-z) = (1 + z^{-1})^{2k} R(z)$ to get $H_0(z)$, where there is a choice of taking zeros from either inside or outside of the unit circle (which will change the phase behavior of the resulting filter) [38]. It turns out that all filters for $k \geq 2$ will meet the bound in Proposition 3.1. and thus lead to continuous wavelets [13]. One such example is shown in Figure 4, where a length 18 filter was designed (which has 9 zeros at $z = -1$) and the corresponding wavelet and its Fourier spectrum are plotted.

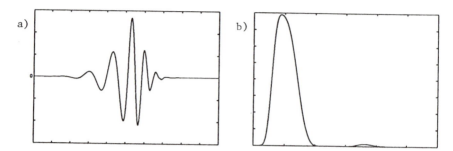

FIGURE 4. Compactly supported orthogonal wavelet designed from a length-18 orthogonal filter having 9 zeros at $z = -1$. (a) Time domain function. (b) Spectrum.

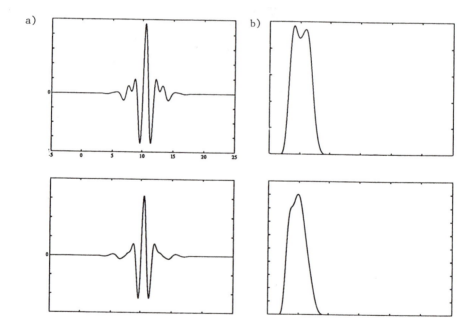

FIGURE 5. Compactly supported symmetric wavelet and its dual. (a) Time domain functions. (b) Spectras.

Alternatively, one can factor $(1 + z^{-1})^{2k} R(z)$ into linear phase factors. This amounts to grouping zeros in quadruples $(\alpha_i, 1/\alpha_i, \alpha_i^*, 1/\alpha_i^*)$, pairs

(α_i, α_i^*) (if they are on the unit circle) or pairs $(\alpha_i, 1/\alpha_i)$ (if they are on the real axis) so as to keep symmetric factors. The two factors correspond to $H_0(z)$ and $H_1(-z) = G_0(z)$, and both have to be regular. This means that the zeros at $z = -1$ have to be carefully allocated so that both filters have similar regularity. An example of a linear phase wavelet and its dual, obtained from a linear phase factorization of the same polynomial as the previous orthogonal example, is shown in Figure 5.

As mentioned earlier (2.2.), "minimal" solutions (that is, having a maximum number of zeros at $z = -1$) can be extended to bigger size, nonminimal solutions using Diophantine equations (see (34)). The added degrees of freedom can then be used to satisfy some other design goals. For example, the discrete filter can now also meet some more traditional digital signal processing requirements, like good stopband attenuation and good passband flatness.

Finally, Figure 6 shows a wavelet based on an orthogonal IIR filter bank with (non-causal) symmetric/antisymmetric filters (see remark (iv) in section 2.1.1.). The allpass filter $A(z)$ in (26) is only second order, but the resulting wavelet looks actually quite good.

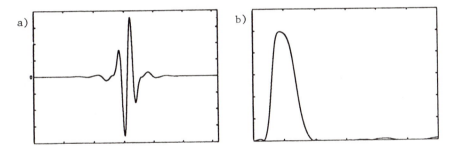

FIGURE 6. Antisymmetric (linear phase) orthogonal wavelet based on IIR filter. (a) Time domain function. (b) Spectrum.

For more design examples, we refer to [13, 8, 38, 55].

3.3. REMARKS

(i) Regularity

It is shown in [13, 8, 38] that arbitrary high regularity can be achieved by increasing the size of the filters appropriately. This also means that the wavelets can have an arbitrarily large number of zero moments, a fact that has proven useful in numerical analysis [5] and could be of importance for signal processing applications as well.

(ii) Numerical conditioning

The design procedure relies on factorization of polynomials, and is thus numerically not well behaved as the size of the filters increases. Furthermore, the resulting filters are given by their impulse response rather than their lattice coefficients (19), leading to implementations which are not robust to quantization (see remark (iii) in section 2.1.1.). If numerical behavior is critical (for example due to coarse quantization), it is worthwhile to rederive the filters in their lattice form (see example below (20)).

4.   THE MULTICHANNEL CASE

In the filter bank literature, the multichannel case has been studied for a number of years [51, 42, 44]. In particular, the case where the filters are obtained by modulation of a single prototype filter has attracted particular attention, because of its resemblance to the short-time Fourier transform (STFT) [37, 32, 53]. All these results can be related, under certain conditions, to wavelets as we will show in this section.

4.1.   MULTICHANNEL FILTER BANKS

4.1.1.   TREE STRUCTURED MULTICHANNEL FILTER BANKS

An easy way to construct multichannel filter banks is obtained by cascading two channel banks appropriately. One case was already seen in Figure 2, where logarithmically spaced frequency analysis was obtained by simply iterating a two channel division on the previous lowpass channel. This is often called a "constant-Q" or constant relative bandwidth filter bank. Another case appears when 2^K equal bandwidth channels are desired. This is simply obtained by a K-step subdivision into 2 channels, that is, the two channel bank is now iterated on both the lowpass and highpass channel. This results in a tree with 2^K leaves, each corresponding to $1/2^K$-th of the original bandwidth, with a subsampling by 2^K. We outline the proof that this is still an orthogonal filter bank if the two channel filter bank is. We show this by induction. To prove it, it is sufficient to show that the dual filter bank (with filters $\tilde{H}_i(z)$) achieves perfect reconstruction. Assume that it holds for the depth-$(K-1)$ tree. The depth-(K) tree is obtained by adding one layer of division into two channels. The dual tree starts with one layer of two channel dual synthesis filters, which cancels the final layer of division, followed by the dual of the depth-$(K-1)$ tree, which cancels by assumption, and thus perfect reconstruction is achieved. If the elementary

bank is not orthogonal, then obviously the dual filter bank will not achieve
perfect reconstruction. Therefore, the depth-(K) tree is orthogonal if and
only if the elementary two channel bank is.

Now, using similar arguments, one can show that *any* tree built from
elementary orthogonal two channel banks will be orthogonal. This gives
rise to so-called *wavelet packets* studied by Coifman and Meyer [9]. The
number of possible trees of a given maximal depth is of the order of $2^{2^{K}-1}$,
that is, a phenomenal number of different tree structured bases. The power
of these wavelet packets is that efficient algorithms can be derived to search
for the "best" basis among all these bases, a feature useful in signal compression [57, 9].

4.1.2. TRUE MULTICHANNEL FILTER BANKS

An N-channel analysis/synthesis system is shown in Figure 7. It is easy to
verify that subsampling a signal $x(n)$ by N followed by upsampling by N
(that is, replacing $x(n), n \bmod N \neq 0$ by 0) produces a signal $y(n)$ with
z-transform $Y(z)$ equal to:

$$Y(z) = \frac{1}{N} \sum_{i=0}^{N-1} X(W^i z), \quad W = e^{-j2\pi/N} \quad j = \sqrt{-1} \tag{52}$$

because of the orthogonality of the roots of unity. Then the output of the
system in Figure 7 becomes, in a similar fashion to (2):

$$\hat{X}(z) = \frac{1}{N} \mathbf{g}^T(z) \cdot \mathbf{H}_m(z) \cdot \mathbf{x}_m(z) \tag{53}$$

where $\mathbf{g}(z) = [G_0(z), \dots, G_{N-1}(z)]$, $\mathbf{x}_m(z) = [X(z), \dots, X(W^{N-1}z)]$ and
the i-th line of $\mathbf{H}_m(z)$ is equal to $[H_i(z), \dots, H_i(W^{N-1}z)]$ (line numbers
start from 0). One can similarly define the equivalent polyphase matrix
(see (16)) which now contains subfilters according to indices taken modulo N.

Assume now that $\mathbf{H}_m(z)$ is paraunitary. It follows from $\mathbf{H}_m(z) \cdot \tilde{\mathbf{H}}_m(z) =
N \cdot \mathbf{I}$ that:

$$1/N \sum_{k=0}^{N-1} H_i(W^k z) \tilde{H}_j(W^k z) = \delta(i - j) \tag{54}$$

that is:

$$< h_i(n + Nl), h_j(n + Nk) >= \delta(l - k)\delta(i - j) \ . \tag{55}$$

Thus, the filter impulse responses and their translates by multiples of N
form an orthonormal set. From $\tilde{\mathbf{H}}_m(z)\mathbf{H}_m(z) = \mathbf{I}$ it follows that the dual

synthesis filters $G_i(z) = \tilde{H}_i(z)$ will produce perfect reconstruction, and thus, the set:

$$\{h_0(n+Nl_0), h_1(n+Nl_1), \ldots h_{N-1}(n+Nl_{N-1})\}, \quad l_i \in Z \qquad (56)$$

is an orthonormal basis for $l^2(Z)$.

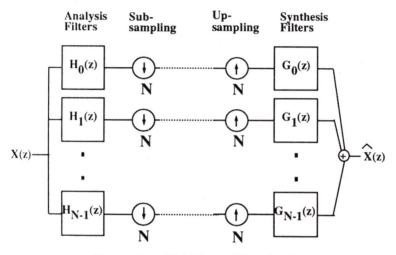

FIGURE 7.   Multichannel filter bank.

How can such paraunitary matrices be designed? Obviously, products of elementary paraunitary matrices (diagonal matrices of delays, unitary matrices) will lead to paraunitary products. More interestingly, such factorizations can be shown to be complete [47]. Three approaches are possible: (i) Product of diagonal matrices of delays and unitary matrices: this is a direct extension of the two channel lattice (19) and uses Givens rotations [17].
(ii) Product of Householder matrices: this is an alternative to the previous lattice and is numerically better conditioned [47].
(iii) Unitary state-space realization: following remark (ii) in section 2.1.1., one can design an arbitrary size paraunitary filter bank by using a unitary $\mathbf{R}$ (see (24)).

4.1.3.   MODULATED FILTER BANKS

Filter banks obtained from a single prototype filter by modulation are attractive both because of their conceptual simplicity (they seem like a "local" Fourier transform) and because of their reduced computational complexity (they can be implemented with FFT algorithms). Two cases of interest are discussed below.

(i) Modulation by roots of unity

Assume that the i-th filter is given by:

$$H_i(z) = H_{proto}(W^i z), \quad i = 0, \ldots, N-1, \quad W = e^{-j2\pi/N} \quad j = \sqrt{-1}. \quad (57)$$

That is, the N filters are equally spaced over the spectrum $[0, 2\pi]$ and $H_{proto}(z)$ is typically a real lowpass filter with a cutoff frequency at $2\pi/2N$ (corresponding to a bandwidth of $2\pi/N$).

This filter bank, subsampled by N, computes a critically sampled discrete short-time Fourier (or Gabor) transform, where the window function is given by the prototype filter. However, it is easy to verify the following negative result [51]:

Proposition 4.1. *There are no finite support orthonormal bases with filters as in (57) (except trivial ones with only N non-zero coefficients).*

The proof consists in noting that the polyphase matrix $\mathbf{H}_p(z)$ corresponding to this filter bank can be written as a Fourier matrix times a diagonal matrix which contains the polyphase components of $H_{proto}(z)$. The determinant of $\mathbf{H}_p(z)$ has to be a delay (for perfect FIR reconstruction), but equals a constant times the product of the polyphase components of $H_{proto}(z)$. Thus, each polyphase component can contain at most one non-zero coefficient. The only orthonormal solutions thus have a prototype filter with polyphase components equal to $a_i z^{-l}$ where a_i is a complex unity norm constant. This result is similar to the Balian-Low theorem for the continuous-time Gabor transform [15].

If IIR filters are allowed, it is possible to design paraunitary uniformly modulated filter banks by cascading a Fourier transform with allpass polyphase components, that is, $\mathbf{H}_p(z)$ is equal to:

$$\mathbf{H}_p(z) = \mathbf{F} \cdot Diag[p_0(z), p_1(z), \ldots, p_{N-1}(z)] \quad (58)$$

where $F_{ij} = e^{-2\pi ij/N}$, $Diag[.]$ is a diagonal matrix and $p_i(z)$ are allpass filters. The lowpass filter equals:

$$\mathbf{H}_{proto} = \sum_{i=0}^{N-1} z^{-i} \cdot p_i(z^N) \quad , \quad (59)$$

and the other filters are modulated versions due to the Fourier transform. With well designed allpass filters $p_i(z)$, this can lead to a good discrete-time orthogonal Gabor transform.

(ii) Cosine modulated filter banks

If one replaces the modulation by roots of unity by appropriate cosines, then one can actually design perfect reconstruction modulated FIR filter banks. This was done in [37, 32] for filters of length $L = 2N$, and that these banks are actually paraunitary was shown in [53]. Recently, the solution was generalized for $L = 2N$ [10] and extended to arbitrary length $L = KN$ [33]. The resulting filter bank imitates a real trigonometric transform with windows overlapping to neighboring blocks, and is thus sometimes called Lapped Orthogonal Transform (LOT). An example of the frequency response of an orthonormal 8 channel filter bank with cosine modulation is given in Figure 8 (from [26]). The filters are of length 16.

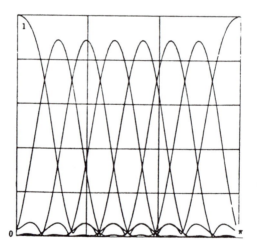

FIGURE 8. Frequency response of an orthonormal 8 channel cosine modulated filter bank.

4.1.4. FILTER BANKS WITH RATIONAL SAMPLING RATES

A final variation involves filter banks with rational sampling rate changes. Until now, such filter banks have been designed by cascading appropriate analysis and synthesis filter banks. For example, a filter bank with a rate 1/3 highpass and a rate 2/3 lowpass can be obtained by synthesizing the 2 lower bands of a three channel analysis bank. The result will be as shown in Figure 9. If all the filter banks are paraunitary, then the overall bank will be paraunitary as well. However, this method is indirect in that it goes through two steps when constructing filter banks with rational sampling rates. Recently, a direct design method has been developed [27] allowing for better filter designs and lower complexity.

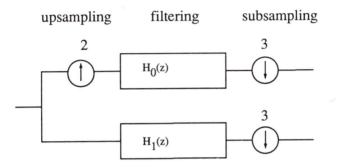

FIGURE 9.   Filter bank with rate 1/3 and 2/3 channels.

4.2.   *WAVELETS DERIVED FROM MULTICHANNEL FILTER BANKS*

Consider the tree structured filter bank case first, and assume that the lowpass filter $h(n)$ is regular and orthogonal to its even translates. Thus, there is a limit function $\phi(x)$ which satisfies a two scale equation as given in (42). But, obviously, $\phi(x)$ satisfies also two scale equations with scale changes by any power of 2 (by iteration). The linear combination is given by the iterated filter $h^{(i)}(n)$ (see (36)):

$$\phi(x) = 2^{i/2} \sum_{k=0}^{L^{(i)}-1} h^{(i)}(k)\phi(2^i x - k) \qquad (60)$$

Then, we can design different "wavelet" bases based on iterated low and highpass filters. Let us take a simple example. Consider the following four filters, corresponding to a four channel filter bank derived from a binary tree:

$$F_0(z) = H_0(z)H_0(z^2) \quad F_1(z) = H_0(z)H_1(z^2) \qquad (61)$$

$$F_2(z) = H_1(z)H_0(z^2) \quad F_3(z) = H_1(z)H_1(z^2) \qquad (62)$$

This corresponds to an orthogonal filter bank as we have seen. Call the impulse responses $f_i(n)$. Then, the following $\phi(x)$ is a scaling function (with scale change by 4):

$$\phi(x) = 2 \sum_k f_0(k)\phi(4x - k) \tag{63}$$

and the following 3 functions are "wavelets":

$$\psi_i(x) = 2 \sum_k f_i(k)\phi(4x - k), \quad i \in \{1, 2, 3\} \tag{64}$$

The set $\{\phi(x - k), \psi_1(x - l), \psi_2(x - m), \psi_3(x - n),\}$ is orthonormal, and $2^j \psi_i(4^j x - l_i), i \in \{1, 2, 3\}, l_i, j \in Z$ is an orthonormal basis for $L^2(R)$ following similar arguments as in the classic, "single" wavelet case (we have simply expanded two successive wavelet spaces into 3 spaces spanned by $\psi_i(x), i \in \{1, 2, 3\}$). Of course, this is a simple variation on the normal wavelet case (note that $\psi_1(x)$ is the usual wavelet). With these methods and the previously discussed concept of wavelet packets, it can be seen how to obtain continuous-time wavelet packets.

The case for general filter banks is very similar. Assume we have a size N filter bank, with a regular lowpass filter. This filter has to be regular with respect to subsampling by N (rather than 2), which amounts, in a similar fashion to Proposition 3.1., to having a sufficient number of zeros at the N-th roots of unity (the aliasing frequencies). The lowpass filter will lead to a scaling function satisfying:

$$\phi(x) = N^{1/2} \sum_k h_0(k)\phi(Nx - k) \tag{65}$$

and the $N - 1$ functions:

$$\psi_i(x) = N^{1/2} \sum_k f_i(k)\phi(Nx - k), \quad i = 1, \ldots N - 1 \tag{66}$$

will form a wavelet basis with respect to scale changes by N.

The problem with scale changes by $N > 2$ is that the resolution steps are even larger between a scale and the next coarser scale than for the typical "octave band" wavelet analysis. A finer resolution change could be obtained for rational scale changes between 1 and 2. In discrete time such finer steps can be achieved with filter banks having rational sampling rates (see section 4.1.4.). The situation is more involved in the continuous time case. In particular, the iterated filter bank method does not lead to regular wavelets. However, solutions not having compact support have been derived by P. Auscher [2].

5. THE MULTIDIMENSIONAL CASE

The multidimensional case is both more interesting and more challenging than the one dimensional one. While the separable case (separable sampling and/or separable filters) is a straightforward extension of the one dimensional case, non-separable sampling and filters broaden the possibilities and can lead to some interesting wavelet designs. However, a number of questions remain to be solved both on design and on regularity.

5.1. *MULTIDIMENSIONAL FILTER BANKS*

This will only be a brief treatment, and we refer to [25, 56, 28] for additional details. Besides the fact that signals and filters are now multidimensional, the major difference with the one dimensional case is that subsampling now uses lattices, meaning, there can be many ways to subsample by N in more than one dimension. More precisely, given a signal on an input lattice, the signal subsampled by N will live on a sublattice of the input, and this sublattice has a "density" which is N times smaller. Such a sublattice is given by all integer linear combinations of m integer basis vectors (where m is the dimension). Let us consider the two dimensional case for simplicity. Calling $\mathbf{D}$ the matrix of the basis vectors for the sublattice, a point (u_1, u_2) in the sublattice corresponds to the following point (n_1, n_2) in the input lattice:

$$\begin{pmatrix} n_1 \\ n_2 \end{pmatrix} = \begin{pmatrix} d_{00} & d_{01} \\ d_{10} & d_{11} \end{pmatrix} \cdot \begin{pmatrix} u_1 \\ u_2 \end{pmatrix} = \mathbf{D} \cdot \mathbf{u} \quad . \tag{67}$$

$\mathbf{D}$ has integer entries and is called the sampling matrix. The subsampling factor is given by $N = Det[\mathbf{D}]$. Note that $\mathbf{D}$ is non-unique, for example, the following 3 matrices are possible representations for the quincunx sublattice (this sublattice consists of all points with indexes (n_1, n_2) where $n_1 + n_2$ is even):

$$\begin{pmatrix} 2 & 1 \\ 0 & 1 \end{pmatrix}, \quad \begin{pmatrix} 1 & 1 \\ 1 & -1 \end{pmatrix}, \quad \begin{pmatrix} 1 & 1 \\ -1 & 1 \end{pmatrix} \quad . \tag{68}$$

These matrices are related through multiplication by a unimodular matrix (an integer matrix with determinant equal to ± 1). A quincunx lattice is shown in Figure 10.

Subsampling according to $\mathbf{D}$ followed by resampling amounts to keeping all samples on the sublattice, and replacing all others by zero. For example,

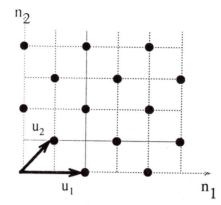

FIGURE 10.   Quincunx sublattice of rectangular lattice.

a two dimensional signal with z-transform $X(z_1, z_2)$ results, after quincunx
subsampling and resampling, in:

$$Y(z_1, z_2) = 1/2 \left[X(z_1, z_2) + X(-z_1, -z_2) \right] \quad . \tag{69}$$

5.1.1.   THE SEPARABLE CASE

When subsampling and filters are separable, then the resulting system is
a trivial extension of the one dimensional case, since it is a tensor product
of one dimensional systems. All the properties will follow from properties
of the individual one dimensional systems. Note that separable sampling
can be written with a diagonal sampling matrix $\mathbf{D}$ (basis vectors of the
sublattice are colinear with the ones of the input lattice), and that separable
filters are convolutions of one dimensional ones. The classic example is 2×2
separable subsampling, given by:

$$\mathbf{D} = \begin{pmatrix} 2 & 0 \\ 0 & 2 \end{pmatrix} \tag{70}$$

and filters:

$$F_0(z_1, z_2) = H_0(z_1)H_0(z_2), \quad F_1(z_1, z_2) = H_0(z_1)H_1(z_2), \tag{71}$$
$$F_2(z_1, z_2) = H_1(z_1)H_0(z_2), \quad F_3(z_1, z_2) = H_1(z_1)H_1(z_2) \quad . \tag{72}$$

This can be implemented in a tree structure, with a splitting in the first
dimension, followed by a splitting of each channel in the other dimension.
The polyphase matrix (which is defined similarly to the one dimensional
case, except that now modulo N becomes modulo a sublattice, leading to
cosets) can be written as:

$$\mathbf{H}_p(z_1, z_2) = \mathbf{H}_p(z_1) \otimes \mathbf{H}_p(z_2) \tag{73}$$

where $\otimes$ is the Kronecker product.

Even if the sublattice is separable, one can use true multidimensional filters. While the computational complexity will be increased, one gains additional freedom. For example, there are linear phase orthogonal FIR filter banks for 2×2 separable subsampling [24, 25], but the filters are non-separable (we know there is no separable solution from the one dimensional case). The reason is that centro-symmetry is sufficient for linear phase behavior in two dimensions, which is less restrictive than what a separable linear phase solution would require.

5.1.2. NON-SEPARABLE CASE

We will consider the quincunx subsampling in more detail because it leads to a two-channel filter bank (shown in Figure 11), and is thus the two dimensional equivalent of the classic two channel filter bank used to construct orthonormal wavelet bases with scaling by 2 in one dimension.

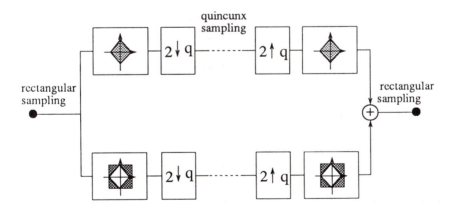

FIGURE 11. Two dimensional filter bank with two channels using quincunx subsampling.

In a similar fashion to (2) and because of (69), the output of an analysis/synthesis system using quincunx subsampling (as shown in Figure 11) is equal to:

$$\hat{X}(z_1, z_2) = \frac{1}{2}[G_0(z_1, z_2) \; G_1(z_1, z_2)]$$

$$\times \begin{bmatrix} H_0(z_1, z_2) & H_0(-z_1, -z_2) \\ H_1(z_1, z_2) & H_1(-z_1, -z_2) \end{bmatrix} \begin{bmatrix} X(z_1, z_2) \\ X(-z_1, -z_2) \end{bmatrix} . \tag{74}$$

Then, a lowpass filter which is of norm 1 and orthogonal to its shifted versions on a quincunx lattice, that is:

$$H_0(z_1, z_2)\tilde{H}_0(z_1, z_2) + H_0(-z_1, -z_2)\tilde{H}_0(-z_1, -z_2) = 2 \tag{75}$$

together with a highpass filter equal to:

$$H_1(z_1, z_2) = -z_1^{-1}\tilde{H}_0(-z_1, -z_2) \tag{76}$$

will yield a paraunitary perfect reconstruction filter bank. All FIR solutions will be of this form (within shifts) [28].

In terms of polyphase components, one can verify that

$$H_{00}(z_1, z_2)\tilde{H}_{00}(z_1, z_2) + H_{01}(z_1, z_2)\tilde{H}_{01}(z_1, z_2) = 1 \tag{77}$$

that is, they have to be power complementary. The similarity of these results with the one dimensional case is striking, and is essentially due to the fact that we have a two channel system. That is why the results carry over to two channel systems in *any* number of dimensions [28].

However, what does not carry over obviously are the design methods from the one dimensional case which are based on factorizations. While cascade structures can be developed that yield desirable filter banks [56, 28] (paraunitary or linear phase), the completeness results are missing, except for very small cases.

5.2.   *WAVELETS IN MULTIPLE DIMENSIONS*

Wavelets in multiple dimensions can again be obtained as limits of iterated filter banks, just as in the one dimensional case. However, the sampling matrix $\mathbf{D}$ (which is not unique) plays a critical role, because iterated subsampling corresponds to taking powers of $\mathbf{D}$. So, for example, considering quincunx subsampling and two of the matrices from (68), we notice:

$$\begin{pmatrix} 1 & 1 \\ 1 & -1 \end{pmatrix}^2 = \begin{pmatrix} 2 & 0 \\ 0 & 2 \end{pmatrix}, \quad \begin{pmatrix} 2 & 1 \\ 0 & 1 \end{pmatrix}^2 = \begin{pmatrix} 4 & 3 \\ 0 & 1 \end{pmatrix} \tag{78}$$

that is, one iteration leads to a separable subsampling, while the other doesn't. Therefore, the same original filter iterated following different sub-

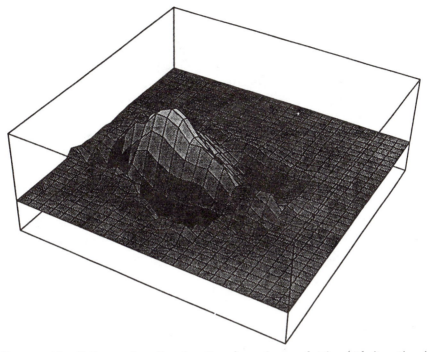

FIGURE 12. Orthogonal scaling function for quincunx lattice (4th iteration is shown).

sampling matrices (but representing the same sublattice) can converge to very different limits. The most striking examples, discovered by Gröchenig and Madych [21] are the equivalent of the Haar basis (simple N-point average as the lowpass filter) which can lead to indicator functions with fractal boundaries (dragons). By construction, they form a self-similar tiling of the plane [21]. Therefore, for a given sublattice like quincunx, the regularity of the iterated filter will depend on the choice of the sampling matrix $\mathbf{D}$.

Let us consider the simplest, separable case first ($\mathbf{D}$ as in (70)). This will lead to a separable wavelet basis with scale changes by 2 in each dimension. It involves one scaling function $\phi(x, y)$ corresponding to $F_0(z_1, z_2)$ in (71) and three wavelets $\psi_i(x, y)$ corresponding to $F_i(z_1, z_2)$, $i = 1, 2, 3$ in (71) [29]. Regularity follows from the regularity of the individual one dimensional filters.

The non-separable cases are more interesting, and we consider the natural two channel case based on the quincunx lattice. Assume we have an orthogonal filter bank (see (75,76)). The iterated lowpass filter can be made regular by placing a sufficient number of zeros at the aliasing frequency $[\omega_1 = \pi, \omega_2 = \pi]$. This is a necessary condition, but not sufficient.

In particular, regularity will depend on the matrix choosen to represent the lattice. Assuming regularity, the limit function defined similarly as in the one dimensional case will be orthogonal with respect to integer shifts, and it will satisfy a two scale equation with respect to a scale change given by the matrix $\mathbf{D}$. The wavelet follows by replacing the lowpass with the highpass filter in the two scale equation. Such a wavelet, together with integer shifts and scales (by $\mathbf{D}$) will form an orthonormal basis of $L^2(R^2)$ (under certain conditions, and for appropriate $\mathbf{D}$'s) [28]. Note that now a filter is regular with respect to a particular matrix representing a lattice.

Figure 12 shows a regular scaling function with compact support designed by putting a maximum number of zeros at $[\pi, \pi]$ in the lowpass filter of an orthogonal filter bank. The design process is now more complicated than in the one dimensional case, due to the absence of a factorization theorem for multidimensional polynomials. Biorthogonal examples can be generated as well. However, a number of questions remain open for both design procedures and regularity of such non-separable filter banks and wavelets [28].

6. IMPLEMENTATION AND APPLICATION CONSIDERATIONS

6.1. NUMERICAL BEHAVIOR

Efficient implementations are often realized with fixed point arithmetic, for example in very large scale integrated silicon circuits. This means that both filter coefficients and results of intermediate computations are coarsely quantized. The latter source of error cannot be eliminated but only minimized with a careful design of the finite precision arithmetic. As far as the quantization of the filter coefficients is concerned, we have seen that lattice structures would remain insensitive in the sense that properties like orthogonality are preserved. But this does not mean that regularity would be maintained (in an exact sense) since the implementation results in a nearby orthogonal filter. It is not expected that this would present a problem with the number of stages that are used in practice.

6.2. COMPUTATIONAL COMPLEXITY

It is well known that the complexity of the discrete wavelet transform is at most twice that of its first stage (due to the subsampling of the later stages), and this results in a complexity of order $O(L)$ per input sample where L is the filter length.

Comparisons with the fast Fourier transform (FFT) are often done in an oversimplified manner, since one would rarely compute an FFT of the total signal, but rather take a short-time Fourier transform (STFT) which has $O(LogM)$ complexity per input sample (M being the number of frequency points, and the window length being a small multiple of M typically). Comparisons are further obscured by the fact that one does not compute comparable things in the first place. It is clear, however, that the block processing structure of the STFT is favored for implementations. Multirate structures, especially unbalanced ones like the logarithmic tree used for the discrete wavelet transform (DWT), are difficult to implement, because various signals have widely different sampling rates, requiring complex multiplexing and resource allocation.

Since the DWT is based on convolution computations, it can actually be sped up by using the FFT [6] to compute the various convolutions. This reduces the complexity from $O(L)$ to $O(LogL)$. Further improvements can be achieved by performing subsampling in frequency domain as well [54].

Note also that complete trees like the ones appearing in wavelet packets (rather than logarithmic trees as in the DWT) raise the complexity to $O(KL)$ where K is the number of leaves.

6.3. *APPLICATIONS*

In signal processing, one of the major applications of filter banks will remain source coding or signal compression. The insight gained with wavelets concerning regularity and zero moments changes the design of filters used in such coding schemes. So far, only preliminary results are available regarding the importance of regularity in coding [1]. It is clear that one would like to avoid very irregular filters, but most filters used in practice are almost regular, and should not present any problem when iterated a few steps. Filters having zero moments have nice theoretical properties as far as compressing smooth functions, and these properties have been used in numerical analysis and signal compression [5, 9, 57].

However, one should keep in mind the ultimate goal which is to minimize bitrate *and* distortion, and subband or wavelet decomposition is only an initial step in this process. From rate distortion theory, it is known that an optimal code will need to look at the whole data set, and that successive approximation is suboptimal except in very particular and limited cases [19]. But, optimal solutions are usually far too complex in practice, and thus, subband or wavelet decompositions can be an interesting compromise between complexity and optimality, especially when multiple resolutions are desired as it is the case for certain applications.

ACKNOWLEDGEMENTS

The author would like to acknowledge the fruitful collaboration with C. Herley and J. Kovačević of Columbia University, and the useful discussions with I. Daubechies (AT&T) and O. Rioul (CNET).

REFERENCES

[1]   M. Antonini, M. Barlaud, P. Mathieu and I. Daubechies, "Image coding using vector quantization in the wavelet transform domain," Proc. IEEE ICASSP, pp.2297-2300, Albuquerque NM, April 1990.

[2]   P.Auscher, "Wavelet bases for $L^2(R)$ with rational dilation factor," in **Wavelets and their Applications**, G. Beylkin et al Eds, Jones and Bartlett, Boston, MA, 1991.

[3]   V. Belevitch, **Classical Network Synthesis**, Holden Day, San Francisco, CA , 1968.

[4]   M. G. Bellanger and J. L. Daguet, "TDM-FDM transmultiplexer: digital polyphase and FFT," IEEE Trans. on Commun., Vol.22, No.9, pp. 1199-1204, Sept. 1974.

[5]   G. Beylkin, R. Coifman and V. Rokhlin, "Fast wavelet transforms and numerical algorithms I," Dept. of Math, Yale Univ. Preprint, 1989.

[6]   R. Blahut, **Fast Algorithms for Digital Signal Processing**, Addison-Wesley, Reading, MA, 1984.

[7]   P. J. Burt and E. H. Adelson, "The Laplacian pyramid as a compact image code," IEEE Trans. on Commun., Vol. 31, No.4, April 1983, pp.532-540.

[8]   A. Cohen, I. Daubechies and J.-C. Feauveau, "Biorthogonal bases of compactly supported wavelets," to appear, Commun. on Pure and Applied Math.

[9]   R. R. Coifman, Y. Meyer, S. Quake and M. V. Wickerhauser "Signal processing and compression with wavelet packets," Dept. of Math, Yale Univ. Preprint, 1990.

[10] R. R. Coifman and Y. Meyer, Note aux Comptes Rendus de l'Académie des Sciences, Paris, to appear.

[11] R. E. Crochiere, S. A. Weber and J. L. Flanagan, "Digital coding of speech in subbands," Bell Syst. Tech. J., Vol.55, pp.1069-1085, Oct.1976.

[12] A. Croisier, D. Esteban and C. Galand, "Perfect channel splitting by use of interpolation, decimation, tree decomposition techniques," Int. Conf. on Information Sciences/Systems, Patras, pp. 443-446, Aug. 1976.

[13] I. Daubechies, "Orthonormal bases of compactly supported wavelets," Commun. on Pure and Applied Mathematics, Vol. XLI, 909-996, 1988.

[14] I. Daubechies, "Orthonormal bases of compactly supported wavelets II: variations on a theme," AT&T preprint 1989.

[15] I. Daubechies, "The wavelet transform, time-frequency localization and signal analysis," IEEE Trans. on Info. Theory, Vol.36, No.5, Sept. 1990, pp.961-1005.

[16] I. Daubechies, **Ten Lectures on Wavelets**, CBMS-NSF Series on Applied Mathematics, SIAM, to appear, 1992.

[17] Z. Doğanata, P. P. Vaidyanathan and T. Q. Nguyen, "General synthesis procedures for FIR lossless transfer matrices, for perfect reconstruction multirate filter bank applications," IEEE Trans. on ASSP, Vol.36, No.10, Oct.1988, pp.1561-1574.

[18] Z. Doğanata and P. P. Vaidyanathan, "Minimal structures for the implementation of digital rational lossless systems," submitted for publication.

[19] W. H. R. Equitz, "Successive refinement of information," Ph. D Thesis, Stanford Univ., 1989.

[20] P. Goupillaud, A. Grossmann and J. Morlet, "Cycle-octave and related transforms in seismic signal analysis," Geoexploration, 23, pp.85-102, 1984.

[21] K. Gröchenig and W. R. Madych, "Multiresolution analysis, Haar bases and self-similar tilings of R^n," submitted for publication, 1990.

[22] A. Grossmann and J. Morlet, "Decomposition of Hardy functions into square integrable wavelets of constant shape," SIAM, J. Math. Anal., Vol.15, No.4, pp.723-736, 1984.

[23] C. Herley and M. Vetterli, "Linear phase wavelets: theory and design," Proc. IEEE ICASSP-91, pp. 2017-2020.

[24] G. Karlsson, M. Vetterli and J. Kovacevic "Non-separable two-dimensional perfect reconstruction filter banks," Proc. of the SPIE Conf. on Visual Communications and Image Processing, Nov. 1988, pp.187-199.

[25] G. Karlsson and M. Vetterli, "Theory of two-dimensional multirate filter banks," IEEE Trans. on ASSP, Vol.38, No.6, pp.925-937, June 1990.

[26] J. Kovačević, D. Le Gall and M. Vetterli, "Image coding with windowed modulated filter banks," Proc. IEEE ICASSP, May 1989, pp.1949-1952.

[27] J. Kovačević and M. Vetterli, "Perfect reconstruction filter banks with rational sampling rate changes," Proc. of IEEE ICASSP, Toronto, May 1991, pp.1785-1788.

[28] J. Kovačević and M. Vetterli, "Multidimensional non-separable filter banks and wavelets bases for R^n," to appear, IEEE Trans. on Info. Theory, Feb. 1992.

[29] S. Mallat, "A theory for multiresolution signal decomposition: the wavelet representation," IEEE Trans. on PAMI, Vol.11, No.7, pp.674-693, 1989.

[30] S. Mallat, "Multiresolution approximations and wavelet orthonormal bases in $L^2(R)$," Trans. of the American Math. Society, Vol. 315, No.1, Sept. 1989, pp.69-87.

[31] S. Mallat, "Multifrequency channel decompositions of images and wavelet models," IEEE Trans. ASSP, Vol.37, No.12, pp.2091-2110, Dec.1989.

[32] H. S. Malvar and D. H. Staelin, "The LOT: transform coding without blocking effects," IEEE Trans. ASSP, Vol.37, No.4, pp.553-559, April 1989.

[33] H. S. Malvar, "Modulated QMF filter banks with perfect reconstruction," Electronics Letters, Vol.26, No.13, June 21 1990, pp.906-907.

[34] Y. Meyer, **Ondelettes**, Vol.1, Hermann, Paris 1990.

[35] F. Mintzer, "Filters for distortion-free two-band multirate filter banks," IEEE Trans. on ASSP, Vol.33, pp.626-630, June 1985.

[36] A. V. Oppenheim and R. W. Schafer, **Discrete Time Signal Processing**, Prentice-Hall, 1989.

[37] J. Princen and A. Bradley, "Analysis/synthesis filter bank design based on time domain aliasing cancellation," IEEE Trans. on ASSP, Vol.34, No. 5, pp.1153-1161, Oct. 1986.

[38] O. Rioul, "A unifying multiresolution theory for the discrete wavelet transform, regular filter banks and pyramid transforms," submitted to IEEE Trans. ASSP, June 1990.

[39] M. J. T. Smith and T. P. Barnwell, "A procedure for designing exact reconstruction filter banks for tree structured sub-band coders," Proc. IEEE ICASSP, San Diego, March 1984.

[40] M. J. T. Smith and T. P. Barnwell, "A unifying framework for analysis/synthesis systems based on maximally decimated filter banks," Proc. IEEE ICASSP-85, pp. 521-524, Tampa, March 1985.

[41] M. J. T. Smith and T. P. Barnwell, "Exact reconstruction for tree-structured subband coders," IEEE Trans. on ASSP, Vol.34, pp. 434-441, June 1986.

[42] M. J. T. Smith and T. P. Barnwell, "A new filter bank theory for time-frequency representation," IEEE Trans. on ASSP, Vol.35, No.3, March 1987, pp. 314-327.

[43] G. Strang, "Wavelets and dilation equations: a brief introduction," SIAM Review, Vol.31, No.4, pp.614-627, Dec. 1989.

[44] P. P. Vaidyanathan, "Theory and design of M-channel maximally decimated quadrature mirror filters with arbitrary M, having perfect reconstruction property," IEEE Trans. on ASSP, Vol.35, No.4, pp.476-492, April 1987.

[45] P. P. Vaidyanathan, "Quadrature mirror filter banks, M-band extensions and perfect-reconstruction technique," IEEE ASSP Magazine, Vol. 4, No. 3, pp.4-20, July 1987.

[46] P. P. Vaidyanathan and P.-Q. Hoang, "Lattice structures for optimal design and robust implementation of two-band perfect reconstruction QMF banks," IEEE Trans. on ASSP, Vol.36, No.1, pp.81-94, Jan. 1988.

[47] P. P. Vaidyanathan, T. Q. Nguyen, Z. Doğanata and T. Saramäki, "Improved technique for design of perfect reconstruction FIR QMF banks with lossless polyphase matrices," IEEE Trans. on ASSP, Vol.37, No.7, July 1989, pp.1042-1056.

[48] P. P. Vaidyanathan and Z. Doğanata, "The role of lossless systems in modern digital signal processing," IEEE Trans. Education, Special issue on Circuits and Systems, Vol. 32, No.3, Aug. 1989, pp.181-197.

[49] M. Vetterli, "Multi-dimensional sub-band coding: some theory and algorithms," Signal Processing, Vol. 6, No.2, pp. 97-112, Feb. 1984.

[50] M. Vetterli, "Splitting a signal into subsampled channels allowing perfect reconstruction," Proc. of the IASTED Conf. on Applied Signal Processing and Digital Filtering, Paris, June 1985.

[51] M. Vetterli, "Filter banks allowing perfect reconstruction," Signal Processing, Vol.10, No.3, April 1986, pp.219-244.

[52] M. Vetterli, "A theory of multirate filter banks," IEEE Trans. ASSP, Vol.35, No.3, pp.356-372, March 1987.

[53] M. Vetterli and D. Le Gall, "Perfect reconstruction FIR filter banks: some properties and factorizations," IEEE Trans. on ASSP, Vol.37, No.7, pp.1057-1071, July 1989.

[54] M. Vetterli and C. Herley, "Wavelets and filter banks: relationships and new results," Proc. IEEE ICASSP, pp.1723-1726, Albuquerque NM, April 1990.

[55] M. Vetterli and C. Herley, "Wavelets and filter banks: theory and design," to appear, IEEE Trans. on ASSP, Sept. 1992.

[56] M. Vetterli, J. Kovačević and D. LeGall, "Perfect reconstruction filter banks for HDTV representation and coding," Image Communication, Vol.2, No.3, Oct.1990, pp.349-364.

[57] M. V. Wickerhauser, "Acoustic signal compression with wave packets," Dept. of Math., Yale Univ. Preprint, 1989.

WAVELETS FOR THE QUINCUNX PYRAMID

JEAN-CHRISTOPHE FEAUVEAU

Matra MS2i,
Laboratoire de Traitement de l'Image
et du Signal
38 Bd Paul Cézanne
78052 Saint Quentin en Yvelines, FRANCE

1. INTRODUCTION

The wavelet theory, as a multiresolution concept, [13, 14, 11, 12, 5] provides powerful tools for signal and image processing. From an algorithmic point of view, it is well known that 1D and 2D multiresolution analysis leads to dyadic pyramidal implementations using filter banks. For introductory details on filter banks, see the previous chapter by Vetterli on "Wavelet and Filter Banks for Discrete-Time Signal Processing" [20] for more details on filter banks and multiresolution analysis.

Before wavelet theory, filter banks were mainly used in speech and image coding systems [18]. The multiresolution interpretation of these filter banks allows their design with the use of associated wavelet criteria to improve sub-band coding results [2].

The natural extension to the multiresolution analysis of images utilizes the tensor product of two 1D analysis. The associated algorithm uses a separable four-channel filter bank: it is the separable pyramid discussed in [20]. But other pyramids also exist, the quincunx pyramid, for instance shows interesting properties: its associated filter bank uses only two channels, the filters are non-oriented and finally, the analysis step of the image is twice thinner than that of the separable pyramid.

This paper links the quincunx pyramid and a new type of multiresolution approximation, and explains a direct method method for building the corresponding filters. The quincunx sampling is introduced in section 5 of Vetterli's chapter. The purpose of this paper is to formulate a theoretical 2D wavelet framework with an associated algorithm which is precisely the quincunx pyramid. Then we derive new results describing a fast implementation of this pyramid. Note that the quincunx pyramid optimized with a wavelet criterion has been already used as the first step of a sub-band coder leading to a compression rate of 100 and preserving image quality very well [3].

This paper is organized as follows: in Section 2, we briefly describe the principle of the quincunx pyramid. In Section 3, we develop a (biorthogonal) wavelet multiresolution analysis related to this pyramid. This approach includes the orthogonal environment developed in [7]. Using the same method as the 1D case, the notion of wavelet related to the quincunx pyramid leads to explaining the associated filter banks as alternated filters. This concept is introduced in Section 4. Finally we develop tools dedicated to the use of 1D and 2D alternated filters with an important application: nxn kernels for quincunx pyramid, known as non-separable [18], can be implemented with only 2xn multiplication/pel. From a computational view-point, it has exactly the same cost as the usual 2D dyadic separable algorithm.

2. THE QUINCUNX PYRAMID

In order to simplify formulas, we use the following notations in the 2D case (to be consistent with the 1D case):

$$\omega = (\omega_1, \omega_2), x = (x_1, x_2) \text{ and } \omega + \pi = (\omega_1 + \pi, \omega_2 + \pi).$$

The basic idea of the quincunx pyramid algorithm is to apply a two-channel filter bank (instead of the 4-channel one used in the dyadic case) using non separable 2D filters : a low-pass filter H_0 and a high-pass one H_1 [18]. Such a system in its "bi-orthogonal" version (described later) is sketched by the decomposition part of Figure 1.

Note that this sub-sampling grid is the one obtained by the $T(x)$ transform on the initial grid Z^2, where $T(x) = T(x_1 + x_2, x_1 - x_2)$.

In this Figure, the transformation $T(x) = T(x_1 + x_2, x_1 - x_2)$ defines the sub-sampling grid of the quincunx pyramid: applied to the basic grid Z^2, it

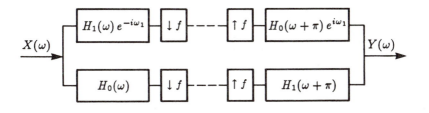

FIGURE 1. Bi-orthogonal filter bank for the quincunx pyramid. In the decomposition part (on the left), the 2D signal is filtered by H_0 and H_1, and then decimated with T:↓T is a 2D downsampling by a factor of 2. We only keep samples (n_1, n_2) with $n_1 + n_2$ even. For the reconstruction part ↑T replaces the dropped samples by zero. The scheme is the same for the 1D bi-orthogonal case when replacing "ω_1" by "ω" and $T(x_1, x_2)$ by $T(x) = 2x$.

performs a pixel decimation by a factor of 2. As a scaling transformation, $T(x)$ keeps the angles unchanged and scales the distances by a factor $\sqrt{2}$. Note that $T^2 = ToT = 2Id$, which is the basic sub-sampling of the dyadic algorithm.

We can verify that the outputs of the decomposition are:

$$Y_0(T(\omega)) = \frac{1}{2}\{H_0(\omega)X(\omega) + H_0(\omega + \pi)X(\omega + \pi)\},$$

$$Y_1(T(\omega)) = \frac{1}{2}\{H_1(\omega)X(\omega) - H_1(\omega + \pi)X(\omega + \pi)\}e^{-i\omega_1}.$$

After reconstruction, we obtain:

$$X'(\omega) = \frac{1}{2}\{H_0(\omega)H_1(\omega + \pi) + H_1(\omega)H_0(\omega + \pi)\}X(\omega)$$

and the aliasing term is cancelled. Up to a delay, the exact reconstruction formula is:

$$H_0(\omega)H_1(\omega + \pi) + H_1(\omega)H_0(\omega + \pi) = 2. \tag{1}$$

To produce the quincunx pyramid, we have to iterate the decomposition on the low-pass filtered output.

Note that H_0, H_1 filters are rotated by $\frac{\pi}{4}$ which yields H_0', H_1' (in order to follow quincunx geometry).

3.   MULTIRESOLUTION ANALYSIS ON QUINCUNX PYRAMID

We have defined the quincunx pyramid as a filter bank algorithm. We know demonstrate how to define a multiresolution analysis using wavelets which

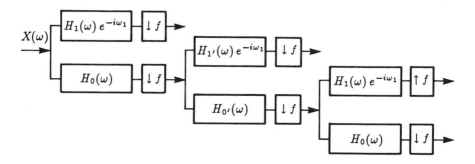

FIGURE 2.   Quincunx pyramid diagram.

admits a quincunx pyramid as a decomposition algorithm.

According to the decimation transformation T defined on Z^2 for the quincunx pyramid, the following notation is defined:

$$f_{j,k}(x) = 2^j f(T^{2j}(x) - k), \text{ for } j \in \frac{1}{2}Z.$$

The application T is the composition of an isometry with a dilation by a factor of $\sqrt{2}$, j is a scaling parameter, while k is a shift parameter in Z^2. Note that for $j \in Z$, we find for $f_{j,k}(x)$ the usual 2D wavelet notation with one scale level inserted between dyadic scales.

We also use the Fourier transform:

$$\hat{f}(\omega) = \int_{R^2} f(x)e^{i\omega x} dx$$

as well as the scalar product in $L^2(R^2)$:

$$<f,g> = \int_{R^2} f(x)\overline{g}(x)dx$$

Using the transformation T, the wavelet representation of an image $I(x)$ consists of writing the following unique decomposition:

$$I(x) = \sum_{j,k \in \Lambda} C_{j,k}\psi_{j,k}(x), \tag{2}$$

where Λ is the Cartesian product $\frac{1}{2}Z \times Z^2$. The function $\psi(x)$ enabling this decomposition is known as a *wavelet*.

The construction of wavelets was a difficult task until S. Mallat [11, 12] proposed to associate them with filter banks in a multiresolution framework. We have to adapt this method to the quincunx pyramid.

Let $H_0(\omega)$ and $H_1(\omega)$ be two 2π-periodic functions such that

$$H_i(0,0) = 1, H_i(\pi,\pi) = 0$$

and

$$H_0(\omega)H_1(\omega) + H_0(\omega + \pi)H_1(\omega + \pi) = 1.$$

We then define two functions $\phi(x)$ and $\phi^*(x)$ by their Fourier transforms:

$$\hat{\phi}(\omega_1 + \omega_2, \omega_1 - \omega_2) = \hat{\phi}(T(\omega)) = H_1(\omega)\hat{\phi}(\omega) = \prod_{n=0}^{+\infty} H_1(T^{-n}(\omega)) \quad (3)$$

and

$$\hat{\phi}^*(\omega_1 + \omega_2, \omega_1 - \omega_2) = \hat{\phi}^*(T(\omega)) = \overline{H}_0(\omega)\hat{\phi}^*(\omega) = \prod_{n=0}^{+\infty} \overline{H}_0(T^{-n}(\omega)), \quad (4)$$

where the $\overline{z}$ indicates the complex conjugate of z. When some additional technical conditions are fulfilled, the functions

$$\hat{\psi}(T(\omega)) = H_0(\omega + \pi)e^{-i\omega_1}\hat{\phi}(\omega) \quad (5)$$

and

$$\hat{\psi}^*(T(\omega)) = \overline{H}_1(\omega + \pi)e^{-i\omega_1}\hat{\phi}^*(\omega) \quad (6)$$

leads to verification of equation (2) when $C_{j,k} = <I, \psi_{i,k}^*>$, for all images $I(x)$ in $L^2(Z^2)$,

$$I(x) = \sum_{(j,k)\in\Lambda} <I, \psi_{j,k}^*> \psi_{j,k}(x). \quad (7)$$

The function $\psi^*(x)$ which enables the computation of the decomposition coefficients is called the *dual wavelet* of the wavelet $\psi(x)$. Thus, the function $\psi^*(x)$ is used for image analysis, while $\psi(x)$ is used for image reconstruction (or synthesis). Let us notice that $\psi^*(x)$ is also a wavelet since the following can be written:

$$I(x) = \sum_{(j,k)\in\Lambda} <I, \psi_{j,k}> \psi_{j,k}^*(x). \quad (8)$$

From the uniqueness of the decomposition, we can easily verify that $<\psi_{j,k}^*, \psi_{j',k'}> = \delta_{j,j'}\delta_{k,k'}$ where $\delta_{n,n'} = 1$ if $n = n'$ and 0 otherwise. This is nothing else than the definition of the bi-orthogonality between the families $(\psi_{j,k})_{(j,k)\in\Lambda}$ and $(\psi_{j,k}^*)_{(j,k)\in\Lambda}$.

The functions $\phi(x)$ and $\phi^*(x)$ are not only a means for a wavelets definition, they actually, define the multiresolution analysis notion. Consider the family of linear applications $(A_j)_{j \in \frac{1}{2} Z}$ defined by

$$A_j(I)(x) = \sum_{-\infty}^{+\infty} < I, \phi^*_{j,k} > \phi_{j,k}(x). \qquad (9)$$

These applications yield continuous projectors which degrade the information contained in $I(x)$ as j decreases, and which provide an increasing better approximation of $I(x)$ as j increases.

More precisely, let V_j be the projection space of A_j. Definition (3) allows us to write the following inclusions :

$$0 \ldots \subset V_j \subset V_{j+\frac{1}{2}} \subset V_{j+1} \subset V_{j+\frac{3}{2}} \ldots \subset L^2(R^2). \qquad (10)$$

This is a multiresolution analysis associated with the resolution factor $\sqrt{2}$ introduced by the transformation T, where $A_j(I)(x)$ can be interpreted as a version of $I(x)$ seen at scale $2^j (j \in \frac{1}{2} Z)$.

The task of the wavelet is therefore to extract the details lost between two consecutive scales. From relations (3) to (6), we can show that for every j in $\frac{1}{2} Z$:

$$A_{j+1}(I)(x) - A_j(I)(x) = \sum_{-\infty}^{+\infty} < I, \psi^*_{j,k} > \psi^*_{j,k}(x), \qquad (11)$$

and we retrieve formula (7), which is nothing more than a simple consequence of the fact that the approximation improves as j increases. All the major ideas defining multiresolution analysis in a 1D orthogonal context are given in [11]. Subsequent efforts have focused on a non-orthogonal version of such multiresolution analysis [4, 9] to make the use of this method easier.

4.   THE ALGORITHM

To use the multiresolution analysis, we have to be able to compute the projections $A_j(I)$ and $(A_{j+1} - A_j)(I)$. By working on coordinates of the bases, we now prove that the quincunx pyramid provides the decomposition algorithm of the multiresolution analysis defined previously.

We write $S_{j,k} =< I, \phi^*_{j,k} >$ and $C_{j,k} =< I, \psi^*_{j,k} >$. The sequence $(S_{j,k})_{k \in Z^2}$ is a discrete representation of the image $I(x)$ at scale 2^j, while $(C_{j,k})_{k \in Z^2}$ provides a discrete representation of the details lost between the scales $2^{j+1/2}$ and 2^j. The increasing property of projectors $(A_j)_{j \in \frac{1}{2} Z}$ shows that there must be a linear transformation from $(S_{j+1,k})_{k \in Z^2}$ to $((S_{j,k})_{k \in Z^2}, (C_{j,k})_{k \in Z^2})$. Let

$$H_0(\omega) = \frac{1}{\sqrt{2}} \sum_{n \in Z^2} h_0(n_1, n_2) e^{i(n_1 \omega_1 + n_2 \omega_2)},$$

and

$$H_1(\omega) = \frac{1}{\sqrt{2}} \sum_{n \in Z^2} h_1(n_1, n_2) e^{i(n_1 \omega_1 + n_2 \omega_2)}.$$

From definitions (3) to (6), we can develop the decomposition and reconstruction formulas between two consecutive scales.

Decomposition:

$$S_{j,(k_1,k_2)} = \sum_{p \in Z^2} S_{j+1/2, T(k_1,k_2)-(p_1,p_2)} h_0(p_1, p_2),$$

$$C_{j,(k_1,k_2)} = \sum_{p \in Z^2} S_{j+1/2, T(k_1,k_2)-(p_1,p_2)+(0,1)} h_1(p_1, p_2).$$

Reconstruction:

$$S_{j+1/2,(p_1,p_2)} = \sum_{p \in Z^2} S_{j,(k_1,k_2)} \overline{h}_0(T(k_1, k_2) - (p_1, p_2))$$

$$+ \sum_{p \in Z^2} C_{j,(k_1,k_2)} \overline{h}_1(T(k_1, k_2) - (p_1, p_2) + (1,0)).$$

Note that the shifts (1,0) and (0,1) appearing in the decomposition and reconstruction formulas are a consequence of the term $e^{i\omega_1}$ in relations (5) and (6). In order to have a better understanding of those formulas, we consider the transformation from the scale index when $j = 0$ and $j = -\frac{1}{2}$.

Note that the filters H_0 and H_1 are usually symmetrical. By hypothesis, the samples $S_{0,k}$ are localized on the 2D grid Z^2, and from decomposition formulas, we can conclude that coefficients $(S_{-1/2,k})_{k \in Z^2}$ are localized on the grid $T(Z^2)$, and obtained through filtering by $H_0(\omega)$. The set of coefficients $(C_{-1/2,k})_{k \in Z^2}$ can be viewed as localized on the grid $T(Z^2)$ and obtained by filtering with $H_1(\omega)e^{-i\omega_1}$, but the true localization is in fact on the grid $T(Z^2) + (1,0)$ after filtering with $H_1(\omega)$. This is exactly the scheme of the filter banks defining the quincunx pyramid described in the previous section.

A simple remark will lead to the Alternated Filtering notion. From a practical point of view, the basic sub-band transformation which provides decomposition is the following: define

$$y_n^0 = \sum_{k \in Z^2} x_{n-k} h_0(k)) \text{ for } n_1 + n_2 \text{ even}$$

and

$$y_n^1 = \sum_{k \in Z^2} x_{n-k} h_1(k)) \text{ for } n_1 + n_2 \text{ odd}.$$

The channel y_n^0 provides the set $(S_{-1/2,k})_{k \in Z^2}$, and the other channel y_n^1 yields $(C_{-1/2,k})_{k \in Z^2}$. Obviously the union of the grids $\{n_1 + n_2 \text{ even}\}$ and $\{n_1 + n_2 \text{ odd}\}$ is exactly Z^2. Therefore, the idea of merging the two channels is natural: the information is preserved and we have a global transformation from $l^2(Z^2)$ onto itself.

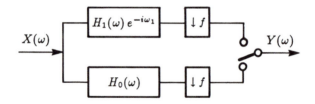

FIGURE 3. 2D Alternated Filter diagram. Notice that the sub-sampling and the delay $e^{-i\omega}$ lead to a merging of the two channels as a single output. The 1D AF diagram is obtained by changing $e^{-i\omega_1}$ by $e^{-i\omega}$ and the decimation T by a sub-sampling by a factor 2.

Using previous notations, let $\underline{H} = (H_0, H_1) \in (L^\infty([0, 2\pi]^2))^2$, the 2D Alternated Filter (AF) associated with $\underline{H}$, is the transformation defined by

$$\mathcal{A}(\underline{H}) : l^2(Z^2) \to l^2(Z^2)$$
$$(x_n) \mapsto (y_n)$$
$$\text{with } y_n = y_n^0 \text{ if } n_1 + n_2 \text{ is even}$$
$$\text{and } y_n = y_n^1 \text{ if } n_1 + n_2 \text{ is odd}.$$

Figure 3 gives a scheme of the Alternated Filter transformation: we organize the two channels produced by filter banks in order to produce a single output (see Figure 4). Of course, for interpretation we have to split this single output into the two chanels. Let us recall that such a notion was previously defined in the 1D case as an alternative to filter

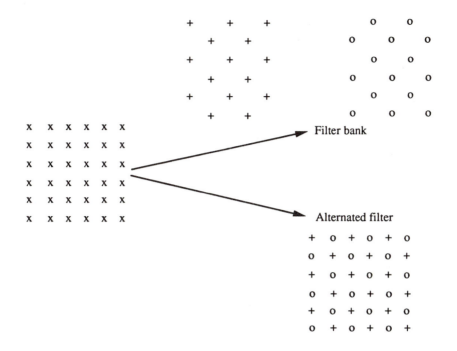

FIGURE 4. Equivalence between filter bank and alternated filter. The input pixels are denoted "x", filtering with H_0 provides pixels "+" and filtering with H_1 provides pixels "o". The shift is introduced by the term $e^{-i\omega_1}$.

banks polyphase representation [8, 9]. In this case, a 1D filter bank defined by two filters $H_0(\omega)$ and $H_1(\omega)e^{-i\omega}$ provides the following 1D Alternated Filter:

$$\mathcal{A}(\underline{H}) : l^2(Z) \to l^2(Z)$$
$$(x_n) \mapsto (y_n)$$

$$\text{with } y_{2n} = \sum_{k \in Z} x_{2n-k} h_0(k)$$

$$\text{and } y_{2n+1} = \sum_{k \in Z} x_{2n+1-k} h_1(k).$$

It is exactly the formalism previously developed for the 2D case.

As we will see below, the Alternated Filter concept is not only a reorganization of filter bank results, it is actually a new way of thinking about filter banks.

5. COMPOSITION, REPRESENTATION AND APPLICATION OF ALTERNATED FILTERS

The goal of this section is to remove the main drawback of the quincunx pyramid: filters used in this pyramid are non-separable and this leads to heavy computations in implementation. This section shows how to reduce nxn convolution kernels of associated filter banks to 2n only multiplications. In fact, each of the two basic kernels are non-separable, but considered together as a 2D Alternated Filter, they can be separated in two 1D Alternated Filter.

In order to produce such results, we first derive tools for Alternated Filters.

COMPOSITION AND REPRESENTATION

Unlike filter bank transformations, 1D (respectively 2D) Alternated Filters have the same kind of input and output, some sequences in $l^2(Z)$ (respectively $l^2(Z^2)$). So it is now possible to compose such transformations. With the convention defined in Section 1 for the 2D case, we now develop a theory which can be read in the 1D or 2D case.

Let $\underline{H}(\omega) = (H_0(\omega), H_1(\omega))$ be a pair of filters, and $\mathcal{A}(\underline{H})$ the associated AF.

Applying the Fourier transform to the equation $(y_n) = \mathcal{A}(\underline{H})((x_n))$ leads to:

$$Y(\omega) = \frac{1}{2}\{X(\omega)(H_0(\omega) + H_1(\omega))$$
$$+ X(\omega + \pi)(H_0(\omega + \pi) - H_1(\omega + \pi))\}.$$

Applied to ω and $\omega + \pi$), this relation can be written equivalently

$$\begin{pmatrix} Y(\omega) \\ Y(\omega + \pi) \end{pmatrix} = \mathcal{R}(\underline{H}) \begin{pmatrix} Y(\omega) \\ Y(\omega + \pi) \end{pmatrix}$$

with the 2π-periodic matrix $\mathcal{R}(\underline{H})$ valued in $\mathcal{M}_{2,2}(C)$:

$$\mathcal{R}(\underline{H}) = \frac{1}{2}\begin{pmatrix} H_0(\omega) + H_1(\omega) & H_0(\omega + \pi) - H_1(\omega + \pi) \\ H_0(\omega) - H_1(\omega) & H_0(\omega + \pi) + H_1(\omega + \pi) \end{pmatrix}.$$

Obviouly AF are stable under composition law if and only if for every $\underline{F}$ and $\underline{G}$, there is a pair of filters $\underline{H}$ such that $\mathcal{R}(\underline{H}) = \mathcal{R}(\underline{G}) \mathrm{x} \mathcal{R}(\underline{F})$. This is true, since a simple computation yields

$$H_0(\omega) = \frac{1}{2}(F_0(\omega)G_0(\omega) + F_1(\omega)G_0(\omega)$$
$$+ F_0(\omega)G_0(\omega + \pi) + F_1(\omega)G_0(\omega + \pi)) \qquad (12)$$

$$H_1(\omega) = \frac{1}{2}(F_0(\omega)G_1(\omega) + F_1(\omega)G_1(\omega)$$
$$+F_1(\omega)G_1(\omega + \pi) - F_0(\omega)G_1(\omega + \pi)) \qquad (13)$$

as the unique solution.

Relations (10) and (11) show that, like time invariant filtering, the AF set is stable under composition (which provides a non-commutative algebra structure on it). Following this analogy, we have defined a representation $\mathcal{R}(\underline{H})$ which expresses AF composition as multiplication of representations, just like the Fourier tranform in time invariant filtering theory. This representation will be a necessary tool for using composition in applications.

For future reference, we note that the perfect reconstruction relation (1) is expressed by $\det(\mathcal{R}(\underline{H})(\omega)) = 1$ for both the 1D and 2D cases. Finally note that the filter bank polyphase approach, which is used for implementation purposes, leads to a representation [16]. Both representations are in fact different, even if some consequences like factorization are very close [17, 19, 8].

APPLICATION

Let us recall that the previous theory developed for AF is valid in the 1D case, but also in the 2D case with the conventions that $\omega = (\omega_1, \omega_2)$ and $\omega + \pi = (\omega_1 + \pi, \omega_2 + \pi)$. We will apply composition for the reduction of computations for 2D filter banks used in the quincunx pyramid. The idea is to factorize, in a composition way, the corresponding 2D AF for easier implementation. We present now two methods for this purpose.

First method. Some 2D AF are obtained by the tensor product of 1D AF. Namely let $\mathcal{A}(\underline{F})$ and $\mathcal{A}(\underline{G})$ be two 1D alternated filters. In order to build a 2D AF, we apply to the image the transformation $\mathcal{A}(\underline{F})$ on rows and the transformation $\mathcal{A}(\underline{G})$ on columns. The global 2D transformation is still a 2D AF, and we can find the filters' components by the representations: the resulting 2D AF is defined by $(H_0(\omega_1, \omega_2), H_1(\omega_1, \omega_2))$ such that

$$\mathcal{R}(\underline{H})(\omega_1, \omega_2) = \mathcal{R}(\underline{G})(\omega_2)\mathcal{R}(\underline{F})(\omega_1).$$

Moreover, we have a perfect reconstruction 2D AF if and only if $\mathcal{H}(\underline{F})$ and $\mathcal{H}(\underline{G})$ are two perfect reconstructions 1D AF. The computation gain is obvious: if each 1D filter F_0, F_1, G_0 or G_1 has an impulse response of length n, the global 2D equivalent kernels have an impulse response containing n^2 terms. So we have reduced computation from n^2 multiplications per pixel to 2n. An example of this "separability" can be found in [9].

Second method.   We can factorize perfect reconstruction 2D AF's by perfect reconstruction elementary 2D AF.

We consider two special classes of irreducible perfect reconstruction elementary 2D AF:

Class 1:
$$(H_0(\omega_1, \omega_2), H_1(\omega_1, \omega_2))$$
$$= (a, b + c(cos(\omega_1) + cos(\omega_2))), \text{ for } (a, b) \in C^{*2} \text{ and } c \in C$$

Class 2:
$$(H_0(\omega_1, \omega_2), H_1(\omega_1, \omega_2))$$
$$= (b + c(cos(\omega_1) + cos(\omega_2)), a), \text{ for } (a, b) \in C^{*2} \text{ and } c \in C$$

It is easy to show that an alternative composition of 2D AF of Classes 1 and 2 leads to a linear phase perfect reconstruction 2D AF (and equivalently to a perfect reconstruction 2D filter bank for the quincunx pyramid). A simple recursive verification shows that a composition of such elementary 2D AF allows one to implement kernels with n^2 multiplication/pel using only 2n multiplications.

For instance, from the representation, we easily verify that:

$$\mathcal{R}(\underline{H}) = \mathcal{R}(1, 1 - 0.5(cos(\omega_1) + cos(\omega_2)))\mathcal{R}(0.5 + 0.25(cos(\omega_1) + cos(\omega_2)), 1)$$

with

$$H_0 = 0.5 + 0.25(cos(\omega_1) + cos(\omega_2))$$

and

$$H_1 = 0.875 - 0.5(cos(\omega_1) + cos(\omega_2)) - 0.25(cos(\omega_1 + \omega_2) + cos(\omega_1 - \omega_2))$$
$$- 0.125(cos(2\omega_1) + cos(2\omega_2)).$$

6.   CONCLUSION

The first aim of this work was to provide for the quincunx pyramid a multiresolution analysis interpretation by a generalization of 2-D wavelets. This theory was developed in the bi-orthogonal framework which allows a large flexibility in the associated filter banks design. Second, following wavelet transformation ideas, we introduced the 1D and 2D Alternated Filter notion, for which we defined analogues of some well known tools existing in time invariant filtering theory. As an application, we have shown that alternated filters allow us to "separate" the non-separable filters associated with the quincunx pyramid.

REFERENCES

[1] E.H. Adelson, E. Simoncelli and Hingorani, "Orthogonal pyramid transforms for image coding", SPIE vol 845 Visual Communication and Image Processing II, pp 50-58, 1987.

[2] M. Antonini, M. Barlaud, P. Mathieu and I. Daubechies, "Image Coding Using Vector Quantization in the Wavelet Transform Domain" IEEE ICASSP, Albuquerque, April 1990.

[3] M. Antonini, M. Barlaud and P. Mathieu, "Image Coding Using Lattice Vector Quantization of Wavelet Coefficients", IEEE ICASSP, Toronto, April 1991.

[4] A. Cohen, I. Daubechies and J. C. Feauveau, "Bi-orthogonal bases of compactly supported wavelets", preprint AT&T Bell Laboratories, 1990.

[5] I. Daubechies, "Ten Lectures on Wavelets", CBMS-NSF Series in Applied Mathematics, SIAM (in press 1991).

[6] D. Esteban and C. Galand, "Application of Quadrature Mirror Filters to split-band voice coding schemes", Proc. of the IEEE International Conf. ASSP, Hartford, Connecticut, May 1977.

[7] J.C. Feauveau, "Analyse multirésolution pour les images avec un facteur de résolution $\sqrt{2}$", Traitement du Signal, September 1990.

[8] J. C. Feauveau, "A new approach for subband processing", Proc. ICASSP, Albuquerque, New Mexico, 1990.

[9] J. C. Feauveau, "Analyse multirésolution par ondelettes non orthogonales et bancs de filtres numériques", Ph. D., Univ. Paris Sud, France, January 1990.

[10] D. Le Gall and A.Tabatabai, "Sub-band coding of digital images using symmetric short kernel filters and arithmetic coding techniques", Proc. ICASSP, pp 761-764, 1988.

[11] S. Mallat, "A theory for multiresolution signal decomposition: the wavelet representation", IEEE PAMI, vol. 2, nb 7, July 1989.

[12]   S. Mallat, "Multiresolution approximation and wavelet orthonormal bases of $L^2(R)$", Trans. of the American Mathematical Society, vol. 315, pp. 69-87, Sept. 1989.

[13]   Y. Meyer, "Principe d'incertitude, bases hilbertiennes et algèbre d'opérateurs", Séminaire Bourbaki, 1985-1986, nb 662.

[14]   Y. Meyer, "Ondelettes, fonctions splines et analyses graduées", CahOCier de Math. du CEREMADE, Univ. Paris-Dauphine, 1987.

[15]   M.J.T. Smith and T.P.Barnwell III, "A procedure for designing exact reconstruction filter bank for tree structured subband coders", IEEE International Conf. ASSP, San Diego, California, March 1984.

[16]   P.P. Vaidyanathan and Phuong-Quan Hoang, "Lattice structure for optimal design and robust implementation of two-channel perfect-reconstruction QMF banks", IEEE ASSP, vol. 36, January 1988.

[17]   P.P. Vaidyanathan and Phuong-Quan Hoang, "Two-channel perfect-reconstruction FIR QMF structures which yield linear-phase analysis and synthesis filters", IEEE ASSP, vol. 37 nb 5, pp 676-690, mai 1989.

[18]   M. Vetterli, "Multi-dimensional Sub-band Coding: Some Theory and Algorithms", Signal Processing, vol. 6, pp 97-112, February 1984.

[19]   M. Vetterli and D. Le Gall, "Perfect reconstruction FIR filter banks: some properties and factorizations", IEEE ASSP, vol. 37, nb 7, July 1989.

[20]   M. Vetterli, "Wavelet and Filter Banks for Discrete-Time Signal Procesing", preceding chapter in this book.

WAVELET TRANSFORM MAXIMA AND MULTISCALE EDGES

STEPHANE MALLAT

SIFEN ZHONG

*Courant Institute of
Mathematical Sciences
New York University, New York, NY*

1. INTRODUCTION

The wavelet transform is particularly well adapted to characterize transient phenomena because it decomposes signals into building blocks that are well localized in space and frequency. For many different types of signals, the important information is carried by singularities and sharp variation points. For example, sharp variation points provide the locations of contours in images. These contours are often the most important image features. This is well illustrated by our ability to recognize objects from a drawing that only outlines edges. The multiscale Canny edge detector [1], which is most often used in computer vision, is closely related to the wavelet transform. For a particular class of wavelets, a Canny edge detection is equivalent to detecting the wavelet transform local maxima. In this chapter we study the properties of these local maxima, and we show that signals can be numerically reconstructed from multiscale edges.

The classical approach to discretize a wavelet transform is to define a

This research was supported by the NSF grant IRI-890331 and Airforce grant AFOSR-90-0040.

frame or an orthogonal basis [3]. As we explain later, when a function is translated, the decomposition coefficients associated with a frame for this function are not translated but completely modified. This is the major inconvenience of wavelet frames for pattern recognition applications. On the contrary, the local maxima of a wavelet transform do translate when the signal is translated. Since the wavelet local maxima can also detect sharp variation points, they define a signal representation that is well adapted for characterizing patterns. It is however important to understand whether the whole signal information can be embedded into these local maxima. This is equivalent to a conjecture made in computer vision by David Marr [10], who predicted that one can reconstruct images from their multiscale edges. Given the mathematical tools of the wavelet theory, we derive an iterative algorithm that recovers one and two dimensional signals from the local maxima of their wavelet transform. We have yet no proof for the convergence of the algorithm although numerically it does converge fast.

Detecting sharp variation points is often not enough for characterizing patterns. Indeed, a sharp variation point might be located at a discontinuity or along a continuous curve or on peaks of different types. It is important to discriminate these different types of sharp variations to analyze the underlying signal structures. For examples, the occlusion of an object by another one, generally creates an intensity discontinuity in the image, whereas a shadow produces a smoother intensity variation due to the diffraction effect. Peaks often correspond to noises or textures. The wavelet transform is particularly well adapted to perform such a discrimination. As we explain, from the evolution across scales of the wavelet transform maxima, we can derive the Lipschitz regularity of a signal or estimate how smooth the signal is.

The reconstruction of signals from the wavelet transform local maxima enables us to develop coding algorithms based on edges. To achieve high compression rates, one cannot afford to code the entire image information. We implement an algorithm that first selects the important edges in the image and then makes an efficient coding of the corresponding data. Examples of image coding with a compression ratio over 30 are shown.

In order to understand the motivations for detecting local maxima in a wavelet transform, we first explain the main idea of a multiscale edge detection and show how it relates to the wavelet transform. We then make a brief review of the wavelet transform properties that are important in the study of these local maxima. Signals are considered as functions of continuous variables. The algorithm that reconstructs signals from the local maxima of their wavelet transform is explained in one dimension and then extended in two dimensions for image applications. The discretization of these algorithms is not described in this chapter, but can be found in another paper [9] where we also give details on the fast numerical implementations.

NOTATION

$\mathbf{L}^2(\mathbf{R})$ denotes the Hilbert space of measurable, square-integrable one-dimensional functions. For $f(x) \in \mathbf{L}^2(\mathbf{R})$ and $g(x) \in \mathbf{L}^2(\mathbf{R})$, the inner product of $f(x)$ with $g(x)$ is written:

$$< g(x) , f(x) > = \int_{-\infty}^{+\infty} g(x) \, f(x) \, dx.$$

The norm (energy) of $f(x) \in \mathbf{L}^2(\mathbf{R})$ is given by

$$\|f\|^2 = \int_{-\infty}^{+\infty} |f(x)|^2 \, dx.$$

We denote the convolution of two functions $f(x) \in \mathbf{L}^2(\mathbf{R})$ and $g(x) \in \mathbf{L}^2(\mathbf{R})$ by

$$f * g(x) = \int_{-\infty}^{+\infty} f(u) \, g(x - u) \, du.$$

We denote by $\mathbf{H}^1(\mathbf{R})$ the Hilbert space of differentiable functions in the sense of Sobolev. The Fourier transform of $f(x) \in \mathbf{L}^2(\mathbf{R})$ is written $\hat{f}(\omega)$ and is defined by

$$\hat{f}(\omega) = \int_{-\infty}^{+\infty} f(x) \, e^{-i\omega x} \, dx.$$

For any function $f(x) \in \mathbf{L}^2(\mathbf{R})$, $f_{2^j}(x)$ denotes the dilation of $f(x)$ by the scale factor 2^j:

$$f_{2^j}(x) = \frac{1}{2^j} f\left(\frac{x}{2^j}\right) .$$

$\mathbf{L}^2(\mathbf{R}^2)$ is the Hilbert space of measurable, square-integrable two dimensional functions. The norm of $g(x,y) \in \mathbf{L}^2(\mathbf{R}^2)$ is given by:

$$\|f\|^2 = \int_{-\infty}^{+\infty} \int_{-\infty}^{+\infty} |f(x,y)|^2 \, dx \, dy .$$

The Fourier transform of $f(x,y) \in \mathbf{L}^2(\mathbf{R}^2)$ is written $\hat{f}(\omega_x, \omega_y)$ and is defined by

$$\hat{f}(\omega_x, \omega_y) = \int_{-\infty}^{+\infty} \int_{-\infty}^{+\infty} f(x,y) \, e^{-i(\omega_x x + \omega_y y)} \, dx \, dy .$$

For any function $g(x, y) \in \mathbf{L}^2(\mathbf{R}^2)$, $f_{2^j}(x, y)$ denotes the dilation of $f(x, y)$ by the scale factor 2^j:

$$f_{2^j}(x, y) \;=\; \frac{1}{2^{2j}}\, f\left(\frac{x}{2^j}, \frac{y}{2^j}\right) \;.$$

2.   MULTISCALE EDGE DETECTION

The sharp variation points of an image intensity are important because they are generally located at the boundaries of the image components. In order to detect the contours of small structures as well as the boundaries of larger objects, several researchers in computer vision have introduced the concept of multiscale edge detection [10,12,13]. The signal is smoothed by a convolution with a smoothing function which is dilated by a scale factor, and the sharp variation points are detected with a first or second order differentiation operator. In one dimension, sharper variation points are local maxima of the absolute value of the first derivative or zero-crossings of the second derivative. Zero-crossings of the second derivative are either a local maxima or a local minima of the absolute value of the first derivative and might thus be sharp or slow variations points (see figure 1). Since we only want to detect the sharpest variation points, the first order derivative method is more effective. Let us now explain the relation between this edge detection procedure and the wavelet transform. To simplify the computer implementation, we only consider scales that vary along a dyadic sequence $\left(2^j\right)_{j \in \mathbf{Z}}$.

Let $\theta(x)$ be the smoothing function that is used to smooth the signal at various scales. In computer vision, $\theta(x)$ is often chosen to be a Gaussian. Let $\psi(x)$ be the derivative of $\theta(x)$:

$$\psi(x) \;=\; \frac{d\theta(x)}{dx} \;. \tag{1}$$

The function $\psi(x)$ is a wavelet [4] because $\int_{-\infty}^{+\infty} \psi(x)\, dx \;=\; 0$. Let $\psi_{2^j}(x) \;=\; \frac{1}{2^j}\psi(\frac{x}{2^j})$. The wavelet transform of a function $f(x) \in \mathbf{L}^2(\mathbf{R})$ at the scale 2^j, at a point x is defined by:

$$W_{2^j} f(x) \;=\; f * \psi_{2^j}(x) \;. \tag{2}$$

The derivative of $f(x)$ smoothed by $\theta_{2^j}(x)$ is proportional to the wavelet transform of $f(x)$. Indeed,

$$W_{2^j} f(x) \;=\; f * (2^j \frac{d\theta_{2^j}}{dx})(x) \;=\; 2^j \frac{d}{dx}(f * \theta_{2^j})(x) \;. \tag{3}$$

The local maxima of $|W_{2^j} f(x)|$ thus indicate the position of the signal sharp variation points at the scale 2^j. When the scale parameter 2^j is small, the

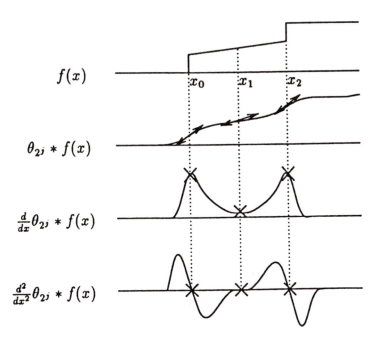

FIGURE 1. The extrema of $\frac{d}{dx} f * \theta_{2^j}(x)$ and the zero-crossings of $\frac{d^2}{dx^2} f * \theta_{2^j}(x)$ are the inflection points of $f * \theta_{2^j}(x)$. We only record the points of abscissa x_0 and x_2 where $|W_{2^j}^1 f(x)|$ is maximum because they locate the sharper variation points of $f(x)$ smoothed at the scale 2^j. The local minima of $|W_{2^j}^1 f(x)|$ at x_1 also corresponds to an inflection point, but it locates a slow variation point.

smoothing of $f(x)$ by $\theta_{2^j}(x)$ is negligible and this edge detection provides the locations of all locally sharper variations of $f(x)$. When the scale 2^j is relatively large, the convolution of $f(x)$ with $\theta_{2^j}(x)$ removes the smaller fluctuations so we only detect the larger amplitude variations of $f(x)$.

This edge detection procedure can easily be extended in two dimensions. The image is smoothed at differents scale 2^j by a convolution with a smoothing kernel $\theta(x, y)$ dilated by 2^j. We then compute the gradient vector $\vec{\nabla}(f * \theta_{2^j})(x, y)$ at each point of the smoothed image. Edges are defined as points where the modulus of the gradient vector is maximum in the direction of the gradient vector. Edge points are inflection points of the surface $f * \theta_{2^j}(x, y)$. Let us relate this edge detection to a two-dimensional wavelet transform. We define

$$\psi^1(x, y) = \frac{\partial \theta(x, y)}{\partial x} \quad \text{and} \quad \psi^2(x, y) = \frac{\partial \theta(x, y)}{\partial y} . \qquad (4)$$

The functions $\psi^1(x, y)$ and $\psi^2(x, y)$ are two dimensional wavelets. Let

$\psi^1_{2^j}(x,y) = \frac{1}{2^{2j}}\psi^1(\frac{x}{2^j}, \frac{y}{2^j})$ and $\psi^2_{2^j}(x,y) = \frac{1}{2^{2j}}\psi^2(\frac{x}{2^j}, \frac{y}{2^j})$. For any function $g(x,y) \in \mathbf{L}^2(\mathbf{R}^2)$, the wavelet transform defined with respect to $\psi^1(x,y)$ and $\psi^2(x,y)$ has two components:

$$W^1_{2^j}f(x,y) = f * \psi^1_{2^j}(x,y) \quad \text{and} \quad W^2_{2^j}f(x,y) = f * \psi^2_{2^j}(x,y) \ . \ (5)$$

Similarly to equation (3), one can easily prove that

$$\begin{pmatrix} W^1_{2^j}f(x,y) \\ W^2_{2^j}f(x,y) \end{pmatrix} = 2^j \begin{pmatrix} \frac{\partial}{\partial x}(f * \theta_{2^j})(x,y) \\ \frac{\partial}{\partial y}(f * \theta_{2^j})(x,y) \end{pmatrix} = 2^j \vec{\nabla}(f * \theta_{2^j})(x,y) \ . \ (6)$$

Hence, the local maxima of the modulus of the gradient vector, in the direction of the gradient vector, can be obtained from the vector whose coordinates are $W^1_{2^j}f(x,y)$ and $W^2_{2^j}f(x,y)$.

A first important problem for the application of multiscale edge detection to signal processing is to understand whether the whole signal information is embedded into these multiscale edges. This issue is important for pattern recognition where one needs to be sure that no important feature of the signal might disappear when representing it with multiscale edges. Also, if one can recover signals from their sharp variation points, then it is possible to process directly these sharp variation points for applications such as denoising or compact coding. This completeness issue has been studied when multiscale edges are obtained from the zero-crossings of the second order derivative of the signal smoothed at various scales. For reviews, the reader is referred to [6] or [7]. One main finding of these analyses [6] is that even if a multiscale zero-crossing representation might be complete for some particular sub-class of signals, it is not stable. This means that a slight perturbation of the position of the edges might correspond to an arbitrary large perturbation of the reconstructed signal. Hence, a precise reconstruction of the original signal for these zero-crossings is not possible in general. When detecting edges from the local maxima of the first order derivative, we can also record the value of the local maxima in order to stabilize the representation. In the following, we see that it is then possible to recover the original signal by using the properties of a wavelet transform.

Another important issue that has been studied in image processing is to understand how to combine the information provided by a multiscale edge detection at different scales, in order to precisely characterize the local behavior of the sharp variation points. A classical theorem proves that this characterization can be obtained by looking at the evolution across scales of the wavelet transform maxima. This is further studied in section 6. In the next section we review the basic properties of the wavelet transform to better understand these local maxima.

3. PROPERTIES OF A WAVELET TRANSFORM IN ONE DIMENSION

We briefly introduce the wavelet transform and explain under what condition a dyadic wavelet transform can be complete and stable. For a thorough presentation the reader is referred to the lecture notes of Daubechies [10] and an advanced functional analysis book of Meyer [11]. Let $\psi(x) \in \mathbf{L}^2(\mathbf{R})$ be a function whose average is zero and $\psi_{2^j}(x) = \frac{1}{2^j} \psi(\frac{x}{2^j})$. The wavelet transform of a function $f(x)$ at the scale 2^j and position x is given by the convolution product:

$$W_{2^j} f(x) = f * \psi_{2^j}(x) . \tag{7}$$

We call the *dyadic wavelet transform* the sequence of functions

$$\left(W_{2^j} f(x) \right)_{j \in \mathbf{Z}} . \tag{8}$$

We denote by $\mathbf{W}$ the dyadic wavelet operator defined by $\mathbf{W}f = \left(W_{2^j} f(x) \right)_{j \in \mathbf{Z}}$. For a multiscale edge detector, the wavelet is the first order derivative of a smoothing function, but this condition does not modify the general properties of a dyadic wavelet transform.

Let us study the completeness and stability of a dyadic wavelet transform. The Fourier transform of $W_{2^j} f(x)$ is

$$\hat{W} f_{2^j}(\omega) = \hat{f}(\omega) \, \hat{\psi}(2^j \omega) . \tag{9}$$

By imposing that there exists two strictly positive constants A and B such that

$$\forall \omega \in \mathbf{R} , \quad A \leq \sum_{j=-\infty}^{+\infty} |\hat{\psi}(2^j \omega)|^2 \leq B , \tag{10}$$

we insure that the whole frequency axis is covered by a dilation of $\hat{\psi}(\omega)$ by the scale factors $\left(2^j \right)_{j \in \mathbf{Z}}$, so that no information on $\hat{f}(\omega)$ is lost. Any wavelet satisfying equation (10) is called a *dyadic wavelet*. We denote $\bar{z}$ the complex conjugate of a complex number z. We can define a reconstructing wavelet $\chi(x)$ whose Fourier transform is given by:

$$\hat{\chi}(\omega) = \frac{\overline{\hat{\psi}(\omega)}}{\sum_{j=-\infty}^{+\infty} |\hat{\psi}(2^j \omega)|^2} \tag{11}$$

The function $f(x)$ is recovered from its dyadic wavelet transform by the following summation:

$$f(x) \;=\; \sum_{j=-\infty}^{+\infty} W_{2^j} f * \chi_{2^j}(x) \;. \tag{12}$$

This equation is proved by computing its Fourier transform and inserting equations (9) and (11). It shows that the dyadic wavelet transform is complete. From equations (9), (10) and by applying the Parseval theorem, we also obtain an energy equivalence equation:

$$A \,\|f\|^2 \;\leq\; \sum_{j=-\infty}^{+\infty} \|W_{2^j} f(x)\|^2 \;\leq\; B \,\|f\|^2 \;. \tag{13}$$

This prove that the dyadic wavelet transform representation is not only complete but also stable. The closer $\frac{A}{B}$ is to 1, the more stable the representation.

Now that we know that a dyadic wavelet transform is complete, let us see how redundant it is and how to express this redundancy. Let $\mathbf{V}$ be the space of the dyadic wavelet transforms $(W_{2^j} f(x))_{j \in \mathbf{Z}}$, for all the functions $f(x) \in \mathbf{L}^2(\mathbf{R})$. Let us denote by $\mathrm{l}^2(\mathbf{L}^2)$ the Hilbert space of all sequences of functions $(g_j(x))_{j \in \mathbf{Z}}$, such that

$$g_j(x) \;\in\; \mathbf{L}^2(\mathbf{R}) \quad \text{and} \quad \sum_{j=-\infty}^{+\infty} \|g_j(x)\|^2 \;<\; +\infty \;.$$

Equation (13) proves that $\mathbf{V}$ is a sub-space of $\mathrm{l}^2(\mathbf{L}^2)$. We explain why $\mathbf{V}$ is different from $\mathrm{l}^2(\mathbf{L}^2)$, in other words, why any sequence of functions $(g_j(x))_{j \in \mathbf{Z}} \in \mathrm{l}^2(\mathbf{L}^2)$ is not a priori the dyadic wavelet transform of some function $f(x) \in \mathbf{L}^2(\mathbf{R})$. We denote by $\mathbf{W}^{-1}$ the operator from $\mathrm{l}^2(\mathbf{L}^2)$ to $\mathbf{L}^2(\mathbf{R})$ defined by:

$$\mathbf{W}^{-1} \,(g_j(x))_{j \in \mathbf{Z}} \;=\; \sum_{j=-\infty}^{+\infty} g_j * \chi_{2^j}(x) \;. \tag{14}$$

The reconstruction formula (14) shows that the *restriction* of $\mathbf{W}^{-1}$ to the wavelet space $\mathbf{V}$ is the inverse of the dyadic wavelet transform operator $\mathbf{W}$. As a consequence, if a sequence $(g_j(x))_{j \in \mathbf{Z}}$ belongs to $\mathbf{V}$, it satisfies

$$\mathbf{W} \left(\mathbf{W}^{-1} \,(g_j(x))_{j \in \mathbf{Z}} \right) \;=\; (g_j(x))_{j \in \mathbf{Z}} \;. \tag{15}$$

If we replace the operators $\mathbf{W}$ and $\mathbf{W}^{-1}$ by their expression given in equations (7) and (14), we obtain:

$$\forall j \in \mathbf{Z} \quad \sum_{l=-\infty}^{+\infty} g_l * K_l, j(x) \;=\; g_j(x) \;, \quad \text{with} \qquad (16)$$

$$K_l, j(x) \;=\; \chi_{2^l} * \psi_{2^j}(x) \;.$$

Equations (16) are not only necessary but also sufficient to prove that $(g_j(x))_{j\in\mathbf{Z}}$ is a dyadic wavelet transform and thus is equal to $(W_{2^j} f(x))_{j\in\mathbf{Z}}$ for some $f(x) \in \mathbf{L}^2(\mathbf{R})$. These equations are called reproducing kernel equations. Equation (15) shows that the reproducing kernel equations are defined by the operator

$$\mathbf{P_V} \;=\; \mathbf{W} \circ \mathbf{W}^{-1} \;. \qquad (17)$$

This operator is an orthogonal projection from $\mathbf{l}^2(\mathbf{L}^2)$ onto the $\mathbf{V}$ space. The operator $\mathbf{P_V}$ which expresses the inner redundancy of a dyadic wavelet transform, is important for reconstructing a function from multiscale edge points.

In practice, since the input signal is measured at a finite resolution, we cannot compute the wavelet transform at an arbitrary fine scale. In order to model this scale limitation, we introduce a smoothing function $\phi(x)$ whose Fourier transform is an aggregation of $\hat{\psi}(2^j\omega)$ at scales 2^j larger than 1:

$$|\hat{\phi}(\omega)|^2 \;=\; \sum_{j=1}^{+\infty} |\hat{\psi}(2^j\omega)|^2 \;. \qquad (18)$$

Let us also define the smoothing operator S_{2^j} such that:

$$S_{2^j} f(x) \;=\; f * \phi_{2^j}(x) \;. \qquad (19)$$

One can prove that for any given scale 2^l, the function $S_{2^l} f(x)$ can be reconstructed from the dyadic wavelet transform of $f(x)$ at scales larger than 2^l: $(W_{2^j} f(x))_{l<j<+\infty}$. Conversely, the dyadic wavelet transform of $f(x)$ for scales larger than 2^l can be computed from $S_{2^l} f(x)$. The function $S_{2^l} f(x)$ can thus be interpreted as the components of $f(x)$ which appear at scales larger than 2^l. We shall suppose that the original signal $f(x)$ has been computed at the scale 1 for normalization purposes. This means that the measured signal is not $f(x)$ but $S_1 f(x)$. In practice, we limit the wavelet decomposition of $S_1 f(x)$ to a finite larger scale 2^J. Since the wavelet transform at the scales smaller than 2^J is characterized by $S_{2^J} f(x)$, we define the *finite dyadic wavelet transform* of $f(x)$ between the scales 1

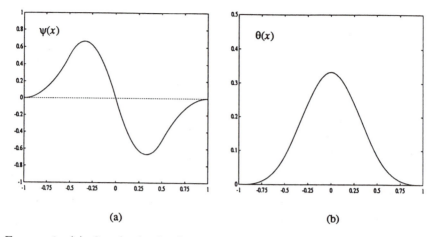

(a)   (b)

FIGURE 2. (a): Graph of a dyadic wavelet of compact support. This wavelet is
a quadratic spline and is continuously differentiable. It is the derivative of the
cubic spline function $\theta(x)$ shown in (b).

and 2^J, as the set of functions

$$\left\{ (W_{2^j} f(x))_{1 \le j \le J} \ , \ S_{2^J} f(x) \right\} \ . \tag{20}$$

Clearly $S_1 f(x)$ is characterized by this finite dyadic wavelet transform.

The discretization of this model is carefully studied in [9,15]. If the
original discrete signal is given by N samples one can compute a uniform
discretization of a finite scale dyadic wavelet transform with an algorithm of
complexity $O(N \ log(N))$. This algorithm is based on a cascade of discrete
convolutions with a low-pass and a band-pass filter. The reconstruction of
the original signal from its dyadic wavelet transform is exact and also requires $O(N \ log(N))$ computations. The wavelet used in most computations
shown in this chapter is given in figure 2(a). This wavelet is the derivative
of the smoothing function shown in figure 2(b). Figure 3(a) is the graph
of a signal $S_1 f(x)$ measured at the scale 1. Figure 3(b) is the finite dyadic
wavelet transform of this signal, between the scales 1 and 2^5. As expected,
the local maxima of the wavelet transform indicate the location of sharp
variation points in the signal. Figure 3(c) gives the locations and values of
the local maxima of the dyadic wavelet transform. In this local maxima
representation, we also keep the signal $S_{2^5} f(x)$ that carries the information
at scales larger than 2^5.

For numerical applications, it is necessary to define a complete and stable
discrete wavelet transform by discretizing the parameter x. One approach
is to build a frame of $\mathbf{L}^2(\mathbf{R})$ with wavelets. However, this generates signal
descriptors that are considerably modified when the signal is translated.

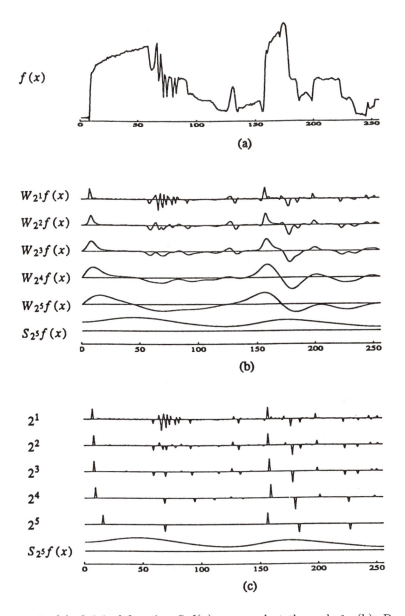

FIGURE 3. (a): Original function $S_1 f(x)$ measured at the scale 1. (b): Dyadic wavelet transform of signal (a) decomposed on 5 scales. This figure gives the graphs of the functions $W_{2^j} f(x)$ for $1 \leq j \leq 5$. The scale increases from top to bottom. At the bottom is the graph of the remaining low-frequencies $S_{2^5} f(x)$. (c): Maxima representation of signal (a). Each Dirac indicates the position and amplitude of a local maximum of the wavelet transform given in (b). The maxima representation also includes the low-frequencies $S_{2^5} f(x)$ shown at the bottom.

Let us denote $\tilde{\psi}_{2^j}(x) = \psi_{2^j}(-x)$. The function $W_{2^j}f(x)$ can also be written as an inner product in $\mathbf{L}^2(\mathbf{R})$:

$$W f_{2^j}(x) = f * \psi_{2^j}(x) = \int_{-\infty}^{+\infty} f(u)\, \tilde{\psi}_{2^j}(u-x)\, du \ , \quad \text{thus}$$

$$W f_{2^j}(x) = <\, f(u)\, , \, \tilde{\psi}_{2^j}(u-x)\, > \ . \tag{21}$$

By imposing a weak condition on $\psi(x)$, Daubechies proved that if the sampling interval a is small enough, the family of functions $\left(\tilde{\psi}_{2^j}(x - a2^j n)\right)_{(n,j)\in\mathbf{Z}^2}$ is a frame of $\mathbf{L}^2(\mathbf{R})$. This means that for any $f(x) \in \mathbf{L}^2(\mathbf{R})$, the inner products $\left(W_{2^j}f(a2^j n) = <\, f(u)\, , \, \tilde{\psi}_{2^j}(u - a2^j n)\, >\right)_{(n,j)\in\mathbf{Z}^2}$ provide a complete and stable characterization of $f(x)$. In other words, $f(x)$ is characterized by sampling uniformly $W_{2^j}f(x)$ at intervals of size $a2^j$, at each scale 2^j. Let us now look at the behavior of these inner products when $f(x)$ is translated. Let $f_\tau(x) = f(x - \tau)$ be a translation of $f(x)$ by τ. Since a wavelet transform at a scale 2^j is given by a convolution product (equation (7)), it is clear that $W_{2^j}f_\tau(x) = W_{2^j}f(x - \tau)$. However, if one samples each of theses functions with an interval $a2^j$, the values obtained may be totally different for $W_{2^j}f(x)$ and $W_{2^j}f_\tau(x)$, unless $\tau = ka2^j$ with $k \in \mathbf{Z}$ (see figure 4). The larger the value a, the more distortions are introduced with translation. In particular, this distortion is maximum for wavelet orthogonal bases.

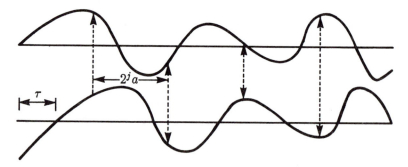

FIGURE 4. This drawing shows that the sampling of a wavelet transform (given by the arrows) can be very different after translating the signal. If $f_\tau(x) = f(x - \tau)$ then $W_{2^j}f_\tau(x) = W_{2^j}f(x - \tau)$, but the sampling does not translate if τ is not proportional to the sampling interval $a2^j$.

We saw in Section 2 that if the wavelet $\psi(x)$ is the first derivative of a smoothing function, then a Canny edge detection is equivalent to an adaptive sampling of $W_{2^j} f(x)$ at the locations where $|W_{2^j} f(x)|$ is locally maximum. Since the function $W_{2^j} f(x)$ is translated when $f(x)$ is translated by τ, its maxima are also translated. Thus an adaptive sampling of the wavelet transform based on local maxima is not modified by translations.

4. RECONSTRUCTION OF SIGNALS FROM THE WAVELET MAXIMA

In this Section, we study the reconstruction of functions from the local maxima of their wavelet transform. We formalize the completeness problem within the wavelet framework and then derive an algorithm to perform the reconstruction. This algorithm is adapted from an algorithm that was designed to reconstruct functions from the zero-crossings of their wavelet transform [7]. Let $f(x) \in \mathbf{L}^2(\mathbf{R})$ and $(W_{2^j} f(x))_{j \in \mathbf{Z}}$ be its dyadic wavelet transform. Since $f(x)$ can be recovered from its dyadic wavelet transform, we first try to reconstruct $(W_{2^j} f(x))_{j \in \mathbf{Z}}$ given the position of the local maxima of each function $|W_{2^j} f(x)|$ and the value of $W_{2^j} f(x)$ at these locations. Clearly, for any scale 2^j, there exists an infinite number of functions $g_j(x)$ which have the same local maxima as $W_{2j} f(x)$. However, any such sequence of functions $(g_j(x))_{j \in \mathbf{Z}}$ is not necessarily the dyadic wavelet transform of some function in $\mathbf{L}^2(\mathbf{R})$. Indeed, we saw in Section 3 that a dyadic wavelet transform must satisfy the reproducing kernel equations (16). Let us recall from Section 3 that the space of all dyadic wavelet transforms is denoted by $\mathbf{V}$. In order to express the conditions given by the maxima of the wavelet transform of $f(x)$, we define the set $\mathbf{\Gamma}$ of all sequences $(g_j(x))_{j \in \mathbf{Z}}$ such that for all scales 2^j, the local maxima of $|g_j(x)|$ and $|W_{2^j} f(x)|$ have the same positions, and the values of $g_j(x)$ and $W_{2^j} f(x)$ are also the same at these locations. The local maxima representation is complete if and only if there exists no dyadic wavelet transform different from $(W_{2j} f(x))_{j \in \mathbf{Z}}$ which has the same local maxima. In other words, the intersection of $\mathbf{\Gamma}$ with $\mathbf{V}$ must be reduced to one element:

$$\mathbf{\Gamma} \bigcap \mathbf{V} = \left\{ (W_{2j} f(x))_{j \in \mathbf{Z}} \right\} . \tag{22}$$

We have no mathematical proof of this statement, but we describe an algorithm that recovers numerically the intersection of $\mathbf{\Gamma}$ with $\mathbf{V}$.

The set $\mathbf{\Gamma}$ is not convex but is close to convex [9]. If instead of only keeping the local maxima of $|W_{2^j} f(x)|$, we keep the position and values of all the extrema of $W_{2^j} f(x)$ then the corresponding set $\mathbf{\Gamma}$ would be convex. A classical technique for recovering the intersection of a convex set with a linear space is to iterate alternative projections on the convex set and the linear space [14]. For any $(g_j(x))_{j \in \mathbf{Z}}$, we can define [9,15] a projection $\mathbf{P_\Gamma}$ on Γ that transforms $(g_j(x))_{j \in \mathbf{Z}}$ into the sequence of functions $(h_j(x))_{j \in \mathbf{Z}} \in \Gamma$ that is the closest to $(g_j(x))_{j \in \mathbf{Z}}$. The functions $h_j(x)$ are smooth deformations of $g_j(x)$ that match the maxima constraints. The deformation is computed in order to minimize the difference between $g_j(x)$ and $h_j(x)$ measured with an $\mathbf{H}^1(\mathbf{R})$ norm [9]. If Γ were convex, then it would guarantee that the projector Γ is non expansive. Let $\mathbf{P_V}$ be the projection on the space $\mathbf{V}$ defined by equation (17). One can prove that this projection is orthogonal with respect to the same Sobolev norm. We define the operator $\mathbf{P} = \mathbf{P_\Gamma} \circ \mathbf{P_V}$ which is a composition of $\mathbf{P_\Gamma}$ and $\mathbf{P_V}$. Clearly any element at the intersection of Γ and $\mathbf{V}$ is a fixed point of $\mathbf{P}$. To compute such a fixed point, we iterate on the operator $\mathbf{P}$. The algorithm is illustrated in figure 5. Let $\mathbf{P}^{(n)}$ be the composition n times of the operator $\mathbf{P}$ and $(g_j(x))_{j \in \mathbf{Z}}$ the initial sequence. The weak convergence of $\mathbf{P}^{(n)}(g_j(x))_{j \in \mathbf{Z}}$ to an element of $\Gamma \cap \mathbf{V}$ is guaranteed if Γ is convex [14]. This is the case when we keep all the local extrema values of $W_{2^j} f(x)$ but not when we only record the local maxima of $|W_{2^j} f(x)|$. The numerical experiments described in the next section shows that even though Γ is not a convex, the algorithm does converge. Moreover, for any initial sequence $(g_j(x))_{j \in \mathbf{Z}}$, it converges to the wavelet transform of $f(x)$ with a good numerical precision. This seems to indicate that the intersection of $\mathbf{V}$ and Γ is indeed reduced to the wavelet transform of $f(x)$.

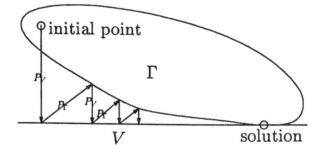

FIGURE 5.   The reconstruction of the wavelet transform of $f(x)$ is done with alternating projections on the set Γ that expresses the constraints on the local maxima and on the space $\mathbf{V}$ of all dyadic wavelet transforms.

5. NUMERICAL RECONSTRUCTION RESULTS IN ONE DIMENSION

Within a discrete framework, if the original signal has N data points, the operators $\mathbf{P_V}$ and $\mathbf{P_\Gamma}$ can be implemented with a complexity of $O(N\ log(N))$. The fast discrete algorithms are described in [9]. Each iteration on the operator $\mathbf{P}$ thus requires $O(N\ log(N))$ computations. After reconstructing the dyadic wavelet transform by iterating on $\mathbf{P}$, we compute the corresponding signal by applying the inverse wavelet transform operator $\mathbf{W^{-1}}$. The error is defined as the difference between this reconstructed signal and the original one. We compute the error to signal ratio with a mean-square measure. Figure 6(b) was reconstructed from the local maxima representation shown in Figure 3(c). The error to signal ratio is $2.9\ 10^{-2}$. We studied the decay of the error to signal ratio as a function of the number of iterations on the operator $\mathbf{P}$. For all the signals that have been tested, the decay of the error is fast during the first 20 iterations. Then it slows down and the remaining errors are concentrated at the finest scale of the wavelet transform. This is due to the structure of the reproducing kernel [9]. The reproducing kernel expresses the redundancy between the functions $W_{2^j} f(x)$ at different scales 2^j. This redundancy is maximum for consecutive scales: 2^j with $2^j + 1$ and $2^j - 1$. At the scale 2^1, there is no information available at the finer scale 2^0, so the correlation constraint is weaker. The behavior of the reconstruction error of the finest scales depends upon the choice of the wavelet $\psi(x)$. For the quadratic wavelet shown in Figure 2(a), it converges to a minimum value approximately equal to 10^{-2}. Let us emphasize again that this remaining error is concentrated at the finer scale and is probably due to the particular situation of the finest scale in a finite reproducing kernel equation. This remaining error can be reduced by choosing a wavelet $\psi(x)$ of smaller support [9]. The Haar wavelet $\psi(x)$ has a support of size 1 and is given by:

$$\psi(x) \;=\; \begin{cases} 1 \;\; if \;\; \dfrac{-1}{2} \le x < 0 \\[2mm] -1 \;\; if \;\; 0 \le x < \dfrac{1}{2} \\[2mm] 0 \;\; otherwise \end{cases} \tag{23}$$

In this case, even the finer scale error decreases constantly although the decay of the error rate is slow. After 500 iterations, the error to signal ratio is $2\ 10^{-3}$. Figure 7 gives the decay of the error to signal ratio during the first 500 iterations for the reconstruction of the signal 3(a) from the wavelet transform maxima obtained with the Haar wavelet. Similar convergence

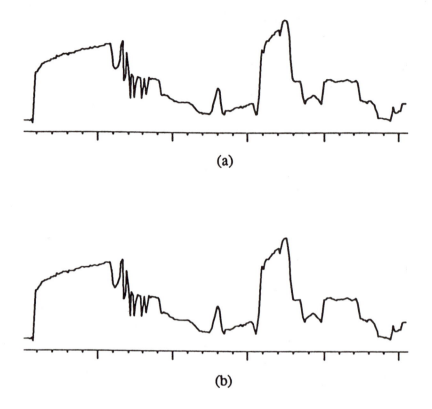

FIGURE 6. (a): Original signal. (b): Signal reconstructed with 20 iterations from the maxima representation given in Figure 3(c). The error to signal ratio of the reconstruction 2.9 10^{-2}.

rates were obtained with all the other types of signals that have been tested. This result is particularly impressive for a step edge: equal to 0 for $x < 0$ and 1 for $x \geq 0$. In this case we only have one local maxima per scale 2^j corresponding to the discontinuity at 0, and this is enough to completely reconstruct the wavelet transform and the signal. A wavelet frame would require an infinite number of coefficients per scale. The local maxima detection is an adaptive sampling that adapts to the complexity of the signal.

The numerical results obtained so far, seem to indicate that the maxima of a dyadic wavelet transform do provide a complete and stable representation if we have no finer scale limitation. We therefore conjecture that for a large class of wavelets, the local maxima of the wavelet transform at all scales $\left(W_{2^j} f(x)\right)_{j \in \mathbf{Z}}$, do provide a complete and stable representation of $f(x)$. The class of wavelets for which this is true remains to be defined.

Although there is a slight instability at the finest scale, this reconstruction algorithm is sufficient for most applications.

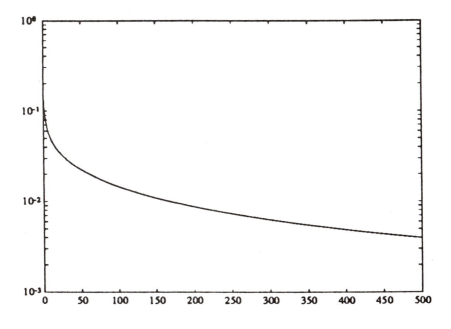

FIGURE 7. Evolution of the error to signal ratio for the reconstruction of signal 3(a) from the local maxima of the wavelet transform computed with a Haar wavelet $\psi(x)$. The abscissa gives the number of iterations on the operator **P**.

6.   CHARACTERIZATION OF SHARP VARIATION POINTS FROM MULTISCALE VALUES

Let us now explain how to describe more precisely the different type of sharp variations that are detected by the wavelet transform local maxima. An important property of the wavelet transform is to characterize the local regularity of functions from the evolution of its value across scales. The regularity of a function at a point x_0, can be measured with a Lipschitz exponent α. Let us recall that a function $f(x)$ is said to be Lipschitz α in x_0 ($0 \leq \alpha \leq 1$), if and only if there exists a constant K such that for all x in a neighborhood of x_0, we have

$$|f(x) - f(x_0)| \leq K |x - x_0|^\alpha . \tag{24}$$

The larger α, the more regular the function. We shall call Lipschitz regularity of $f(x)$ at x_0 the superior bound of all α that satisfy property (24).

A discontinuity for example has a Lipschitz regularity $\alpha = 0$. The function $f(x)$ is uniformly Lipschitz α over an interval $]a, b[$ if and only if it satisfies equation (24) for all $(x, x_0) \in]a, b[^2$. On can prove that if a function is uniformly Lipschitz α over an interval if and only if its primitive is uniformly Lipschitz $\alpha + 1$ over the same interval. This enables us to extend the notion of Lipschitz exponents to distributions by saying that a distribution is uniformly Lipschitz α in an interval $]a, b[$ if its primitive is Lipschitz $\alpha + 1$ in the same interval. For example, a Dirac function centered at x_0 is Lipschitz $\alpha = -1$ in the neighborhood of x_0. Indeed, its primitive is bounded but discontinuous at x_0 and is therefore uniformly Lipschitz $\alpha = 0$ in the neighborhood of x_0. The following theorem proves that the Lipschitz exponent of a function can be measured from the evolution across scales of the absolute value of the wavelet transform. We suppose that the wavelet $\psi(x)$ is continuously differentiable and has a decay at infinity which is $O(\frac{1}{1 + x^2})$.

Theorem 1.   *Let $f(x) \in \mathbf{L}^2(\mathbf{R})$, $f(x)$ is uniformly Lipschitz α in $]a, b[$ if and only if there exists a constant $K > 0$ such that for all $x \in]a, b[$, the wavelet transform satisfies*

$$|W_{2^j} f(x)| \leq K (2^j)^\alpha . \tag{25}$$

The proof can be found in [5]. From equation (25) we derive that

$$log_2 |W_{2^j} f(x)| \leq log_2(K) + \alpha j .$$

This theorem proves that a Lipschitz exponent is equivalent to an upper bound on the decay of the absolute value of the wavelet transform and thus on the wavelet transform local maxima. A mathematical and numerical study on the characterization of Lipschitz exponents from the decay of the wavelet transform maxima can be found in [8]. The wavelet transform local maxima provide simple strategies to detect and characterize the singularities of a signal. An application to the suppression of white noises from one-dimensional and two-dimensionals signals is described [8]. In practice, we are limited by the scale of measurement of our signal that we normalized to 1. We do not have access to $f(x)$ but to $S_1 f(x)$. This means that we can at most estimate the Lipschitz regularity of the signal $f(x)$ from the behavior of its wavelet transform up to the scale 1.

Often the signal singularities are blurred due to some diffusion process. We thus get smoothed "singularities" and it is important to estimate this smoothing factor. For example, shadows do not produce sharp discontinuities of the image intensity, but relatively smooth variations because of the diffraction effect. We model smooth variations as singularities convolved with a Gaussian of a given variance. This is equivalent to running a reverse heat equation on the signal, and looking at how long it takes to create a singularity at the point where the signal had a smooth variation. The Gaussian variance is a measurement of the time when the singularity occurs. Let us write our signal $f(x)$ as the convolution of a signal $h(x)$ which has a singularity Lipschitz α in x_0, with a Gaussian of variance σ:

$$f(x) = h * g_\sigma(x) \quad \text{where} \quad g_\sigma(x) = \frac{1}{\sqrt{2\pi}\sigma} exp(-\frac{x^2}{2\sigma^2}) .$$

We saw that the wavelet transform of $f(x)$ can be written

$$W_{2^j} f(x) = 2^j \frac{d}{dx} (f * \theta_{2^j})(x) = 2^j \frac{d}{dx} (h * g_\sigma * \theta_{2^j})(x) . \quad (26)$$

Let us suppose that the function $\theta(x)$ is close to a Gaussian function. In this case we have:

$$\theta_{2^j} * g_\sigma(x) \approx \theta_{s_0}(x) \quad \text{with} \quad s_0 = \sqrt{2^{2j} + \sigma^2} . \quad (27)$$

Equation (26) can thus be rewritten:

$$W_{2^j} f(x) = 2^j \frac{d}{dx} (h * \theta_{s_0})(x) = \frac{2^j}{s_0} W_{s_0} h(x) , \quad (28)$$

where $W_{s_0} h(x)$ is the wavelet transform of $h(x)$ at the scale s_0:

$$W_{s_0} h(x) = h * \psi_{s_0}(x) . \quad (29)$$

Hence, the wavelet transform at the scale 2^j of a singularity smoothed by a Gaussian of variance σ , is equal to the wavelet transform of the non-smoothed singularity at the scale $s_0 = \sqrt{2^{2j} + \sigma^2}$. Since $h(x)$ is locally Lipschitz α, similarly to Theorem 1, one can prove that there exists a constant $K > 0$ such that $|W_s h(x)| \leq K s^\alpha$, for any scale $s > 0$. By inserting this inequality in equation (28), we obtain

$$|W_{2^j} f(x)| \leq K 2^j s_0^{\alpha-1} , \quad \text{with} \quad s_0 = \sqrt{2^{2j} + \sigma^2} . \quad (30)$$

We can thus estimate the decay of the wavelet transform maxima of $f(x)$ as a function of the Lipschitz regularity α and the smoothing variance σ.

Figure 8 gives several examples of singularities smoothed by Gaussians of different variances. The decay of the maxima are clearly affected by the different Lipschitz exponents as well as the variance of the Gaussian smoothing. For a non smooth step edge, the maxima values are constant across scales because $\alpha = 0$ whereas for a Dirac distribution, the maxima amplitude increase by a factor 2 when the scale decreases from $2^j + 1$ to 2^j

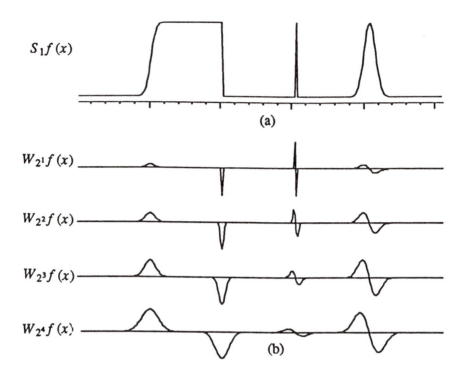

FIGURE 8. (a): The signal $S_1 f(x)$ is built from a function $f(x)$ which has four singularities respectively characterized by $(\alpha = 0, \sigma = 0), (\alpha = 0, \sigma = 3), (\alpha = -1, \sigma = 0)$ and $(\alpha = -1, \sigma = 4)$.
(b): The behavior of the local maxima across scales depend upon the Lipschitz regularity α and the smoothing scale σ.

because $\alpha = -1$. Let us now explain how to compute numerically the Lipschitz regularity α and the smoothing scale σ of a signal sharp variation. We suppose that the sharp variation that occurs at a point x_0 produces a local maxima of the wavelet transform, whose amplitude at the scale 2^j is a_j for $1 \le j \le l$. The finest scale is normalized to 1. We compute the three values K, σ and α so that the inequality of equation (30) is as close as possible to an equality for each maxima a_j. These values are obtained by minimizing the following expression:

$$\sum_{j=1}^{l} \left(log_2|a_j| - log_2(K) - j - \frac{\alpha - 1}{2} log_2(\sigma^2 + 2^{2j}) \right)^2 . \quad (31)$$

We shall then say that our signal in x_0, at the scale 1, has a sharp variation of regularity Lipschitz α, smoothed at the scale σ. The value K gives the amplitude of the sharp variation. If the signal is multiplied by a constant

λ then K is also multiplied by λ, but σ and α are not affected. On the contrary, if the signal is smoothed by a Gaussian of variance σ_0 (and integral 1), then K and α are not affected but σ becomes $\sqrt{\sigma^2 + \sigma_0^2}$. This shows clearly that the parameters α, σ and K describe different properties of the sharp variation that occurs in x_0. When computing the values of σ and α from the evolution of the maxima across scales in Figure 8, we have a numerical error of less than 10%, which is mainly due to the fact that the wavelet we use is not the derivative of a Gaussian but of some smoothing function $\theta(x)$ which only approximates a Gaussian (see Figure 2(b)).

This analysis is valid only if the sharp variation in x_0 dominates locally the other local variations of the signal, so that it can be considered as isolated in a first approximation. For example at the abscissa 160 of Figure 3(a), the signal clearly has a sharp variation that dominates its neighborhood and the wavelet maxima produced by this sharp variation dominates over a large range of scales. On the contrary this is not the case for the sharp variations between the abscissa 60 and 80. None of these sharp variations can be considered as isolated and the value of the wavelet transform at scales larger than 2^1 reflects the interaction between these sharp variation points. Here, the signal irregularities cannot be considered as isolated edges but rather as texture variations. The differentiation between edges and texture variations is a difficult ill-defined problem in computer vision and this issue appears naturally when trying to interpret the evolution of the wavelet transform maxima across scales.

7. WAVELET TRANSFORM OF IMAGES AND LOCAL MAXIMA

We explained in Section 2 that in two dimensions, a multiscale edge detection can be reformalized through a wavelet transform defined with respect to two wavelets $\psi^1(x, y)$ and $\psi^2(x, y)$. We first study the general properties of such a two-dimensional wavelet transform. We denote $\psi^1_{2^j}(x, y) = \frac{1}{2^{2j}} \psi^1(\frac{x}{2^j}, \frac{y}{2^j})$ and $\psi^2_{2^j}(x, y) = \frac{1}{2^{2j}} \psi^2(\frac{x}{2^j}, \frac{y}{2^j})$. The wavelet transform of a function $g(x, y) \in \mathbf{L}^2(\mathbf{R}^2)$, at the scale 2^j, has two components defined by:

$$W^1_{2^j} f(x, y) \;=\; f * \psi^1_{2^j}(x, y) \quad \text{and} \quad W^2_{2^j} f(x, y) \;=\; f * \psi^2_{2^j}(x, y) \; . \; (32)$$

We call the *two-dimensional wavelet transform* of $f(x, y)$ the set of functions

$$\mathbf{W} f \;=\; \left(W^1_{2^j} f(x, y) \, , \, W^2_{2^j} f(x, y) \, \right)_{j \in \mathbf{Z}} \; . \tag{33}$$

Let $\hat{\psi}^1(\omega_x, \omega_y)$ and $\hat{\psi}^2(\omega_x, \omega_y)$ be the Fourier transforms of $\psi^1(x, y)$ and $\psi^2(x, y)$. The Fourier transforms of $W_{2^j}^1 f(x, y)$ and $W_{2^j}^2 f(x, y)$ are respectively given by:

$$\hat{W}_{2^j}^1 f(\omega_x, \omega_y) \;=\; \hat{f}(\omega_x, \omega_y)\, \hat{\psi}^1(2^j \omega_x, 2^j \omega_y) \quad \text{and} \qquad (34)$$

$$\hat{W}_{2^j}^2 f(\omega_x, \omega_y) \;=\; \hat{f}(\omega_x, \omega_y)\, \hat{\psi}^2(2^j \omega_x, 2^j \omega_y) \; . \qquad (35)$$

To insure that a dyadic wavelet transform is a complete and stable representation of $f(x, y)$, similarly to the one dimensional case, we impose that the two-dimensional Fourier plane is covered by the dyadic dilations of $\hat{\psi}^1(\omega_x, \omega_y)$ and $\hat{\psi}^2(\omega_x, \omega_y)$. This means that there exists two strictly positive constants A and B such that

$$\forall (\omega_x, \omega_y) \in \mathbf{R}, \quad A \le \sum_{j=-\infty}^{+\infty} \left(|\hat{\psi}^1(2^j \omega_x, s^j \omega_y)|^2 \right.$$
$$\left. + |\hat{\psi}^2(2^j \omega_x, 2^j \omega_y)|^2 \right) \le B \; . \qquad (36)$$

Let $\chi^1(x, y)$ and $\chi^2(x, y)$ be the two functions whose Fourier transforms are defined by:

$$\hat{\chi}^1(\omega_x, \omega_y) \;=\; \frac{\overline{\hat{\psi}}^1(\omega_x, \omega_y)}{\sum_{j=-\infty}^{+\infty} \left(|\hat{\psi}^1(2^j \omega_x, 2^j \omega_y)|^2 \;+\; |\hat{\psi}^2(2^j \omega_x, 2^j \omega_y)|^2 \right)} \quad \text{and} \qquad (37)$$

$$\hat{\chi}^2(\omega_x, \omega_y) \;=\; \frac{\overline{\hat{\psi}}^2(\omega_x, \omega_y)}{\sum_{j=-\infty}^{+\infty} \left(|\hat{\psi}^1(2^j \omega_x, 2^j \omega_y)|^2 \;+\; |\hat{\psi}^2(2^j \omega_x, 2^j \omega_y)|^2 \right)} \; . \qquad (38)$$

We can derive from equations (34-38) that $f(x, y)$ is reconstructed from its dyadic wavelet transform with the following summation:

$$f(x, y) \;=\; \sum_{j=-\infty}^{+\infty} \left(W_{2^j}^1 f * \chi_{2^j}^1(x, y) \;+\; W_{2^j}^2 f * \chi_{2^j}^2(x, y) \right) \; . \qquad (39)$$

This equation defines an inverse wavelet operator $\mathbf{W}^{-1}$. As in one dimension, one can also prove that any sequence of two dimensional functions $\left(g_j^1(x, y), g_j^2(x, y) \right)_{j \in \mathbf{Z}}$ is not a priori the dyadic wavelet transform of some two-dimensional function $f(x, y)$. In order to be a dyadic wavelet transform, such a sequence must satisfy reproducing kernel equations similar to equations (16). The set of all sequences of functions which are the dyadic wavelet transform of some function in $\mathbf{L}^2(\mathbf{R}^2)$ define a Hilbert space that we denote by $\mathbf{V}$. This Hilbert space is strictly included in the space of all sequences of $\mathbf{L}^2(\mathbf{R}^2)$ functions. The operator

$$\mathbf{P_V} \;=\; \mathbf{W} \circ \mathbf{W}^{-1} \;, \tag{40}$$

is an orthogonal projection from the space of all sequences of $\mathbf{L}^2(\mathbf{R}^2)$ functions onto $\mathbf{V}$.

Images are measured at a finite resolution so we cannot compute the wavelet transform at a scale below the limit set by this resolution. Like in one dimension, in order to model the limitation of resolution, we introduce a smoothing function $\phi(x, y)$ whose Fourier transform satisfies

$$|\hat{\phi}(\omega_x, \omega_y)|^2 \;=\; \sum_{j=1}^{+\infty} (\; |\hat{\psi}^1(2^j\omega_x, 2^j\omega_y)|^2 \;+\; |\hat{\psi}^2(2^j\omega_x, 2^j\omega_y)|^2 \;) \;. \tag{41}$$

We also define a smoothing operator S_{2^j} such that:

$$S_{2^j} f(x, y) \;=\; f * \phi_{2^j}(x, y) \;. \tag{42}$$

One can prove that at any given scale 2^l, from $S_{2^l} f(x, y)$ we can compute the wavelet transform of $f(x, y)$ at scales larger than 2^l, $\big(W_{2^j}^1 f(x, y)$, $W_{2^j}^2 f(x, y) \big)_{l<j<+\infty}$, and conversely $S_{2^l} f(x, y)$ can be recovered from this sub-part of the dyadic wavelet transform. We suppose that the input image is measured at the scale 1 and can be written $S_1 f(x, y)$. The finite dyadic wavelet transform of $S_1 f(x, y)$ up to the scale 2^J is given by:

$$\Big\{ \big(W_{2^j}^1 f(x, y) \,, \; W_{2^j}^2 f(x, y) \;\big)_{1 \le j \le J} \,, \; S_{2^J} f(x, y) \Big\} \;. \tag{43}$$

The discretization of this model has been studied in [9,15]. If the original signal $S_1 f(x, y)$ is uniformly sampled, one can compute a uniform sampling of the two-dimensional dyadic wavelet transform. If the image has N pixels, the wavelet transform is computed with $O(N \; log(N))$ operations. The reconstruction of the original image from its wavelet transform has also a complexity of $O(N \; log(N))$.

In Section 2, we explained that multiscale sharp variation points can be obtained from a dyadic wavelet transform, if the two wavelets $\psi^1(x, y)$ and $\psi^2(x, y)$ are respectively the partial derivatives along x and y of a two dimensional smoothing function $\theta(x, y)$. To implement such a wavelet transform with a fast discrete algorithm, we need to relax slightly this constraint. We choose two wavelets $\psi^1(x, y)$ and $\psi^2(x, y)$ that can be written as separable products of functions of the x and y variables. More precisely, given a one-dimensional dyadic wavelet $\psi(x)$, we define the two wavelets $\psi^1(x, y)$ and $\psi^2(x, y)$ by:

$$\psi^1(x, y) \;=\; \psi(x) \, \xi(y) \quad \text{and} \quad \psi^2(x, y) \;=\; \xi(x) \, \psi(y) \;, \tag{44}$$

where $\xi(x)$ is a one-dimensional smoothing function. The function $\xi(x)$ is chosen so that the wavelet transform can be implemented with a fast pyramidal algorithm [9,15]. Since $\psi(x) = \frac{d\theta(x)}{dx}$, these two wavelets can be rewritten:

$$\psi^1(x,y) = \frac{\partial \theta^1(x,y)}{\partial x} \quad \text{and} \quad \psi^2(x,y) = \frac{\partial \theta^2(x,y)}{\partial y} \quad, \quad \text{with} \quad (45)$$

$$\theta^1(x,y) = \theta(x)\,\xi(y) \quad \text{and} \quad \theta^2(x,y) = \xi(x)\,\theta(y) \ .$$

We cannot choose $\xi(x)$ equal to $\theta(x)$ but it can be close to $\theta(x)$. This means that the two smoothing functions $\theta^1(x,y)$ and $\theta^2(x,y)$ are not equal but are close. The numerical experiments given in this chapter are computed with two wavelets that are derived from the one-dimensional quadratic spline wavelet $\psi(x)$ given in figure 2(a). Similarly to equation (6), we derive from (44) that the wavelet transform can be rewritten:

$$W_{2^j}^1 f(x,y) = 2^j \frac{\partial}{\partial x} (f * \theta_{2^j}^1)(x,y) \quad \text{and} \quad (46)$$

$$W_{2^j}^2 f(x,y) = 2^j \frac{\partial}{\partial y} (f * \theta_{2^j}^2)(x,y) \ . \quad (47)$$

The dyadic wavelet transform $W_{2^j}^1 f(x,y)$ and $W_{2^j}^2 f(x,y)$ are respectively the partial derivatives along the horizontal and vertical directions of $f(x,y)$ smoothed at the scale 2^j. In practice, we can suppose that the two functions $\theta^1(x,y)$ and $\theta^2(x,y)$ are close enough to be approximated by a unique smoothing function $\theta(x,y)$. The two components of the wavelet transform are thus approximately proportional to the two components of the gradient vector $\vec{\nabla}(f * \theta_{2^j})(x)$. Figure 9 shows the dyadic wavelet transform of a circle image decomposed between the scales 1 and 2^4.

At each scale 2^j, the modulus of the gradient vector is proportional to

$$M_{2^j} f(x,y) = \sqrt{|W_{2^j}^1 f(x,y)|^2 + |W_{2^j}^2 f(x,y)|^2} \ . \quad (48)$$

The angle of the gradient vector with the horizontal direction is given by:

$$A_{2^j} f(x,y) = \arctan\left(\frac{W_{2^j}^2 f(x,y)}{W_{2^j}^1 f(x,y)} \right) \ . \quad (49)$$

The modulus and angle images of the circle are also shown in Figure 9.

Like in a Canny edge detection, the sharper variation points of $f * \theta_{2^j}(x,y)$ are obtained from the local maxima of $M_{2^j} f(x,y)$ along the gradient direction given by $A_{2^j} f(x,y)$. We record the position of each of these local maxima as well the values of the modulus $M_{2^j} f(x,y)$ and

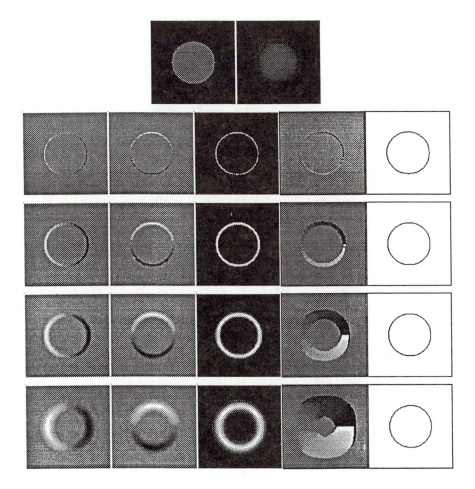

FIGURE 9. The original image $S_1 f(x, y)$ is at the top left and next to it the low-frequency image $S_{2^4} f(x, y)$. The first column from the left gives the images $\left(W_{2^j}^1 f(x, y)\right)_{1 \le j \le 4}$ and the scale increases from top to bottom. The second column displays $\left(W_{2^j}^2 f(x, y)\right)_{1 \le j \le 4}$. Black, grey and white pixels indicate respectively negative, zero and positive sample values. The third column displays the modulus images $(M_{2^j} f(x, y))_{1 \le j \le 4}$, black pixels indicate zero values whereas white ones correspond to the highest value. The fourth column gives the angle images $(A_{2^j} f(x, y))_{1 \le j \le 4}$, and we can see that the angle value turns from π to $-\pi$ along the circle contour. The fifth column displays in black the position of the local maxima of $(M_{2^j} f(x, y))_{1 \le j \le 4}$ in the direction given by the corresponding angle images $A_{2^j} f(x, y)$.

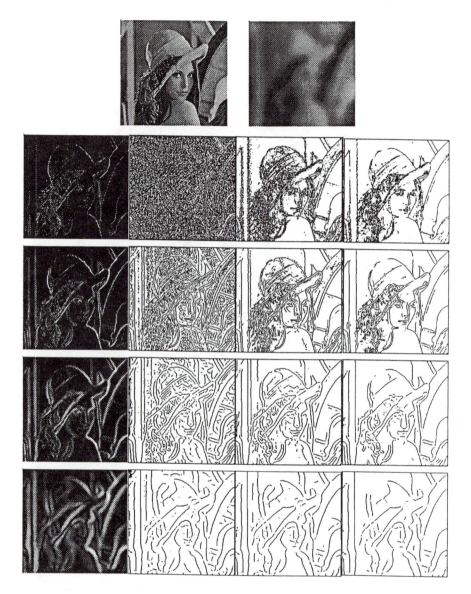

FIGURE 10. The original image $S_1 f(x, y)$ is at the top left and next to it the low-frequency image $S_{2^4} f(x, y)$. The first column gives the modulus images $M_{2^j} f(x, y)$ for $1 \leq j \leq 4$. The second column displays the position of the local maxima of $M_{2^j} f(x, y)$, for $1 \leq j \leq 4$. The third and fourth columns display the position of the local maxima whose amplitudes are respectively larger than two thresholds set to 4 and 8. The local maxima corresponding to light textures variations are removed by the thresholding.

the angle $A_{2^j} f(x, y)$ at the corresponding locations. In Figure 9, the local maxima are located at the border of the circle. The first column of Figure 10 gives an other example of modulus images $M_{2^j} f(x, y)$ for the Lena image shown at the top. The second column gives the corresponding position of the maxima. At fine scales there are many maxima created by the image noise. Most of these maxima have a small modulus value. The third and fourth columns display the maxima whose modulus are larger than thresholds respectively equal to 4 and 8. The edge points with a high modulus value are the sharper intensity variations of the image.

8.   RECONSTRUCTION OF TWO-DIMENSIONAL SIGNALS FROM THE WAVELET MAXIMA

The algorithm that reconstructs images from the local maxima of their wavelet transform is a simple extension of the one-dimensional algorithm. Let $g(x, y) \in \mathbf{L}^2(\mathbf{R}^2)$ and $\left(W_{2^j}^1 f(x, y), W_{2^j}^2 f(x, y) \right)_{j \in \mathbf{Z}}$ be its dyadic wavelet transform. For each scale 2^j, we detect the maxima of $M_{2^j} f(x, y)$ along the direction of the gradient given by the angle image $A_{2^j} f(x, y)$. We record the position of each maxima as well as the value of $M_{2^j} f(x, y)$ and $A_{2^j} f(x, y)$ at the corresponding location. It is possible to reconstruct the dyadic wavelet transform of $f(x, y)$ given these maxima if and only if there is no other wavelet transform having the same maxima (position, amplitude and angle). Let us formalize this statement. For any pair of functions $(g_j^1(x, y), g_j^2(x, y))$, we define the modulus and angle images respectively as:

$$M_j(x, y) = \sqrt{|g_j^1(x, y)|^2 + |g_j^2(x, y)|^2} \quad \text{and}$$

$$A_j(x, y) = \arctan\left(\frac{g_j^1(x, y)}{g_j^2(x, y)}\right) .$$

Let us denote by Γ the set of all sequences of functions $\left(g_j^1(x, y), g_j^2(x, y) \right)_{j \in \mathbf{Z}}$ such that for all $j \in \mathbf{Z}$, the two pairs of functions $(g_j^1(x, y), g_j^2(x, y))$ and $(W_{2^j}^1 f(x, y), W_{2^j}^2 f(x, y))$ have the same maxima. By this we mean that the local maxima of $M_j(x, y)$ in the direction defined by $A_j(x, y)$ have the same location as the maxima of $M_{2^j} f(x, y)$ in the direction given by $A_{2^j} f(x, y)$. Moreover, at these locations the modulus and the angle values are the same. Let us also recall that we denote by $\mathbf{V}$ the space of all possible dyadic wavelet transforms. The reconstruction is possible if and only if the intersection of $\mathbf{V}$ with Γ is reduced to one element which is the wavelet transform of $f(x, y)$:

$$\mathbf{V} \bigcap \Gamma = \left\{ \left(W_{2^j}^1 f(x, y) , W_{2^j}^2 f(x, y) \right)_{j \in \mathbf{Z}} \right\} \quad . \tag{50}$$

As in the one-dimensional case, we have no mathematical proof of this statement, but we describe an alternative projection algorithm to compute numerically the intersection of $\mathbf{V}$ and Γ. One can define [9] a projector $\mathbf{P_\Gamma}$ that transforms any sequence $\left(g_j^1(x, y) , g_j^2(x, y) \right)_{j \in \mathbf{Z}}$ of a two-dimensional function into a new sequence $\left(h_j^1(x, y) , h_j^2(x, y) \right)_{j \in \mathbf{Z}}$ which is in Γ. The projector $\mathbf{P_\Gamma}$ makes a non-linear deformation of $(g_j^1(x, y) , g_j^2(x, y))$ so that the resulting functions $(h_j^1(x, y) , h_j^2(x, y))$ have the same maxima (position, modulus, angle) as $(W_{2^j}^1 f(x, y) , W_{2^j}^2 f(x, y))$. The difference between the two pairs of functions $(g_j^1(x, y) , g_j^2(x, y))$ and $(h_j^1(x, y) , h_j^2(x, y))$ is minimized with respect to an $\mathbf{H}^1(\mathbf{R}^2)$ norm. We also mentioned briefly in the previous section that we can define an orthogonal projector $\mathbf{P_V}$ on $\mathbf{V}$, given by equation (40). The algorithm iterates on the projectors $\mathbf{P_V}$ and $\mathbf{P_\Gamma}$ to reach the intersection of $\mathbf{V}$ and Γ. Since Γ is not a convex, the convergence is not guaranteed a priori. However, the numerical results show that it does converge fast.

9. NUMERICAL RECONSTRUCTION OF IMAGES FROM THE WAVELET MAXIMA

If the original image has N pixels, the implementation of both $\mathbf{P_V}$ and $\mathbf{P_\Gamma}$ requires $O(N \log(N))$ operations [9,15]. The upper left image of Figure 12 is the original Lena image whereas the upper right is the reconstructed image from the maxima representation with 8 iterations on the projectors $\mathbf{P_\Gamma}$ and $\mathbf{P_V}$. These two images are visually identical on a good quality image display. The error to signal ratio of this reconstruction is 6×10^{-2}. Like in the one-dimensional case, the error is concentrated in the finest scale of the reconstructed image. This is also due to the weaker constraint imposed by the reproducing kernel on the finer scale. The decay rate of this fine scale error depends upon the size of the support of the wavelets $\psi^1(x, y)$ and $\psi^2(x, y)$ [9]. If instead of choosing two wavelets $\psi^1(x, y)$ and $\psi^2(x, y)$ that are derived from the quadratic spline wavelet $\psi(x)$ of Figure 2(a), we choose two wavelets built from the Haar wavelet $\psi(x)$ given by equation (23), the decay of the finer scale error is faster. Figure 11 gives the evolution of the error to signal ratio when reconstructing the Lena image from the wavelet maxima obtained with wavelets derived from the one-dimensional Haar wavelet. After 10 iterations, the error to signal ratio is approximately $3. \ 10^{-2}$. After the first 20 iterations, the fine scale error

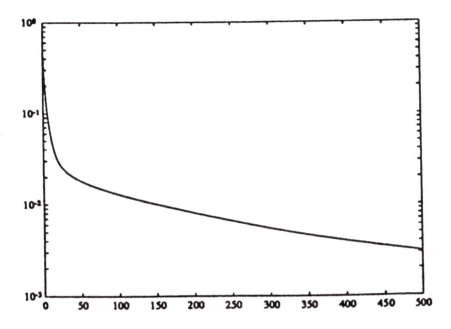

FIGURE 11. This graph gives the error to signal ratio when reconstructing the Lena image from its maxima representation, as a function of the number of iterations on the operator **P**. After the first 20 iterations, most of the remaining error is concentrated at the finest scale.

decreases slowly but steadily and reaches $3 \ 10^{-3}$ with 500 iterations. After 300 iterations, the error made on the value of any pixel in the reconstructed image is always smaller than 0.5. Since the pixel values of the original image are coded with integers between 0 and 255, we can recover exactly this image with a round-off operation. The reconstruction algorithm has been tested for a large collection of images including special two-dimensional functions such as Diracs, sinusoidal waves, step edges, Brownian noises... For all these experiments, the error to signal ratio behaves similarly to Figure 11. Let us emphasize that for image processing applications, we need at most 10 iterations for reconstructing an image with no perceivable distortions, whether we choose two-dimensional wavelets derived from the Haar wavelet or from the quadratic spline wavelet. Similar results have also been obtained for wavelets of larger supports although the fine scale errors increase with the size of the support. These numerical results seem to indicate that if we do not have a finest scale effect, then the local maxima of a two-dimensional wavelet transform provide a complete and stable representation. Let us emphasize that this is only a conjecture based on numerical results. However, for image processing applications, the numerical precision of the reconstruction algorithm is sufficient even when we limit the number of iterations below 10.

The stability of the convergence enables us to slightly perturbate the local maxima representation and reconstruct a close image. The lower left and lower right images in Figure 12 are reconstructed from the local maxima shown in the last two columns of Figure 10. By thresholding the local maxima based on their modulus values, we suppressed the local maxima produced by the image noise and the light textures. As expected, these textures have disappeared in the reconstructed images but the sharp variations are not affected. The higher the threshold, the more textures disappear. In Section 11 we study more carefully how to select the maxima that can be removed from the representation without affecting too much the quality of the reconstructed image.

10. LOCAL MAXIMA CURVES

Sharp variations of two dimensional signals are often not isolated but belong to curves in the image plane. These curves are more meaningful than the edge points by themselves because they generally are the boundaries of the image structures. We thus reorganize the maxima representation into chains of local maxima to recover these edge curves. To chain an edge point with its neighbors, we use the fact that the orientation of the gradient angle given by $A_{2^j} f(x, y)$, is approximately perpendicular to the tangent of the edge curve that goes through this point. This is only an approximation because the two wavelets $\psi^1(x, y)$ and $\psi^2(x, y)$ are derived from two smoothing functions $\theta^1(x, y)$ and $\theta^2(x, y)$ which are slightly different. Along a given chain, we also impose that the value of $M_{2^j} f(x, y)$ varies smoothly which means that the intensity profile along the boundary varies smoothly. We thus chain together two local maxima whose respective position is approximately perpendicular to the direction indicated by $A_{2^j} f(x, y)$, only if their modulus value $M_{2^j} f(x, y)$ is close enough. With this chaining procedure, we build a representation that is a set of chains of maxima at each scale 2^j.

Image edges might correspond to very different types of sharp variations. We saw in one dimension that the Lipschitz regularity of a singularity is characterized by the decay across scales of the absolute value of the wavelet transform. The same result is valid for the decay of the wavelet transform modulus $M_{2^j} f(x, y)$. Let us suppose that the two wavelets $\psi^1(x, y)$ and $\psi^2(x, y)$ are continuously differentiable and that their decay at infinity is $O(\frac{1}{(1 + x^2)(1 + y^2)})$. One can prove that a function $f(x, y)$ is uniformly

FIGURE 12. The upper left is the original image. The upper right image is reconstructed from the maxima representation shown in the second column of Figure 10. This reconstruction is performed with 8 iterations and the error to signal ratio is $6 \ 10^{-2}$. The lower left and lower right images have been reconstructed from the maxima representations shown respectively in the fourth and third column of Figure 10. The light textures have disappeared but the strong edges and textures remain unchanged.

Lipschitz α in all points of an open set of $\mathbf{R}^2$ if and only if there exists a strictly positive constant K such that for all points (x,y) of this open set, the modulus $M_{2^j}f(x,y)$ of the wavelet transform satisfies

$$M_{2^j}f(x,y) \ \leq \ K \ (2^j)^\alpha \ . \tag{51}$$

If the two smoothing functions $\theta^1(x,y)$ and $\theta^2(x,y)$ are closely approximated by a two-dimensional Gaussian, then we can also estimate any Gaussian blurr with the same type of computations as in one dimension. The numerical procedure to compute the Lipschitz regularity α, the smoothing variance σ and the constant K are the same as in one dimension. This characterization of edge types is important for pattern recognition algorithms. As we already mentioned, we can for example discriminate occlusions from shadows by looking whether the corresponding edge is an intensity discontinuity or a smoother variation. In general, we believe that an edge detection should not be a binary process that labels the image pixels as edge points or non edge points but a procedure that characterizes precisely the different types of image variations.

11. COMPACT IMAGE CODING FROM MULTISCALE EDGES

An important problem in image processing is to code images with a minimum number of bits for transmission or storage. To obtain high compression rates in image coding we cannot afford to code all the information available in the image. It is necessary to remove part of the image components that are not important for the visualization. A major problem is to identify the "important" information that we need to keep. Since edges provide meaningful features for image interpretation, it is natural to represent the image information with an edge based representation in order to select the important information for coding, as suggested by Carlsson [2]. In this section, we describe a compact coding algorithm based on the wavelet maxima representation. This compact image coding algorithm involves two steps. First we select the edge points that we consider important for the visual image quality. Then we make an efficient coding of this edge information. The selection of the most important edge curves can require sophisticated algorithms if we take into account the image context. For example, in the Lena image, it is important not to introduce distortions around the eyes because these are very visible for a human observer. In the following, we do not introduce such context information for the selection. We use a wavelet decomposition on three scales. This means that we must

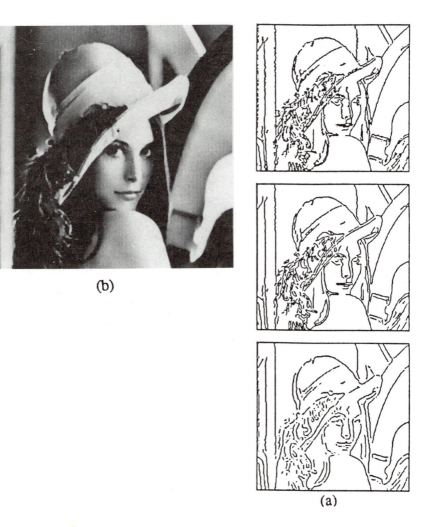

(b)

(a)

FIGURE 13. (a): Edge points that are selected for the image coding at the scales 2^1, 2^2 and 2^3, from top to bottom. The edge chains at the scale 2^2 are selected based on their length and average modulus value. For the scales 2^1 and 2^3 we keep the edge points that propagate to a local maxima that has been kept at the scale 2^2. (b): Image reconstructed from the local maxima shown in 13(a).

code the edges that appear at the scales 2^1, 2^2 and 2^3 plus the remaining low-frequency information provided by $S_{2^3} f(x, y)$. The selection is first performed at the scale 2^2 because at the finer scale 2^1 the edges are too much contaminated by noises. The boundaries of the important coherent structures often generate long edge curves. We thus first remove any edge curve whose length is smaller than a given length threshold. Among the remaining curves, we select the ones that correspond to the sharpest discontinuities in the image. This is done by removing the edge curves, along which the average value of the wavelet transform modulus $M_{2^j} f(x, y)$ is smaller than a given amplitude threshold. The middle edge image of Figure 13(a) shows the remaining edges at the scale 2^2. We then only keep at the scales 2^1 and 2^3 the edge points that propagate at the scale 2^2 to an edge point that has been kept. The top and bottom edge images of Figure 13(a) display the remaining edge points at these two scales. The image 13(b) was reconstructed from the corresponding maxima representation. The light textures and some small image components have disappeared in the reconstructed image since the corresponding edge points were removed from the representation. However, the reconstructed image is still sharp because the important intensity discontinuities are not perturbed.

Once the selection is done, we must code efficiently the remaining information. This requires coding the position, modulus value and angle value of each point along the edge curves at the scales 2^1, 2^2 and 2^3, plus the low frequency image $S_{2^3} f(x, y)$. To save bits, we only code the geometry of the edge curves at the scale 2^2 and use these to approximate the edge curves at the other scales. At the scale 2^2, the edge chains are coded by recording the position of the first point of each chain, and then coding the increment between the position of one edge point to the next one along the chain. Carlsson [2] showed that this requires on average 1.3 bit per point along the chain, with an entropy coding. We know that the direction of the gradient image intensity at the edge locations is approximately orthogonal to the tangent of the edge curves. We thus do not code the angle values $A_{2^j} f(x, y)$, but approximate it from the angle value of the edge tangent that is computed with the encoding of the edge curves. The values of the modulus $M_{2^j} f(x, y)$ at the scales 2^1, 2^2 and 2^3, along the edge curves, are recorded with a simple predictive coding using a coarse quantization of the prediction values. The low-frequency image information is encoded in an image whose size is 2^3 smaller than the original one and whose grey level values are coded on 6 bits.

We give in Figure 14(a), 14(b) and 14(c) three examples of images coded with this algorithm. The same length and amplitude thresholds were used for each of these images to select the edge chains at the scale 2^2. For each

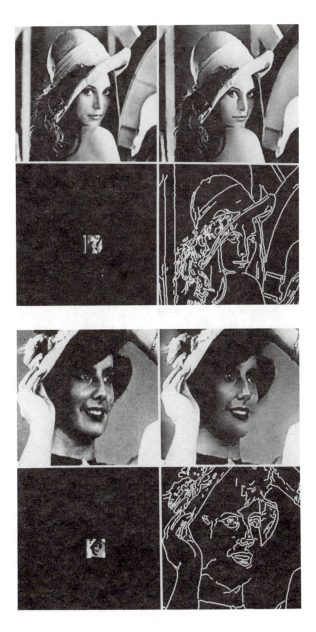

FIGURE 14. The top left of (a), (b) and (c) gives the original image of 256 by 256 pixels. The top right is the reconstructed image from the coded multiscale representation. Image (a) requires 0.30 bits per pixel, image (b) 0.24 bits per pixel and image (c) 0.20 bits per pixel. The bottom right gives the edge curves that are encoded at the scale 2^2. The bottom left shows the reduced image that carries the remaining low-frequency information.

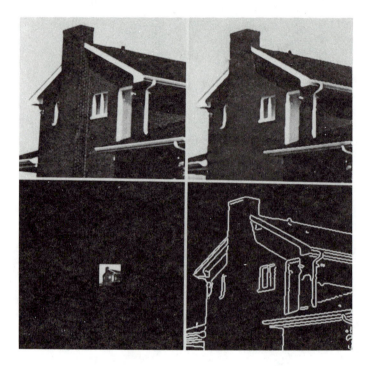

FIGURE 14 c.

example, we display at the top left the original image, at the top right the image reconstructed after the coding, at the lower right the edge map at the scale 2^2 that is encoded and at the bottom left the reduced image that carries the low-frequency information. The original images have 256 by 256 pixels. We need 0.30 bits per pixels to code Figure 14(a), 0,24 bits per pixel to code Figure 14(b) and 0.20 bits per pixel to code Figure 14(c). The compression rate varies with the number of edge points that remain after the selection operation. This type of coding removes the image textures, however it does not produce any distortion such as Gibbs phenomena. For the Lena image, the distortions are particularly visible around the eyes because too much edge points have been removed in this region by our simple selection algorithm. Although a lot of information has been removed in the coded images, they remain sharp and are not much degraded.

This compact coding algorithm is a feasibility study and we believe that it can be considerably improved both at the selection and the coding stages. For applications to images where textures are important, we are now extending this technique by developing an algorithm that makes a specific coding of the texture after this edge based coding. Distortions of textures are generally much less visible than distortion of edges. Hence, a separate coding of these two types of features can be adapted to the specificity of the visual perception as shown by Carlsson [2].

12. CONCLUSION

We saw that multiscale edges can be obtained from the local maxima of a wavelet transform and we explained how to reconstruct the original signal from these local maxima. The reconstruction algorithm can restore images and thus gives a verification of Marr's conjecture on the completeness of multiscale edges. It converges quickly and the numerical computations are mostly sequences of discrete convolutions. A real time implementation can be done on a pipe-lined hardware architecture. The mathematical study of the completeness problem remains an open question. We also proved that one can characterize different types of edge points from the evolution of the wavelet maxima amplitude across scales. The wavelet maxima representation is a new reorganization of the image information which enables us to develop an algorithm that only processes multiscale edges. In particular, we described a compact image coding procedure that selects the important visual information before coding. Other applications of the wavelet transform maxima to singularity characterization and signal denoising are described in [8].

REFERENCES

[1] Canny, J., "A Computational approach to edge detection," *IEEE Trans. Pattern Analysis and Machine Intelligence,* vol. 8, pp. 679–698, 1986.

[2] Carlsson, S., "Sketch based coding of grey level images," *Signal Processing, North Holland,* vol. 15, pp. 57–83, July 1988.

[3] Daubechies, *Ten lectures on wavelets,* CBMS-NSF Series in App. Math., SIAM, 1991.

[4] Grossmann, A., and Morlet, J., "Decomposition of Hardy functions into square integrable wavelets of constant shape," *SIAM J. Math.*, vol. 15, pp. 723–736, 1984.

[5] Holschneider, M. and Tchamitchian, P., "Regularite locate de la fonction non-differentiable de Rieman," in *Les ondelettes en 1989*, ed. P. G. Lemarie, notes in Mathematics, Springer-Verlag, 1989.

[6] Hummel, R. and Moniot, R., "Reconstruction from zero-crossing in space-space," *IEEE Trans. on Acoustic Speech and Signal Processing*, vol. 37, no. 12, Dec. 1989.

[7] Mallat, S., "Zero-crossings of a wavelet transform," /it to appear in IEEE Trans. on Information Theory, vol. 37, No. 4, July, 1991, p. 1019–1033.

[8] Mallat, S. and Hwang, W. L., "Singularity detection and processing with wavelets," *NYU, Computer Science Tech. Report 549*, To appear in IEEE Trans on Information Theory, March 1991.

[9] Mallat, S. and Zhong, S., "Characterization of signals from multiscale edges," *NYU, Computer Science Tech. Report,* To appear in IEEE Trans. on Pattern Analysis and Machine Intelligence, 1991.

[10] Marr, D., in *Vision,* W. H. Freeman and Company, 1982.

[11] Meyer, Y., in *Ondelettes et Operateurs,* Hermann, 1988.

[12] Rosenfeld, A. and Thurston, M., "Edge and curve detection for visual scene analysis," *IEEE Trans. on Computers,* 1971 vol. C-20.

[13] Witkin, A., "Scale space filtering," *Proc. Int. Joint Conf. Artificial Intelligence,* 1983.

[14] Youla, D. and Webb, H., "Image restoration by the method of convex projection," *IEEE Trans. Medical Imaging,* vol. 1, pp. 81–101, Oct. 1982.

[15] Zhong, S., *Edges representation from wavelet transform maxima,* Ph.D. Thesis, New York University, Sept. 1990.

WAVELETS AND DIGITAL SIGNAL PROCESSING

ALBERT COHEN

AT&T Bell Laboratories
Murray Hill, New Jersey

INTRODUCTION

Orthonormal bases of wavelets have recently found many interesting applications in different domains such as functional analysis and digital signal processing. In those two fields, the wavelet theory seems to be an efficient alternative to Fourier analysis. This classic method suffers from a lack of localization of the analyzing function, namely the complex exponential.

Attempts to circumvent the disadvantages of the Fourier transform are not new but many of them were made independently, either by mathematicians for a better characterization of functional spaces, or by engineers to process non-stationary signals in an appropriate way.

In functional analysis, wavelets can be viewed as a natural result of the combined works of Haar (1911) and Littlewood-Paley (1930), who both introduced a multiscale approach, using the dyadic scaling of a single mother function: The Haar function, i.e. $h(x) = \chi_{[0;\,\frac{1}{2}]}(x) - \chi_{[\frac{1}{2};\,1]}(x)$ and the band-pass filter used in Littlewood-Paley theory, i.e. $\hat{\psi}(\omega) = \chi_{[-2\pi;\,-\pi]}(\omega) + \chi_{[\pi;\,2\pi]}(\omega)$ are two "limit cases" of wavelets, the first well localized in space, the second in frequency. The general constructions made by Meyer ([13]) and Daubechies ([9]) allow the analyzing function to be well localized in both domains.

In digital signal processing, several algorithms have been developed since the fifties in order to apply multiscale methods to discrete data. Here, the Fast Wavelet Transform algorithm (FWT) introduced by Mallat ([12]) appears as a particular case of subband coding schemes allowing perfect reconstruction. Such pyramidal algorithms had been introduced by Smith and Barnwell ([2]) in 1983 without the help of wavelet bases. Discrete wavelet transforms can thus be defined without any input from the continuous theory mentioned above.

Our main goal in this paper is to show the usefulness of the continuous approach for digital signal processing. We shall insist on the regularity of the wavelet and the importance of this property for the FWT algorithm, in particular for the reconstruction stage.

We start with a brief review of multiscale analysis which can be considered as a unifying framework for the introduction of orthonormal bases of wavelets as well as exact reconstruction subband coding schemes.

I. MULTISCALE ANALYSIS

We use the notations of Meyer in [13]. A multiscale analysis is thus an increasing sequence of closed subspaces $\{V_j\}_{j\in\mathbb{Z}}$ which approximates $L^2(\mathbb{R})$

$$\{0\} \rightarrow \ldots \subset V_{-1} \subset V_0 \subset V_1 \subset \ldots \rightarrow L^2(\mathbb{R}) \qquad (1)$$

and satisfies the following properties:

$$f(x) \in V_j \iff f(2x) \in V_{j+1} \qquad (2)$$

There exists a function $\phi(x)$ in V_0 such that the set $\qquad$ (3)
$\{\phi(x-k)\}_{k\in\mathbb{Z}}$ is an orthonormal basis of V_0 .

The "scaling function" $\phi(x)$ satisfies the well known two scale difference equation,

$$\phi(x) = 2\sum_{n\in\mathbb{Z}} h_n \, \phi(2x - n) \qquad (4)$$

where $\{h_n\}_{n\in\mathbb{Z}}$ is at least in $\ell^2(\mathbb{Z})$. This equation plays a crucial part in the theory for several reasons.

a) The associated wavelet can be obtained by changing the coefficients in (4). If we define

$$\psi(x) = 2\sum_{n\in\mathbb{Z}} (-1)^n \, \overline{h}_{1-n} \, \phi(2x - n) \qquad (5)$$

then $\{\psi(x-k)\}$ is an orthonormal basis for W_0, the orthogonal complement of V_0 in V_1. The family $\{\psi_k^j\} = \{2^{j/2}\,\psi(2^j x - k)\}_{j,k\in\mathbb{Z}}$ is consequently an orthonormal basis for $L^2(\mathbb{R})$. This is actually the only general process to build wavelet bases.

b) Equations (4) and (5) are the key to the FWT algorithm. If the data to be analyzed are discrete, one assumes that they represent the approximation of a continuous signal $f(x)$ in a space V_j whose scale corresponds to the discretization step.

The starting point of the algorithm is thus the sequence of sampling coefficients $\{S_k^j\}_{k\in\mathbb{Z}} = \{2^{j/2}\,\langle f \mid \phi_k^j\rangle\}_{k\in\mathbb{Z}}$. The approximations $\{S_k^{j'}\}_{k\in\mathbb{Z}}$ and detail coefficients $\{D_k^{j'}\}_{k\in\mathbb{Z}} = \{2^{j'/2}\,\langle f \mid \psi_k^{j'}\rangle\}_{k\in\mathbb{Z}}$ at lower scales can be computed by using a convolution with the discrete filter h_n and its conjugate $\overline{g}_n = (-1)^n\,h_{1-n}$, followed by a decimation by a factor of two.

The reconstruction stage uses the same filters to interpolate the decimated data, so that the algorithm can be summarized by the following formulas:

$$S_k^j = \sum_{n\in\mathbb{Z}} \overline{h}_{n-2k}\, S_n^{j+1} \tag{6}$$

$$D_k^j = \sum_{n\in\mathbb{Z}} \overline{g}_{n-2k}\, S_n^{j+1} \tag{7}$$

$$S_k^{j+1} = 2\sum_{n\in\mathbb{Z}} h_{k-2n}\, S_n^j + 2\sum_{n\in\mathbb{Z}} g_{k-2n}\, D_n^j\,. \tag{8}$$

Note that the functions f, ϕ and ψ have been introduced artificially. In fact, only the coefficients h_n are used here. However they must satisfy an identity of perfect reconstruction, which holds automatically when this filter comes from a multiscale analysis:

$$2\sum_{k\in\mathbb{Z}} h_k\, \overline{h}_{k+2\ell} = \delta_{0,\,\ell}\,. \tag{9}$$

This condition was first introduced by Smith and Barnwell in [2] as a special case of exact reconstruction subband coders. One of its interesting aspects is that the filter $\{h_n\}_{n\in\mathbb{Z}}$ can have finite impulse response.

The particular set of FIR (finite impulse response) filters satisfying (9) and having real coefficients has been completely characterized by I. Daubechies in [9] for the construction of compactly supported wavelets.

c) In fact, the scaling function and the wavelet can be constructed directly from the filter $\{h_n\}_{n\in\mathbb{Z}}$.

- In the Fourier domain, define the transfer function

$$m_0(\omega) = \sum_{n \in \mathbb{Z}} h_n \, e^{-in\omega} \, . \tag{10}$$

The two scale difference equation (4) and the exact reconstruction condition (9) can be expressed by

$$\hat{\phi}(2\omega) = m_0(\omega) \, \hat{\phi}(\omega) \tag{11}$$

and

$$|m_0(\omega)|^2 + |m_0(\omega + \pi)|^2 = 1 \, . \tag{12}$$

It is also possible to assume (see [3]) that $\hat{\phi}(0) = 1$. We thus have

$$m_0(0) = 1 \quad \text{and} \quad m_0(\pi) = 0 \, .$$

By iteration of (11), one gets, if $m_0(\omega)$ is regular enough,

$$\hat{\phi}(\omega) = \prod_{k=1}^{+\infty} m_0 \left(2^{-k} \omega \right) \, . \tag{13}$$

The functions ϕ and ψ are thus completely characterized by the data of the filter $\{h_n\}_{n \in \mathbb{Z}}$.

- In the time domain, the scaling function can be viewed as a fixed point of the operator T defined by

$$T(f) = 2 \sum_{n \in \mathbb{Z}} h_n \, f(2x - n) \, . \tag{14}$$

A natural construction method for ϕ and ψ consists of the iterative application of this operator to some well chosen starting point.

We will now focus on this particular process. Its implementation and the conditions for its convergence are the key to our initial problem: The role of the functions ϕ and ψ in the Fast Wavelet Transform.

II.   THE CASCADE ALGORITHM

Let us start with the indicator function of the interval $\left[-\frac{1}{2}; \frac{1}{2} \right]$. If we apply the operator T iteratively, it is clear that the result after j steps is a piecewise constant function on the intervals $\left[2^{-j} \left(k - \frac{1}{2} \right); 2^{-j} \left(k + \frac{1}{2} \right) \right]$ as shown in Figure 1.

Note that this piecewise constant function can be regarded as a sequence $\{s_k^j\}$ in $\ell^2(2^{-j}\mathbb{Z})$. The action of T on this sequence is expressed

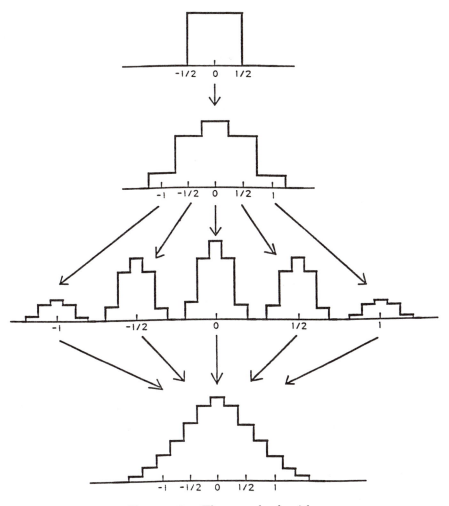

FIGURE 1.   The cascade algorithm

by

$$s_k^{j+1} = 2 \sum_{n \in \mathbb{Z}} h_n \, s_{k-2^j n}^j \, . \tag{15}$$

Using the Fourier expansion

$$S_j(\omega) = \sum_{k \in \mathbb{Z}} s_k^j \, e^{-i2^{-j}k\omega} \tag{16}$$

(15) can also be written as

$$S_{j+1}(\omega) = 2m_0(\omega/2) \, S_j(\omega/2) \tag{17}$$

and thus, one has

$$S_j(\omega) \;=\; 2^j \prod_{k=1}^{j} m_0 \left(2^{-j}\omega\right). \tag{18}$$

We see here that $S_{j+1}(\omega)$ can also be obtained by another relation

$$S_{j+1}(\omega) \;=\; 2S_j(\omega)\, m_0(2^{-j-1}\omega) \tag{19}$$

which can be expressed in the space domain by

$$s_n^{j+1} \;=\; 2\sum_{k\in\mathbb{Z}} h_{n-2k}\, s_k^j\,. \tag{20}$$

Although it leads to the same result, this new formula is more interesting than (15). It shows that the cascade algorithm is local (see Figure 1) and that it identifies exactly with the interpolation process (8) used in the reconstruction of a discrete signal from its lower scale coefficients.

This last remark justifies the interest in a continuous approach for digital signal processing. In many applications such as compression, coding or approximation, one would like the reconstruction from a low resolution component to be as smooth as possible. Note that "smoothness" is difficult to define when dealing with discrete data. Here, it means that the cascade converges to a regular function ϕ.

The regularity of the scaling function and of the wavelet (which are, in general, equal) are thus important for signal processing tasks which involve the reconstruction from coarse scales. We believe that the wavelet approach is of no use in subband filtering when the goal is only the analysis of the signal by a multiscale decomposition.

Before we take a closer look at the convergence of the cascade algorithm, we shall illustrate it with some examples.

- Figures 2 and 3 show the results after 1, 3 and 8 iterations for two different filters, both associated with compactly supported scaling functions. However ϕ is $\mathcal{C}^2$ in the first case and not even continuous in the second situation. Note that the regularity is already visible after the third iteration.

- In Figure 4, we have used a 3 time dilated version of the Haar filter which satisfies the conditions for exact reconstruction but does not lead to a multiscale analysis. Indeed, the function

$$\phi(x) \;=\; \frac{1}{3}\, \chi_{[0;\,3]}(x) \tag{21}$$

clearly satisfies the two scale difference equation but it translates are not orthonormal. The algorithm actually converges to (21) but only in the distribution sense and such a filter is not adapted for a nice reconstruction process. We shall see that the orthonormality of the family $\{\phi(x-k)\}_{k\in\mathbb{Z}}$ is a minimum requirement on exact reconstruction filters to have a smooth cascade.

- Figure 5 emphasizes this last point: Some filter may lead to a smooth scaling function but with very irregular results at each iteration.

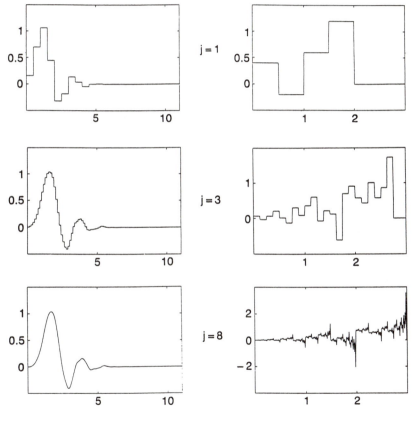

FIGURE 2.   FIGURE 3.

Cascades leading to orthonormal bases of wavelets.

FIGURE 2. ϕ is in $\mathcal{C}^2$ and the algorithm converges uniformly in $\mathcal{C}^2$.

FIGURE 3. ϕ is not continuous but the algorithm converges in L^2.

A. COHEN

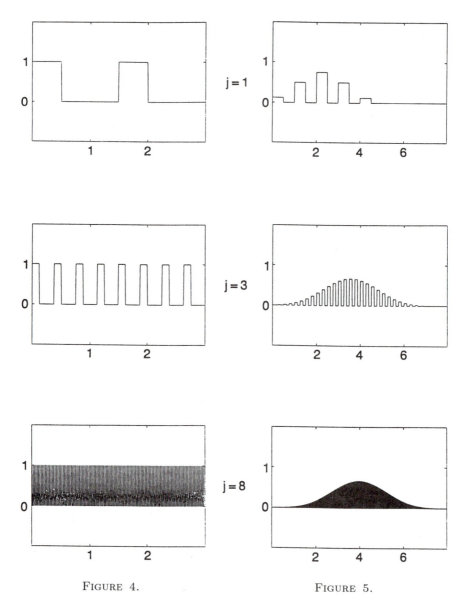

FIGURE 4.   FIGURE 5.

Cases of bad convergence in the cascade algorithm. The algorithm does
not converge in L^2 but only in the distribution sense.

III. CONVERGENCE OF THE CASCADE: SOME RESULTS

First, we remark that the sequences s_k^j are also the regular sampling values, at the points $2^{-j}k$, of the band-limited function $u_j(x)$ defined by

$$\hat{u}_j(\omega) = \prod_{k=1}^{j} m_0(2^{-k}\omega)\, \chi_{[-2^j\pi;\, 2^j\pi]}(\omega)\,. \tag{22}$$

We shall use this sequence u_j for two reasons:

- These approximations of ϕ are $\mathcal{C}^\infty$ functions (unlike the piecewise constant functions). This will help show the convergence of the discrete derivative schemes in the cascade.

- The family $\{u_j(x-k)\}_{k\in\mathbb{Z}}$ is already an orthonormal basis. Indeed, this is true for $u_0(x) = \frac{\sin x}{x}$ and, using (12), we find:

$$
\begin{aligned}
\langle u_j(x)|u_j(x-k)\rangle &= \frac{1}{2\pi}\int_{\mathbb{R}} |\hat{u}_j(\omega)|^2 e^{ik\omega}\, d\omega \\
&= \frac{1}{2\pi}\int_{-2^j\pi}^{2^j\pi} \prod_{k=1}^{j} |m_0(2^{-k}\omega)|^2\, e^{ik\omega}\, d\omega \\
&= \frac{1}{2\pi}\int_{-2^{j-1}\pi}^{2^{j-1}\pi} \prod_{k=1}^{j-1} |m_0(2^{-k}\omega)|^2 \\
&\qquad \times \left(|m_0(2^{-j}\omega)|^2 + |m_0(2^{-j}\omega+\pi)|^2\right) e^{ik\omega}\, d\omega \\
&= \langle u_{j-1}(x)|u_{j-1}(x-k)\rangle \\
&= \delta_{0,k}\,.
\end{aligned}
$$

The pointwise convergence of $\hat{u}_j(\omega)$ to $\hat{\phi}(\omega)$ is trivial. Note that it implies $\|\phi\|_{L^2(\mathbb{R})} \le 1$ (by Fatou's lemma). The scaling function is thus in L^2 but, as we mentioned above, $\{\phi(x-k)\}$ is not necessarily an orthonormal basis. For this last requirement, we need the L^2 convergence of the sequence u_j to ϕ. The exact characterization of the filters such that this is achieved has been solved by different approaches (see [3], [7], [8], [11]). We present here the most general result. The following definition will be useful.

Definition. *A compact set K congruous to $[-\pi;\, \pi]$ modulo 2π is a finite union of closed intervals $\{I_n\}_{n=0...N}$ of total measure 2π and such that there exists a set of integers $\{k_n\}_{n=0...N}$ satisfying*

$$\cup_{n=0}^{N} (I_n + 2k_n\pi) = [-\pi;\, \pi]\,. \tag{23}$$

Figure 6 shows an example of such a set.

FIGURE 6. Compact set congruous to $[-\pi; \pi]$ modulo 2π.

Theorem 1. *Suppose that the transfer function $m_0(\omega)$ has a Hölder exponent $\alpha > 0$.*

Then the infinite product (13) generates an orthonormal basis $\{\phi(x-k)\}_{k\in\mathbb{Z}}$ and the cascade converges in $L^2(\mathbb{R})$ if and only if $m_0(\omega)$ satisfies the following hypothesis:

(H) *There exist a compact set K congruous to $[-\pi; \pi]$ which contains a neighborhoods of the origin, such that, for all $n > 0$, $m_0(2^{-n}\omega)$ does not vanish on K.*

Proof. We first note that since $m_0(\omega)$ has a global Hölder exponent, $\hat{\phi}(\omega)$ is a continuous function.

Suppose that $\{\phi(x-k)\}_{k\in\mathbb{Z}}$ is an orthonormal basis. This can be expressed by the identity

$$\sum_{\ell\in\mathbb{Z}} |\hat{\phi}(\omega + 2\ell\pi)|^2 = 1 . \tag{24}$$

One can thus find an integer m such that for all ω in $[-\pi; \pi]$,

$$\sum_{\ell=-m}^{m-1} |\hat{\phi}(\omega + 2\ell\pi)|^2 > \frac{1}{2} . \tag{25}$$

Consequently for all ω in $[-\pi; \pi]$, there exists an integer k_ω such that $|\hat{\phi}(\nu + 2k_\omega\pi)| > \frac{1}{\sqrt{4m}}$ on a neighborhood V_ω of ω in $[-\pi; \pi]$. Note that $k_0 = 0$. Let us choose a finite sequence $\{\omega_n\}_{n=0...N}$ such that $\omega_0 = 0$ and the union of the V_{ω_n} covers $[-\pi; \pi]$, and definite $W_0 = \overline{V}_0$, $W_n = \overline{V_{\omega_n} \setminus (\cup_{k=0}^{n-1} W_k)}$, $I_n = W_n + 2k_{\omega_n}\pi$. Then $K = \cup_{n=0}^{N} I_n$ is a possible choice to check (H): K is congruous to $[-\pi; \pi]$, contains V_0 which is a neighborhood of the origin and $\hat{\phi}(\omega)$ $(= \prod_{k=1}^{+\infty} m_0(2^{-k}\omega))$ does not vanish on K.

Suppose now that (H) is satisfied. We shall use a third sequence v_j of approximations for ϕ, defined by

$$\hat{v}_j(\omega) = \prod_{k=1}^{j} m_0(2^{-k}\omega) \, \chi_{2^j K}(\omega) \tag{26}$$

which converges pointwise to $\hat{\phi}(\omega)$, since K contains a neighborhood of the origin. Since K is congruous to $[-\pi; \pi]$, one checks easily that for u_j as in (22), $s_k^j = u_j(2^{-j}k) = v_j(2^{-j}k)$ and that $\langle v_j(x-k)|v_j(x) \rangle = \langle u_j(x-k)|u_j(x) \rangle = \delta_{0,k}$.

Multiplying (26) by $\hat{\phi}(2^{-j}\omega)$, we find

$$\hat{\phi}(2^{-j}\omega) \, \hat{v}_j(\omega) = \hat{\phi}(\omega) \, \chi_{2^j K}(\omega) \, . \tag{27}$$

We now fully use (H) which implies that $\chi_{2^j K}(\omega) \, (\hat{\phi}(2^{-j}\omega))^{-1}$ is a bounded function. Since $\hat{\phi}$ is square-integrable, v_j converges to ϕ in L^2 (by Lebesgue theorem) and $\{\phi(x-k)\}_{k \in \mathbb{Z}}$ is an orthonormal basis.

Remarks. It is necessary to make some comments on the hypothesis (H) which may seem mysterious and difficult to satisfy.

- Since K is a compact set and $m_0(0) = 1$, one has to check $m_0(2^{-n}\omega) \neq 0$ only for a finite number of integers n.

- A trivial case where (H) holds is when $m_0(\omega)$ does not vanish on $\left[-\frac{\pi}{2}; \frac{\pi}{2} \right]$. Here, $K = [-\pi; \pi]$ and $u_j = v_j$ are a convenient choice.

- With a little more work, one finds that, if $m_0(\omega)$ does not vanish on $\left[-\frac{\pi}{3}; \frac{\pi}{3} \right]$, (H) is still satisfied.

- If $m_0(\omega)$ has a finite number of zeros in $[-\pi; \pi]$, then (H) has an arithmetic interpretation. (H) is not satisfied if and only if $|m_0(\omega)|$ is 1 on a cyclic orbit different from $\{0\}$, for the transformation $\omega \to 2\omega[2\pi]$ (see [5], [7], [8]).

This last remark explains the behavior of the cascade in the case of the dilated Haar System (Figure 4). After each iteration, the frequencies $\frac{2^j \pi}{3}$ and $\frac{-2^j \pi}{3}$ reproduce with the same weight since $\left| m_0 \left(\frac{2\pi}{3} \right) \right| = \left| m_0 \left(-\frac{2\pi}{3} \right) \right| = 1$. The L^2 convergence is thus a basic requirement on the filter. Yet, it is not sufficient to characterize "nice wavelets", as shown in Figure 3.

We now give a result which ensures more regular convergence.

Theorem 2. *Suppose that (H) is satisfied and that, for some $\alpha \geq 0$,*

$$\int_{\mathbb{R}} (1 + |\omega|)^{\alpha} \, |\hat{\phi}(\omega)| \, d\omega < +\infty \, . \tag{28}$$

Then ϕ is in $\mathcal{C}^\alpha$ and v_j converges to ϕ uniformly in $\mathcal{C}^\alpha$.

Proof.   It is well known that (28) implies that ϕ has global Hölder exponent superior to α.

Multiplying (27) by $(1+|\omega|)^\alpha$, we use the same argument as in Theorem 1. From the Lebesgue theorem, $(1+|\omega|)^\alpha v_j(\omega)$ converges to $(1+|\omega|)^\alpha \hat{\phi}(\omega)$ in $L^1(\mathbb{R})$. Consequently, u_j tends to ϕ uniformly in $\mathcal{C}^\alpha$.

This result tells us that if (H) is satisfied and ϕ is smooth, then the iterations in the cascade will also have a smooth aspect.

The next step is to characterize the filters leading to a scaling function which satisfies (28).

IV.   REGULARITY: A SPECTRAL APPROACH AND AN ASYMPTOTIC RESULT

To study the regularity from a spectral point of view, we shall use three classes of functions:

- $\mathcal{C}^\alpha$, the Hölder space of exponent α.

- $\mathcal{R}^\alpha$, the set of functions f such that $\hat{f}(\omega)(1+|\omega|)^\alpha$ is in $L^1(\mathbb{R})$.

- $\mathcal{D}^\alpha$, the set of functions f such that $\hat{f}(\omega)(1+|\omega|)^\alpha$ is in $L^\infty(\mathbb{R})$.

It is clear that $\mathcal{D}^{\alpha+1+\epsilon} \subset \mathcal{R}^\alpha \subset \mathcal{C}^\alpha$. If ϕ has compact support, we also have

$$\phi \in \mathcal{C}^{\alpha+\frac{1}{2}+\epsilon} \Rightarrow \phi \in \mathcal{R}^\alpha \quad \text{and} \quad \phi \in \mathcal{C}^\alpha \Rightarrow \phi \in \mathcal{D}^\alpha . \tag{29}$$

Let us recall the method used by I. Daubechies in [9] to estimate the decay of $\hat{\phi}(\omega)$. It uses the factorization

$$m_0(\omega) \;=\; \left(\frac{1+e^{i\omega}}{2} \right)^N \, p(\omega) \tag{30}$$

where N is the order of cancellation in π. From the identity $\frac{\sin \omega}{\omega} = \prod_{k=1}^{+\infty} \cos(2^{-k}\omega)$, we obtain

$$|\hat{\phi}(\omega)| \;=\; \left| \frac{2 \sin(\omega/2)}{\omega} \right|^N \prod_{k=1}^{+\infty} |p(2^{-k}\omega)|$$

$$=\; \left| \frac{2 \sin(\omega/2)}{\omega} \right|^N \prod_{1} (\omega) . \tag{31}$$

The growth of $\prod_1(\omega)$ can be controlled. Define

$$B_j = \operatorname*{Sup}_{\omega \in \mathbb{R}} \left| \prod_{k=0}^{j-1} p(2^k \omega) \right| = \operatorname*{Sup}_{\omega \in \mathbb{R}} |p_j(\omega)| \qquad (32)$$

and

$$b_j = \frac{\log(B_j)}{j \log 2} . \qquad (33)$$

If we take from $\prod_1(\omega)$ the factors such that $|2^{-k}\omega| > 1$ and group them in packets of size j, the rest of $\prod_1(\omega)$ will be bounded and we thus get

$$\prod_1(\omega) \leq CB_j \frac{\log\ (|\omega| + 1)}{j \log 2} = C(1 + |\omega|)^{b_j} . \qquad (34)$$

It follows that if $\alpha < N - b_j$ for some j, then ϕ is in $\mathcal{D}^\alpha$, $\mathcal{R}^{\alpha-1}$ and $\mathcal{C}^{\alpha-1}$. Let us now define

$$b = \operatorname*{Inf}_{j>0}\ b_j . \qquad (35)$$

By subadditivity ($B_{m+n} \leq B_m B_n$ and $b_{kn} \leq b_n$), b is also the limit of b_j when j goes to $+\infty$. The quantity b is called the "critical exponent" of the filter $m_0(\omega)$. It is helpful to optimize the above estimation of the decay of $\hat{\phi}(\omega)$ in the following sense:

Theorem 3. *If $\alpha < N - b$, then ϕ is in $\mathcal{D}^\alpha$, $\mathcal{R}^{\alpha-1}$ and $\mathcal{C}^{\alpha-1}$. Suppose that $|p(\pi)| > |p(0)| = 1$ and that $m_0(\omega)$ satisfies (H). Then, if $\alpha > N - b$, ϕ does not belong to $\mathcal{D}^\alpha$, $\mathcal{C}^\alpha$, $\mathcal{R}^{\alpha-1}$.*

Proof. The first part is trivial since b is the infimum of all b_j. The proof of the second part is rather technical and we shall give only its main aspects. The reader can consult [4] or [5] for a complete demonstration.

Consider the points ω_j such that $p_j(\omega_j) = B_j$. Since $|p(\omega + \pi)| > |p(\omega)|$ in a neighborhood of the origin, the same holds for $p_j(\omega)$. This shows that ω_j stays superior to some constant $C_1 > 0$.

Note that as $p_j(\omega)$ has period 2π, ω_j can be chosen in the compact set K which is involved in the hypothesis (H).

We thus have:

$$\prod_1(\omega_j) \geq |\hat{\phi}(\omega_j)| > C_2 > 0 , \qquad (36)$$

and

$$\prod_1(2^j \omega_j) = |p_j(\omega_j)| \prod_1(\omega_j) > C_3(1 + |2^j \omega_j|)^{b_j} . \qquad (37)$$

This shows that the growth of $\prod_1(\omega)$ is at least $|\omega|^b$ for some points. To show that ϕ is not in $\mathcal{D}^\alpha$ when $\alpha > N - b$ we have to multiply (37) by the factor $\frac{\sin(2^{j-1}\omega_j)}{2^{j-1}\omega_j}$, which could unfortunately vanish. To avoid this problem, we remark that the derivative of $p_j(\omega)$ satisfies

$$|p_j'(\omega)| \leq C_4\, 2^j B_j \tag{38}$$

(using the Leibniz formula and the convergence of b_j to b). Consequently, (37) holds also on intervals of constant size surrounding the points $2^j\omega_j$. It is thus possible to choose a sequence λ_j close to $2^j\omega_j$ and satisfying

$$\hat{\phi}(\lambda_j) > C_5(1 + |\lambda_j|)^{b_j - N} . \tag{39}$$

This concludes the proof of Theorem 3.

This result characterizes the exact decay of $\hat{\phi}(\omega)$ with the help of the critical exponent b. Note that the Hölder regularity cannot be exactly evaluated by this method since we only have $\mathcal{C}^\alpha \subset \mathcal{D}^\alpha \subset \mathcal{C}^{\alpha-1-\epsilon}$ (in the case of compactly supported functions). Better estimates were found in [10] by Daubechies and Lagarias who use infinite products of matrices. However this technique becomes unpracticable in the case of high regularity which requires filters with a large number of coefficients. In this last case, our spectral approach can be very interesting since the Hölder exponent of ϕ and the rate of decay of $\hat{\phi}$ are asymptotically equivalent.

Consider the family of compactly supported wavelets $\{\psi_N\}_{N>0}$ built in [9], from the family of filters defined by

$$|m_0^N(\omega)|^2 = \left(\cos^2\left(\frac{\omega}{2}\right)\right)^N P_N\left(\sin^2\left(\frac{\omega}{2}\right)\right) \tag{40}$$

with

$$P_N(y) = \sum_{j=0}^{N-1} \binom{N-1+j}{j} y^j = |p^N(\omega)|^2 . \tag{41}$$

Theorem 4.   *Let $\alpha(N)$ be the Hölder exponent of ψ_N and ϕ_N. Then,*

$$\lim_{N \to +\infty} \frac{\alpha(N)}{N} = 1 - \frac{\log 3}{2\log 2} \simeq 0.2075 . \tag{42}$$

Proof.   It is clear that $m_0^N(\omega)$ satisfies the hypothesis of Theorem 3 i.e. (H) with $K = [-\pi;\,\pi]$ and $|p^N(\pi)| > |p(0)| = 1$. Let $b(N)$ be the associated critical exponent.

We first remark that since $\left|p_j^N\left(\frac{2\pi}{3}\right)\right| = \left|p^N\left(\frac{2\pi}{3}\right)\right|^j$, then

$$b(N) \geq \frac{\log\left|p^N\left(\frac{2\pi}{3}\right)\right|}{\log 2} = \frac{\log(P_N(3/4))}{2\log 2}. \qquad (43)$$

If we simply minorate $P_N(y)$ by its last term, we obtain

$$\log\left(P_N\left(\frac{3}{4}\right)\right) > \log\left(\binom{2(N-1)}{N-1}\left(\frac{3}{4}\right)^{N-1}\right)$$

$$> \log\left(\frac{3^{N-1}}{2N}\right). \qquad (44)$$

Consequently, we have

$$\liminf_{N\to+\infty}\left(\frac{\alpha(N)}{N}\right) = 1 - \liminf\left(\frac{b(N)}{N}\right) \leq 1 - \frac{\log 3}{2\log 2}. \qquad (45)$$

The other inequality is more delicate. We shall use here a method developed independently by M. Volkner in [15] and J. P. Conze and the author in [6]. It is based on the following lemma:

Lemma.

$$y \leq 3/4 \quad\Rightarrow\quad P_N(y) = |p^N(\omega)|^2 \leq 3^{N-1} \qquad (46)$$

$$\frac{3}{4} \leq y \leq 1 \quad\Rightarrow\quad P_N(y)P_N(4y(1-y))$$

$$= |p^N(\omega)p^N(2\omega)|^2 \leq 3^{2(N-1)}. \qquad (47)$$

Proof. From the definition of $P_N(y)$, we see that, for all $y \geq \frac{1}{2}$,

$$P_N(y) = \sum_{j=0}^{N-1}\binom{N-1+j}{j}\left(\frac{1}{2}\right)^j(2y)^j \leq P_N\left(\frac{1}{2}\right)(2y)^{N-1}. \qquad (48)$$

As $P_N(y)$ grows between 0 and 1, and $P_N\left(\frac{1}{2}\right) = 2^N\left|m_0^N\left(\frac{\pi}{2}\right)\right|^2 = 2^{N-1}$, we have

$$P_N(y) \leq (\max(2;4y))^{N-1}. \qquad (49)$$

It is then trivial to check that $(\max(2;4y))^{N-1}$ satisfies the two inequalities (46) and (47) wanted for $P_N(y)$.

We now conclude the proof of Theorem 4. Dividing the product $p_j^N(\omega)$ into packets of one or two factors in order to apply either (46) or (47), we obtain

$$\left| p_j^N(\omega) \right|^2 \leq (3^{N-1})^{j-1} \, \sup \left| p^N(\omega) \right|^2 . \tag{50}$$

This implies $b(N) \leq (N-1)\frac{\log 3}{2\log 2}$ and finally

$$\lim_{N \to \infty} \frac{\alpha(N)}{N} = 1 - \lim \frac{b(N)}{N} = 1 - \frac{\log 3}{2\log 2} \simeq 0.2075 . \tag{51}$$

CONCLUSION

We have shown that the wavelet theory can be very useful in the design of subband coders. A filter associated with a regular scaling function will often be chosen if one wants to avoid the reconstruction from the low scales containing high frequencies. Theorem 3 indicates that this regularity cannot be ensured only by cancellation rules on the filter coefficients, as the critical exponent b may counterbalance the value of N.

Finally, Theorem 4 shows that to get a wavelet that is r times differentiable, approximately $10r$ taps are necessary since the filter $m_0^N(\omega)$ has length $2N$. This leads to a "trade-off" between the smoothness of the reconstruction and the amount of computations in the FWT algorithm.

REFERENCES

[1]   I. Daubechies, "Ten Lectures on Wavelets", CBMS-NSF Series in Applied Mathematics, SIAM (in press 1991).

[2]   T. P. Barnwell and M. J. T. Smith, "Exact reconstruction techniques for tree structured subband coders", IEEE ASSP, Vol. 34, pp. 434–441, 1986.

[3]   A. Cohen, "Ondelettes, analyses multiresolutions et filtres miroirs en quadrature", Annales de l'Institut Henri Poincaré, Analyse non lineaire, vol. 7, no. 5, pp. 439–459, 1990.

[4]   A. Cohen, "Construction de bases d'ondelettes α-Hölderiennes", Revista Matematica Iberoamericana, 1991.

[5]   A. Cohen, "Ondelettes, analysis multirésolutions et traitement numérique du signal", Ph.D. dissertation, Université Paris IX Dauphine, 1990.

[6] A. Cohen and J. P. Conze, "régularité des bases d'ondelettes et mesures ergodiques", to appear at Comptes rendus de l'Académie des Sciences, Paris, 1991.

[7] A. Cohen, "Two scale difference equations and bases of translates", to be published.

[8] J. P. Conze and A. Gaugi, "Fonctions harmoniques pour un opérateur de transition", preprint math. dept., Université de Rennes I, 35042 France, 1989.

[9] I. Daubechies, "Orthonormal bases of compactly supported wavelets", Comm. in Pure and Applied Math., Vol. 41, pp. 909–996, 1988.

[10] I. Daubechies and J. Lagarias, "Two scale difference equations", Part I and II, SIAM Journ. Math. Anal., 1990.

[11] W. Lawton, "Necessary and sufficient conditions for constructing orthonormal wavelet bases", J. Math. Phys., Vol. 32, No. 1, pp. 57–61, January 1991

[12] S. Mallat, "A theory for multiresolution approximation the wavelet representation", IEEE PAMI, Vol. 2, no. 7, 1989.

[13] Y. Meyer, "Ondelettes et Operateurs", tomes I, II et III, ed. Hermann, Paris, 1990.

[14] Y. Meyer, "Wavelets and Operators", Analysis at Urbana, Vol. 1, edited by E. Berkson, N. T. Peck and J. Uhl, London Math. Society Lecture Notes, Series 1, 1987.

[15] H. Volkner, "On the regularity of wavelets", preprint, Dept. of Math., University of Wisconsin – Milwaukee, P.O. Box 413, Milwaukee, WI 53201, 1990.

RIDGE AND SKELETON EXTRACTION FROM THE WAVELET TRANSFORM

PH. TCHAMITCHIAN
B. TORRÉSANI

Centre de Physique Théorique
CNRS-Luminy, Case 907 F-13288 Marseille
Cedex 09, FRANCE

I. INTRODUCTION

The starting point of standard signal analysis procedures is the computation of time-frequency representations of the analyzed signal, as for instance spectrograms, scalograms, or bilinear representations such as Wigner or smoothed Wigner distributions. Most of the algorithms that have been developed in that context are essentially based on the study of the time-frequency "energy localization", where the representation in the bilinear case or its squared-modulus in the linear case are interpreted as an energy density in the time-frequency plane. Among the linear representations, the Gabor representation [Ga] (or sliding window Fourier transform) based on time and frequency translations, has been the most popular for a long time. More recently, Grossmann and Morlet [Gr-Mo] have proposed an alternative representation, called the wavelet transform, which basically has the same structure [Gr-Mo-Pa1]. The frequency translations are replaced by dilations (the dilation parameter is interpreted as a reference frequency, divided out by the frequency shift). Both methods have been applied to signal analysis, with comparable performance levels, in different contexts.

We consider here the application of wavelets to specific problems of signal analysis, namely the problems of extraction and characterization of amplitude and frequency modulated signals in the asymptotic (i.e. high frequency) limit. More specifically, we are interested not in the asymptotic signals themselves, for which standard techniques based on the Hilbert transform give excellent results, but rather in composite signals of asymptotic type, which can be defined as sums of asymptotic signals. Notice that such a composite signal of asymptotic type has no reason to be asymptotic; moreover, the Hilbert transform is unable to give precise information about the characteristics of the components, such as frequency and amplitude modulation laws. Nevertheless, the first test of the algorithms presented will be performed on asymptotic signals.

The plan of the paper is as follows. After briefly recalling in Section II the basic definitions and properties of wavelet analysis, we describe in Section III the behaviour of wavelet analysis in the asymptotic limit for asymptotic wavelets. We then show in Section IV how, for Gaussian wavelets, the asymptotic assumption on the analyzing function can be relaxed. Algorithms are described, which allow the extraction of instantaneous frequencies (and spectral lines), and in some cases the separation of frequency-modulated components in a signal. Applications to matched filtering are discussed in Section V, and Section VI is devoted to conclusions.

II.   CONTINUOUS WAVELET ANALYSIS

We describe in this Section the basic properties of continuous wavelet analysis which we use. For more details on such techniques, we refer to [Da1], [Gr-Mo], [Gr-Mo-Pa1], [Gr-Mo-Pa2] and [Gr-KM-Mo].

We restrict our wavelet analysis to the real Hardy space:

$$H^2(\mathbf{R}) = \left\{ f \in L^2(\mathbf{R}); \hat{f}(\omega) = 0 \; for \; \omega \leq 0 \right\}.$$

This is sufficient for our purpose. The wavelet analysis of $L^2(\mathbf{R})$ follows from easy modifications.

Let us start with two functions g and h in $H^2(\mathbf{R})$ such that the following admissibility condition holds: if

$$c_{g,h} = \int_0^{+\infty} \frac{\overline{\hat{g}(\omega)}\hat{h}(\omega)}{\omega} \, d\omega \,, \qquad (II-1)$$

then

$$0 \; < \; |c_{g,h}| \; < \; \infty \qquad (II-2)$$

where our convention for the Fourier transform is

$$\hat{f}(\omega) = < f, \varepsilon_\omega >_{L^2} = \int_{-\infty}^{+\infty} f(t) e^{-i\omega t} dt$$

with $\varepsilon_\lambda(t) = e^{i\lambda t}$. We shall call g the analyzing wavelet and h the reconstructing wavelet. The wavelets $g_{(b,a)}$ and $h_{(b,a)}$ associated with these
functions are:

$$g_{(b,a)}(t) = \frac{1}{a} g\left(\frac{t-b}{a}\right) \qquad (II-3)$$

(with the same definition for $h_{(b,a)}$), where the translation and dilation
parameters run over the Poincaré half plane $\mathbf{H} = \{(b,a) \in \mathbf{R}^2; a > 0\}$.
The admissibility condition then implies the following resolution of the
identity: to each $s \in H^2(\mathbf{R})$ one can associate the following family of
wavelet coefficients

$$T_s(b,a) = < s, g_{(b,a)} >_{L^2} . \qquad (II-4)$$

These are simply the L^2 products of $s(t)$ by the dilated and translated
wavelets. When there is no possible confusion, the subscript s of T_s will be
omitted. We then have a reconstruction formula:

$$s(t) = \frac{1}{c_{g,h}} \int_0^{+\infty} \int_{-\infty}^{+\infty} T_s(b,a) \, h_{(b,a)}(t) \, \frac{da \, db}{a}, \qquad (II-5)$$

which expresses $s(t)$ as a sum of dilated and translated wavelets $h_{(b,a)}$,
the equality being understood in the L^2 sense. The coefficients of the
decomposition are the corresponding wavelet coefficients.

It is convenient in practice to use wavelets having good localization properties in both the direct space and the Fourier space; in particular, g can
be chosen so that both g and $\hat{g}$ have exponential decay at infinity (and/or
the same for h and $\hat{h}$). Moreover this allows one to interpret the dilation
parameter a as the inverse of a frequency shift parameter (up to a mul-
tiplicative constant, characteristic of the analyzing wavelet), and then to
make the connection with the Gabor coefficients.

Note that it is often useful for numerical applications to use a simplified
version of the reconstruction formula:

$$s(t) = \frac{1}{k_g} \int_0^{+\infty} T_s(t,a) \, \frac{da}{a} \qquad (II-6)$$

which expresses $s(t)$ as a sum over the positive frequencies with logarithmic
measure. This is equivalent to formally taking $h = \delta$ in equation $(II-5)$

where the admissibility condition $(II - 1)$ has the form:

$$0 < |k_g| = \left| \int_0^{+\infty} \frac{\overline{\hat{g}(\omega)}}{\omega} d\omega \right| < \infty. \qquad (II - 7)$$

Assuming that $c_{g,g} < \infty$, one then has $T_s \in L^2(\mathbf{H}, \frac{da\, db}{a})$. However, the wavelet coefficients do not span the whole $L^2(\mathbf{H}, \frac{da\, db}{a})$, but a subspace of it, called the **reproducing kernel space**:

$$\mathcal{H}_{g,h} = \left\{ F \in L^2(\mathbf{H}, \frac{da\, db}{a}); K_{g,h}F = F \right\}. \qquad (II - 8)$$

The **reproducing kernel** is given by:

$$K_{g,h}(b, a; b', a') = \frac{1}{c_{g,h}} < g_{(b,a)}, h_{(b',a')} > \qquad (II - 9)$$

with respect to the measure $\frac{da\, db}{a}$ on the Poincaré half plane. It is worth noticing that $K_{g,h}$ is a projection operator, which is orthonormal if and only if $h = g$.

To understand the information provided by the wavelet coefficients, the standard technique [Gr-KM-Mo] codes them into two images, $|T_s(b, a)|$ and $\Psi(b, a) = Arg\, [T_s(b, a)]$. Thanks to the Plancherel equality:

$$||s||^2 = \frac{1}{c_{g,g}} \int |T_s(b, a)|^2 \frac{da\, db}{a}, \qquad (II - 10)$$

the $\frac{1}{c_{g,g}} |T_s(b, a)|^2$ is interpreted as an energy density in the Poincaré half plane (we have assumed here that $0 < c_{g,g} < \infty$), and is often referred to as a scalogram. The reader will find in [Gr-KM-Mo] a description of the usual conventions and interpretations of the images. We will focus mainly on the study of the argument $\Psi(b, a)$ of the wavelet coefficients.

III. ASYMPTOTIC WAVELETS

We will now describe the behaviour of the wavelet coefficients in the asymptotic limit. An asymptotic signal is a signal such that the variations of its amplitude are slow compared with the variations of its phase. Assuming that the signal and the analyzing wavelet (and then all the wavelets) are asymptotic leads to an approximate expression for the corresponding coefficients. As we will see, this allows the specification of some particular sets of curves in the time-scale half-plane called the ridge, which describes the frequency modulation law of the signal and the wavelet curves. One then gets a simple geometrical description of the transform (that we call the factorization property): essentially, the restriction of the transform to the ridge (the so-called skeleton) reproduces the analytic signal of the analyzed signal, while the restriction to a given wavelet curve reproduces the corresponding wavelets.

We first describe the computation of the wavelet coefficients in the asymptotic limit, and introduce the notions of ridge and wavelet curves; we then derive simple formulas, leading to algorithms for the extraction of the ridge.

III–1 ASYMPTOTIC WAVELET COEFFICIENTS

Let $s \in L^2(\mathbf{R})$ be an asymptotic locally monochromatic signal. Let Z_s be its analytic signal, obtained by a linear filtering which suppresses the negative frequencies. Let us express Z_s in its exponential form:

$$Z_s(t) = A_s(t) \exp\left(i\phi_s(t)\right). \qquad (III-1)$$

Then clearly

$$s(t) = Re[Z_s(t)] = A_s(t) \cos\left(\phi_s(t)\right). \qquad (III-2)$$

Let us assume now that s is asymptotic in the following sense: $\phi'_s(t) > 0$ and for some fixed (small) positive real number ϵ

$$\left|\frac{1}{\phi'_s(t)}\right| \left|\frac{A'_s(t)}{A_s(t)}\right| < \epsilon \,,$$

$$\left|\frac{\phi''_s}{\phi'^2_s}\right| < \epsilon \qquad (III-3)$$

for all t in the domain under consideration. Let $g \in H^2(\mathbf{R})$ be an analytic analyzing wavelet. One then has:

$$T(b, a) = <s, g_{(b,a)}> = \frac{1}{2} <Z_s, g_{(b,a)}> \qquad (III-4)$$

Assume that the analyzing wavelet g is asymptotic, and set:

$$g(t) = A_g(t) \exp\left(i\phi_g(t)\right). \qquad (III-5)$$

Then all the dilated and translated wavelets $g_{(b,a)}(t)$ are asymptotic too (since they are of constant shape). One then has the following wavelet coefficient expression.

Theorem 1.   *Let $s \in L^2(\mathbf{R})$ and $g \in H^2(\mathbf{R})$ be asymptotic functions. Denote by t_s the stationary points of*

$$\Phi_{(b,a)}(t) = \phi_s(t) - \phi_g\left(\frac{t-b}{a}\right). \qquad (III-6)$$

Then $T(b,a)$ is essentially given by the contribution of such points to the scalar product $(III-4)$. If moreover

$$\Phi''_{(b,a)}(t_s) \neq 0, \qquad (III-7)$$

then the first order term in the asymptotic development of $T(b,a)$ reads:

$$T(b,a) = \frac{Z_s(t_s)\overline{g\left(\frac{t_s-b}{a}\right)}}{Corr(b,a)} + O(\epsilon) \qquad (III-8)$$

where

$$Corr(b,a) = \sqrt{\frac{2}{\pi}}\,\sqrt{a^2\left|\Phi''_{(b,a)}(t_s)\right|}\,e^{-i\frac{\pi}{4}Sgn\Phi''_{(b,a)}(t_s)}\,. \qquad (III-9)$$

Proof.   Under the assumptions of the theorem, $T(b,a)$ takes the form of a rapidly oscillating integral:

$$T(b,a) = \frac{1}{2a}\int_{-\infty}^{+\infty} A_s(t)\,A_g\left(\frac{t-b}{a}\right)e^{i\left[\phi_s(t)-\phi_g\left(\frac{t-b}{a}\right)\right]}\,dt, \qquad (III-10)$$

which can be approximated by means of the stationary phase method. The essential contribution to $(III-10)$ is provided by the stationary points of the argument $\Phi_{(b,a)}(t)$ of the integrand, i.e. the points t_s such that:

$$\phi'_s(t_s) = \frac{1}{a}\,\phi'_g\left(\frac{t-b}{a}\right)\,. \qquad (III-11)$$

Let a domain $\Omega \subset \mathbf{H}$ be such that for any $(b,a) \in \Omega$, there is associated an unique first order stationary point t_s (i.e. such that equation $(III-7)$ holds). Then by a first order development of the modulus of the integrand, and a second order development of its argument, one is led to a Gaussian integral, whose solution is given by equation $(III-8)$ and equation $(III-9)$.

It is interesting to notice that such an expression involves the evaluation of the analytic signal only at the stationary points. This property will be crucial in the following. Note also that any stationary point t_s is actually a function $t_s(b,a)$ on the half plane $\mathbf{H}$.

Using the same techniques, one can also derive an explicit expression for the first corrective term in the asymptotic development of $T(b, a)$, involving the first derivatives of the amplitudes of the signal and the translated and dilated wavelets. Due to the vanishing at infinity of the analyzing wavelet, this term is easily shown to vanish in the considered case, which proves the quality of the approximation. The next corrective term (involving second derivatives of the amplitudes) is in general nonzero. From now on, the wavelet coefficients will always be replaced by their first order approximation.

In equation $(III - 8)$, it clearly appears that some particular sets of points will play an important role in the understanding of the wavelet transform. These sets of points are basically the points $(b, a) \in \Omega$ for which $t_s(b, a) = b$ (ridge of the transform) or $t_s(b, a) = C^{st}$ (wavelet curves). Equation $(III - 8)$ basically states that the wavelet transform basically behaves like the analytic signal of the analyzed signal on the ridge, while when restricted to a given wavelet curve, it behaves like the corresponding dilated and translated wavelets. We will refer to such a property as the factorization property of the asymptotic wavelet transform.

In the three next Sections, we give more details on the ridge and wavelet curves. Additional comments and numerical results will be presented in Section $III - 5$.

III–2   THE RIDGE OF THE WAVELET TRANSFORM

Let us introduce now the notion of **ridge** of the transform.

Definition 2.   The ridge of the wavelet transform of s is the set of points $(b, a) \in \Omega$ such that $t_s(b, a) = b$.

It immediatly follows from the definition of stationary points that on the ridge:

$$a = a_r(b) = \frac{\phi'_g(0)}{\phi'_s(b)}. \qquad (III - 12)$$

The ridge is then a curve $\mathcal{R} = \{(b, a) \in \Omega; a = a_r(b)\} \subset \Omega$ in the domain Ω, from which one easily recovers the frequency modulation law. Moreover, equation $(III - 8)$ and equation $(III - 9)$ simplify on the ridge to:

$$T(b, a_r(b)) = \sqrt{\frac{\pi}{2}} \frac{e^{i\frac{\pi}{4} Sgn[\Phi''_{(b,a_r(b))}(b)]}}{\sqrt{a_r(b)^2 \left|\Phi''_{(b,a_r(b))}(b)\right|}} \overline{g(0)} Z_s(b). \qquad (III - 13)$$

The restriction of the wavelet transform of $s(t)$ to the associated ridge gives, up to a corrective function of b, the analytic signal $Z_s(t)$ of $s(t)$. To evaluate this corrective function, note that owing to equation $(III - 12)$, for any $(b, a) \in \Omega$:

$$\Phi''_{(b,a)}(t) = -\frac{a'_r(t)\phi'_g(0)}{a_r(t)^2} - \frac{1}{a^2}\phi''_g\left(\frac{t-b}{a}\right) \qquad (III - 14)$$

and then,

$$\Phi''_{(b,a_r(b))}(b) = -\frac{1}{a_r(b)^2}\left[a'_r(b)\phi'_g(0) + \phi''_g(0)\right] . \qquad (III - 15)$$

Let $Corr(t) = g(\bar{0})/Corr(t, a_r(t))$. One then has the following result on the wavelet coefficients:

Proposition 3. *Under the assumptions of theorem 1, the restriction of the wavelet coefficients to the ridge is given by:*

$$T(t, a_r(t)) = Corr(t) \, Z_s(t) \qquad (III - 16)$$

where the function $Corr(t)$ is completely determined by the analyzing wavelet and the ridge of the transform.

Notice that once the ridge is known, one is then able to recover directly $A_s(t)$ from the wavelet coefficients $T(b, a)$.

The restriction of the wavelet transform to its ridge is called the **skeleton** of the transform. The ridge and the skeleton on Ω give sufficient information to characterize the transform on Ω. We introduce the corresponding **skeleton operator** $\mathcal{F}^{\mathcal{R}}_0 : L^2(\mathbf{R}) \to L^2(\mathbf{R})$, defined by:

$$\left[\mathcal{F}^{\mathcal{R}}_0 s\right](t) = \sqrt{\frac{2}{\pi}} \frac{\sqrt{a^2 \left|\Phi''_{(t,a_r(t))}(t)\right|}}{g(0)} e^{-i\frac{\pi}{4}Sgn[\Phi''_{(t,a_r(t))}(t)]} \, T(t, a_r(t)).$$

$$(III - 17)$$

This allows the recovery of the analytic signal from the wavelet transform. A slight generalisation of this operator will be useful in the matched filtering context (see Section V).

Remark. The phase shift $\pm\frac{\pi}{4}$ is here a global phase shift in Ω, since we have assumed that for any $(b, a) \in \Omega$, $\Phi''_{(b,a)}(t_s) \neq 0$.

III–3   THE WAVELET CURVES

Consider $(b_0, a_r(b_0)) \in \Omega$. We now introduce the notion of **wavelet curve**.

Definition 4.   The wavelet curve through $(b_0, a_r(b_0))$ is the connected component of $(b_0, a_r(b_0))$ of the set of points $(b, a) \in \Omega$ such that $t_s(b, a) = b_0$.

Let $(b, a) \in \Omega$ be such a point. This implies that:

$$\phi'_g \left(\frac{b_0 - b}{a} \right) = \frac{a}{a_r(b_0)} \; \phi'_g(0). \qquad (III - 18)$$

The wavelet curves are then completely determined by the analyzing wavelet, or more precisely by its frequency modulation law. For instance, in the case of fixed-frequency wavelets, $\phi'_g(0) = \omega_0$, the wavelet curve through $(b_0, a_r(b_0))$ is the line $a = a_r(b_0)$. This property is a characteristic property of the fixed-frequency analyzing wavelets.

Restricting the wavelet transform to a given wavelet curve yields:

$$[T(b, a)]_{t_s(b,a) = b_0} = \sqrt{\frac{\pi}{2}} \frac{e^{i \frac{\pi}{4} Sgn[\Phi''_{(b,a)}(b_0)]}}{\sqrt{a^2 \left| \Phi''_{(b,a)}(b_0) \right|}} Z_s(b_0) \; \overline{g \left(\frac{b_0 - b}{a} \right)}. \qquad (III - 19)$$

So the behaviour of the wavelet transform restricted to a given wavelet curve is governed by that of the corresponding dilated and translated wavelets (the dilation parameter a being a function of the translation parameter b). Once more, $|T(b, a)|$ is obtained from $\overline{g(\frac{b_0 - b}{a})}$ via a perturbation by the correction function $\sqrt{a^2 \left| \Phi''_{(b,a)}(b_0) \right|}$. The argument of $T(b, a)$ restricted to a given wavelet curve will be discussed in the next Section.

It is worth noticing that one now has a nice geometrical picture of the wavelet transform in the time-scale half plane. When restricted to the ridge, it is essentially determined by the analytic signal of the analyzed signal, and when restricted to a given wavelet curve, it is essentially determined by the wavelets, in both cases up to a corrective function, completely determined by the ridge and the analyzing wavelet. This is the so-called **factorization** property of the transform, that will be frequently used in the sequel.

III–4   RIDGE EXTRACTION AND TIME-FREQUENCY ENERGY LOCALIZATION

An important question is that of the extraction of the ridge from the $T(b, a)$ coefficients. The most usual and natural information comes from the squared modulus of the coefficients, interpreted as an energy density in the time-scale half-plane $\mathbf{H}$. When the analyzing wavelet is assumed to be maximum at $t = 0$, equation $(III - 8)$ shows that $|T(b,a)|$ is locally maximum at $b = t_s$, if one neglects the influence of the corrective function in the denominator of equation $(III - 8)$. Taking this corrective function into account shows that the maximality of $T(b,a)$ at $b = t_s$ only holds in an approximate sense, which is exact if and only if the ridge $a_r(b)$ is a linear function of b, i.e. for hyperbolic frequency modulated signals and $\phi''_g(0) = 0$.

The ridge of the transform can be extracted in a much more precise way from the phase of the transform. Let

$$\Psi(b,a) = \arg\left[T(b,a)\right]. \qquad (III - 20)$$

Then we have the following property for $\Psi(b,a)$.

Proposition 5. *Under the assumptions of Theorem 1, the argument of the wavelet transform has the two following properties on Ω:*

$$\frac{\partial \Psi(b,a)}{\partial a} = 0 \text{ on the ridge,}$$

and

$$\left[\frac{d\Psi(b,a)}{db}\right]_{t_s(b,a)=b_0} = \frac{\phi'_g(0)}{a} \text{ on the intersection with the ridge.}$$

Proof. By definition of $t_s(b,a)$

$$\frac{\partial \Psi(b,a)}{\partial a} = -\frac{t_s - b}{a^2}\, \Phi'_{(b,a)}\left(\frac{t_s - b}{a}\right) \qquad (III - 21)$$

and then vanishes on the intersection of the lines $b = C^{st}$ and the ridge. Moreover

$$\left[\frac{d\Psi(b,a)}{db}\right]_{t_s(b,a)=b_0} = \frac{1}{a}\,\phi'_g\left(\frac{t_s - b}{a}\right) + \left[\frac{da}{db}\right]_{t_s(b,a)=b_0} \frac{t_s - b}{a^2}\,\phi'_g\left(\frac{t_s - b}{a}\right).$$
$$(III - 22)$$

Hence the instantaneous frequency of the restriction of the wavelet transform to a fixed wavelet curve equals the central frequency of the dilated wavelet at the intersection with the ridge.

These two properties yield two algorithms for the ridge extraction, which are quite useful for numerical computations. Numerical applications of these algorithms will be presented in a forthcoming paper [De-E-Gu-KM-Tc-To].

Remark: Note that in the case of an asymptotic fixed-frequency analyzing wavelet, the second property does not allow the extraction of the ridge. Indeed, in that case, $t_s(b, a) = b_0$ is equivalent to $a = a_r(b_0)$, and one has that $\frac{\partial \Psi}{\partial b}(b, a) = \frac{\phi_g'(0)}{a}$. This illustrates the fact that the algorithms presented here are based on necessary conditions (see Proposition 5) and not on necessary and sufficient conditions.

III–5 *SOME ADDITIONAL REMARKS AND COMMENTS*

The main results of the present Section can be summarized as follows:

Let $T_s(b, a)$ be the wavelet transform of the asymptotic locally monochromatic signal $s(t)$ with respect to the progressive asymptotic wavelet $g(t)$. Then $T_s(b, a)$ is non-negligible only in a neighborhood of a curve in $\Omega \in \mathbf{H}$ (the ridge) directly related to the frequency modulation law of the analyzed signal.

$T_s(b, a)$ (and then $s(t)$) is completely characterized by its restriction to such a curve (i.e. its skeleton).

The ridge can be extracted directly from the phase of $T_s(b, a)$.

Let us then consider as a numerical example a function $s(t) = a(t) \cos(\phi(t))$ with $a(t)$ a triangular function and $\phi(t) = \frac{2\pi\nu_0}{\alpha} \log(1 + \alpha t)$, so that the instantaneous frequency yields: $\nu(t) = \frac{\nu_0}{1+\alpha t}$, and ν_0 is large enough to ensure the asymptotic character of the signal. Figure 1 represents such a signal, and the modulus and phase of its wavelet transform are partly plotted in Figures 2-a and 2-b (coded in grey levels).

Notice the energy localization around the ridge; notice also that the lines of constant phase (see in particular the null phase lines, white in Figure 2-b) are curved, in such a way that their top lies at their intersection with the ridge (property similar to the first statement of Proposition 5). Finally, Figure 3 represents the skeleton of the wavelet transform of such a signal, to be compared with Figure 1. This is in perfect agreement with Proposition 3, apart from the neighborhoods of the angular points of the $a(t)$ amplitude, where the stationary phase approximation fails, and where the skeleton is a smoothed version of the signal.

In such a very simple case, the wavelet transform allows the characterization of the frequency and amplitude modulation laws of the signal. Standard methods based on the Hilbert transform can do so as well. But the main interest of wavelet analysis in that context is that it also applies to linear combinations of asymptotic signals (which not need be asymptotic themselves) for which standard methods in general fail. Indeed, consider an analytic signal of the form $Z_s(t) = \sum_{k=1}^{N} a_k(t) \exp[i\phi_k(t)] = A_s(t) \exp[i\Phi_s(t)]$. From $Z_s(t)$ one can directly recover $A_s(t)$ and $\Phi_s(t)$, but one is in general much more interested in the $a_k(t)$ and $\phi_k(t)$ functions. Clearly, if the individual ridges of the asymptotic components are located in different regions in **H**, the wavelet transform allows their separation as follows:

Due to the fast decay of both g and $\hat{g}$, the wavelet transforms of the asymptotic components are numerically non-negligible in disjoint domains in the half-plane.

Restricting to such a domain, one then can extract the ridge and the frequency modulation law of the corresponding asymptotic component.

Restricting the wavelet transform to the ridge yields the corresponding skeleton, and then the amplitude modulation law of the corresponding component.

As a numerical illustration, consider the sum of two signals identical to the previous one, with different values of ν_0 (see Figure 4).

Clearly the instantaneous frequency of such a signal computed via Hilbert transform is an oscillating function, because of the beats phenomena. However, choosing wavelets with a sufficiently small width in the Fourier space allows the separation of the two components, via individual ridge extractions (see [De-E-Gu-KM-Tc-To] for a more systematic presentation). Indeed, Figure 5 shows the skeleton of one of the two components of the signal, which is quite close to the original signal presented in Figure 1.

Finally, let us stress that there are situations where our extraction methods fail partly or completely, i.e. situations in which the interactions between components are too strong. The prototype of such situations is the case where the instantaneous frequency functions of two components intersect.

This is the case in Figure 6 where the signal is the sum of that of Figure 1 and a monochromatic function.

Figure 7 represents the skeleton of the first component, which is perturbed in a neighborhood of the value of t for which the instantaneous frequency equals the frequency of the second component. The perturbation is local here, but there are many examples for which this is not the case, such as musical examples (clarinet sounds in [De], [De-E-Gu-KM-Tc-To]).

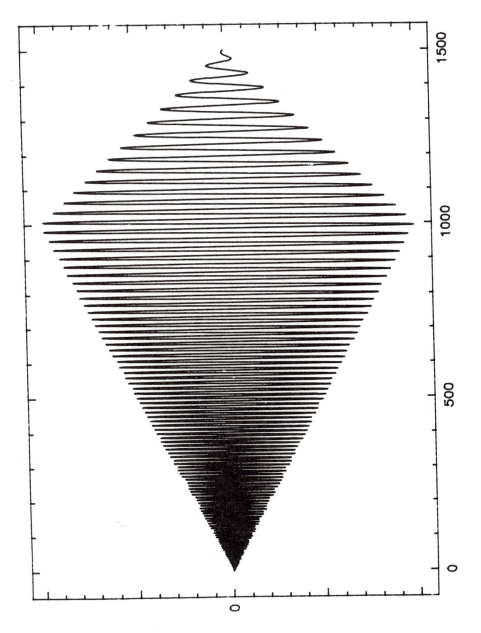

FIGURE 1. Reference signal: hyperbolic frequency modulation law and triangular amplitude modulation law.

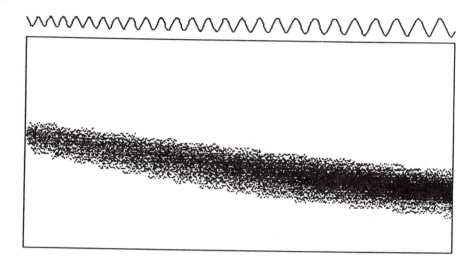

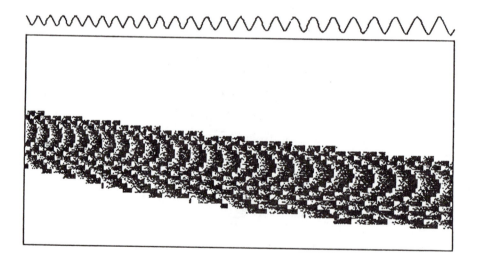

FIGURE 2. Wavelet transform of the reference signal of Figure 1 (the analyzing wavelet is the Morlet wavelet). Figure 2-a: modulus; Figure 2-b: phase.

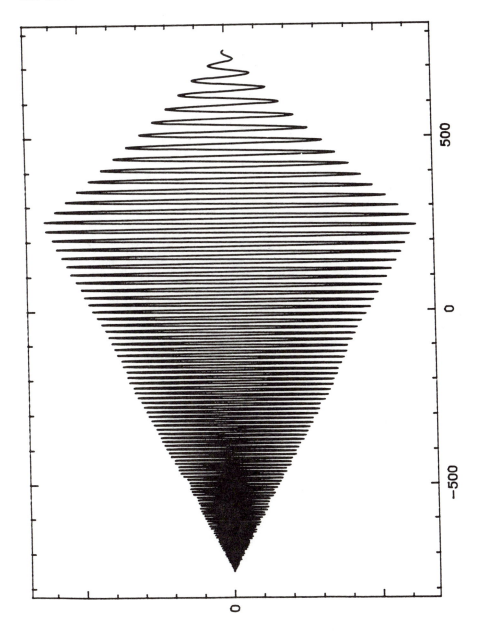

FIGURE 3. Skeleton of the wavelet transform of the reference signal.

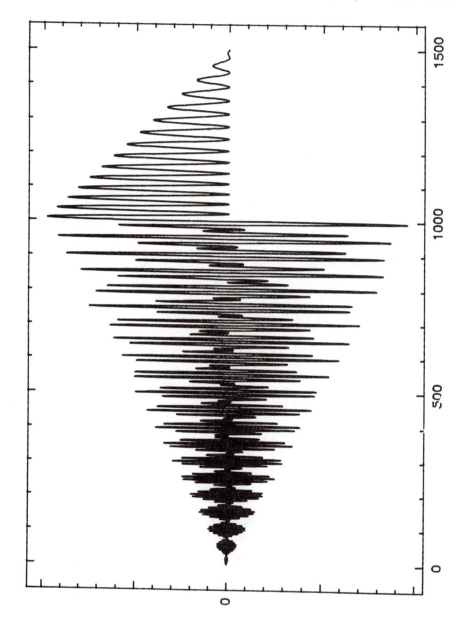

FIGURE 4. Sum of two copies of the reference signal of Figure 1, one of which is
uniformly frequency-shifted.

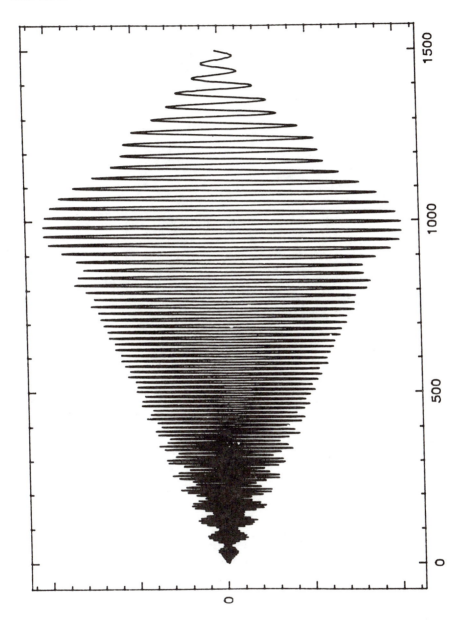

FIGURE 5. Skeleton of the wavelet transform of the signal of Figure 4.

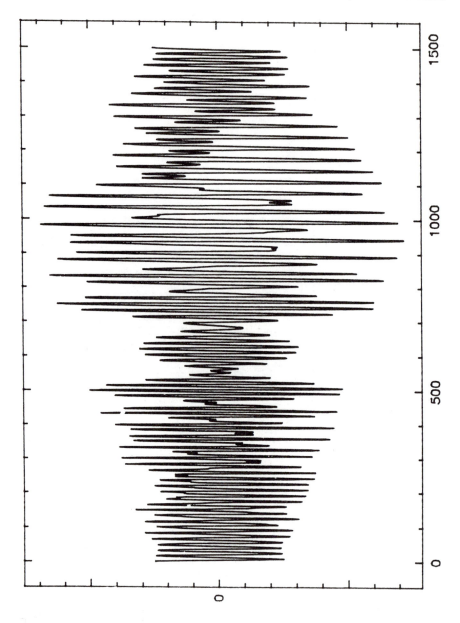

FIGURE 6. Sum of the reference signal and a monochromatic signal, with intersecting ridges.

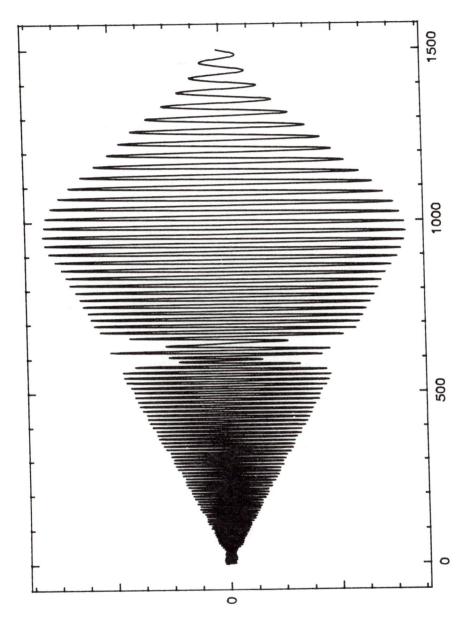

FIGURE 7.   Skeleton of the signal of Figure 6.

IV.   NON-ASYMPTOTIC WAVELETS

It must be noticed that the asymptotic assumption made on the analyzing functions in the last section is not very natural in the wavelet philosophy. Indeed, the frequency localization of these functions is enforced, but the time localization is destroyed. This has moreover a bad consequence on the numerical point of view, in the sense that the computational time is increased considerably.

In the case where the analyzing wavelet has a Gaussian amplitude, it is however possible to slightly generalize the stationary phase calculus to non-asymptotic wavelets (or Gabor curves), the signal still being assumed asymptotic. The notions of ridge and wavelet curves can still be introduced, and have the same intuitive meaning. The only changes concern the explicit expressions of the coefficients, and the ridge extraction algorithms.

The plan of this section is essentially the same as the previous one, apart from the fact that we will not present numerical results, which would be completely analogous to the previous ones.

IV–1   WAVELET COEFFICIENTS

Consider an analytic analyzing wavelet $g \in L^2(\mathbf{R})$, such that its amplitude $A_g(t)$ is a Gaussian function, which can be supposed without loss of generality to be centered and with unit variance. Assume moreover that $\hat{g}(\omega) < \epsilon$ for $\omega \leq 0$. The stationary phase method no longer yields accurate estimates, but can be slightly modified as follows. Equation $(III - 10)$ can now be approximated by:

$$T(b,a) \approx \frac{A_s(t_s)e^{i\Phi_{(b,a)}(t_s)}}{2a} \int_{-\infty}^{+\infty} A_g\left(\frac{t-b}{a}\right) e^{i\frac{(t-t_s)^2}{2}\Phi''_{(b,a)}(t_s)} dt.$$

$$(IV-1)$$

This new integral, up to a rotation in the complex plane, reduces to a Gaussian integral. Here, $t_s = t_s(b,a)$ is still defined by equation $(III-11)$.

A straightforward computation then yields the following analogue of Theorem 1.

Theorem 1'.   *Let $s \in L^2(\mathbf{R})$ be an asymptotic signal, and $g \in H^2(\mathbf{R})$ an analytic analyzing wavelet. Denote by t_s the stationary points of*

$$\Phi_{(b,a)}(t) = \phi_s(t) - \phi_g\left(\frac{t-b}{a}\right).$$

Then $T(b,a)$ is essentially given by the contribution of such points to the *scalar product* $(III - 4)$. *If moreover*

$$\Phi''_{(b,a)}(t_s) \neq 0,$$

then the first order term in the asymptotic development of $T(b, a)$ *reads:*

$$T(b, a) = |T(b, a)| \, e^{i\Psi(b,a)}, \qquad (IV - 2 - a)$$

$$|T(b, a)| = \sqrt{\frac{\pi}{2}} \, \frac{A_s(t_s) e^{-\frac{1}{2} \frac{(t_s - b)^2 \Phi''_{(b,a)}(t_s)^2}{a^2 \Phi''_{(b,a)}(t_s)^2 + \frac{1}{a^2}}}}{\left[1 + a^4 \Phi''_{(b,a)}(t_s)^2\right]^{\frac{1}{4}}} + O(\epsilon), \qquad (IV - 2 - b)$$

$$\Psi(b, a) = \phi_s(t_s) - \phi_g\left(\frac{t_s - b}{a}\right) + \frac{1}{2} \frac{(t_s - b)^2 \Phi''_{(b,a)}(t_s)}{1 + a^4 \Phi''_{(b,a)}(t_s)^2}$$

$$+ \frac{1}{2} \arctan\left[a^2 \Phi''_{(b,a)}(t_s)\right] + O(\epsilon) \qquad (IV - 2 - c)$$

Contrary to the case described in Section III the first corrective term in the asymptotic development, proportional to $A'_s(t_s)$, is not identically zero in Ω. This term is:

$$T_{(1)}(b, a) = \frac{A'_s(t_s)}{2a} e^{i\Phi_{(b,a)}(t_s)} \int_{-\infty}^{+\infty} (t - t_s) A_g\left(\frac{t - b}{a}\right) e^{i\frac{(t - t_s)^2}{2} \Phi''_{(b,a)}(t_s)} \, dt, \qquad (IV - 3)$$

and can be shown to equal

$$T_{(1)}(b, a) = |T_{(1)}(b, a)| \, e^{i\Psi_{(1)}(b,a)}, \qquad (IV - 4 - a)$$

$$|T_{(1)}(b, a)| = \sqrt{\frac{\pi}{2}} \, \frac{A'_s(t_s) e^{-\frac{1}{2} \frac{(t_s - b)^2 \Phi''_{(b,a)}(t_s)^2}{a^2 \Phi''_{(b,a)}(t_s)^2 + \frac{1}{a^2}}}}{\left[1 + a^4 \Phi''_{(b,a)}(t_s)^2\right]^{\frac{3}{4}}} \, |t_s - b|, \qquad (IV - 4 - b)$$

$$\Psi_{(1)}(b, a) = \Phi_s(t_s) - \Phi_g\left(\frac{t_s - b}{a}\right) + \frac{1}{2} \frac{(t_s - b)^2 \Phi''_{(b,a)}(t_s)}{1 + a^4 \Phi''_{(b,a)}(t_s)^2}$$

$$+ \frac{3}{2} \arctan\left[a^2 \Phi''_{(b,a)}(t_s)\right]. \qquad (IV\text{-}4\text{-}c)$$

Notice that $T_{(1)}(b, a)$ is proportional to $|t_s - b|$. This is not the case for the second corrective term $T_{(2)}(b, a)$ of the asymptotic expansion.

IV-2 RIDGE AND SKELETON

The ridge of the transform is still defined by Definition 2, and still describes the frequency modulation law according to equation $(III - 12)$. The skeleton of the transform, i.e. the restriction of the wavelet transform to the associated ridge has a slightly different form than that given by equation $(III - 13)$:

$$T(b, a_r(b)) = \sqrt{\frac{\pi}{2}} \frac{e^{\frac{i}{2} \arctan\left[a_r(b)^2 \Phi''_{(b,a_r(b))}(b)\right]}}{\left[1 + a^4 \Phi''_{(b,a_r(b))}(b)^2\right]^{\frac{1}{4}}} e^{-i\phi_g(0)} Z_s(b) . \qquad (IV-5)$$

One easily checks on such an expression that Proposition 3 still holds. Notice also that from equation $(IV - 4 - b)$

$$T_{(1)}(b, a_r(b)) = 0. \qquad (IV-6)$$

So the first corrective term of the asymptotic development vanishes identically on the ridge.

IV-3 WAVELET CURVES

Consider now the wavelet curves, as defined in definition 4, i.e. as connected sets of points $(b, a) \in \Omega$ with a constant stationary point. The restriction of the wavelet transform to such a curve is essentially governed by the corresponding translated and dilated wavelets (its expression is obtained by setting $t_s = b_0$ in equation $(IV - 4)$, for some fixed b_0). Notice that in this case, the deviation from the dilated and translated wavelets is bigger than in the completely asymptotic case.

IV-4 RIDGE AND SKELETON EXTRACTION

The extraction of the ridge of the transform from the zeros of $\left[\frac{\partial \Psi}{\partial a}\right]_{b=b_0}$ is no longer valid. Indeed, it is not difficult to check that $\left[\frac{\partial \Psi}{\partial a}\right]_{b=b_0}$ involves a (in general nonzero) term proportional to $\left[\frac{\partial t_s}{\partial a}\right]_{b=b_0}$. The main reason for this is that the phase of $T(b, a)$ is now perturbed by the contribution of the modulus of the wavelet. Nevertheless, this phase is still a useful quantity for the ridge extraction, since on a fixed wavelet curve

$$\left[\frac{d\Psi}{db}\right]_{t_s(b,a)=b_0} = \frac{1}{a}\left[\phi'_g(0) + \frac{1}{2} \frac{\phi'''_g(0)}{1 + \left[\phi_g(0)a'_r(b) + \phi''_g(0)\right]^2}\right] \qquad (IV-7)$$

on the intersection with the ridge. Then for any analyzing wavelet whose central frequency is an inflection point (i.e. $\phi'''_g(0) = 0$), equation $(III - 21)$

is still valid, yielding an algorithm for the extraction of the ridge. Proposition 5 then has to be replaced in that case by:

Proposition 5'. *Under the assumptions of Theorem 1', and assuming moreover that the analyzing wavelet is such that $\phi_g'''(0) = 0$, the argument of the wavelet transform has the following property on Ω:*

$$\left[\frac{d\Psi(b, a)}{db}\right]_{t_s(b,a)=b_0} = \frac{\phi_g'(0)}{a} \text{ on the intersection with the ridge.}$$

Such a property is very useful for numerical computations. Although Theorem 1' is a generalisation of Theorem 1, we insist on the fact that this is no longer the case for the corresponding algorithms. When the wavelet is asymptotic, the above mentioned property is numerically irrelevant (since it holds everywhere and not only on the ridge), while for a non-asymptotic wavelet, it is a good starting point for ridge extraction. In other words, the "more asymptotic" the wavelet is, the less precise the algorithm is.

V. APPLICATIONS TO MATCHED FILTERING

The computation of the wavelet coefficients $T(b, a)$ of $s(t)$ can actually be thought of as band-pass filtering of the signal, which localizes it into cells (in the time-frequency plane) of constant relative bandwidth $\frac{\Delta\nu}{\nu}$. It is then natural to combine such filters to build other filters by hand, adapted to special kinds of signals. This is the purpose of matched filtering, the matched filter being defined as the linear filter which optimizes the signal to noise ratio of the output (see e.g. [Pap], [Bl]).

We will briefly describe here how to model filters close to matched filters, in the particular case of asymptotic locally monochromatic signals . We will describe two algorithms, called the ridge coincidence algorithm (proposed in [To 1]) and the ridge rectification algorithm (proposed in [E-To]) respectively, whose purpose is detection and time-location of a-priori well known asymptotic locally monochromatic components $s(t)$ in a signal, imbedded in an input signal $\tilde{s}(t)$. The output signal exhibits a sharp peak at the estimated date, like for instance the autocorrelation function of $s(t)$.

Both algorithms basically follow the same philosophy. They first use the wavelets to localize the signal into time-frequency cells of constant relative bandwidth, and the contributions of such cells are summed up in different ways. Finally, the same operations are performed with time-shifted copies of the cells. The output signal is in both cases different than the autocorrelation function of $s(t)$, but has similar characteristics.

Let us stress that one can build in the same way many slightly different versions of such algorithms; our goal in this Section is not to present algorithms in a final form, but rather to illustrate how wavelet analysis can be used to get approximations of the matched filter. Let us also point out the simple geometrical interpretation of these algorithms in the time-frequency half-plane.

V–1 THE RIDGE COINCIDENCE ALGORITHM [E-Gr-KM-To],[To 1]

Let $s \in L^2(\mathbf{R})$ be an asymptotic locally monochromatic signal, and let $g \in H^2(\mathbf{R})$ be an analytic analyzing wavelet, written in the form given by equation $(III - 5)$. Denote by:

$$\mathcal{R} = \{(b, a) \in \Omega; a = a_r(b)\} \qquad (V - 1)$$

the ridge of the wavelet transform $T(b, a)$ of $s(t)$. It is convenient to introduce here the τ-shifted skeleton operator $\mathcal{F}^{\mathcal{R}}_\tau : L^2(\mathbf{R}) \to L^2(\mathbf{R})$. If $\tilde{s} \in L^2(\mathbf{R})$, and $T_{\tilde{s}}$ is its wavelet transform:

$$\left[\mathcal{F}^{\mathcal{R}}_\tau.\tilde{s}\right](t) = \sqrt{\frac{2}{\pi}} \frac{\sqrt{a_r(t)^2 \left|\Phi''_{(t,a_r(t))}(t)\right|}}{\overline{g(0)}} \, e^{-i\frac{\pi}{4}Sgn[\Phi''_{(t,a_r(t))}(t)]} T_{\tilde{s}}(t + \tau, a_r(t)).$$

$$(V - 2)$$

At $\tau = 0$, $\mathcal{F}^{\mathcal{R}}_\tau$ associates with $\tilde{s}$ its wavelet transform restricted to the $\mathcal{R}$ ridge, and corrected by a (local) normalisation function and phase shift, needed to compensate the terms introduced by the stationary phase evaluation of the integrals. For $\tau \neq 0$, the wavelet transform is now restricted to a τ-shifted copy of the $\mathcal{R}$ ridge. Clearly, when $\tilde{s}(t) = s(t - \tau_0)$, $\mathcal{R}^\tau$ coincides with the ridge of $\tilde{s}$, so that both functions

$$\mathcal{SC}(\tau) = <\mathcal{F}^{\mathcal{R}}_\tau.\tilde{s}, Z_s> \qquad (V - 3)$$

$$\mathcal{SE}(\tau) = ||\mathcal{F}^{\mathcal{R}}_\tau.\tilde{s}||^2 \qquad (V - 4)$$

are maximal at $\tau = \tau_0$ (to see this, just take for $\mathcal{F}^{\mathcal{R}}_\tau.\tilde{s}$ its expression provided by the stationary phase argument in Section III, and evaluate the scalar products). $\mathcal{SC}$ is called the **skeleton correlation** function, and $\mathcal{SE}$ is the **skeleton energy** function.

Consider now $\tilde{s}(t)$ of the form:

$$\tilde{s}(t) = \sum_{i=1}^{N} A_i s(t - \tau_i) + r(t) \qquad (V - 5)$$

for some real numbers A_i and τ_i, and some deterministic or random $r(t)$. $\tilde{s}(t)$ contains N shifted copies of $s(t)$, and the corresponding ridge is a multi-component ridge, which contains in particular N copies of the ridge of $s(t)$. Assume moreover that the τ_i parameters are large enough to ensure that these components of the ridge are non-interacting. Then $\mathcal{SC}(\tau)$ is still locally maximal (more precisely it exhibits sharp peaks) at $\tau = \tau_i, i = 1, ..N$, at least as soon as the wavelet coefficients $T_r(b + \tau, a_r(b))$ are small compared with all the $A_i T_s(b, a_r(b))$. This in a first step allows the determination of the time delay parameters τ_i, and in a second step the determination of the A_i parameters, by using the factorization property described in Section IV. In the opposite case, additional local maxima appear, one says that the false alarm probability becomes more important.

Note that the local maxima of $\mathcal{SC}(\tau)$ correspond to the cases where the ridges of $\tilde{s}(t)$ coincide with the theoretical ridge of $s(t)$. This explains the name given to the algorithm.

V–2   THE RIDGE RECTIFICATION ALGORITHM [E],[E-To]

We now describe another algorithm dedicated to matched filtering. The main idea of this algorithm is to transform a signal with a given ridge into another signal, with a ridge as close as possible to a vertical ridge:

$$\mathcal{R}_{b_0} = \{(b, a) \in \Omega; b = b_0\} \qquad (V - 6)$$

for some fixed b_0. Then, we use the simplified reconstruction formula to construct the output of the algorithm from the modified wavelet coefficients. We describe here the generic structure of the algorithm in the particular case of hyperbolic frequency modulation laws, for which it is known that representations associated with the affine group are optimal in terms of energy localization in the time-scale plane.

The ridge rectification algorithm is as follows.

In a first step, pick a fixed number of dilation parameters $a_i, i = 1, ..N$, chosen in such a way that the corresponding relative bandwidth

$$\frac{\Delta a}{a} = 2\,\frac{a_i - a_{i-1}}{a_i + a_{i-1}} \qquad (V - 7)$$

is constant. To these dilation parameters are then associated the proper frequencies of the corresponding dilated wavelets:

$$\nu_i = \frac{\phi_g'(0)}{2\pi a_i}, \qquad (V - 8)$$

and then the time delays t_i such that

$$\nu_i = \frac{1}{2\pi}\phi_s'(t_i). \qquad (V-9)$$

Fix a reference time T, and set:

$$\theta_i = T - t_i. \qquad (V-10)$$

In a second step, compute the wavelet transform of the input signal for these fixed values of the dilation parameter, and time-translate the i-voice by a time-delay equal to θ_i. One then gets modified wavelet coefficients, which no longer satisfy the reproducing kernel equation. This in particular means that one does not have any control a priori on the signal reconstructed from these modified coefficients. Nevertheless, a ridge extraction gives a vertical ridge. Another modification is then necessary.

In a third step, to achieve coherence between all the voices, the output of each voice is multiplied by a corrective function

$$Corr_k(t) = \sqrt{\frac{2}{\pi}}\, \frac{\sqrt{a_k^2 \Phi_{(t-\theta_k,a_k)}''(t_s)}}{g(0)}, \qquad (V-11)$$

and a phase shift

$$\delta\phi_k = -2\pi\phi_s(t_s(t-\theta_k,a_k)) + Cst \qquad (V-12)$$

which are completely determined by the a-priori known ridge.

Finally, the modified outputs of each voice are summed in a coherent way:

$$W(t) = \sum_{i=1}^{N} Corr_k(t)e^{-\delta\phi_k}T_s(t-\theta_k,a_k), \qquad (V-13)$$

or in a partially coherent way

$$E(t) = \sum_{i=1}^{N} |Corr_k(t)T_s(t-\theta_k,a_k)|^2. \qquad (V-14)$$

It is then not difficult to show that if the receiver is matched to the frequency modulation law of the incoming signal, i.e. if the incoming signal is hyperbolically frequency modulated, both $W(t)$ and $E(t)$ are maximal at $t = T$. If the input signal is now of the form described in equation $(V-5)$, the output signal will exhibit sharp peaks at all the $t = T + \tau_i$, yielding estimates for the τ_i delays.

VI. CONCLUSION

We have described here the behaviour of the wavelet coefficients in the limit where the analyzed signal is rapidly oscillating. In that case, standard asymptotic techniques yield an approximate expression for the coefficients, and it is then possible to build algorithms for instantaneous frequency extraction and matched filtering. It must be stressed that the same analysis can be carried out starting from Gabor analysis instead of wavelet analysis, leading to similar algorithms (such an analysis is developed in [De-E-Gu-KM-Tc-To]. Let us also point out that in the particular case where the instantaneous frequency is constant (i.e. for spectral lines), the extraction algorithms presented here have been improved, and have given very precise results [Gu-KM-Ma]. Finally, let us summarize the limitations of our approach. The basic idea we always used here is that time-frequency methods allow the separation of independent components in the analyzed signal; it is then possible to process them separately. However, all the material described here fails as soon as these independent components are interacting, that is to say that they are localized in intersecting regions in the time-frequency plane. In such a case, the algorithms presented here yield biased estimates for instantaneous frequencies. We refer to [De-E-Gu-KM-Tc-To] for a detailed discussion of these aspects and for numerical applications. However, let us stress that there is some hope that new decompositions recently introduced in [To3] and [DD-Mu-To], generalizing wavelet decompositions and Gabor decompositions, could help to enlarge the application domain of the methods we described here.

Let us finally make a short comment on sampled signals. The signals that have to be processed numerically are obviously sampled signals, and there is no way to associate with it a (continuous) frequency modulation law without additional assumption. Then, our methodology can be applied only with the implicit assumption that the sampled signal is modeled by an asymptotic signal, having well-defined (i.e. continuously defined) phase and amplitude.

This paper is a report on joint works with N. Delprat, B. Escudié, A. Grossmann, P. Guillemain and R. Kronland-Martinet. We are happy to thank them for the pleasure of collaboration.

REFERENCES

[Bl]   R. E. Blahut, Principles and practice of information theory, Addison-Wesley publishing company, AW series in electrical and computer engineering (1987).

[Co-Gr-Tc]   J. M. Combes, A. Grossmann, Ph. Tchamitchian Eds., Wavelets, Proceedings of the first conference, Marseille (1987), Springer.

[Cop]   E. Copson, Asymptotic expansions, Cambridge University Press (1965).

[Da1]   I. Daubechies,The wavelet transform, time-frequency localization and signal analysis, IEEE Trans. Inf. Th. 36 (1991), pp. 961-1005

[Da2]   I. Daubechies,Ten lectures on wavelets, CBMS-NSF series in Applied Mathematics, SIAM (in Press 1991)

[De-E-Gu-KM-Tc-To]   N. Delprat, B. Escudié, P. Guillemain, R. Kronland-Martinet, P. Tchamitchian, B. Torrésani, Asymptotic wavelet and Gabor analysis; extraction of instantaneous frequencies, Preprint CPT-91/P.2512, submitted to IEEE Trans. Inf. Th., special issue on Wavelets and multiresolution analysis.

[DD-Mu-To]   M. Duval-Destin, M. A. Muschietti, B. Torrésani, From continuous wavelets to wavelet packets, in preparation.

[E]   B. Escudié, Wavelet analysis of asymptotic signals: A tentative model for bat sonar receiver, to appear in [Me2].

[E-Gr-KM-To]   B. Escudié, A. Grossmann, R. Kronland-Martinet, B. Torrésani, Analyse en ondelettes de signaux asymptotiques: emploi de la phase stationnaire, proceedings of the GRETSI conference, Juan-les-Pins (1989).

[E-To]   B. Escudié, B. Torrésani, Wavelet representation and time-scaled matched receiver for asymptotic signals, Proceedings of the conference EUSIPCO V, Barcelona (1990), North Holland, 305-308.

[Ga]   D. Gabor, Theory of communication, J. Inst. Elec. Eng. 903 (1946), 429.

[Gr-KM-Mo]   A. Grossmann, R. Kronland-Martinet, J. Morlet, Reading and understanding continuous wavelet transform, in [Co-Gr-Tc].

[Gr-Mo]   A. Grossmann, J. Morlet, Decomposition of Hardy functions into square integrable wavelets of constant shape, SIAM J. of Math. An. 15 (1984), 723.

[Gr-Mo-Pa1]   A. Grossmann, J. Morlet, T. Paul, Transforms associated with square integrable group representations I, J. Math. Phys. 27 (1985), 2473.

[Gr-Mo-Pa2] A. Grossmann, J. Morlet, T. Paul, Transforms associated with square integrable group representations II, Ann. Inst. H. Poincaré, 45 (1986) 293.

[Gu] P. Guillemain, Analyse et modélisation de signaux acoustiques: application de la transformée en ondelettes, DEA, LMA-Preprint (1990) Marseille.

[Gu-KM-Ma] P. Guillemain, R. Kronland-Martinet, B. Martens, Application de la transformee en ondelettes a la spectroscopie RMN, Note-LMA 112, Marseille (1989).

[Ko-Ge-dG] K. Kodera, R. Gendrin, C. de Gilledary, Analysis of time-varying signals with small BT values, IEEE Trans. ASSP 26 (1978), 64.

[Me1] Y. Meyer, Ondelettes et opérateurs I: Ondelettes (1989), Hermann.

[Me2] Y. Meyer Ed., Wavelets and applications, Proceedings of the second wavelet conference, Marseille (1989), Masson/Springer, to appear.

[Pap] A. Papoulis, Probability, random variables and stochastic processes, McGraw-Hill series in electrical engineering, 2nd edition (1989).

[Pi-Ma] B. Picinbono, W. Martin, Représentation des signaux par amplitude et phase instantanées, Annales des Télécommunications 38 (1983), 179-190.

[To1] B. Torrésani, Wavelet analysis of asymptotic signals: Ridge and Skeleton of the transform, to appear in [Me2].

[To2] B. Torrésani, Wavelets associated with representations of the affine Weyl-Heisenberg group, Preprint CPT-90/P.2390, Marseille, to appear at J. Math. Phys. (May 1991).

[To3] B. Torrésani, Time-frequency representations: wavelet packets and optimal decompositions, Preprint CPT-90/P.2466, Marseille, to appear at Ann. Inst. H. Poincaré.

Wavelet Analysis and Signal Processing

RONALD R. COIFMAN

YVES MEYER

VICTOR WICKERHAUSER

Yale University,
New-Haven, Connecticut

Wavelet Analysis consists of a versatile collection of tools for the analysis and manipulation of signals such as sound and images,as well as more general digital data sets. The user is provided with a collection of standard libraries of waveforms, which can be chosen to fit specific classes of signals. These libraries come equipped with fast numerical algorithms enabling realtime implementation of a variety of signal processing tasks, such as data compression, extraction of parameters for recognition and diagnostics, transformation and manipulation of data. The process of analysis of data is usually started by comparing acquired segments of data with stored waveforms.

As can be seen at the top portion Figure 1, representing a segment of a recording of the word armadillo, voice signals consist of modulated oscillations of small duration and varying frequencies and intensity. Figures (3–16) represent a variety of waveforms selected from different libraries, as well as illustrations of analysis tasks performed on them as described in the figure captions.

A general signal (Figure 1 or Figure 2 for example) is a superposition of different structures occurring on different time scales at different times (or spatial scales at different locations). One purpose of analysis is to separate and sort these structures. The example of music (or voice) can be used

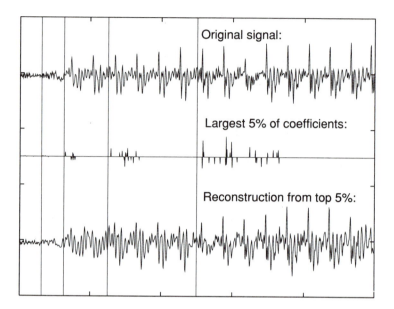

FIGURE 1. The first 1024 samples ($\frac{1}{8}$ second) of the word armadillo, are plotted on the top part. The library of local sine waveforms is then used to select the combination of windows of highest efficiency (lowest entropy). Expansion coefficients are then ordered by window in decreasing order. The top 5% are plotted in the center and used to reconstruct a compressed form of the signal which is plotted below.

to illustrate some of these ideas. A musical note can be described by four basic parameters, intensity (or amplitude), frequency, time duration, time position. Wavelet packets or trigonometric wave forms are indexed by the same parameters, plus others corresponding to choice of library (we can think of a library as a musical instrument, i.e. the recipe used to generate all the waveforms, notes, in the library).

The process of analysis compares a sound (or other signals) with all elements of a given library, picks up large correlations (or notes which are closest to segments of the signal) and uses these to rebuild the signal with a minimal number of waveforms. The result provides an economical transcription, which if ordered by decreasing intensity sorts the main features in order of importance. (In Figure 1 the signal has been segmented in windows, the top 5% of the expansion kept, and used to rebuild the signal in the bottom half).[1]

---

[1] See Figures 8–15 for examples of analysis of various typical examples.

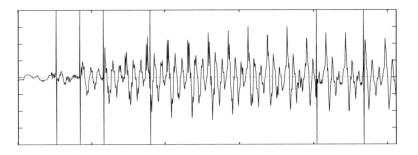

FIGURE 2. Automatic segmentation of a voice recording (armadillo) by using least entropy windowing in the local sine library. The windows are selected to obtain optimal efficiency in representing the signal. We see different patterns fall into distinct windows.

This realization of the signal in terms of the "best basis" providing efficient superpositon in terms of oscillatory modes on different time scales, can be used to compress signals for digital transmission and storage.

Of more importance for applications is the ability to compute and manipulate data in compressed parameters. This feature is particularly important for recognition and diagnostic purposes.

As an illustration consider a hypothetical diagnostic device for heartbeats, in which fifty consecutive beats are recorded. We would like to use this data as a statistical foundation for detection of significant changes in the next batch of beats. Theoretically this can be done by factor analysis (or Karhunen-Loéve bases); unfortunately the computation involving raw data is too large to be useful. The ability to efficiently compress the recorded data in terms of a single statistical best basis (or the representation of the data in terms of an adapted efficient coordinate system) enables us to perform a factor analysis (if needed) and to compute the deviation of the next few heartbeats from their predecessors, thereby detecting significant changes on the fly.

This procedure, in which we first compress a large data set of measurements in order to compute with the compressed parameters, can reduce dramatically the time needed to compute and manipulate data, it generalizes the usual transform methods (like FFT) by building an adapted fast transform for various classes of data or of operations on that data. (As an example the data could consist of a three dimensional atmospheric pressure map, and the computation would involve the evolution of the pressure. In this case it is natural to break the computation up as a sum of interactions on different scales, and some limited interaction between adjacent scales; such breakup is automatic if the pressure map is expressed in the wavelet basis, which in this case is also the natural choice for compression of the data.) Demo software is available on anonymous ftp from Yale [6].

DEFINITIONS OF MODULATED WAVEFORM LIBRARIES

We now introduce the concept of a "Library of orthonormal bases". For the sake of exposition we restrict our attention to two classes of numerically useful waveforms, introduced recently [2][3].

We start with trigonometric waveform libraries. These are localized sine transforms (LST) associated to covering by intervals of $\mathbf{R}$ (more generally, of a manifold).

We consider a cover $\mathbf{R} = \bigcup_{-\infty}^{\infty} I_i$, where $I_i = [\alpha_i \alpha_{i+1})$, $\alpha_i < \alpha_{i+1}$; write $\ell_i = \alpha_{i+1} - \alpha_i = |I_i|$ and let $p_i(x)$ be a window function supported in $[\alpha_i - \ell_{i-1}/2, \alpha_{i+1} + \ell_{i+1}/2]$ such that

$$\sum_{-\infty}^{\infty} p_i^2(x) = 1$$

and

$$p_i^2(x) = 1 - p_i^2(2\alpha_{i+1} - x) \quad \text{for} \quad x \quad \text{near} \quad \alpha_{i+1}$$

then the functions

$$S_{i,k}(x) = \frac{2}{\sqrt{2\ell i}} p_i(x) \sin[(2k+1)\frac{\pi}{2\ell_i}(x - \alpha_i)]$$

form an orthonormal basis of $L^2(\mathbf{R})$ subordinate to the partition p_i. The collection of such bases for all partitions $\{p_i\}$ forms a library of orthonormal bases. See Figure 3.

It is easy to check that if H_{I_i} denotes the space of functions spanned by $S_{i,k}$ $k = 0, 1, 2, \ldots$ then $H_{I_i} + H_{I_{i+1}}$ is spanned by the functions

$$P(x)\frac{1}{\sqrt{2(\ell_i + \ell_{i+1})}} \sin[(2k+1)\frac{\pi}{2(\ell_i + \ell_{i+1})}(x - \alpha_i)]$$

where

$$P^2 = p_i^2(x) + p_{i+1}^2(x)$$

is a "window" function covering the interval $I_i \cup I_{i+1}$.

Another new library of orthonormal bases called the Wavelet packet library can be constructed. This collection of modulated wave forms corresponds roughly to a covering of "frequency" space. This library contains the wavelet basis, Walsh functions, and smooth versions of Walsh functions called wavelet packets. See Figure 5.

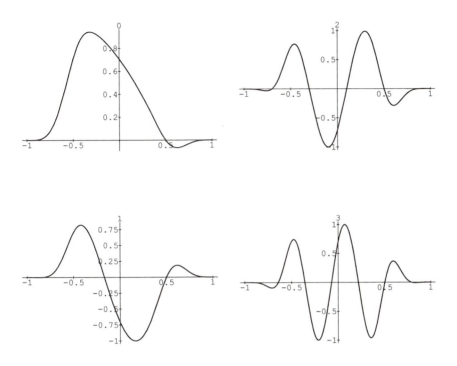

FIGURE 3.

We'll use the notation and terminology of [4], whose results we shall assume.

We are given an exact quadrature mirror filter $h(n)$ satisfying the conditions of Theorem (3.6) in [4], p. 964, i.e.

$$\sum_n h(n - 2k)h(n - 2\ell) = \delta_{k,\ell}, \quad \sum_n h(n) = \sqrt{2}.$$

We let $g_k = \bar{h}_{1-k}(-1)^k$ and define the operations F_i on $\ell^2(\mathbf{Z})$ into "$\ell^2(2\mathbf{Z})$"

$$F_0\{s_k\}(i) = 2\sum s_k h_{k-2i} \qquad (1.0)$$

$$F_1\{s_k\}(i) = 2\sum s_k g_{k-2i}.$$

The map $\mathbf{F}(s_k) = F_0(s_k) \oplus F_1(s_k) \in \ell^2(2\mathbf{Z}) \oplus \ell^2(2\mathbf{Z})$ is orthogonal and

$$F_0^* F_0 + F_1^* F_1 = I \qquad (1.1)$$

We now define the following sequence of functions.

$$\begin{cases} W_{2n}(x) &= \sqrt{2}\sum h_k W_n(2x - k) \\ W_{2n+1}(x) &= \sqrt{2}\sum g_k W_n(2x - k). \end{cases} \qquad n = 0, 1, 2, \ldots \quad (1.2)$$

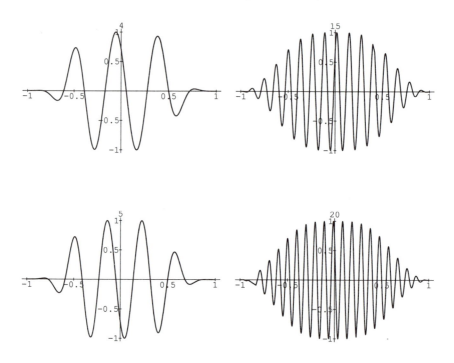

FIGURE 4.

Clearly the function $W_0(x)$ can be identified with the scaling function φ in [4] and W_1 with the basic wavelet ψ.

Let us define $m_0(\xi) = \frac{1}{\sqrt{2}} \sum h_k e^{-ik\xi}$ and

$$m_1(\xi) = -e^{i\xi}\bar{m}_0(\xi + \pi) = \frac{1}{\sqrt{2}} \sum g_k e^{ik\xi}$$

Remark.   The quadrature mirror condition on the operation $\mathbf{F} = (F_0, F_1)$ is equivalent to the unitarity of the matrix

$$\mathcal{M} = \left[\begin{array}{cc} m_0(\xi) & m_1(\xi) \\ m_0(\xi + \pi) & m_1(\xi + \pi) \end{array} \right]$$

Taking the Fourier transform of (1.2) when $n = 0$ we get

$$\hat{W}_0(\xi) = m_0(\xi/2)\hat{W}_0(\xi/2)$$

i.e.,

$$\hat{W}_0(\xi) = \prod_{j=1}^{\infty} m_0(\xi/2^j)$$

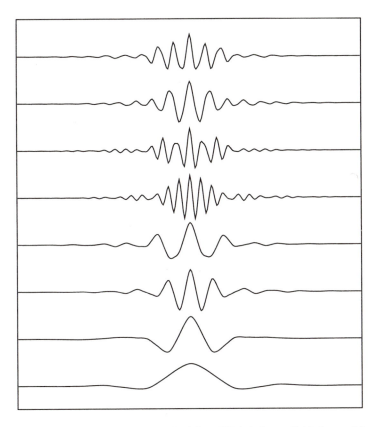

FIGURE 5. A few wavelet packets, $W_0(x) \cdots W_7(x)$ from C-18 formed by using the relations (1.2). These waveforms are mutually orthogonal; moreover, each of them is orthogonal to all of its integer translates and dyadic rescaled versions. The full collection of these wavelet packets (including translates and rescaled versions) provides us with a library of "templates" or "notes" which are matched "efficiently" to signals for analysis and synthesis.

and

$$\hat{W}_1(\xi) = m_1(\xi/2)\hat{W}_0(\xi/2) = m_1(\xi/2)m_0(\xi/4)m_0(\xi/2^3) \cdots$$

More generally, the relations (1.2) are equivalent to

$$\hat{W}_n(\xi) = \prod_{j=1}^{\infty} m_{\varepsilon_j}(\xi/2j) \tag{1.3}$$

and $n = \sum\limits_{j=1}^{\infty} \varepsilon_j 2^{j-1} (\varepsilon_j = 0 \text{ or } 1)$.

The functions $W_n(x - k)$ form an orthonormal basis of $L^2(\mathbf{R}^n)$.

We define a <u>library</u> of wavelet packets to be the collection of functions of the form $W_n(2^\ell x - k)$ where $\ell, k \in \mathbf{Z}, n \in N$. Here, each element of the library is determined by a scaling parameter ℓ, a localization parameter k and an oscillation parameter n. (The function $W_n(2^\ell x - k)$ is roughly centered at $2^{-\ell}k$, has support of size $\approx 2^{-\ell}$ and oscillates $\approx n$ times).

We have the following simple characterization of subsets forming orthonormal bases.

Proposition. *Any collection of indices (ℓ, n) such that the intervals $[2^\ell n, 2^\ell n + 1)$ form a disjoint cover up to measure 0 of $[0, \infty)$ gives rise to an orthonormal basis of L^2.*[2]

Motivated by ideas from signal processing and communication theory we were led to measure the "distance" between a basis and a function in terms of the Shannon entropy of the expansion. More generally, let H be a Hilbert space.

Let $v \in H$, $\|v\| = 1$ and assume

$$H = \oplus \sum H_i$$

is an orthogonal direct sum. We define

$$\varepsilon^2(v, \{H_i\}) = -\sum \|v_i\|^2 \ell n \|v_i\|^2$$

as a measure of distance between v and the orthogonal decomposition.

ε^2 is characterized by the Shannon equation which is a version of Pythagoras' theorem.

Let

$$H = \oplus \left(\sum H^i\right) \oplus \left(\sum H_j\right)$$
$$= H_+ \oplus H_-$$

H^i and H_j give orthogonal decompositions $H_+ = \sum H^i, H_- = \sum H_j$. Then

$$\varepsilon^2(v; \{H^i, H_j\}) = \varepsilon^2(v, \{H+, H_-\}$$
$$+ \|v_+\|^2 \varepsilon^2 \left(\frac{v_+}{\|v_+\|}, \{H^i\}\right) + \|v_-\|^2 \varepsilon^2 \left(\frac{v_-}{\|v_-\|}, \{H_j\}\right)$$

This is Shannon's equation for entropy (if we interpret, as in quantum mechanics, $\|P_{H_+} v\|^2$ to be the "probability" of v being in the subspace H_+).

---

[2]We can think of this cover as an even covering of frequency space by windows roughly localized over the corresponding intervals.

This equation enables us to search for a smallest entropy space decomposition of a given vector.

In fact, for the example of the first library restricted to covering by dyadic intervals we can start by calculating the entropy of an expansion relative to a local trigonometric basis for intervals of length one, then compare the entropy of an adjacent pair of intervals to the entropy of an expansion on their union. Pick the expansion of minimal entropy and continue until a minimum entropy expansion is achieved (see Figures 1 and 2).

In practice, discrete versions of this scheme can be implemented in $CN \log N$ computations (where N is the number of discrete samples $N = 2^L$.)

For voice signals and images this procedure leads to remarkable compression algorithms; see below.

Of course, while entropy is a good measure of concentration or efficiency of an expansion, various other information cost functions are possible, permitting discrimination and choice between various special function expansion.

Other possible libraries can be constructed. The space of frequencies can be decomposed into pairs of symmetric windows around the origin ,on which a smooth partition of unity is constructed. This and other constructions were obtained by one of our students, E. Laeng [5].

Higher dimensional libraries can also be easily constructed,(as well as libraries on manifolds) leading to new and direct analysis methods for linear transformations.

We will describe an algorithm to produce a rectangle in which coefficients are grouped by frequency, since this is simpler and since the transformation to the other form is evident. For definiteness, consider a function defined at 8 points $\{x_1, \ldots, x_8\}$, i.e., a vector in $\mathbf{R}^8$. We may develop the (periodicised) wavelet packet coefficients of this function by filling out the following rectangle:

x_1	x_2	x_3	x_4	x_5	x_6	x_7	x_8
s_1	s_2	s_3	s_4	d_1	d_2	d_3	d_4
ss_1	ss_2	ds_1	ds_2	sd_1	sd_2	dd_1	dd_2
sss_1	dss_1	sds_1	dds_1	ssd_1	dsd_1	sdd_1	ddd_1

A rectangle of wavelet packet coefficients.

Each row is computed from the row above it by one application of either F_0 or F_1, which we think of as "summing" (s) or "differencing" (d) operations, respectively. Thus, for example the subblock $\{ss_1, ss_2\}$ is obtained

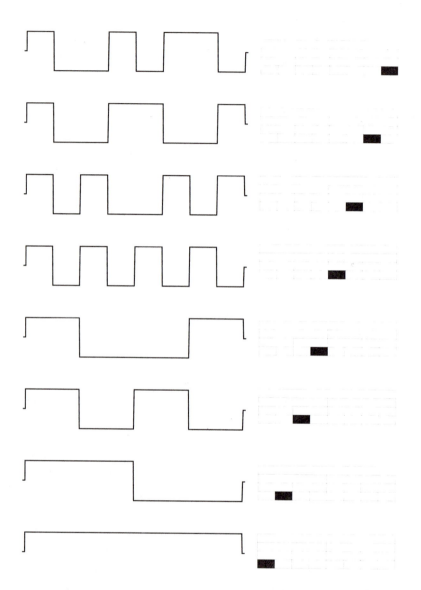

FIGURE 6. Wavelet packets are generated by the same algorithm in which different QMF are selected, generating the shapes in Figure 5. Wavelet packets from the deepest level have the largest scale. Coefficients computed in the shaded box represent a correlation between the initial signal $x_1, \ldots, x_8$ and the waveform described on the left. Here we take the Haar QMF, i.e. $s_1 = \frac{1}{\sqrt{2}}(x_1 + x_2) \quad d_1 = \frac{1}{\sqrt{2}}(x_1 - x_2), \ldots$ etc. The waveforms produced are the classical Walsh functions.

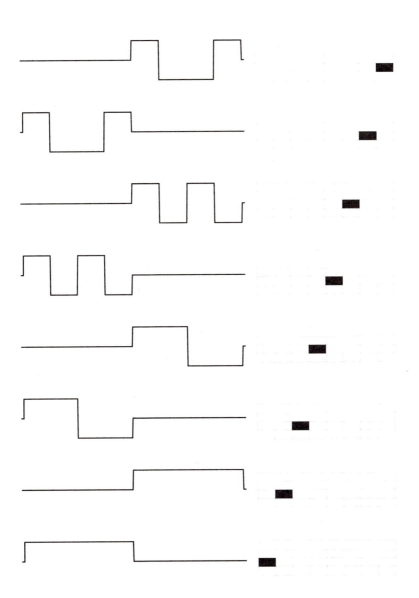

FIGURE 7. Wavelet packets from intermediate levels have a shorter time duration than the Walsh functions.

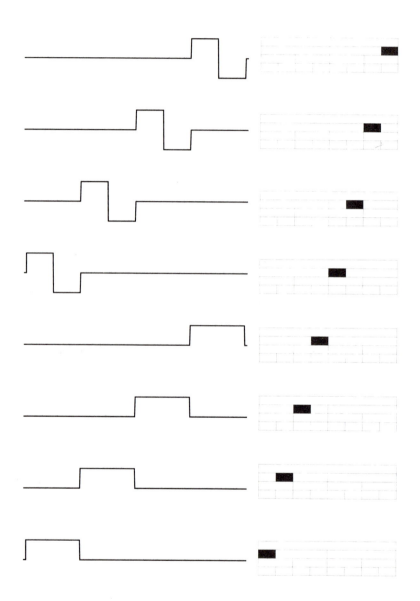

FIGURE 8. Wavelet packets from the first level have the smallest time duration.

by convolution-decimation of $\{s_1, s_2, s_3, s_4\}$ with F_0, while $\{ds_1, ds_2\}$ comes from similar convolution-decimation with F_1. In the simplest case, where we use the Haar filters $h = \{\frac{1}{\sqrt{2}}, \frac{1}{\sqrt{2}}\}$ and $g = \{\frac{1}{\sqrt{2}}, -\frac{1}{\sqrt{2}}\}$, we have in particular $ss_1 = \frac{1}{\sqrt{2}}(s_1 + s_2)$, $ss_2 = \frac{1}{\sqrt{2}}(s_3 + s_4)$, $ds_1 = \frac{1}{\sqrt{2}}(s_1 - s_2)$, and $ds_2 = \frac{1}{\sqrt{2}}(s_3 - s_4)$. The two daughter s and d subblocks on the $n + 1$st row are determined by their mutual parent on the nth row, which conversely is determined by them through the adjoint anticonvolution.

Reconstructing the nth row from the $n + 1$st row consists of applying F_0^* to the left daughter and F_1^* to the right daughter, then summing the images into the parent. In this manner, we generated the graphs of the functions which are included in fig 3. We used a rectangle of size 1024×10 to obtain 1024-point approximations. We filled the rectangle with 0's except for a single 1, then applied the deconvolutions F_0^* and F_1^* up to 10 times in various orders, so as to generate a vector of length 1024. This vector approximates one of the 10240 wavelet packets in R^{1024}. The details of this reconstruction determine the frequency, scale, and location parameters.

From this rectangle, we may choose subsets of N coefficients which correspond to orthonormal bases for $\mathbf{R}^N$. For example, the subset corresponding to the labelled boxes in the figure below is the wavelet basis.

				d_1	d_2	d_3	d_4
		ds_1	ds_2				
sss_1	dss_1						

The wavelet basis.

The two figures below give other orthonormal basis subsets.

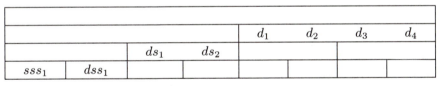

ss_1	ss_2	ds_1	ds_2	sd_1	sd_2	dd_1	dd_2

A subband basis.

s_1	s_2	s_3	s_4				
						dd_1	dd_2
				ssd_1	dsd_1		

An orthonormal basis subset.

The boxes of coefficients in the rectangle have a natural binary tree structure. Each box is a direct sum of its two children. Call a subset of the rectangle a *graph* if it contains only whole boxes and each column of the rectangle has exactly one element. We have the following relation between dyadic coverings and orthonormal bases.

Proposition. *Every graph is an orthonormal basis subset.*

The number of graphs may be counted by induction. If $N = 2^L$, let A_L be the number of graphs in the coefficient rectangle of N columns and L rows. Then $A_0 = 1$ and we have the relation $A_{L+1} = 1 + A_L^2$, which implies that $A_{L+1} > 2^{2^L} = 2^N$.

This last algorithm is beautifully suited for a best basis selection algorithm. By comparing the information cost of two "children" with their parent box we can, starting from the bottom of the rectangle, replace each node of the tree by the least costly combination.

If entropy is taken as information cost, the Shannon equation guarantees that we will end up with a basis with minimum entropy.

A simple variant on this selection algorithm permits the construction of a statistical best basis. Here we start with a collection of vectors X_n $n = 1, \ldots, N$ in $\mathbf{R}^d$ (for example, recording of successive distinct heartbeats). We construct the average vector $\bar{X} = \frac{1}{N} \sum_{n=1}^{N} X_n$ and would like to select in our library a basis which is most efficient, on the average, in compressing all vectors $\tilde{X}_n = X_n - \bar{X}$.

This is easily achieved by repeating the preceding search where, in each node of the tree, we compare the total cost (or entropy) of the node to the cost of its children. (Where by total cost we mean the sum of entropies of all vectors contributing to the node, or some other measure of information.)

Of course this procedure is related to the Karhunen-Loeve expansion, in which we find in $\mathbf{R}^d$ the most efficient basis by diagonalizing the autocovariance matrix

$$M_{ij} = \frac{1}{N} \sum_{n=1}^{N} \tilde{X}_n(i) \tilde{X}_n(j) \quad i = 1, \ldots, d \; j = 1, \ldots, d.$$

Intuitively we think of the various sample vectors as forming an ellipsoidal cloud centered at $\bar{X}$, with the principal axis of this ellipsoid being the eigenvectors of M, pointing in the direction of maximum variance. This Karhunen-Loeve basis is the most efficient basis for capturing on the average most of the energy of a random sample. Ideally, to analyze the fluctuations of $\tilde{X}_n$ we should compute this basis and expand each sample in it. Numerically this task is too expensive when dealing with raw data. It is advisable to first find the statistical best basis within a library thereby compressing the data. In this new coordinate system one can compute the Karhunen-Loeve basis and compare its entropy to that of the best basis. If the two entropies are close, we have an indication that we selected the correct library for compression of the collection $\tilde{X}_n$.

The reduction in complexity achieved by finding a best basis provides a significant speedup in computation time, not only for finding the $K - L$ basis, but also for obtaining a fast algorithm to compute correlations with the $K - L$ basis. Normally to compute an expansion in $K - L$ will take d^2 computations (if $N \simeq d$) by proceeding through the best basis. This can be reduced to $Cd \log d$.

Returning to the specific example of heartbeats, we could design a diagnostic tool as follows. We compress fifty consecutive heartbeats (to a desired accuracy) in their statistical best basis. Next, we compute a $K - L$ basis which we use to analyze the variance of the next batch (of course, it may be simpler and also sufficient to skip the $K - L$ construction).

We should also mention that the $K - L$ basis, while optimal in the L^2 sense, may be quite lacking in efficiency in other norms. This fact for the case of trigonometric expansions has led us to pick wavelets for more local questions. We are therefore proposing the best basis selection with different norm criteria as a tool for obtaining more flexible coding and compressions.

BEST BASIS FOR NUMERICAL COMPUTATIONS

As seen in the preceding paragraph, expressing the raw data in a statistical best basis provides a transform permitting faster manipulation and computation with the data. The paper on wavelets and numerical algorithms in this volume provides illustrations of this fact where, rather than diagonalize a matrix (generally, a numerically expensive procedure), one chooses to express it in an appropriate wavelet basis obtaining a banded version and fast computation. In that case two procedures arise. The first is a mere coordinate change to a well chosen basis. The second, or so-called nonstandard form, consists in compressing a matrix $[a_{ij}]$ as if it were an image by finding a best 2-dimensional basis to compress it. Since the two dimensional bases of wavelet-packets or trigonometric waveforms can be obtained as a product of one-dimensional versions, this procedure amounts to a separation of variables and translates efficient compression into fast computation of linear transformations.

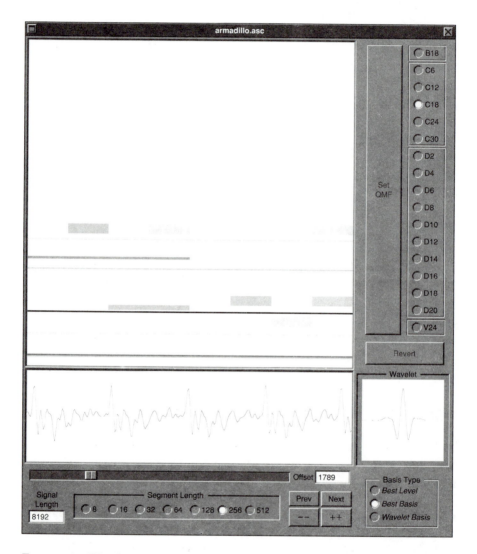

FIGURE 9. Wavelet packet analysis of a segment of "armadillo." The shaded rectangles measure the correlation with selected elements of the best basis, they are located above the portion of the signal with which they correlate. They have a base corresponding to duration and a height centered at the main frequency. See Appendix I for a description of the screen display.

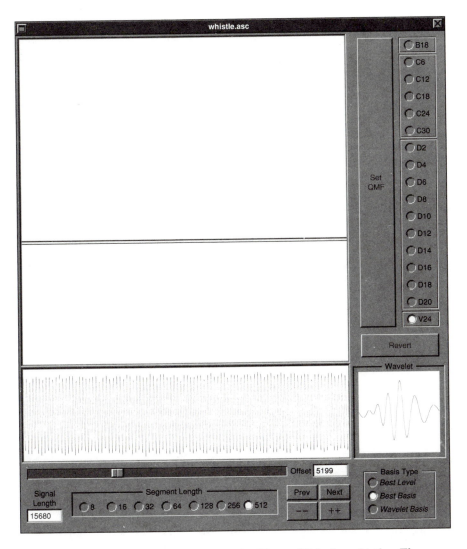

FIGURE 10. Best basis analysis, using the library V24 of a whistle. There are essentially two basic frequencies along the whole signal.

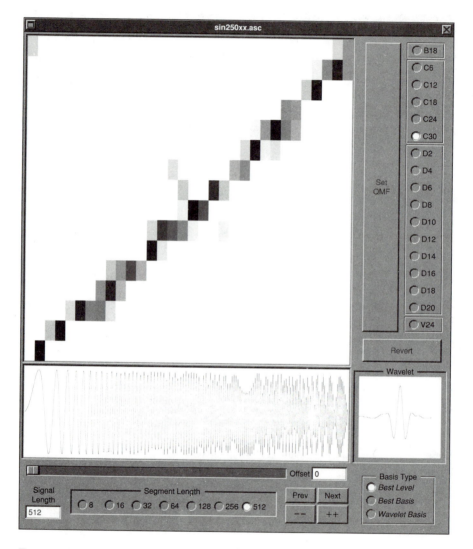

FIGURE 11. This is a plot of $\sin(250\,\pi x^2)$, a chirp up to the Nyquist frequency.

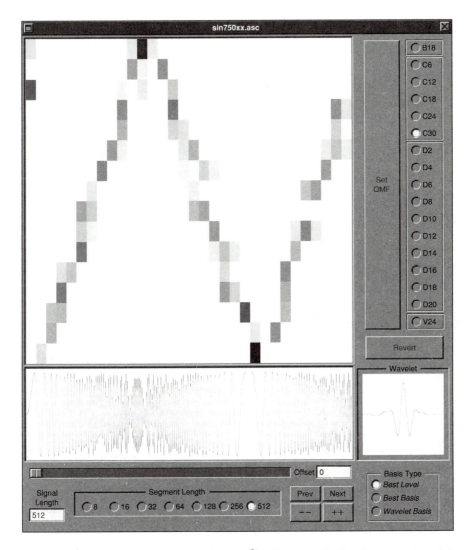

FIGURE 12. This is a plot of $\sin(750\ \pi x^2)$, showing aliasing between 1 and 3 times the Nyquist frequency.

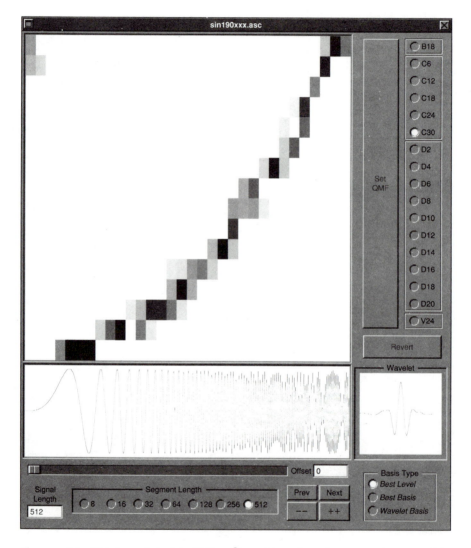

FIGURE 13. This is a plot of $\sin(250\,\pi x^3)$, a chirp whose frequency increases like a parabola.

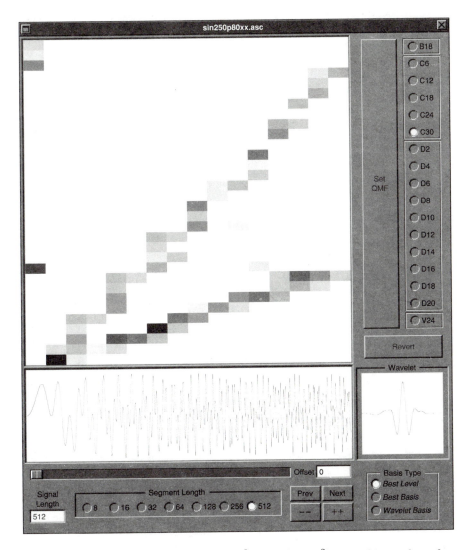

FIGURE 14. This plot shows $\sin(250\,\pi x^2) + \sin(80\,\pi x^2)$, two chirps whose frequencies increase at different rates.

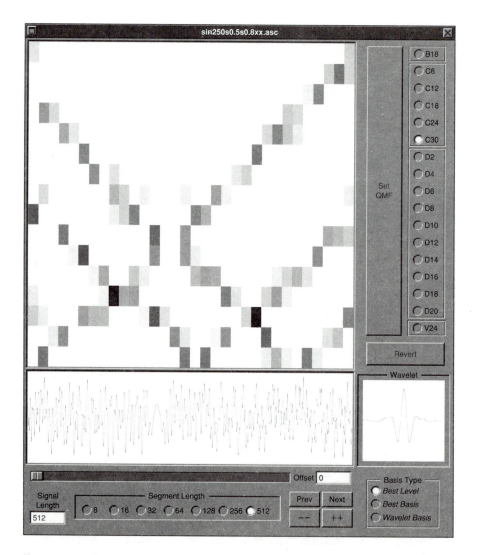

FIGURE 15. Let $y = x - 0.5$ and $z = x - 0.8$. This plot shows the superposition $\sin(250\ \pi x^2) + \sin(250\ \pi y^2) + \sin(250\ \pi z^2)$, three chirps of parallel increasing frequency which are 0.5 and 0.8 intervals out of phase.

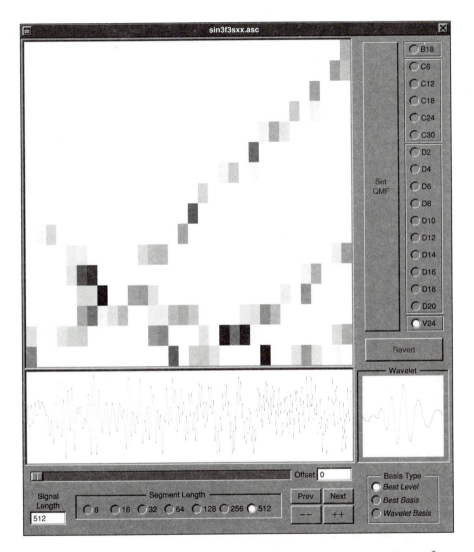

FIGURE 16. Let $y = x - 0.5$, and let $z = x - 0.8$. This plot of $\sin(250\ \pi x^2) +$ $\sin(190\ \pi y^2) + \sin(120\ \pi z^2)$, showing three chirps increasing at different rates and 0.5 and 0.8 intervals out of phase, respectively.

APPENDIX 1 USING THE WAVELET PACKET LABORATORY

READING A SIGNAL FROM A FILE

WPLab can read text files which have the extension ".asc" appended to their names, and which contain just ASCII floating-point numbers.

Selecting the "Open" item from the application's main menu brings up a browser panel which allows the user to select any single file with the proper extension. The entire file is read into a double-precision floating-point array in memory, and the array is padded with zeroes up to an integer multiple of the longest available segment length.

The number of samples in the signal is displayed in the "Signal Length" text field. The signal file name becomes the title for the main window, and also for the miniwindow upon miniaturization.

DISPLAYING A SIGNAL SEGMENT

Segments of the signal file are plotted in the rectangular view near the bottom of the window.

WPLab can display segments whose lengths are a power of 2, and starting at arbitrary offsets. Use the "Segment Length" radio buttons to select this length, and any combination of the offset form, slider, and buttons to set the index of the first displayed sample.

The buttons "Prev", "Next", "++", and "–" adjust the offset. Their actions, respectively, are to subtract a segment length, add a segment length, add 1, and subtract 1. The program does the best it can given the signal length.

If a newly selected segment length is too long for the current offset, the offset is decreased to accommodate it.

CHOOSING A QUADRATURE MIRROR FILTER

There are 17 quadrature mirror filters (QMFs) available for wavelet packet analysis. They are identified by a letter ("B", "C", "D", or "V"), followed by (finite) impulse response length. For example, the Haar filters $(\sqrt{2}), (\sqrt{2}), (\sqrt{2}), -(\sqrt{2})$ are designated $D2$.

Preview a filter by clicking on its radio button. In the small "Wavelet" window will appear a plot of the mother wavelet associated to that QMF. This action enables the "Set QMF" and "Revert" buttons.

Roughly speaking, longer filters produce smoother wavelets and wavelet packets with better frequency resolution.

Click on the "Set QMF" button to register your choice and update the Phase Plane. This disables the "Set QMF" and "Revert" buttons.

Click on the "Revert" button to cancel previewing filters and return to the last registered QMFs. This will disable the "Set QMF" and "Revert" buttons.

THE PHASE PLANE REPRESENTATION OF A SIGNAL

The large square view contains that portion of the phase plane affected by the plotted segment. WPLab draws a rectangle in the phase plane for every modulated waveform in the basis chosen to represent the signal.

Each modulated waveform can be assigned 4 attributes: amplitude a, timescale s, frequency f, and position p. In a musical note, these correspond to loudness, duration, pitch, and the instant it is played.

Suppose that the signal segment has length $N = 2^n$. Coefficient (a, s, f, p) is displayed as the rectangle $[p2^s, (p+1)2^s] \times [f2^{(n-s)}, (f+1)2^{(n-s)}]$, shaded in proportion to a^2.

Because of the Heisenberg uncertainty principle, position and frequency cannot both be specified to arbitrary precision. The uncertainty of the frequency is 2^s, and the uncertainty in position is $2^{(n-s)}$. Thus each rectangle or "Heisenberg box" has area $2^n = N$. Since the total area of the displayed section of the phase plane is N^2, there are exactly N Heisenberg boxes in a disjoint cover of the section.

A library consists of all possible Heisenberg boxes, and bases from the library consist of certain disjoint covers of the phase plane by such rectangles.

CHOOSING A BASIS

Wavelet Basis: this forces a display of the wavelet basis constructed with the given mother wavelet.

Best Level: this forces all of the Heisenberg boxes to have the same time scale. In particular, they must be congruent. There are (log N) such bases for a segment of length N, and the one displayed has minimum entropy.

Best Basis: this minimizes entropy over all bases corresponding to disjoint dyadic covers of the segment. There are more than 2^N such bases for a segment of length N.

PRINTING THE WINDOW

The entire contents of the key window may be printed at full scale with the "Print" menu item.

REFERENCES

[1]   R. Coifman, *Adaptive multiresolution analysis, computation, signal processing and operator theory*, ICM 90 (Kyoto)

[2]   R. Coifman, Y. Meyer, S. Quake and V. Wickerhauser *Signal processing and compression with wavelet packets*, Numerical Algorithms Research Group, Yale University (1990)

[3]   R. Coifman and Y. Meyer *Remarques sur l'analyse de Fourier à fenêtre*, C. R. Acad. Sci. Paris **312** info série I (1991) pp. 259–261

[4]   I. Daubechies *Orthonormal bases of compactly supported wavelets*, Communications on Pure and Applied Mathematics **XLI** pp. 909–996 (1988)

[5]   E. Laeng *Une base orthonormale de $L^2(\mathbf{R})$, dont les éléments sont bien localisés dans l'espace de phase et leurs supports adaptés à toute partition symétrique de l'espace des fréquences*, C. R. Acad. Sci. Paris **311** info série I (1990) pp. 677–680

[6]   Available by anonymous ftp, ceres.math.yale.edu info InterNet address 130.132.23.22

III
NUMERICAL ANALYSIS

WAVELETS IN NUMERICAL ANALYSIS

G. BEYLKIN

University of Colorado at Boulder
Program in Applied Mathematics
Campus Box 526 Boulder, Colorado

R. COIFMAN[1]

Yale University, Department of
Mathematics, P.O.Box 2155 Yale Station,
New Haven, Connecticut

V. ROKHLIN[2]

Yale University, Department of
Computer Science, P.O.Box 2158 Yale
Station, New Haven, Connecticut

I   INTRODUCTION

The use of wavelet based algorithms in numerical analysis is superficially similar to other transform methods, in which, instead of representing a vector or an operator in the usual way it is expanded in a wavelet basis, or it's matrix representation is computed in this basis. It turns out, however, that because of the localization of wavelet bases in both space and wave number domains, wavelet expansions organize transformations efficiently in terms of proximity on a given scale (wave number) and interactions between different neighbouring scales. Such organization of transformations (both linear and non-linear) has been a powerful tool in Harmonic Analysis and usually referred to as Littlewood-Paley, and Calderón-Zygmund theories (see e.g. [1]).

Initially, the relations between computation and Calderón-Zygmund theory was described in [2], [3], [4], where the Fast Multipole algorithm for

[1] Research partially supported by ONR grant N00014-88-K0020

[2] Research partially supported by ONR grants N00014-89-J1527, N00014-86-K0310 and IBM grant P00038436

computing potential interactions has made explicit many of the ingredients of Calderón-Zygmund theory. In that paper a fast algorithm of order N to compute all sums

$$p_j = \sum_{i=1}^{N} \frac{g_i g_j}{|x_i - x_j|} \quad \text{where} \quad x_i \in \mathbf{R}^3 \quad i,j = 1,\ldots,N$$

is constructed. Naively it would seem to be impossible to do this calculation in less than N^2 computations, since this is the number of interactions. It was observed that the far field effect of a cloud of charges located in a box can be described, to any accuracy, by the effect of a single multipole at the center of the box, requiring only a few numbers (Taylor coefficients of the field at the center of external boxes removed from the source). All boxes were organized in a dyadic hierarchy enabling an efficient $O(N)$ algorithm. This algorithm is $O(N)$ independently of the configuration of the charges, therefore providing a substantial improvement over FFT, even though the problem is to evaluate a convolution.

Wavelet based algorithms provide a systematic elegant generalization of the fast multipole method, in which the geometric and cancellation structure of the basis functions (which may be thought of as multipoles) provide for automatic adaptability and economy in computation [5]. Another novel aspect of transform analysis appearing naturally in connection with wavelets is the so-called non-standard form, in which a transformation is analyzed as a combination of successive contributions from different scales. We start with an initial smooth or blurred input and output vector which is then upgraded successively in input and output to higher and higher resolution, in much the same way as the pyramid scheme in image processing. This non-standard form corresponds algebraically to an imbedding of the original vector of length N into a $2N - 2$ dimensional space where all scales are uncoupled and in which the original transformation becomes sparse, followed by a projection into N dimensional space, in which scale interactions are introduced.

Even in standard form, i.e. the usual matrix realization of an operator in the wavelet basis, we gain a remarkable insight about operator compressions. In fact, already for the case of the Haar basis we see that the numerical manipulations needed to convert a given matrix to the Haar basis, involve a succesion of difference operations between neighbouring columns thus taking advantage of smoothness to reduce numerical complexity and, thereby, providing a general method for selecting an orthonormal basis for numerical compression of operators [5], [6], [7].

The class of operators which can be efficiently treated with wavelet bases includes Calderón-Zygmund and pseudo-differential operators; these operators are well behaved under translations and dilations (they satisfy translation and scale invariant size estimates). The numerical implementations described in this paper are the beginning of a program for the conversion of pseudo-differential calculus into a numerical tool. The main idea here is the conversion of smoothing operators (error terms in pseudo-differential calculus) into sparse matrices with a small number of significant entries. Various numerical examples and applications are described in [5], [6], [8].

II PRELIMINARY REMARKS

Computing in the Haar basis, $h_{j,k}(x) = 2^{-j/2} h(2^{-j}x - k)$ $j, k \in \mathbf{Z}$, where

$$h(x) = \begin{cases} 1 & \text{for } 0 < x < 1/2 \\ -1 & \text{for } 1/2 \le x < 1 \\ 0 & \text{elsewhere.} \end{cases} \tag{2.1}$$

offers a glimpse of the algorithms that we will review.

First, we note that the decomposition of a function into the Haar basis is an order N procedure. Given $N = 2^n$ "samples" of a function, which can for simplicity be thought of as values of scaled averages

$$s_k^0 = 2^{n/2} \int_{2^{-n}k}^{2^{-n}(k+1)} f(x)dx, \tag{2.2}$$

of f on intervals of length 2^{-n}, we obtain the Haar coefficients

$$d_k^{j+1} = \frac{1}{\sqrt{2}}(s_{2k-1}^j - s_{2k}^j) \tag{2.3}$$

and averages

$$s_k^{j+1} = \frac{1}{\sqrt{2}}(s_{2k-1}^j + s_{2k}^j) \tag{2.4}$$

for $j = 0, \ldots, n-1$ and $k = 0, \ldots, 2^{n-j-1} - 1$. It is easy to see that evaluating the whole set of coefficients d_k^j, s_k^j in (2.3), (2.4) requires $2(N-1)$ additions and $2N$ multiplications.

Second, we note that in two dimensions, there are two natural ways to construct Haar systems. The first is simply the tensor product $h_{j,j',k,k'}(x,y) = h_{j,k}(x)h_{j',k'}(y)$, so that each basis function $h_{j,j',k,k'}(x,y)$ is supported on a rectangle. This basis leads to what we call the standard representation of an operator.

The second basis is defined by the set of three kinds of basis functions supported on squares: $h_{j,k}(x)h_{j,k'}(y)$, $h_{j,k}(x)\chi_{j,k'}(y)$, and $\chi_{j,k}(x)h_{j,k'}(y)$, where $\chi(x)$ is the characteristic function of the interval $(0,1)$ and $\chi_{j,k}(x) = 2^{-j/2}\chi(2^{-j}x - k)$. This basis leads to what we call the non-standard representation of an operator (the terminology will become clear later).

Third, we note that if we consider an integral operator

$$T(f)(x) = \int K(x,y)f(y)dy, \qquad (2.5)$$

and expand its kernel in a two-dimensional Haar basis we find (for a wide class of operators) that the decay of entries as a function of the distance from the diagonal is faster in these representations than that in the original kernel. This decay depends on the number of vanishing moments of the functions of the basis. The Haar functions have only one vanishing moment, $\int h(x)dx = 0$, and for this reason the gain in the decay is insufficient to make computing in the Haar basis practical.

To have a faster decay, it is necessary to use a basis in which the elements have several vanishing moments. This is accomplished by wavelets. In particular, the orthonormal bases of compactly supported wavelets constructed by I. Daubechies [9] following the work of Y. Meyer [10] and S. Mallat [11] prove to be very useful. We outline here the properties of compactly supported wavelets and refer for the details to [9] and [12].

The orthonormal basis of compactly supported wavelets of $\mathbf{L}^2(\mathbf{R})$ is formed by the dilation and translation of a single function $\psi(x)$

$$\psi_{j,k}(x) = 2^{-j/2}\psi(2^{-j}x - k), \qquad (2.6)$$

where $j, k \in \mathbf{Z}$. The function $\psi(x)$ has a companion, the scaling function $\varphi(x)$, and these functions satisfy the following relations:

$$\varphi(x) = \sqrt{2}\sum_{k=0}^{L-1} h_k\varphi(2x - k), \qquad (2.7)$$

$$\psi(x) = \sqrt{2}\sum_{k=0}^{L-1} g_k\varphi(2x - k), \qquad (2.8)$$

where

$$g_k = (-1)^k h_{L-k-1}, \qquad k = 0,\ldots, L-1 \qquad (2.9)$$

and

$$\int_{-\infty}^{+\infty} \varphi(x)dx = 1. \qquad (2.10)$$

In addition, the function ψ has M vanishing moments

$$\int_{-\infty}^{+\infty} \psi(x)x^m dx = 0, \qquad m = 0, \ldots, M - 1. \qquad (2.11)$$

The number of coefficients L in (2.7) and (2.8) is related to the number of vanishing moments M. For the wavelets in [9] $L = 2M$. If additional conditions are imposed (see [5] for an example), then the relation might be different, but L is always even.

The decomposition of a function into the wavelet basis is an order N procedure. Given the coefficients s_k^0, $k = 0, 1, \ldots, N - 1$ as "samples" of the function f, the coefficients s_k^j and d_k^j on scales $j \geq 1$ are computed at a cost proportional to N via

$$s_k^j = \sum_{n=0}^{n=L-1} h_n s_{n+2k}^{j-1}, \qquad (2.12)$$

and

$$d_k^j = \sum_{n=0}^{n=L-1} g_n s_{n+2k+1}^{j-1}, \qquad (2.13)$$

where s_k^j and d_k^j are viewed as periodic sequences with the period 2^{n-j}.

We note that the Haar system is a degenerate case of Daubechies's wavelets. There is, however, a different way to construct orthonormal bases which generalize the Haar system and yields basis functions with several vanishing moments. We will discuss this construction in Section III in greater detail, since it approaches the problem of vanishing moments directly and does not require any prior knowledge of the wavelet theory.

To discuss the standard and non-standard representations, and compression of operators in Sections IV-VI, we use Daubechies's bases. Effectively, these representations yield two schemes for the numerical evaluation of integral operators. The first uses the standard representation and leads to numerical schemes which are, generally, of order $N \log(N)$, even for such simple operators as multiplication by a function. Another class of algorithms is obtained using the non-standard representation, which leads to numerical schemes of order N. Also, the non-standard representation leads to a proof of the celebrated "$T(1)$ theorem" of David and Journé (see [13]) (necessary and sufficient conditions for a Calderon-Zygmund operator to be bounded in $\mathbf{L}^2(\mathbf{R})$), and to uniform estimates for the error of the numerical algorithms.

The non-standard forms of many basic operators, such as fractional derivatives, Hilbert and Riesz transforms, etc., may be computed explicitly [8]. In Section VII we give an example of constructing the non-standard form for differential operators.

In Section VIII we show how to multiply two standard forms, and in Section IX describe a fast iterative algorithm for constructing the generalized inverse. This algorithm as well as several examples of Section X contain the beginning of the program for conversion of the pseudo-differential calculus into a numerical tool.

III   BASES WITH VANISHING MOMENTS

Let us describe very simple bases for $\mathbf{L}^2([0,1])$ which are composed of functions with several vanishing moments. Our construction generalizes the Haar basis and does not require any prior knowledge of the wavelet theory [6], [14].

Using the notion of multiresolution analysis [15], [16], we define $\mathbf{V}_j^M$ to be a space of piecewise polynomial functions,

$$\mathbf{V}_j^M = \{f : \quad \text{the restriction of } f \text{ to the interval } (2^j n, 2^j (n+1)) \text{ is}$$
$$\text{a polynomial of degree less than } M, \text{ for } n = 0, \ldots, 2^{-j} - 1,$$
$$\text{and } f \text{ vanishes elsewhere}\},$$

$$(3.1)$$

where M is a positive integer and $j = 0, -1, -2, \ldots$. The space $\mathbf{V}_j^M$ has dimension $2^{-j}M$,

$$\mathbf{V}_0^M \subset \mathbf{V}_{-1}^M \subset \cdots \subset \mathbf{V}_j^M \subset \cdots,$$

and

$$\mathbf{L}^2([0,1]) = \overline{\bigcup_{j \leq 0} \mathbf{V}_j}.$$

We define the $2^{-j}M$-dimensional space $\mathbf{W}_j^M$ to be the orthogonal complement of $\mathbf{V}_j^M$ in $\mathbf{V}_{j-1}^M$,

$$\mathbf{V}_{j-1}^M = \mathbf{V}_j^M \bigoplus \mathbf{W}_j^M,$$

and obtain

$$\mathbf{L}^2([0,1]) = \mathbf{V}_0^M \bigoplus_{j \leq 0} \mathbf{W}_j^M. \qquad (3.2)$$

If functions $h_1, \ldots, h_M : [0,1] \to \mathbf{R}$ form an orthogonal basis for $\mathbf{W}_0^M$, then the orthogonality of $\mathbf{W}_0^M$ to $\mathbf{V}_0^M$ implies that the first M moments of $h_1, \ldots, h_M$ vanish,

$$\int_0^1 h_i(x)\, x^m\, dx = 0, \qquad m = 0, 1, \ldots, M-1.$$

The $2M$-dimensional space $\mathbf{W}_{-1}^M$ is spanned by the $2M$ orthogonal functions $h_1(2x),\ \ldots,h_M(2x),\ h_1(2x-1),\ldots,h_M(2x-1)$, of which M are supported on the interval $[0,\frac{1}{2}]$ and M on $[\frac{1}{2},1]$. In general, the space $\mathbf{W}_j^M$ is spanned by $2^{-j}M$ functions obtained from $h_1,\ldots,h_M$ by translation and dilation. There is some freedom in choosing the functions $h_1,\ldots,h_M$ within the constraint that they be orthogonal; by requiring normality and additional vanishing moments, we specify them uniquely (up to sign).

First let us construct M functions $f_1,\ldots,f_M\ :\ \mathbf{R}\to\mathbf{R}$ supported on the interval $[-1,1]$ and such that

1. The restriction of f_i to the interval $(0,1)$ is a polynomial of degree $M-1$.

2. The function f_i is extended to the interval $(-1,0)$ as an even or odd function according to the parity of $i+M-1$.

3. The functions $\{f_i\}_{i=1}^{i=M}$ are orthonormal,

$$\int_{-1}^{1} f_i(x)\,f_l(x)\,dx = \delta_{il}, \quad i,l=1,\ldots,M.$$

4. The function f_i has vanishing moments,

$$\int_{-1}^{1} f_i(x)\,x^m\,dx = 0, \qquad m=0,1,\ldots,i+M-2.$$

Properties 1 and 2 imply that there are M^2 polynomial coefficients that determine the functions $f_1,\ldots,f_M$, while properties 3 and 4 provide M^2 constraints. It turns out that the equations uncouple to give M nonsingular linear systems that may be solved to obtain the coefficients, yielding the functions uniquely (up to sign).

We now determine $f_1,\ldots,f_M$ constructively by starting with $2M$ functions which span the space of polynomials of degree less than M on the interval $(0,1)$ and on $(-1,0)$, then orthogonalize M of them, first to the functions $1,x,\ldots,x^{M-1}$, then to the functions $x^M,x^{M+1},\ldots,x^{2M-1}$, and finally among themselves. We define $f_1^1,f_2^1,\ldots,f_M^1$ by the formula

$$f_m^1(x) = \begin{cases} x^{m-1}, & x\in(0,1), \\ -x^{m-1}, & x\in(-1,0), \\ 0, & \text{otherwise}, \end{cases}$$

and note that the $2M$ functions $1,x,\ldots,x^{M-1},f_1^1,f_2^1,\ldots,f_M^1$ are linearly independent.

1. By the Gram-Schmidt process we orthogonalize f_m^1 with respect to $1, x, \ldots, x^{M-1}$, to obtain f_m^2, for $m = 1, \ldots, M$. This orthogonality is preserved by the remaining orthogonalizations, which only produce linear combinations of the f_m^2.

2. The following sequence of steps yields $M - 1$ functions orthogonal to x^M, of which $M - 2$ functions are orthogonal to x^{M+1}, and so forth, down to one function which is orthogonal to x^{2M-2}. First, if at least one of f_m^2 is not orthogonal to x^M, we reorder the functions so that it appears first, $\langle f_1^2, x^M \rangle \neq 0$. We then define $f_m^3 = f_m^2 - a_m \cdot f_0^2$ where a_m is chosen so $\langle f_m^3, x^M \rangle = 0$ for $m = 2, \ldots, M$, achieving the desired orthogonality to x^M. Similarly, we orthogonalize to $x^{M+1}, \ldots, x^{2M-2}$, each in turn, to obtain $f_1^2, f_2^3, f_3^4, \ldots, f_M^{M+1}$ such that $\langle f_m^{m+1}, x^i \rangle = 0$ for $i \leq m + M - 2$.

3. Finally, we do Gram-Schmidt orthogonalization on $f_M^{M+1}, f_{M-1}^M, \ldots, f_1^2$, in that order, and normalize to obtain $f_M, f_{M-1}, \ldots, f_1$.

It is easy to see that functions $\{f_m\}_{m=1}^{m=M}$ satisfy properties 1-4 of the previous paragraph. Defining $h_1, \ldots, h_M$ as

$$h_m(x) = 2^{1/2} f_m(2x - 1), \qquad m = 1, \ldots, M,$$

we obtain

$$\mathbf{W}_j^M = \text{linear span } \{h_{m,j}^n : \quad h_{m,j}^n(x) = 2^{-j/2} h_m(2^{-j}x - n), \\ m = 1, \ldots, M; \; n = 0, \ldots, 2^{-j} - 1\}. \qquad (3.3)$$

Let $\{u_m\}_{m=1}^{m=M}$ denote an orthonormal basis for $\mathbf{V}_0^M$. Combining it with (3.3) (in view of (3.2)) we obtain an orthonormal basis of $\mathbf{L}^2([0, 1])$. We refer to this basis as the *multi-wavelet* basis of order M.

It is easy to see that the orthonormal set

$$\{h_{m,j}^n : \quad h_{m,j}^n(x) = 2^{-j/2} h_m(2^{-j}x - n), \; m = 1, \ldots, M; \; n \in Z\}.$$

is an orthonormal basis of $\mathbf{L}^2(\mathbf{R})$.

Algorithms, various numerical examples and applications utilizing bases of this Section are described in [6] and [14].

We now outline the construction of bases for $\mathbf{L}^2[0, 1]^d$ and $\mathbf{L}^2(\mathbf{R}^d)$, for any dimension d. We describe this extension by giving the basis for $\mathbf{L}^2([0, 1]^2)$, which is illustrative of the construction for any finite-dimensional space. Let us define the space $\mathbf{V}_j^{M,2}$ as

$$\mathbf{V}_j^{M,2} = \mathbf{V}_j^M \times \mathbf{V}_j^M, \qquad j = 0, -1, -2, \ldots,$$

where $\mathbf{V}_j^M$ is given in (3.1), and the space $\mathbf{W}_j^{M,2}$ as the orthogonal complement of $\mathbf{V}_j^{M,2}$ in $\mathbf{V}_{j-1}^{M,2}$,

$$\mathbf{V}_{j-1}^{M,2} = \mathbf{V}_j^{M,2} \bigoplus \mathbf{W}_j^{M,2}.$$

The space $\mathbf{W}_0^{M,2}$ is spanned by the orthonormal basis

$$\{u_i(x)h_l(y),\ h_i(x)u_l(y),\ h_i(x)h_l(y):\ i,l = 1,\ldots,M\}.$$

Among these $3M^2$ basis elements each element $v(x,y)$ has vanishing moments,

$$\int_0^1 \int_0^1 v(x,y)\, x^i\, y^l\, dx\, dy = 0, \qquad i,l = 0,1,\ldots,M-1.$$

The space $\mathbf{W}_j^{M,2}$ is spanned by dilations and translations of the $v(x,y)$ and the basis of $\mathbf{L}^2([0,1]^2)$ consists of these functions and the low-order polynomials $\{u_i(x)u_l(y):\ i,l = 1,\ldots,M\}$.

IV   THE NON-STANDARD FORM

The two-dimensional multi-wavelet basis described in Section III (with the functions of the basis supported on squares) requires $3M^2$ different combinations of one-dimensional basis functions, where M is the number of vanishing moments. On the other hand, the two-dimensional bases obtained using compactly supported wavelets [9] require only three such combinations. Thus, we will use Daubechies' bases to review the non-standard form [5].

The wavelet basis induces a multiresolution analysis on $\mathbf{L}^2(\mathbf{R})$ [15], [16], i.e., the decomposition of the Hilbert space $\mathbf{L}^2(\mathbf{R})$ into a chain of closed subspaces

$$\ldots \subset \mathbf{V}_2 \subset \mathbf{V}_1 \subset \mathbf{V}_0 \subset \mathbf{V}_{-1} \subset \mathbf{V}_{-2} \subset \ldots \tag{4.1}$$

such that

$$\bigcap_{j\in\mathbf{Z}} \mathbf{V}_j = \{0\}, \quad \overline{\bigcup_{j\in\mathbf{Z}} \mathbf{V}_j} = \mathbf{L}^2(\mathbf{R}). \tag{4.2}$$

By defining $\mathbf{W}_j$ as an orthogonal complement of $\mathbf{V}_j$ in $\mathbf{V}_{j-1}$,

$$\mathbf{V}_{j-1} = \mathbf{V}_j \bigoplus \mathbf{W}_j, \tag{4.3}$$

the space $\mathbf{L}^2(\mathbf{R})$ is represented as a direct sum

$$\mathbf{L}^2(\mathbf{R}) = \bigoplus_{j\in\mathbf{Z}} \mathbf{W}_j. \tag{4.4}$$

On each fixed scale j, the wavelets $\{\psi_{j,k}(x)\}_{k \in \mathbf{Z}}$ form an orthonormal basis of $\mathbf{W}_j$ and the functions $\{\varphi_{j,k}(x) = 2^{-j/2}\varphi(2^{-j}x - k)\}_{k \in \mathbf{Z}}$ form an orthonormal basis of $\mathbf{V}_j$.

If there is the coarsest scale n, then the chain of the subspaces (4.1) is replaced by

$$\mathbf{V}_n \subset \ldots \subset \mathbf{V}_2 \subset \mathbf{V}_1 \subset \mathbf{V}_0 \subset \mathbf{V}_{-1} \subset \mathbf{V}_{-2} \subset \ldots, \quad \mathbf{L}^2(\mathbf{R}) = \mathbf{V}_n \bigoplus_{j \leq n} \mathbf{W}_j.$$
$$(4.5)$$

If there are finitely many scales, then without loss of generality we set the scale $j = 0$ to be the finest scale. Instead of (4.5) we then have

$$\mathbf{V}_n \subset \ldots \subset \mathbf{V}_2 \subset \mathbf{V}_1 \subset \mathbf{V}_0, \quad \mathbf{V}_0 \subset \mathbf{L}^2(\mathbf{R}). \tag{4.6}$$

In numerical realizations the subspace $\mathbf{V}_0$ is finite dimensional.

Let T be an operator

$$T : \mathbf{L}^2(\mathbf{R}) \to \mathbf{L}^2(\mathbf{R}), \tag{4.7}$$

with the kernel $K(x, y)$. We define projection operators on the subspace $\mathbf{V}_j$, $j \in \mathbf{Z}$,

$$P_j : \mathbf{L}^2(\mathbf{R}) \to \mathbf{V}_j, \tag{4.8}$$

as follows

$$(P_j f)(x) = \sum_k \langle f, \varphi_{j,k} \rangle \varphi_{j,k}(x). \tag{4.9}$$

Expanding T in a "telescopic" series, we obtain

$$T = \sum_{j \in \mathbf{Z}} (Q_j T Q_j + Q_j T P_j + P_j T Q_j), \tag{4.10}$$

where

$$Q_j = P_{j-1} - P_j \tag{4.11}$$

is the projection operator on the subspace $\mathbf{W}_j$. If there is the coarsest scale n, then instead of (4.10) we have

$$T = \sum_{j=-\infty}^{n} (Q_j T Q_j + Q_j T P_j + P_j T Q_j) + P_n T P_n, \tag{4.12}$$

and if the scale $j = 0$ is the finest scale, then

$$T_0 = \sum_{j=1}^{n} (Q_j T Q_j + Q_j T P_j + P_j T Q_j) + P_n T P_n, \tag{4.13}$$

where $T \sim T_0 = P_0 T P_0$ is a discretization of the operator T on the finest scale.

The non-standard form is a representation (see [5]) of the operator T as a chain of triplets

$$T = \{A_j, B_j, \Gamma_j\}_{j \in \mathbf{Z}} \tag{4.14}$$

acting on the subspaces $\mathbf{V}_j$ and $\mathbf{W}_j$,

$$A_j : \mathbf{W}_j \to \mathbf{W}_j, \tag{4.15}$$

$$B_j : \mathbf{V}_j \to \mathbf{W}_j, \tag{4.16}$$

$$\Gamma_j : \mathbf{W}_j \to \mathbf{V}_j. \tag{4.17}$$

The operators $\{A_j, B_j, \Gamma_j\}_{j \in \mathbf{Z}}$ are defined as $A_j = Q_j T Q_j$, $B_j = Q_j T P_j$ and $\Gamma_j = P_j T Q_j$.

The operators $\{A_j, B_j, \Gamma_j\}_{j \in \mathbf{Z}}$ admit a recursive definition (see [5]) via the relation

$$T_j = \begin{pmatrix} A_{j+1} & B_{j+1} \\ \Gamma_{j+1} & T_{j+1} \end{pmatrix}, \tag{4.18}$$

where operators $T_j = P_j T P_j$,

$$T_j : \mathbf{V}_j \to \mathbf{V}_j. \tag{4.19}$$

If there is a coarsest scale n, then

$$T = \{\{A_j, B_j, \Gamma_j\}_{j \in \mathbf{Z}: j \leq n}, T_n\}, \tag{4.20}$$

where $T_n = P_n T P_n$. If the number of scales is finite, then $j = 1, 2, \ldots, n$ in (4.20) and the operators are organized as blocks of the matrix (see Figures 1 and 2).

We will now make the following observations:

1). The map (4.15) implies that the operator A_j describes the interaction on the scale j only, since the subspace $\mathbf{W}_j$ is an element of the direct sum in (4.4).

2). The operators B_j, Γ_j in (4.16) and (4.17) describe the interaction between scale j and all coarser scales. Indeed, the subspace $\mathbf{V}_j$ contains all the subspaces $\mathbf{V}_{j'}$ with $j' > j$ (see (4.1)).

3). The operator T_j is an "averaged" version of the operator T_{j-1}.

The operators A_j, B_j and Γ_j are represented by the matrices α^j, β^j and γ^j,

$$\alpha^j_{k,k'} = \int \int K(x,y) \, \psi_{j,k}(x) \, \psi_{j,k'}(y) \, dx dy, \tag{4.21}$$

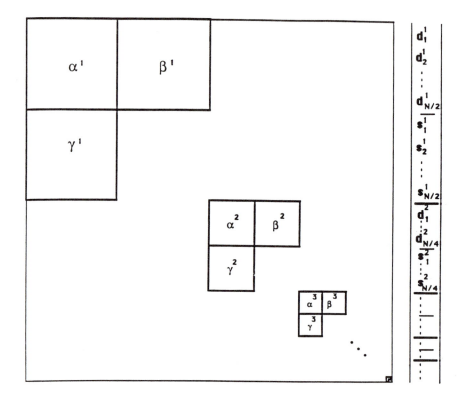

FIGURE 1. Structure of the non-standard form of a matrix. Submatrices α, β and γ on different scales are the only nonzero submatrices. In fact, most of the entries of these submatrices can be set to zero given the desired accuracy.

$$\beta_{k,k'}^{j} = \int \int K(x, y)\, \psi_{j,k}(x)\, \varphi_{j,k'}(y)\ dx dy, \qquad (4.22)$$

and

$$\gamma_{k,k'}^{j} = \int \int K(x, y)\, \varphi_{j,k}(x)\, \psi_{j,k'}(y)\ dx dy. \qquad (4.23)$$

The operator T_j is represented by the matrix s^j,

$$s_{k,k'}^{j} = \int \int K(x, y)\, \varphi_{j,k}(x)\, \varphi_{j,k'}(y)\ dx dy. \qquad (4.24)$$

Given a set of coefficients $s_{k,k'}^{0}$ with $k, k' = 0, 1, \ldots, N - 1$, repeated application of the formulae (2.12), (2.13) produces

$$\alpha_{i,l}^{j} = \sum_{k,m=0}^{L-1} g_k g_m s_{k+2i,m+2l}^{j-1}, \qquad (4.25)$$

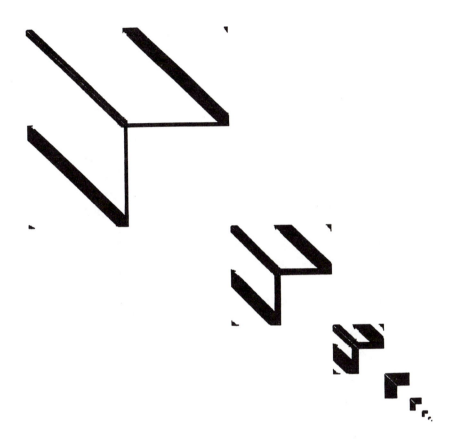

FIGURE 2. An example of a matrix in the non-standard form, $A_{ij} = 1/(i-j), i \neq j$. Entries above the threshold of 10^{-7} are shown black. Note that the width of the bands does not grow with the size of the matrix.

$$\beta_{i,l}^{j} = \sum_{k,m=0}^{L-1} g_k h_m s_{k+2i,m+2l}^{j-1}, \tag{4.26}$$

$$\gamma_{i,l}^{j} = \sum_{k,m=0}^{L-1} h_k g_m s_{k+2i,m+2l}^{j-1}, \tag{4.27}$$

$$s_{i,l}^{j} = \sum_{k,m=0}^{L-1} h_k h_m s_{k+2i,m+2l}^{j-1}, \tag{4.28}$$

with $i,l = 0,1,\ldots,2^{n-j}-1$, $j = 1,2,\ldots,n$. Clearly, formulae (4.25) – (4.28) provide an order N^2 scheme for the evaluation of the elements of all matrices $\alpha^j, \beta^j, \gamma^j$ with $j = 1,2\ldots,n$.

To compute the coefficients $s_{k,k'}^0$, we refer to [5], where wavelet-based quadratures for the evaluation of these coefficients are developed. Also, we refer to [5] for a fast algorithm (order N) for constructing the non-standard form for operators with known singularities, and to [8] for the direct evaluation of non-standard forms of several basic operators (see Section VII).

V   THE STANDARD FORM

The standard form is obtained by representing

$$\mathbf{V}_j = \bigoplus_{j'>j} \mathbf{W}_{j'}, \tag{5.1}$$

and considering for each scale j the operators $\{B_j^{j'}, \Gamma_j^{j'}\}_{j'>j}$,

$$B_j^{j'} : \mathbf{W}_{j'} \to \mathbf{W}_j, \tag{5.2}$$

$$\Gamma_j^{j'} : \mathbf{W}_j \to \mathbf{W}_{j'}. \tag{5.3}$$

If there is the coarsest scale n, then instead of (5.1) we have

$$\mathbf{V}_j = \mathbf{V}_n \bigoplus_{j'=j+1}^{j'=n} \mathbf{W}_{j'}. \tag{5.4}$$

In this case, the operators $\{B_j^{j'}, \Gamma_j^{j'}\}$ for $j' = j+1,\ldots,n$ are as in (5.2) and (5.3) and, in addition, for each scale j there are operators $\{B_j^{n+1}\}$ and $\{\Gamma_j^{n+1}\}$,

$$B_j^{n+1} : \mathbf{V}_n \to \mathbf{W}_j, \tag{5.5}$$

$$\Gamma_j^{n+1} : \mathbf{W}_j \to \mathbf{V}_n. \tag{5.6}$$

FIGURE 3. An example of a matrix in the standard form, $A_{ij} = 1/(i-j), i \neq j$. Different "finger" bands represent "interactions" between scales.

(In this notation, $\Gamma_n^{n+1} = \Gamma_n$ and $B_n^{n+1} = B_n$). If there are finitely many scales and $\mathbf{V}_0$ is finite dimensional, then the standard form is a representation of $T_0 = P_0 T P_0$ as

$$T_0 = \{A_j, \{B_j^{j'}\}_{j'=j+1}^{j'=n}, \{\Gamma_j^{j'}\}_{j'=j+1}^{j'=n}, B_j^{n+1}, \Gamma_j^{n+1}, T_n\}_{j=1,\dots,n}. \qquad (5.7)$$

The operators (5.7) are organized as blocks of the matrix (see Figure 3).

If the operator T is a Calderón-Zygmund or a pseudo-differential operator then, for a fixed accuracy, all the operators in (5.7) (except T_n) are banded. As a result, the standard form has several "finger" bands which correspond to the interaction between different scales. For a large class of operators (pseudo-differential, for example), the interaction between different scales characterized by the size of the coefficients of "finger" bands, decays as the distance $j' - j$ between the scales increases. Therefore, if the scales j and j' are well separated, then for a given accuracy, the operators $B_j^{j'}, \Gamma_j^{j'}$ can be neglected.

There are two ways of computing the standard form of a matrix. First consists in applying the one-dimensional transform (see (2.12) and (2.13)) to each column (row) of the matrix and, then, to each row (column) of the result. Alternatively, one can compute the non-standard form and then apply the one-dimensional transform to each row of all operators B^j and each column of all operators Γ_j. We refer to [5] for details.

VI COMPRESSION OF OPERATORS

If the operator T is a Calderon-Zygmund or a pseudo-differential operator, then by using the wavelet basis with M vanishing moments, we force operators $\{A_j, B_j, \Gamma_j\}_{j \in \mathbf{Z}}$ to decay roughly as $1/d^{M+1}$, where d is a distance from the diagonal. For example, let the kernel satisfy the conditions

$$|K(x,y)| \leq \frac{1}{|x-y|}, \tag{6.1}$$

$$|\partial_x^M K(x,y)| + |\partial_y^M K(x,y)| \leq \frac{C_0}{|x-y|^{1+M}} \tag{6.2}$$

for some $M \geq 1$. Then by choosing the wavelet basis with M vanishing moments, the coefficients $\alpha_{i,l}^j, \beta_{i,l}^j, \gamma_{i,l}^j$ of the non-standard form (see (4.21) - (4.23)) satisfy the estimate

$$|\alpha_{i,l}^j| + |\beta_{i,l}^j| + |\gamma_{i,l}^j| \leq \frac{C_M}{1 + |i-l|^{M+1}}, \tag{6.3}$$

for all

$$|i-l| \geq 2M. \tag{6.4}$$

If, in addition to (6.1), (6.2),

$$|\int_{I \times I} K(x,y)\, dx dy | \leq C|I| \tag{6.5}$$

for all dyadic intervals I (this is the "weak cancellation condition", see [10]), then (6.3) holds for all i, l.

If T is a pseudo-differential operator with symbol $\sigma(x, \xi)$ defined by the formula

$$T(f)(x) = \sigma(x, D)f = \int e^{ix\xi}\, \sigma(x, \xi)\hat{f}(\xi)\, d\xi = \int K(x,y)f(y)\, dy, \tag{6.6}$$

where K is the distributional kernel of T, then assuming that the symbols σ of T and σ^* of T^* satisfy the standard conditions

$$| \partial_\xi^\alpha \partial_x^\beta \sigma(x, \xi) | \leq C_{\alpha,\beta}(1+ | \xi |)^{\lambda-\alpha+\beta} \tag{6.7}$$

$$| \partial_\xi^\alpha \, \partial_x^\beta \, \sigma^*(x, \xi) | \leq C_{\alpha, \beta} (1 + | \, \xi \, |)^{\lambda - \alpha + \beta}, \tag{6.8}$$

we have the inequality

$$|\alpha_{i,l}^j| + |\beta_{i,l}^j| + |\gamma_{i,l}^j| \leq \frac{2^{\lambda j} \, C_M}{(1 + |i - l|)^{M+1}}, \tag{6.9}$$

for all integer i, l.

Suppose now that we approximate the operator T_0 by the operator T_0^B obtained from T_0 by setting to zero all coefficients of matrices α^j, β^j and γ^j outside of bands of width $B \geq 2M$ around their diagonals. We obtain

$$\|T_0^B - T_0\| \leq \frac{C}{B^M} \log_2 N, \tag{6.10}$$

where C is a constant determined by the kernel K. In most numerical applications, the accuracy ε of calculations is fixed, and the parameters of the algorithm (in our case, the band width B and order M) have to be chosen in such a manner that the desired precision of calculations is achieved. If M is fixed, then we choose B so that

$$B \geq \left(\frac{C}{\varepsilon} \log_2 N \right)^{1/M}. \tag{6.11}$$

In other words, T_0 has been approximated to precision ε with its truncated version, which can be applied to arbitrary vectors for a cost proportional to $N \left((C/\varepsilon) \log_2 N \right)^{1/M}$, which for all practical purposes does not differ from N.

A more detailed investigation [5] permits the estimate (6.10) to be replaced with the estimate

$$\|T_0^B - T_0\| \leq \frac{C}{B^M}, \tag{6.12}$$

making the application of the operator T_0 to an arbitrary vector with arbitrary fixed accuracy into a procedure of order N. Obtaining this uniform estimate leads to a proof of

Theorem (G. David, J.L. Journé) Suppose that the operator

$$T(f) = \int K(x, y) \, f(y) \, dy \tag{6.13}$$

satisfies the conditions (6.1), (6.2), (6.5). Then a necessary and sufficient condition for T to be bounded on L^2 is that

$$\beta(x) = T(1)(x), \tag{6.14}$$

$$\gamma(y) = T^*(1)(y) \tag{6.15}$$

belong to dyadic $B.M.O.$, i.e. satisfy condition

$$\sup_J \frac{1}{|J|} \int_J |\beta(x) - m_J(\beta)|^2 dx \le C, \tag{6.16}$$

where J is a dyadic interval and

$$m_J(\beta) = \frac{1}{|J|} \int_J \beta(x) dx. \tag{6.17}$$

Again we refer to [5] for details.

VII   THE OPERATOR d/dx IN WAVELET BASES

As an example, we construct the non-standard form of the operator d/dx [8]. The matrix elements α_{il}^j, β_{il}^j, and γ_{il}^j of A_j, B_j, and Γ_j, where $i, l, j \in \mathbf{Z}$ for the operator d/dx are easily computed as

$$\alpha_{il}^j = 2^{-j} \int_{-\infty}^{\infty} \psi(2^{-j}x - i)\, \psi'(2^{-j}x - l)\, 2^{-j} dx = 2^{-j}\alpha_{i-l}, \tag{7.1}$$

$$\beta_{il}^j = 2^{-j} \int_{-\infty}^{\infty} \psi(2^{-j}x - i)\, \varphi'(2^{-j}x - l)\, 2^{-j} dx = 2^{-j}\beta_{i-l}, \tag{7.2}$$

and

$$\gamma_{il}^j = 2^{-j} \int_{-\infty}^{\infty} \varphi(2^{-j}x - i)\, \psi'(2^{-j}x - l)\, 2^{-j} dx = 2^{-j}\gamma_{i-l}, \tag{7.3}$$

where

$$\alpha_l = \int_{-\infty}^{+\infty} \psi(x - l)\, \frac{d}{dx}\psi(x)\, dx, \tag{7.4}$$

$$\beta_l = \int_{-\infty}^{+\infty} \psi(x - l)\, \frac{d}{dx}\varphi(x)\, dx, \tag{7.5}$$

and

$$\gamma_l = \int_{-\infty}^{+\infty} \varphi(x - l)\, \frac{d}{dx}\psi(x)\, dx. \tag{7.6}$$

Moreover, using (2.7) and (2.8) we have

$$\alpha_i = 2 \sum_{k=0}^{L-1} \sum_{k'=0}^{L-1} g_k\, g_{k'}\, r_{2i+k-k'}, \tag{7.7}$$

$$\beta_i = 2 \sum_{k=0}^{L-1} \sum_{k'=0}^{L-1} g_k \, h_{k'} \, r_{2i+k-k'}, \tag{7.8}$$

and

$$\gamma_i = 2 \sum_{k=0}^{L-1} \sum_{k'=0}^{L-1} h_k \, g_{k'} \, r_{2i+k-k'}, \tag{7.9}$$

where

$$r_l = \int_{-\infty}^{+\infty} \varphi(x-l) \, \frac{d}{dx} \varphi(x) \, dx, \quad l \in \mathbf{Z}. \tag{7.10}$$

Therefore, the representation of d/dx is completely determined by r_l in (7.10) or in other words, by the representation of d/dx on the subspace $\mathbf{V}_0$.

Rewriting (7.10) in terms of $\hat{\varphi}(\xi)$, where

$$\hat{\varphi}(\xi) = \frac{1}{\sqrt{2\pi}} \int_{-\infty}^{+\infty} \varphi(x) \, e^{ix\xi} \, dx, \tag{7.11}$$

we obtain

$$r_l = \int_{-\infty}^{+\infty} |\hat{\varphi}(\xi)|^2 (i\xi) e^{-il\xi} \, d\xi. \tag{7.12}$$

The following proposition [8] reduces the computation of the coefficients r_l to solving a system of linear algebraic equations.

1. *If the integrals in (7.10) or (7.12) exist, then the coefficients r_l, $l \in \mathbf{Z}$ in (7.10) satisfy the following system of linear algebraic equations*

$$r_l = 2 \left[r_{2l} + \frac{1}{2} \sum_{k=1}^{L/2} a_{2k-1} (r_{2l-2k+1} + r_{2l+2k-1}) \right], \tag{7.13}$$

and

$$\sum_l l \, r_l = -1, \tag{7.14}$$

where the coefficients a_{2k-1},

$$a_{2k-1} = 2 \sum_{i=0}^{L-2k} h_i \, h_{i+2k-1}, \quad k = 1, \ldots, L/2. \tag{7.15}$$

2. *If $M \geq 2$, then equations (7.13) and (7.14) have a unique solution with a finite number of non-zero r_l, namely, $r_l \neq 0$ for $-L+2 \leq l \leq L-2$*

and

$$r_l = -r_{-l}, \tag{7.16}$$

Solving equations (7.13), (7.14), we present the results for Daubechies' wavelets with $M = 2, 3, 4, 5$. For further examples we refer to [8].

1. $\quad M = 2$

$$a_1 = \frac{9}{8}, \quad a_3 = -\frac{1}{8},$$

and

$$r_1 = -\frac{2}{3}, \quad r_2 = \frac{1}{12},$$

We note, that the coefficients $(-1/12, 2/3, 0, -2/3, 1/12)$ of this example can be found in many books on numerical analysis as a choice of coefficients for numerical differentiation.

2. $\quad M = 3$

$$a_1 = \frac{75}{64}, \quad a_3 = -\frac{25}{128}, \quad a_5 = \frac{3}{128},$$

and

$$r_1 = -\frac{272}{365}, \quad r_2 = \frac{53}{365}, \quad r_3 = -\frac{16}{1095}, \quad r_4 = -\frac{1}{2920}.$$

3. $\quad M = 4$

$$a_1 = \frac{1225}{1024}, \quad a_3 = -\frac{245}{1024}, \quad a_5 = \frac{49}{1024}, \quad a_7 = -\frac{5}{1024},$$

and

$$r_1 = -\frac{39296}{49553}, \quad r_2 = \frac{76113}{396424}, \quad r_3 = -\frac{1664}{49553},$$

$$r_4 = \frac{2645}{1189272}, \quad r_5 = \frac{128}{743295}, \quad r_6 = -\frac{1}{1189272}.$$

4. $\quad M = 5$

$$a_1 = \frac{19845}{16384}, \quad a_3 = -\frac{2205}{8192}, \quad a_5 = \frac{567}{8192}, \quad a_7 = -\frac{405}{32768}, \quad a_9 = \frac{35}{32768},$$

and

$$r_1 = -\frac{957310976}{1159104017}, \quad r_2 = \frac{265226398}{1159104017}, \quad r_3 = -\frac{735232}{13780629},$$

$$r_4 = \frac{17297069}{2318208034}, \quad r_5 = -\frac{1386496}{5795520085}, \quad r_6 = -\frac{563818}{10431936153},$$

$$r_7 = -\frac{2048}{8113728119}, \quad r_8 = -\frac{5}{18545664272}.$$

Remark 1. If $M = 1$, then equations (7.13) and (7.14) have a unique solution but the integrals in (7.10) or (7.12) may not be absolutely convergent. For the Haar basis ($h_1 = h_2 = 2^{-1/2}$) $a_1 = 1$ and $r_1 = -1/2$ and we obtain the simplest finite difference operator $(1/2, 0, -1/2)$. In this case the function φ is not continuous and

$$\hat{\varphi}(\xi) = \frac{1}{\sqrt{2\pi}} \frac{\sin \frac{1}{2}\xi}{\frac{1}{2}\xi} e^{i\frac{1}{2}\xi}.$$

Remark 2. For the coefficients $r_l^{(n)}$ of d^n/dx^n, $n > 1$, the system of linear algebraic equations is similar to that for the coefficients of d/dx. This system (and (7.13)) may be written in terms of

$$\hat{r}(\xi) = \sum_l r_l^{(n)} e^{il\xi}, \tag{7.17}$$

as

$$\hat{r}(\xi) = 2^n \left(|m_0(\xi/2)|^2 \, \hat{r}(\xi/2) + |m_0(\xi/2 + \pi)|^2 \, \hat{r}(\xi/2 + \pi) \right), \tag{7.18}$$

where m_0 is the 2π-periodic function

$$m_0(\xi) = 2^{-1/2} \sum_{k=0}^{k=L-1} h_k e^{ik\xi}, \tag{7.19}$$

and h_k are the wavelet coefficients. Considering the operator M_0 on 2π-periodic functions

$$(M_0 f)(\xi) = |m_0(\xi/2)|^2 \, f(\xi/2) + |m_0(\xi/2 + \pi)|^2 \, f(\xi/2 + \pi), \tag{7.20}$$

we rewrite (7.18) as

$$M_0 \hat{r} = 2^{-n} \hat{r}, \tag{7.21}$$

so that $\hat{r}$ is an eigenvector of the operator M_0 corresponding to the eigenvalue 2^{-n}. Thus, finding the representation of the derivatives in the wavelet basis is equivalent to finding trigonometric polynomial solutions of (7.21) and vice versa [8].

Remark 3. While theoretically it is well understood that the operators with homogeneous symbols have an explicit diagonal preconditioner in wavelet bases, the numerical evidence illustrating this fact is of interest, since it represents one of the advantages of computing in the wavelet bases.

For operators with a homogeneous symbol the bound on the condition number depends on the particular choice of the wavelet basis (by the condition number we understand the ratio of the largest singular value to the smallest singular value above the theshold of accuracy). After applying such a preconditioner, the condition number κ_p of the matrix of the preconditioned operator is uniformly bounded with respect to the size of the matrix. We recall that the condition number controls the rate of convergence of a number of iterative algorithms; for example the number of iterations of the conjugate gradient method is $O(\sqrt{\kappa_p})$.

We present here two tables illustrating such preconditioning applied to the standard form of the second derivative [5], [8]. In the following examples the standard form of periodized second derivative D_2 of size $N \times N$, where $N = 2^n$, is preconditioned by the diagonal matrix P,

$$D_2^p = PD_2P,$$

where $P_{il} = \delta_{il}2^j$, $1 \le j \le n$, and where j is chosen depending on i, l so that $N - N/2^{j-1} + 1 \le i, l \le N - N/2^j$, and $P_{NN} = 2^n$.

The following tables compare the original condition number κ of D_2 and κ_p of D_2^p.

N	κ	κ_p
64	0.14545E+04	0.10792E+02
128	0.58181E+04	0.11511E+02
256	0.23272E+05	0.12091E+02
512	0.93089E+05	0.12604E+02
1024	0.37236E+06	0.13045E+02

TABLE 1. Condition numbers of the matrix of periodized second derivative (with and without preconditioning) in the basis of Daubechies' wavelets with three vanishing moments $M = 3$.

N	κ	κ_p
64	0.10472E+04	0.43542E+01
128	0.41886E+04	0.43595E+01
256	0.16754E+05	0.43620E+01
512	0.67018E+05	0.43633E+01
1024	0.26807E+06	0.43640E+01

TABLE 2. Condition numbers of the matrix of periodized second derivative (with and without preconditioning) in the basis of Daubechies' wavelets with six vanishing moments $M = 6$.

VIII MULTIPLICATION OF MATRICES IN THE STANDARD FORM

We now turn to the multiplication of matrices of Calderón-Zygmund and pseudo-differential operators in the standard form. We show that such multiplication requires at most $O(N \log^2 N)$ operations and that we may control the width of the "finger" bands by setting to zero all the entries in the product below the threshold of accuracy.

Let us compute $T = \tilde{T}\hat{T}$, where

$$\tilde{T} = \{\tilde{A}_j, \{\tilde{B}_j^{j'}\}_{j'=j+1}^{j'=n}, \{\tilde{\Gamma}_j^{j'}\}_{j'=j+1}^{j'=n}, \tilde{B}_j^{n+1}, \tilde{\Gamma}_j^{n+1}, \tilde{T}_n\}_{j=1,\ldots,n} \qquad (8.1)$$

and

$$\hat{T} = \{\hat{A}_j, \{\hat{B}_j^{j'}\}_{j'=j+1}^{j'=n}, \{\hat{\Gamma}_j^{j'}\}_{j'=j+1}^{j'=n}, \hat{B}_j^{n+1}, \hat{\Gamma}_j^{n+1}, \hat{T}_n\}_{j=1,\ldots,n}. \qquad (8.2)$$

Since the standard form is a representation in an orthonomal basis (which is the tensor product of one-dimensional bases), the result of the multiplication of two standard forms is also a standard form in the same basis. Thus, the product T must have a representation

$$T = \{A_j, \{B_j^{j'}\}_{j'=j+1}^{j'=n}, \{\Gamma_j^{j'}\}_{j'=j+1}^{j'=n}, B_j^{n+1}, \Gamma_j^{n+1}, T_n\}_{j=1,\ldots,n}. \qquad (8.3)$$

Due to the block structure of the corresponding matrix, each element of (8.3) is obtained as a sum of products of the corresponding blocks of $\tilde{T}$

and $\hat{T}$. For example,

$$\Gamma_1^2 = \tilde{\Gamma}_1^2 \hat{A}_1 + \tilde{A}_2 \hat{\Gamma}_1^2 + \sum_{j'=3}^{j'=n+1} \tilde{B}_2^{j'} \hat{\Gamma}_1^{j'}. \tag{8.4}$$

If the operators $\tilde{T}$ and $\hat{T}$ are Calderón-Zygmund or pseudo-differential operators, then all the blocks of (8.1) and (8.2) (except for $\tilde{T}_n$ and $\hat{T}_n$) are banded and it is clear that Γ_1^2 is banded. This example is generic for all operators in (8.3) except for B_j^{n+1}, Γ_j^{n+1}, $(j = 1, \ldots, n)$ and T_n. The latter are dense due to the terms involving $\tilde{T}_n$ and $\hat{T}_n$. It is easy now to estimate the number of operations necessary to compute T. It takes no more than $O(N \log^2 N)$ operations to obtain T, where $N = 2^n$.

If, in addition, when the scales j and j' are well separated, the operators $B_j^{j'}$, $\Gamma_j^{j'}$ can be neglected for a given accuracy (as in the case of pseudo-differential operators), then the number of operations reduces asymptotically to $O(N)$.

We note, that we may set to zero all the entries of T below the threshold of accuracy and, thus, prevent the widening of the bands in the product. On denoting $\tilde{T}_\epsilon$ and $\hat{T}_\epsilon$ the approximations to $\tilde{T}$ and $\hat{T}$ obtained by setting all entries that are less than ϵ to zero, and assuming (without a loss of generality) $||\tilde{T}|| = ||\hat{T}|| = 1$, we obtain using the result of [5]

$$||\tilde{T} - \tilde{T}_\epsilon|| \le \epsilon, \quad ||\hat{T} - \hat{T}_\epsilon|| \le \epsilon, \tag{8.5}$$

and, therefore,

$$||\tilde{T}\hat{T} - (\tilde{T}_\epsilon \hat{T}_\epsilon)_\epsilon|| \le \epsilon + \epsilon(1 + \epsilon) + \epsilon(1 + \epsilon)^2. \tag{8.6}$$

The right hand side of (8.6) is dominated by 3ϵ. For example, if we compute T^4 then we might lose one significant digit.

IX   FAST ITERATIVE CONSTRUCTION OF THE GENERALIZED INVERSE

The fast multiplication algorithm of Section VIII gives a second life to a number of iterative algorithms. As an example, we consider an iterative construction of the generalized inverse. In order to construct the generalized inverse $A^\dagger$ of the matrix A we use the following result [17]:

.   Let σ_1 be the largest singular value of the $m \times n$ matrix A. Consider the sequence of matrices X_k

$$X_{k+1} = 2X_k - X_k A X_k \tag{9.1}$$

with

$$X_0 = \alpha A^*, \tag{9.2}$$

where A^* is the adjoint matrix and α is chosen so that the largest eigenvalue of $\alpha A^* A$ is less than two. Then the sequence X_k converges to the generalized inverse $A^\dagger$.

When this result is combined with the fast multiplication algorithm of Section VIII, we obtain an algorithm for constructing the generalized inverse in at most $O(N \log^2 N \log R)$ operations, where R is the condition number of the matrix. (By the condition number we understand the ratio of the largest singular value to the smallest singular value above the threshold of accuracy).

The details of this algorithm (in the context of computing in wavelet bases) will be described in [18]. We note that throughout the iteration (9.1), it is necessary to maintain the "finger" band structure of the standard form of matrices X_k. Hence, the standard form of both the operator and its generalized inverse must admit such structure. We note that the pseudodifferential operators satisfy this condition.

Size $N \times N$	SVD	FWT Generalized Inverse	L_2-Error
128×128	20.27 sec.	25.89 sec.	$3.1 \cdot 10^{-4}$
256×256	144.43 sec.	77.98 sec.	$3.42 \cdot 10^{-4}$
512×512	1,155 sec. (est.)	242.84 sec.	$6.0 \cdot 10^{-4}$
1024×1024	9,244 sec. (est.)	657.09 sec.	$7.7 \cdot 10^{-4}$
$\cdots$	$\cdots$	$\cdots$	$\cdots$
$2^{15} \times 2^{15}$	9.6 years (est.)	1 day (est.)	

TABLE 3.

We now present an example. The table above contains timings and accuracy comparison of the construction of the generalized inverse via the singular value decomposition (SVD), which is $O(N^3)$ procedure, and via the iteration (9.1)–(9.2) in the wavelet basis using Fast Wavelet Transform (FWT). The computations were performed on Sun Sparc workstation and we used a routine from LINPACK for computing the singular value decomposition. For tests we used the following full rank matrix

$$A_{ij} = \begin{cases} \frac{1}{i-j} & i \neq j \\ 1 & i = j \end{cases},$$

where $i, j = 1, \ldots, N$. The accuracy theshold was set to 10^{-4}, i.e., entries of X_k below 10^{-4} were systematically removed after each iteration.

X   SOME PRELIMINARY RESULTS AND DIRECTIONS OF RESEARCH

In this section we describe several iterative algorithms indicating that numerical functional calculus with operators can be implemented efficiently (at least for pseudo-differential operators). Numerical results and relative performance of these algorithms will be reported separately.

Remark on iterative computation of the projection operator on the null space.

We present here a fast iterative algorithm for computing P_{null} for a wide class of operators compressible in the wavelet bases.

Let us consider the following iteration

$$X_{k+1} = 2X_k - X_k^2 \qquad (10.1)$$

with

$$X_0 = \alpha A^* A, \qquad (10.2)$$

where A^* is the adjoint matrix and α is chosen so that the largest eigenvalue of $\alpha A^* A$ is less than two.

Then $I - X_k$ converges to P_{null}. This can be shown either directly or by combining an invariant representation for $P_{null} = I - A^*(AA^*)^\dagger A$ with the iteration (9.1)–(9.2) to compute the generalized inverse $(AA^*)^\dagger$. The fast multiplication algorithm makes the iteration (10.1)–(10.2) fast for a wide class of operators (with the same complexity as the algorithm for the generalized inverse). The important difference is, however, that (10.1)-(10.2) does not require compressibility of the inverse operator but only of the powers of the operator.

Remark on iterative computation of a square root of an operator.

Let us describe an iteration to construct both $A^{1/2}$ and $A^{-1/2}$, where A is, for simplicity, a self-adjoint and non-negative definite operator. We consider the following iteration

$$Y_{l+1} = 2Y_l - Y_l X_l Y_l, \qquad (10.3)$$

$$X_{l+1} = \frac{1}{2}(X_l + Y_l A), \qquad (10.4)$$

with

$$Y_0 = \frac{\alpha}{2}(A + I),$$

$$X_0 = \frac{\alpha}{2}(A + I), \qquad (10.5)$$

where α is chosen so that the largest eigenvalue of $\frac{\alpha}{2}(A+I)$ is less than $\sqrt{2}$.

The sequence X_l converges to $A^{1/2}$ and Y_l to $A^{-1/2}$. By writing $A = V^*DV$, where D is a diagonal and V is a unitary, it is easy to verify that both X_l and Y_l can be written as $X_l = V^*P_lV$ and $Y_l = V^*Q_lV$, where P_l and Q_l are diagonal and

$$Q_{l+1} = 2Q_l - Q_lP_lQ_l,$$

$$P_{l+1} = \frac{1}{2}(P_l + Q_lD), \qquad (10.6)$$

with

$$Q_0 = \frac{\alpha}{2}(D + I),$$

$$P_0 = \frac{\alpha}{2}(D + I). \qquad (10.7)$$

Thus, the convergence need to be checked only for the scalar case, which we leave to the reader. If the operator A is a pseudo-differential operator, then the iteration (10.3)–(10.4) leads to a fast algorithm due to the same considerations as in the case of the generalized inverse in Section IX.

Remark on fast algorithms for exponential, sine and cosine of a matrix

The exponential of a matrix (or an operator), as well as sine and cosine functions are among the first to be considered in any calculus of operators. In this section we present a fast algorithm for computing the exponential, cosine and sine functions of a matrix. Again, as in the case of the generalized inverse, we use previously known algorithms (see e.g. [19]) which obtain completely different complexity estimates when we use them in conjunction with the wavelet representations. We do not study these algorithms in detail since our main goal in this paper is limited to pointing out the advantages of computing in the wavelet bases.

The algorithm for the exponential is based on the identity

$$\exp(A) = \left[\exp(2^{-L}A)\right]^{2^L}. \qquad (10.8)$$

First, $\exp(2^{-L}A)$ is computed by, for example, using the Taylor series. The number L is chosen so that the largest singular value of $2^{-L}A$ is less than one. At the second stage of the algorithm the matrix $2^{-L}A$ is squared L times to obtain the result.

Similarly, sine and cosine of a matrix can be computed using the elementary double-angle formulas. On denoting

$$Y_l = \cos(2^{l-L}A) \tag{10.9}$$
$$X_l = \sin(2^{l-L}A), \tag{10.10}$$

we have for $l = 0, \ldots, L-1$

$$Y_{l+1} = 2Y_l^2 - I \tag{10.11}$$
$$X_{l+1} = 2Y_lX_l,, \tag{10.12}$$

where I is the identity. Again, we choose L so that the largest singular value of $2^{-L}A$ is less than one, compute the sine and cosine of $2^{-L}A$ using the Taylor series, and then use (10.11) and (10.12).

Ordinarily, such algorithms require at least $O(N^3)$ operations, since a number of multiplications of dense matrices has to be performed [19]. Fast multiplication algorithm of Section VIII reduces complexity to not more than $O(N \log^2 N)$ operations.

To acheive such perfomance it is necessary to maintain the "finger" band structure of the standard form throughout the iteration. Whether it is possible to do depends on the particular operator and, usually, can be verified analytically.

Unlike the algorithm for the generalized inverse, the algorithms of this remark are not self-correcting. Thus, it is necessary to maintain sufficient accuracy initially so as to obtain the desired accuracy after all the multiplications have been performed.

REFERENCES

[1] Y. Meyer, Ondelettes et Operateurs, Hermann, Paris, 1990.

[2] V. Rokhlin, *Rapid Solution of Integral Equations of Classical Potential Theory*, Journal of Computational Physics, vol. 60, 2, 1985.

[3] L. Greengard and V. Rokhlin, *A Fast Algorithm for Particle Simulations*, Journal of Computational Physics, 73(1), **325**, 1987.

[4] J. Carrier, L. Greengard and V. Rokhlin *A Fast Adaptive Multipole Algorithm for Particle Simulations*, Yale University Technical Report, YALEU/DCS/RR-496 (1986), SIAM Journal of Scientific and Statistical Computing, 9 (4), 1988.

[5] G. Beylkin, R. R. Coifman and V. Rokhlin, *Fast wavelet transforms and numerical algorithms I*. Yale University Technical Report YALEU/DCS/RR-696, August 1989, Comm. on Pure and Applied Math., vol. **XLIV**, 141-183, 1991.

[6] B. Alpert, G. Beylkin, R. R. Coifman and V. Rokhlin, *Wavelets for the fast solution of second-kind integral equations*, Technical report, Department of Computer Science, Yale University, New Haven, CT, 1990.

[7] R. R. Coifman, Y. Meyer and V. Wickerhauser, *Wavelet Analysis and Signal Processing*, this volume.

[8] G. Beylkin, *On the representation of operators in bases of compactly supported wavelets*, preprint, to appear in SIAM J. on Numerical Analysis.

[9] I. Daubechies, *Orthonormal Bases of Compactly Supported Wavelets*, Comm. Pure and Applied Math., **XL1**, 1988.

[10] Y. Meyer, *Wavelets and Operators*, Analysis at Urbana, vol.1, edited by E. Berkson, N.T. Peck and J. Uhl, London Math. Society, Lecture Notes Series 137, 1989.

[11] S. Mallat, *Review of Multifrequency Channel Decomposition of Images and Wavelet Models*, Technical Report 412, Robotics Report 178, NYU (1988).

[12] I. Daubechies, Ten Lectures on Wavelets, CBMS-NSF Series in Applied Mathematics, SIAM, (1991 in press).

[13] R. Coifman and Yves Meyer, *Non-linear Harmonic Analysis, Operator Theory and P.D.E.*, Annals of Math Studies, Princeton, 1986, ed. E. Stein.

[14] B. Alpert, *Sparse Representation of Smooth Linear Operators*, PhD thesis, Yale University, 1990.

[15] Y. Meyer, Ondelettes et functions splines. Technical Report, Séminaire EDP, Ecole Polytechnique, Paris, France, 1986.

[16] S. Mallat, *Multiresolution approximation and wavelets*, Technical report, GRASP Lab, Dept. of Computer and Information Science, University of Pennsylvania.

[17] G. Schulz, *Iterative Berechnung der reziproken Matrix*, Z. Angew. Math. Mech. 13, 57-59, 1933.

[18]   G. Beylkin, R. R. Coifman and V. Rokhlin, *Fast wavelet transforms and numerical algorithms II.*, in progress.

[19]   R. C. Ward, *Numerical computation of the matrix exponential with accuracy estimates*, SIAM. J. Numer. Anal. vol.14, 4, 600-610, 1977.

Construction of Simple Multiscale Bases for Fast Matrix Operations

BRADLEY K. ALPERT

Lawrence Berkeley Laboratory
and
Department of Mathematics
University of California
Berkeley, California

Several common matrix operations often dominate the computer time in scientific computations in which they occur. These operations include the application of a matrix to a vector (matrix-vector multiplication), the application of the inverse of a matrix to a vector, and the addition or multiplication of two matrices. For arbitrary $n \times n$-matrices, these operations each require computer time of order $O(n^2)$ to order $O(n^3)$, and typical scientific computations require many repetitions of these operations. The resulting computation times are often prohibitive and sparse matrix techniques have been developed to reduce these costs in cases where the matrices involved are sparse, *i.e.*, in which most of the matrix elements are zero.

In recent years a number of numerical algorithms have been developed ([7], [10], [11], [14]) in which operators of the type arising in potential theory are represented with sparse constructions. Through various schemes in which a dense (non-sparse) matrix is represented with relatively few elements, these algorithms effectively achieve application of a dense $n \times n$-matrix to an arbitrary vector in order $O(n)$ operations. Yet more recently, algorithms have been constructed ([3], [6], [8]) which explicitly transform dense matrices representing integral operators to sparse matrices in order $O(n)$ or $O(n \log n)$ operations. The application of these sparse matrices to vectors is similarly fast.

Perhaps of even greater interest is the fact that the inverses of these sparse matrices, when they exist, are also sparse and can be obtained in order $O(n)$ or $O(n \log^2 n)$ operations. As a result, the application of the inverse of a dense matrix to a vector can be made fast, and a variety of integral equations can be solved rapidly.

The transformation of dense matrices to sparse ones is accomplished by a coordinate transformation, in which the revised coordinates, or basis, is a basis of wavelets. For these applications the essential properties of the basis is that it consists of functions that (generally) are

(1) non-zero on finite intervals of various lengths, and

(2) orthogonal to low-order polynomials.

These two properties combine so that "smooth" operators are transformed to sparse matrices. The defining property of wavelet bases, that a basis should consist of a single basic shape which is identical on all scales, is not essential for these applications. In this chapter we outline the construction of bases which retain the multiscale structure of wavelets, but in which the requirement of a single basic shape is discarded. The additional flexibility thereby obtained enables construction of simple bases for $L^2[0,1]$, as well as bases for the finite-dimensional space of functions defined on a set of points $\{x_1, \ldots, x_n\} \subset [0,1]$. In addition, we construct bases (in one dimension) for which the transformed matrices can be factored into sparse lower and upper-triangular matrices. The latter construction leads to the rapid solution of certain first-kind integral equations not treated by other algorithms.

In Section 1. we present a sketch of the construction of these classes of bases (details may be found in the references). In Section 2. we state the fundamental analytical properties of the bases which lead to the sparse representation of integral operators. In Section 3. we present several numerical examples, and in Section 4. we give a few concluding remarks.

1. CONSTRUCTION OF SIMPLE MULTISCALE BASES

We begin this section with the construction in Subsection 1.1. of a class of bases for $L^2[0,1]$. The class is indexed by k, a positive integer, which denotes the number of vanishing moments of the basis functions: we say a basis $\{b_1, b_2, b_3, \ldots\}$ from this class is of *order* k if

$$\int_0^1 b_i(x)\, x^j\, dx = 0, \qquad j = 0, \ldots, k-1,$$

for each b_i with $i > k$. We will see that in addition to several vanishing moments, most basis functions b_i are non-zero only on small subintervals of $[0, 1]$. In Subsection 1.2. we show how a slightly different point of view leads to a class of bases for functions defined on a discrete set of points $\{x_1, \ldots, x_n\}$. Finally, in Subsection 1.3. we show how the constructions of bases defined in Subsections 1.1. and 1.2. can be revised to yield bases supporting sparse factorizations.

1.1.   BASES FOR $L^2[0, 1]$

We employ the multi-resolution analysis framework developed by Mallat [12] and Meyer [13], and discussed in detail by Daubechies [9]. For $m = 0, 1, 2, \ldots$ and $i = 0, 1, \ldots, 2^m - 1$ we define a half-open interval $I_{m,i} \subset [0, 1)$ by the formula

$$I_{m,i} = \left[2^{-m}i, 2^{-m}(i+1)\right). \tag{1}$$

For a fixed m, the dyadic intervals $I_{m,i}$ are disjoint and their union is $[0, 1)$; also $I_{m,i} = I_{m+1,2i} \cup I_{m+1,2i+1}$. Now we suppose that k is a positive integer and for $m = 0, 1, 2, \ldots$ and $i = 0, 1, \ldots, 2^m - 1$ we define a space $S^k_{m,i}$ of piecewise polynomial functions,

$$S^k_{m,i} = \{f: \quad f: \mathbf{R} \to \mathbf{R}, \text{ the restriction of } f \text{ to the interval} \tag{2}$$
$$I_{m,i} \text{ is a polynomial of degree less than } k, \text{ and}$$
$$f \text{ vanishes elsewhere}\}$$

and we further define the space S^k_m by the formula

$$S^k_m = S^k_{m,0} \oplus S^k_{m,1} \oplus \cdots \oplus S^k_{m,2^m-1}.$$

It is apparent that for each m and i the space $S^k_{m,i}$ has dimension k, the space S^k_m has dimension $2^m k$, and

$$S^k_{m,i} \subset S^k_{m+1,2i} \oplus S^k_{m+1,2i+1};$$

thus

$$S^k_0 \subset S^k_1 \subset \cdots \subset S^k_m \subset \cdots.$$

For $m = 0, 1, 2, \ldots$ and $i = 0, 1, \ldots, 2^m - 1$, we define the k-dimensional space $R^k_{m,i}$ to be the orthogonal complement of $S^k_{m,i}$ in $S^k_{m+1,2i} \oplus S^k_{m+1,2i+1}$,

$$S^k_{m,i} \oplus R^k_{m,i} = S^k_{m+1,2i} \oplus S^k_{m+1,2i+1}, \qquad R^k_{m,i} \perp S^k_{m,i},$$

and we further define the space R^k_m by the formula

$$R^k_m = R^k_{m,0} \oplus R^k_{m,1} \oplus \cdots \oplus R^k_{m,2^m-1}.$$

Now we have $S_m^k \oplus R_m^k = S_{m+1}^k$, so we inductively obtain the decomposition

$$S_m^k = S_0^k \oplus R_0^k \oplus R_1^k \oplus \cdots \oplus R_{m-1}^k. \tag{3}$$

Suppose that functions $h_1, \ldots, h_k : \mathbf{R} \to \mathbf{R}$ form an orthogonal basis for R_0^k. Since R_0^k is orthogonal to S_0^k, the first k moments of $h_1, \ldots, h_k$ vanish,

$$\int_0^1 h_i(x)\, x^j\, dx = 0, \qquad j = 0, 1, \ldots, k-1.$$

The space $R_{m,i}^k$ then has an orthogonal basis consisting of the k functions $h_1(2^m x - i), \ldots, h_k(2^m x - i)$, which are non-zero only on the interval $I_{m,i}$; furthermore, each of the functions has k vanishing moments. Introducing the notation $h_{m,i}^j$ for $j = 1, \ldots, k$, $m = 0, 1, 2, \ldots$, and $i = 0, 1, \ldots, 2^m - 1$, by the formula

$$h_{m,i}^j(x) = h_j(2^m x - i), \qquad x \in \mathbf{R},$$

we obtain from decomposition (3) the formula

$$\begin{aligned} S_m^k = S_0^k \oplus \text{linear span } \{ h_{m,i}^j : \quad & j = 1, \ldots, k; \\ & m = 0, 1, 2, \ldots; \\ & i = 0, 1, \ldots, 2^m - 1 \}. \end{aligned} \tag{4}$$

An explicit construction of $h_1, \ldots, h_k$ is given in [3].

We define the space S^k to be the union of the S_m^k, given by the formula

$$S^k = \bigcup_{m=0}^{\infty} S_m^k, \tag{5}$$

and observe that $\overline{S^k} = L^2[0,1]$. In particular, S^1 contains the Haar basis for $L^2[0,1]$, which consists of functions piecewise constant on each of the intervals $I_{m,i}$. Here the closure $\overline{S^k}$ is defined with respect to the L^2-norm. We let $\{u_1, \ldots, u_k\}$ denote any orthogonal basis for S_0^k; in view of (4) and (5), the orthogonal system

$$\begin{aligned} B_k = \quad & \{ u_j : \quad j = 1, \ldots, k \} \\ & \cup \{ h_{m,i}^j : \quad j = 1, \ldots, k; \ m = 0, 1, 2, \ldots; \ i = 0, \ldots, 2^m - 1 \} \end{aligned}$$

spans $L^2[0,1]$; we refer to B_k as the *multi-wavelet basis of order k* for $L^2[0,1]$.

In [3] it is shown that B_k may be readily generalized to bases for $L^2(\mathbf{R})$, $L^2(\mathbf{R}^d)$, and $L^2[0,1]^d$.

1.2. BASES FOR DISCRETELY DEFINED FUNCTIONS

The bases B_k described in Subsection 1.1. can be revised somewhat to yield bases for the n-dimensional space of functions defined on a set of points $X = \{x_1, \ldots, x_n\} \subset [0,1]$, where $x_1 < \cdots < x_n$. In the following development, for simplicity we assume that $n = 2^l k$, where l and k are positive integers. Analogous to the intervals $I_{m,i}$ defined in (1), for $m = 0, 1, \ldots, l$ and $i = 0, 1, \ldots, 2^m - 1$ we define the sets $X_{m,i}$ by the formula

$$X_{m,i} = \left\{ x_{n2^{-m}i+1}, \, x_{n2^{-m}i+2}, \ldots, x_{n2^{-m}(i+1)} \right\}.$$

We again assume that k is a positive integer and for $m = 0, 1, \ldots, l$ and $i = 0, 1, \ldots, 2^m - 1$ we define the k-dimensional space $U_{m,i}^k$ by the formula

$$U_{m,i}^k = \{f : \; f : X \to \mathbf{R}, \text{ the restriction of } f \text{ to the set}$$
$$X_{m,i} \text{ is a polynomial of degree less than } k,$$
$$\text{and } f \text{ vanishes elsewhere}\}$$

and we further define the $2^m k$-dimensional space U_m^k by the formula

$$U_m^k = U_{m,0}^k \oplus U_{m,1}^k \oplus \cdots \oplus U_{m,2^m-1}^k.$$

Clearly, as for $S_{m,i}^k$, we have the inclusions

$$U_{m,i}^k \subset U_{m+1,2i}^k \oplus U_{m+1,2i+1}^k$$

and

$$U_0^k \subset U_1^k \subset \cdots \subset U_l^k.$$

For $m = 0, \ldots, l-1$ and $i = 0, 1, \ldots, 2^m - 1$, we define the k-dimensional space $T_{m,i}^k$ (analogous to $R_{m,i}^k$) to be the orthogonal complement of $U_{m,i}^k$ in $U_{m+1,2i}^k \oplus U_{m+1,2i+1}^k$,

$$U_{m,i}^k \oplus T_{m,i}^k = U_{m+1,2i}^k \oplus U_{m+1,2i+1}^k, \qquad T_{m,i}^k \perp U_{m,i}^k,$$

and we further define the space T_m^k by the formula

$$T_m^k = T_{m,0}^k \oplus T_{m,1}^k \oplus \cdots \oplus T_{m,2^m-1}^k.$$

Now we have $U_m^k \oplus T_m^k = U_{m+1}^k$, so we inductively obtain the decomposition

$$U_l^k = U_0^k \oplus T_0^k \oplus T_1^k \oplus \cdots \oplus T_{l-1}^k. \tag{6}$$

This decomposition parallels (3). In both decompositions the constituent subspaces, which reflect various scales, are naturally spanned by "locally" supported basis elements, each orthogonal to low-order polynomials. Although the spaces R_m^k are spanned by the translates and dilates of only

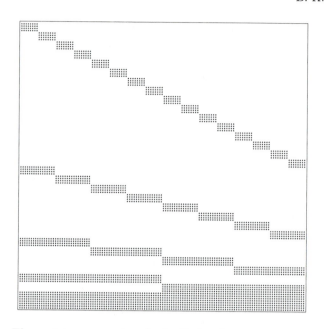

FIGURE 1. The matrix represents a basis C_k for the space of functions defined on a set of points $\{x_1, \ldots, x_n\}$, where $n = 128$ and $k = 4$. Each row denotes one basis vector, with the dots depicting non-zero elements. All but the final k vectors have k vanishing moments.

k basis functions $h_1, \ldots, h_k$, the spaces T_m^k do not have the same scale-invariance. Nevertheless, a basis for the space U_l^k, which reflects the hierarchical decomposition (6), can be constructed by constructing bases for each of the k-dimensional spaces $T_{m,i}^k$ and combining them with a basis for U_0^k. We refer to such a basis as C_k; the construction is illustrated in Figure 1.

It is noteworthy that the construction of C_k requires only n/k orthogonalizations of $2k \times 2k$-matrices and is therefore an order $O(n)$ procedure (see [6] for this construction).

1.3. BASES SUPPORTING LU FACTORIZATION

In Section 2. we will see that certain integral operators, whose kernels are smooth (and non-oscillatory) except at diagonal singularities, are represented as sparse matrices in the coordinates B_k or C_k. Also, the inverses of these matrices, when they exist, are sparse. It would be convenient if in addition these matrices could be factored into sparse lower and upper-triangular matrices.

Integral operators expanded in bases B_k and C_k do not directly admit sparse LU-factorizations. Bases with this property do exist, however, and we outline their construction next. We construct bases (analogous to C_k) for the space U_l^k of functions defined on $\{x_1, \ldots, x_n\}$; similar bases exist for the spaces S_m^k.

The idea of the construction is to create subspaces of U_l^k of various scales, analogous to the T_m^k, but whose basis vectors are supported on *separated* sets of points in X; the construction decomposes the spaces T_m^k. For the following definitions, we assume that the points $x_1, \ldots, x_n$ are equispaced. For $m = 1, \ldots, l-1$ and $i = 1, 3, 5, \ldots, 2^m - 1$, we define the space $\tilde{T}_{m,i}^k$ by the formula

$$\tilde{T}_{m,i}^k = T_{m,i}^k \oplus T_{m+1,2i}^k \oplus T_{m+2,4i}^k \oplus \cdots \oplus T_{l-1,2^{l-1-m}i}^k,$$

we define $\tilde{T}_m^k$ by the formula

$$\tilde{T}_m^k = \tilde{T}_{m,1}^k \oplus \tilde{T}_{m,3}^k \oplus \tilde{T}_{m,5}^k \oplus \cdots \oplus \tilde{T}_{m,2^m-1}^k,$$

and we define $\tilde{T}_0^k$ by the formula

$$\tilde{T}_0^k = T_{0,0}^k \oplus T_{1,0}^k \oplus \cdots \oplus T_{l-1,0}^k.$$

We now have the analogue of (6), namely

$$U_l^k = U_0^k \oplus \tilde{T}_0^k \oplus \cdots \oplus \tilde{T}_{l-1}^k, \tag{7}$$

which expresses our intended decomposition. As is the case for (6), which gives rise to the basis C_k, a basis D_k of the space U_l^k that reflects the structure in (7) can be constructed by constructing bases for the $\tilde{T}_{m,i}^k$. The basic properties of such a basis will be discussed in Section 2.. At this point we mention, however, that one such basis, illustrated in Figure 2, can be obtained by a straightforward reordering of the elements of C_k.

2. ANALYTICAL PROPERTIES OF THE BASES

We now describe how the wavelet-like bases B_k, C_k, and D_k whose construction is outlined in Section 1. produce sparse matrix representations for many integral operators. The main idea is the following: a function which is analytic except at a finite set of singularities can be well approximated on intervals separated from the singularities by low-order polynomials. An interval separated from the singularities is a line segment in the complex plane (generally lying on the real axis) whose length is less than its distance to the nearest singularity. We say that a function is well approximated by low-order polynomials on such intervals if the relative error of Chebyshev interpolation on each interval decays exponentially and uniformly in the

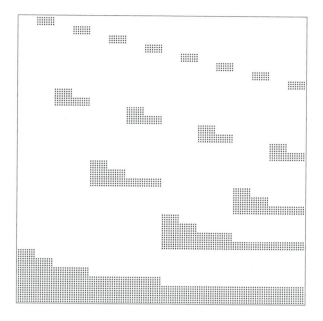

FIGURE 2. The matrix represents a basis D_k, for $n = 128$ and $k = 4$ (compare with Figure 1). The limited interaction between basis functions on one scale and between scales leads to sparse LU factorizations of integral operators represented in these bases.

order of the interpolations. The singularities are assumed to be poles or branch points. An illustration of this notion is given in Figure 3.

The ability to locally approximate by low-order polynomials any function with a finite number of singularities leads to the efficiency of function representation in the bases B_k, C_k, and D_k. For such a function f, if the order k is chosen based on the required precision, a basis element of B_k whose interval of support is separated from the singularities of f (and is orthogonal to polynomials of degree less than k) receives negligible projection from f. Consider now an integral operator $\mathcal{K}$,

$$(\mathcal{K}g)(x) = \int_0^1 K(x, t)\, g(t)\, dx,$$

whose kernel K is an analytic function of x and t except at $x = t$, where it is singular. (Here the function $K(x, t)$, for fixed x or fixed t, is the analogue of the function f.) We can approximate the integral with the trapezoidal rule, which results in a system of linear equations. For a positive integer n and equispaced $x_1, \ldots, x_n$ on $[0, 1]$, we define an $n \times n$ matrix $L = L(n)$

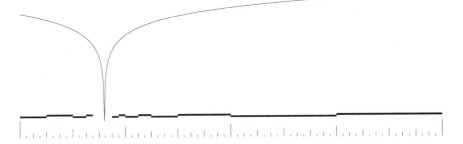

FIGURE 3. The function $f(x) = \ln|x - .2|$ is graphed on the interval $[0, 1]$, which is divided into dyadic subintervals. On each subinterval separated from the singularity (indicated by solid line segments), the function can be represented by a polynomial of order 7 to six-digit accuracy.

with elements L_{ij} given by the formula

$$L_{ij} = \begin{cases} K(x_i, x_j)/n & i \neq j \\ 0 & i = j. \end{cases}$$

The sequence of matrices $\{L(n)\}_{n \in \mathbf{Z}^+}$ converges to $\mathcal{K}$. Although the rate of convergence is low, a change to endpoint-corrected quadratures [4] can be used to greatly accelerate the convergence. Each matrix L can be subdivided into squares separated from the diagonal, as shown in Figure 4, in which the elements can be approximated by low-order polynomials. As a result, transformation of L to the basis C_k yields a sparse matrix. In particular, given a precision ϵ, the order k can be chosen so that, up to ϵ, the number of non-negligible elements in the matrices $\{L(n)\}$ grows linearly in n.

Next we consider two integral operators represented as sparse matrices in the basis C_k, possessing kernels K_1 and K_2 which are analytic except at diagonal singularities. The kernel K_3 of their product, given by the formula

$$K_3(x, t) = \int_0^1 K_2(x, y) K_1(y, t) \, dy,$$

is itself analytic except for $x = t$, and therefore has a sparse representation in the basis C_k. The product matrix can be obtained for a cost of order $O(n)$ simply by multiplying the matrices of the two operators! This fact makes the Schulz method, a Newton-iteration-like scheme for inverting matrices, useful in practice [6].

All invertible integral operators of the above type have asymptotically sparse inverses (the number of non-negligible elements of the $n \times n$-matrix representing the inverse operator is proportional to n). In particular, a

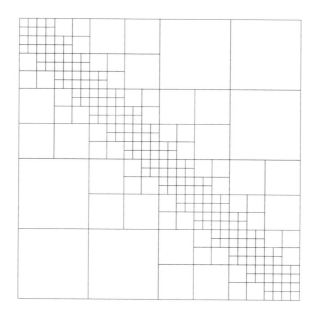

FIGURE 4. The matrix represents a discretized integral operator with a kernel that is singular along the diagonal. The matrix is divided into submatrices of rank k (to high precision) and transformed to a sparse matrix with $O(n)$ elements. Here $n/k = 32$.

large variety of second-kind integral equations, given by the formula

$$f(x) - \int_0^1 K(x,t)\, f(t)\, dt = g(x),$$

where f is unknown, can be solved rapidly in coordinates C_k by explicitly computing an inverse matrix. For other operators, however, the number of non-negligible elements in the inverse matrix for moderate values of n may be nearly n^2. This lack of sparseness occurs when inverting operators from first-kind integral equations, given by the formula

$$\int_0^1 K(x,t)\, f(t)\, dt = g(x).$$

One solution to this problem would be to obtain an operator's decomposition into a product of lower and upper-triangular matrices, if these matrices are sparse. Operators represented in the bases D_k have such sparse factorizations, due to two properties:

(1) Two basis vectors b_1 and b_2 of the space $\tilde{T}_m^k$ which are non-zero on different sets of points have negligible interaction $b_1\, L\, b_2{}^T$. This

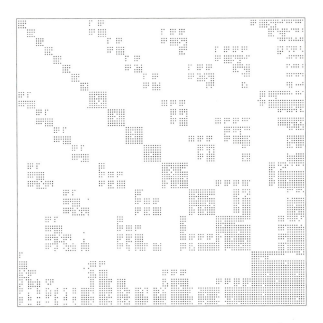

FIGURE 5. The non-negligible matrix elements ($\epsilon = 2 \times 10^{-3}$) are shown for a matrix representing an integral operator in the basis D_k. Note the matrix can be factored into lower and upper-triangular matrices without fill-in (additional elements) provided that pivoting is done only within each block.

follows from the fact that $L\,b_2{}^T$ is negligible outside the $3 \cdot 2^{l-m}k$ points centered on the $2^{l-m}k$ points where b_2 is non-zero.

(2) The interaction of basis vectors of different subspaces has "controlled growth" as one moves up the hierarchy. This property is somewhat complicated to state precisely [2], but is illustrated in Figure 5.

The number of non-negligible elements in $n \times n$-matrices representing integral operators in the basis D_k is potentially of order $O(n \log^2 n)$, but in numerical experiments we have observed slower growth.

Analytical properties of the wavelet-like bases are stated and proved in greater detail in [5].

3. NUMERICAL EXAMPLES

In this section we give several numerical examples using the bases defined in Section 1.. We consider a class of second-kind integral equations with logarithmic kernel,

$$f(x) - p(x) \int_0^1 \log |x - t|\, f(t)\, dt = g_m(x), \qquad x \in [0, 1], \qquad (8)$$

where the right hand side g_m is chosen so that the solution f is given by the formula $f(x) = \sin(mx)$. The integration can be performed explicitly, yielding the formula

$$
\begin{aligned}
\int_0^1 \log|x - t|\,\sin(mt)\,dt = {}& \log(x) - \cos(m)\log(1 - x) \\
& - \cos(mx)[\mathrm{Ci}(mx) - \mathrm{Ci}(m(1 - x))] \\
& - \sin(mx)[\mathrm{Si}(mx) + \mathrm{Si}(m(1 - x))],
\end{aligned}
$$

where Ci and Si are the cosine integral and sine integral, respectively (see, e.g., [1], p. 231). Equation (8) clearly requires quadratures with increasing resolution as m increases; for our examples we let $n = m$, which corresponds to 2π points per oscillation of the right hand side g_m. We use quadratures derived from the trapezoidal rule with endpoint corrections [4], [15]; these quadrature rules converge rapidly.

We consider cases with three different values of the coefficient $p(x)$, namely $p(x) = 1$, $p(x) = 100$, and $p(x) = 1 + \frac{1}{2}\sin(100x)$. The first two cases can be solved using the bases of this chapter, while the third case requires a minor elaboration. For the case $p(x) = 1$, both the operator and its inverse are sparse in the bases C_k and D_k, so either the inverse or the LU factorization can be obtained cheaply. The second case, in which $p(x) = 100$, is poorly-conditioned and the inverse is not as sparse as the operator. For this case, it is dramatically cheaper to obtain the LU factorization.

The third case, with an oscillatory coefficient p, is solved by the construction [6] of bases in which the basis elements are orthogonal to the functions $p(x)^{1/2}x^j$, for $j = 0, \ldots, k-1$, rather than simply the moments x^j. This construction leads to the sparse matrix representation of an integral operator with a highly oscillatory coefficient. As we will see, the inverse is also sparse.

Table 1 presents the numerical results for various coefficients $p(x)$ and various right hand sides $g_m(x)$. For each problem instance, the dense matrix representing the integral operator was transformed to wavelet coordinates C_k reordered for sparse LU factorization. The smallest matrix elements were discarded according to an element threshold determined by the chosen precision ϵ, leaving a sparse matrix representing the integral operator. Second, the inverse matrix, which is also sparse, was computed by the Schulz method. As an alternative method, the LU factorization is obtained by direct Gaussian elimination. Third, the solution is obtained and the error is measured, for each of the two methods: application of the inverse to the right hand side and forward and back-substitution using the LU factorization. The table presents the "bandwidths," computation

TABLE 1. The integral equation $f(x) - p(x) \int_0^1 \log |x - t| f(t)\, dt = g_m(x)$, for which an explicit solution is known, is solved for various coefficients $p(x)$ and various values of m (see text). Precision $\epsilon = 10^{-3}$ and order $k = 4$ were used. The "bandwidths" N_1, N_2, and N_3 denote the average number of elements/row in the sparse matrices representing the operator, its inverse, and its LU factorization, while t_1, t_2, and t_3 denote the time in seconds required to compute these matrices. The condition number of the operator matrix is denoted by κ, and e_2 and e_3 denote the relative L^2 errors of the solution obtained by inversion and LU factorization.

	Operator			Inverse			LU Factorization		
n, m	N_1	t_1	κ	N_2	t_2	e_2	N_3	t_3	e_3
$p(x) = 1$									
64	27.7	3	5	31.3	30	.235E−3	27.7	0	.237E−3
128	31.0	7	4	34.2	80	.169E−3	31.1	1	.171E−3
256	30.6	16	4	33.6	173	.161E−3	30.6	3	.161E−3
512	27.5	33	3	30.2	306	.130E−3	27.5	7	.130E−3
1024	21.7	64	3	24.4	380	.597E−3	21.7	13	.164E−3
2048	15.5	115	3	18.1	487	.479E−3	15.5	15	.117E−3
4096	9.7	199	3	10.6	463	.415E−3	9.7	13	.107E−3
8192	6.0	357	3	7.3	549	.354E−3	6.0	17	.104E−3
$p(x) = 100$									
64	29.1	3	411	63.3	180	.442E−2	29.6	0	.802E−2
128	33.3	8	393	113.0	812	.374E−2	34.3	1	.445E−2
256	34.6	17	386	169.2	3133	.468E−2	35.9	3	.521E−2
512	32.2	38	350	202.1	8991	.337E−2	33.7	6	.352E−2
1024	28.6	78					29.9	13	.433E−2
2048	22.9	159					23.6	28	.599E−2
4096	15.9	316					16.3	47	.516E−2
8192	10.7	628					10.8	32	.382E−2
$p(x) = 1 + \frac{1}{2}\sin(100x)$									
64	36.2	3	4	41.3	52	.228E−2	36.3	0	.228E−2
128	40.8	8	4	47.0	150	.209E−3	40.8	1	.210E−3
256	40.5	18	4	47.3	343	.177E−3	40.5	4	.177E−3
512	34.7	36	4	40.9	574	.125E−3	34.7	9	.125E−3
1024	26.6	69	3	32.5	840	.134E−3	26.7	15	.134E−3
2048	18.7	124	3	22.5	858	.597E−3	18.7	22	.117E−3
4096	12.2	221	3	14.2	910	.529E−3	12.2	26	.100E−3
8192	7.2	394	3	8.4	893	.461E−3	7.2	21	.913E−4

times, and errors associated with the methods, based on our FORTRAN implementation run on a Sun Sparcstation 1+.

We make several observations:

(1) The "bandwidths" of the operator, its inverse, and its LU factorization *decrease* with increasing matrix size. In other words, in the range of matrix sizes tabulated, the number of matrix elements grows more slowly than the matrix dimension n.

(2) The operator matrix in wavelet coordinates is computed in time that grows nearly linearly in n. The inverse matrix and LU factorization are each computed in time which grows sublinearly in n, due to the decreasing "bandwidths" as n increases.

(3) The solution accuracies are generally within the specified precision. Exceptions are attributable to quadrature errors (for $n = 64$) and to poor conditioning of the underlying problem (for $p(x) = 100$). When confronted with large condition numbers, one must obtain the integral operator to higher accuracy than the accuracy required in the solution (as is well known).

(4) The times to compute the LU factorizations are much less than those for the inverses, even when the sparsities are comparable. In the case of the poorly-conditioned problem, the inverse matrices were not very sparse (large sizes exhausted available memory and were not computed), yet LU factorizations remained sparse and inexpensive to compute.

(5) The ability to solve problems requiring an 8192-point discretization in a few minutes on a Sparcstation using these methods should be compared to an estimated 78 days for Gauss-Jordan inversion and 26 days for LU decomposition of an 8192×8192-matrix. One dense matrix-vector multiplication of that size requires roughly 19 minutes.

We summarize these observations by remarking that the LU factorization of an integral operator represented in the wavelet coordinates described above is a highly effective method for the numerical solution of integral equations.

4. SUMMARY

In this chapter we have described a class of bases in which integral operators are represented as sparse matrices. More generally, a dense matrix with elements that are a smooth function of their indices is transformed to a sparse matrix in these bases (to high precision). The sparseness results from the fact that the typical basis element has "local" support and is orthogonal to low-order polynomials. A locally smooth function is therefore concisely represented in these bases.

These sparse matrices can be manipulated rapidly. Matrix operations, including application of a matrix to a vector, application of its inverse to a vector, and matrix-matrix multiplication, scale roughly linearly with the dimension of the matrices. We have presented examples demonstrating the effectiveness of these methods for the solution of a variety of integral equations. We anticipate that these bases will be applied successfully to other matrix computations, including some arising from parabolic and hyperbolic partial differential equations.

ACKNOWLEDGEMENTS

The author would like to thank Vladimir Rokhlin for essential cooperation in all aspects of the research summarized in this chapter. This research was supported in part by the Applied Mathematical Sciences subprogram of the Office of Energy Research, U.S. Department of Energy under Contract DE-AC03-76SF00098. The author's current address is National Institute of Standards and Technology, 325 Broadway, Boulder, CO 80303.

REFERENCES

[1]   M. Abramowitz and I. A. Stegun. *Handbook of Mathematical Functions*. National Bureau of Standards, Washington, DC, 1972.

[2]   B. Alpert. Bases for sparse LU factorization of integral operators (in preparation).

[3]   B. Alpert. A class of bases in $\mathcal{L}^2$ for the sparse representation of integral operators. Technical report LBL-30092, Lawrence Berkeley Laboratory, University of California, Berkeley, CA, 1990.

[4]   B. Alpert. Rapidly-convergent quadratures for integral operators with singular kernels. Technical report LBL-30091, Lawrence Berkeley Laboratory, University of California, Berkeley, CA, 1990.

[5]   B. Alpert. *Sparse Representation of Smooth Linear Operators*. Ph.D. thesis, Yale University, December, 1990. Technical report RR-814, Department of Computer Science.

[6]   B. Alpert, G. Beylkin, R. Coifman, and V. Rokhlin. Wavelets for the fast solution of second-kind integral equations. Technical report RR-

837, Department of Computer Science, Yale University, New Haven, CT, 1990. To appear, *SIAM Journal on Scientific and Statistical Computing*.

[7]   B. Alpert and V. Rokhlin. A fast algorithm for the evaluation of Legendre expansions. *SIAM Journal on Scientific and Statistical Computing*, 12:158–179, 1991.

[8]   G. Beylkin, R. Coifman, and V. Rokhlin. Fast wavelet transforms and numerical algorithms I. *Communications in Pure and Applied Mathematics*, XLIV:141–183, 1991.

[9]   I. Daubechies. Orthonormal bases of compactly supported wavelets. *Communications on Pure and Applied Mathematics*, XLI:909–996, 1988.

[10] L. Greengard and V. Rokhlin. A fast algorithm for particle simulations. *Journal of Computational Physics*, 73:325–348, 1987.

[11] L. Greengard and J. Strain. The fast Gauss transform. *SIAM Journal on Scientific and Statistical Computing*, 12:79–84, 1991.

[12] S. Mallat. Multiresolution approximation and wavelets. Technical report, GRASP Lab., Department of Computer and Information Science, University of Pennsylvania.

[13] Y. Meyer. Ondelettes et functions splines. Technical report, Séminaire EDP, Ecole Polytechnique, Paris, France, 1986.

[14] S. O'Donnell and V. Rokhlin. A fast algorithm for the numerical evaluation of conformal mappings. *SIAM Journal on Scientific and Statistical Computing*, 10:475–487, 1989.

[15] V. Rokhlin. End-point corrected trapezoidal quadrature rules for singular functions. *Computers and Mathematics with Applications*, 20:51–62, 1990.

NUMERICAL RESOLUTION OF NONLINEAR PARTIAL DIFFERENTIAL EQUATIONS USING THE WAVELET APPROACH

J. LIANDRAT[1]

IMST
12 avenue général Leclerc
13003 Marseille, France

V. PERRIER

Laboratoire d'analyse numérique
Centre d'Orsay, Bat 425
91405 Orsay Cedex, France

PH. TCHAMITCHIAN

CPT-CNRS Luminy Case 907-13288
Marseille, France and Faculté Saint
Jérôme Université Aix - Marseille III,
Marseille, France

1. INTRODUCTION

Numerical resolution of nonlinear multidimensional partial differential equations are challenging due to the tremendously high number of degrees of freedom classically required for a satisfactory approximation of the solution.

[1]Research was supported in part by the National Aeronautics and Space Administration under NASA Contract No. NAS1-18605 while the author was in residence at the Institute for Computer Applications in Science and Engineering (ICASE), NASA Langley Research Center, Hampton, VA 23665 and aslo by AFOSR grant No. 90-0093.

227

In standard methods, the number of degrees of freedom is a predetermined number of points or elementary functions that determines once and for all the space in which the solution is approximated. Such approaches are problematic, or at least not optimal, when singularities develop in the solution. To address this phenomenon, which occurs locally in the space domain and eventually in the time domain (examples are the occurence of shock waves), a domain decomposition is often proposed. The problems are the detection and tracking of the different regions. In many methods, numerical difficulties are introduced with the definition of new projection or extrapolation operators or are linked to the requirement of extra conditions on the boundaries of the different domains.

Our goal in this paper is to derive a fast algorithm that can easily and dynamically in time and space define the set to search for an aproximate solution. The method proposed is based on orthonormal families of wavelets. It uses extensively the localization properties of the wavelets in the Fourier and space domains and the fast algorithms (tree algorithms) connected to their construction.

The first Section of this paper identifies some preliminary properties of orthonormal wavelets. In the second Section, the algorithm is derived in the general framework of nonlinear, constant coefficient and diffusion partial differential equations. Numerical tests are presented in the last Section in the case of the regularized Burgers equation in one dimension and with periodic boundary conditions.

2. PRELIMINARY PROPERTIES OF ORTHONORMAL WAVELETS

The natural framework for the construction of orthonormal bases of wavelets is given by a multiresolution analysis. This new concept, elaborated by S. Mallat and Y. Meyer, is also well adapted to the approximation of functions for numerical purposes.

The existence of fast wavelet transforms based on the class of tree algorithms, is also a key point (see [Mallat 1988] and [Daubechies 1990]).

For main definitions and properties of orthonormal wavelets, the reader must refer to I. Daubechies series of lectures [Daubechies 1991] and to Y. Meyer's book ([Meyer 1990]). We will only recall the more specific properties needed for our purposes.

2.1. MULTIRESOLUTION ANALYSIS

We consider a multiresolution analysis of $L^2(\mathbb{R})$ using the notations and definitions of Y. Meyer (i.e. the family of closed subspaces $V_j, j \in \mathbb{Z}$ and the Riesz basis of V_0, $g(x - k), k \in \mathbb{Z}$).

The multiresolution analysis used in this paper is *r-regular* meaning that the function $g(x)$ has the following property:

$$|\partial^\alpha g(x)| \leq C_p(1 + |x|)^{-p} \text{ for } \alpha \leq r, \text{ all } x \in \mathbb{R}, \text{ and all integers } p. \quad (2.1)$$

This can be interpreted as a quantification of the localization properties of the function g in the physical space ($\alpha = 0$) and in the Fourier space ($\alpha \geq 1$)

2.2. WAVELETS AND TREE ALGORITHMS

Using a Gram algorithm, an orthonormal basis of V_j, invariant under translation, can be constructed. It yields $\phi_{jk}(x) = 2^{j/2}\phi(2^j x - k), k \in \mathbb{Z}$. ϕ is called, according to S. Mallat, the scaling function. It satifies the same regularity property (2.1) as g. Numerically, a scaling function is similar to a low pass function.

If one introduces the space W_j defined as $V_j \oplus W_j = V_{j+1}$, then one obtains the following theorem proved by S. Mallat and Y. Meyer:

Theorem 1. **(Construction of the wavelet family)** *There exists a function ψ of W_0 such that $\{\psi(x - k), k \in \mathbb{Z}\}$ is an orthonormal basis of W_0. ψ has the same regularity properties (2.1) as g, and ψ is an oscillating function:*

$$\int x^k \psi(x)dx = 0 \quad \text{for } 0 \leq k \leq r. \quad (2.2)$$

Then the family of functions $\{\psi_{jk}(x) = 2^{j/2}\psi(2^j x - k), k \in \mathbb{Z}\}$ is an orthonormal family of W_j and, as $L^2(\mathbb{R}) = \oplus_{j \in \mathbb{Z}} W_j$, $\{\psi_{jk}(x), j \in \mathbb{Z}, k \in \mathbb{Z}\}$ is a hilbertian basis of $L^2(\mathbb{R})$. This family is an orthonormal wavelets family. Combined with (2.1) applied to ψ for $\alpha > 0$, equation (2.2) characterizes the localization of the wavelet generating function in the Fourier space. Numerically, a wavelet is very close to a band limited function.

A very attractive property of wavelets depends on the hierarchical approach that leads directly to the fast algorithms. Indeed, the algorithms used to compute the wavelet coefficients, i.e. $< U, \psi_{jk} >$, where $< ., . >$ stands for the standard $L^2(\mathbb{R})$ scalar product, are tree algorithms. Thanks to the scaling covariance property of the transform, if

$U(x) = \sum_k c_{j+1,k}\phi_{j+1,k}$, one has $< U, \psi_{jl} >= \sum_k c_{j+1,k}\Xi(2l - k)$ and
$< U, \phi_{jl} >= \sum_k c_{j+1,k}\Xi'(2l - k)$ where $\Xi(i)$ and $\Xi'(i)$ are two discrete
filters. This allows a fast computation of the wavelet coefficients at every
scale.

More generally, this property is available for the computation of the
scalar product $< U, \theta_{jk} >$ as soon as θ_{jk} is orthogonal to the set generated
by $\psi_{j'l}$, $j' \geq j + p$. Indeed, with this hypothesis one has $< U, \theta_{jk} >=$
$\sum_l c_{j'l}\Xi_{jj'}(2^{j'-j}l - k)$ where $\Xi_{jj'}(i)$ is a discrete filter.

2.3. APPROXIMATION PROPERTIES OF WAVELETS

The adaptative philosophy of our algorithm is based on the following properties of the wavelet transform [Meyer 1990]:

Theorem 2.   (Characterization of the Sobolev Space $H^s(\mathbb{R})$)   *If
f belongs to $H^{-r}(\mathbb{R})$, V_j, $j \in \mathbb{Z}$ is a r-multiresolution analysis of $L^2(\mathbb{R})$
and $-r \leq s \leq r$ then f belongs to $H^s(\mathbb{R})$ if and only if $\Pi_{V_0}(f) \in L^2(\mathbb{R})$
and $||\Pi_{W_j}(f)||_2 = \varepsilon_j 2^{-js}, j \in \mathbb{N}$, where $\varepsilon_j \in l^2(\mathbb{N})$. ($\Pi_X(f)$ stands for the
orthogonal projection of f on X)*

Theorem 3.   (Characterization of Lipschitz functions)   *A func-
tion f belonging to $L^2(\mathbb{R})$ is α Lipschitz with $0 < \alpha < 1$ for all points of an
open interval if and only if for all (j, k) such that $k2^{-j}$ is in the interval
one has:*

$$| < f, \psi_{jk} > | = O(2^{-j(\alpha+1/2)})$$

From a numerical point of view, one only deals with a finite number
of data samples, for instance, point values or discrete wavelet coefficients.
Access to the asymptotic behaviours is then not available directly. Usually, the analyzed function f is replaced by one of its projections, $\tilde{f}$ on a
space V_{j_M}. If the projection is orthogonal, than the above theorems provide
information on the decreasing of the coefficients for $j \leq j_M - 1$. When a
collocation projection of f is performed, the result depends on the choice of
the used set of points. In the case of spline wavelets of even order m and the
set of points $k2^{-j_M}, k \in \mathbb{Z}$, (see [Perrier 1991] and [Liandrat et al. 1990]),
an estimate comparable to one of the orthogonal projection is obtained.

2.4. WAVELETS AND ELLIPTIC OPERATORS

Wavelets are not, as the Fourier functions e^{ikx}, eigenfunctions of differential operators. This is due to the mixing in frequencies imposed by the

localization in the physical space. However, despite the constraint of the
Heisenberg uncertainty principle, the r *regular* wavelets also have good
localization in the Fourier space. Consequently, the wavelets leave the differential operators nearly diagonal: i.e., they are not too widely spread out
by the differential operators.

In the case of constant coefficient elliptic differential operators, the following theorem holds (we define the symbol of the differential operator
$L = \sum a_\alpha D^\alpha$ as $\sigma(\omega) = \sum a_\alpha(i\omega)^\alpha$ where $D = \partial/\partial x$ and $i^2 = -1$.):

Theorem 4. **(Action of elliptic operators on wavelets)** *If L is
a constant coefficients elliptic differential operator of second order with
$\sigma(\omega) > 0$ for all $\omega \in \mathbb{R}$, if ψ_{jk}, $(j,k) \in \mathbb{Z}^2$, is a family of r regular
wavelets satisfying (2.1), then the family $\theta_{jk} = (L^{-1})^*(\psi_{jk})$ is a wavelet
family satisfying:*

$$\int x^k \theta(x)dx = 0 \quad \text{for } 0 \leq k \leq r$$

and

$$|\partial^\alpha \theta_{jk}(x)| \leq C'_p \frac{2^{j(\alpha+1/2)}}{\sigma(2^j)}(1 + |2^j x - k|)^{-p} \tag{2.3}$$

for $\alpha \leq r$, all $x \in \mathbb{R}$ and all integers $p \leq 1 + r + \alpha$.

(L^* stands for the adjoint of the operator L)

From a numerical point of view, the localization of $\hat{\theta}_{jk}$ in the Fourier
space makes the θ_{jk} belong, up to a given precision, to a space $V_{j'}$ with
$j' - j = p$ where p is a positive integer. The value of p depends *only* on the
wavelet family, on the operator L and on the precision but is *independant*
of j.

In the physical space, the numerical localization of θ_{jk} "follows" those
of ψ_{jk}.

These two properties are the key points of our algorithm.

3. DESCRIPTION OF THE ADAPTIVE ALGORITHM

3.1. GENERAL DESCRIPTION

Let us consider the general evolution equation:

$$\begin{cases} \frac{\partial U}{\partial t} + LU + G(U) = 0 \\[2mm] U(0,t) = U(1,t) \\[2mm] U = U_0 \text{ for } t = 0 \\[2mm] t \geq 0, 0 \leq x \leq 1. \end{cases} \tag{3.1}$$

where L is a linear, constant coefficients, elliptic operator with $\sigma(\omega) \geq 0$ for all $\omega \in \mathbb{R}$ and G a function of U and its derivatives. An approximated solution U_M, belonging to a trial space of wavelets X_M, is sought as a solution of the equation:

$$\begin{cases} \frac{\partial U_M}{\partial t} + L_M U_M = G_M \\[2mm] U_M(x,0) = U_{0_M}. \end{cases} \tag{3.2}$$

Here L_M is an approximation of L in X_M, G_M stands for $\Pi_{X_M}(G(U_M))$ and U_{0_M} for $\Pi_{X_M}(U_0)$.

Following the notations of the Section 2.4, we introduce the space Y_M as: $Y_M = [(I+CL)^{-1}]^*(X_M)$ and the functions $\theta_{jk} = [(I+CL)^{-1}]^*\psi_{jk}$ with I the identity operator and C a strictly positive constant to be defined later.

Using ψ_{jk} as trial functions and θ_{jk} as test functions, the application of the classical method of weighted residuals (see [Gottlieb and Orzag 1977] and [Canuto et al. (1987)]) provides the following set of equations:

$$\begin{cases} < \frac{\partial U_M}{\partial t}, \theta_{jk} > + < L_M U_M, \theta_{jk} > = < G_M, \theta_{jk} > \\[2mm] U_M(x,0) = U_{0_M}(x). \end{cases} \tag{3.3}$$

The approximation L_M is chosen such that $< L_M U_M, \theta_{jk} > = < LU_M, \theta_{jk} >$. It gives $L_M = P_{X_M} L \Pi_{X_M}$ where P_{X_M} is the projection onto X_M and parallel to the $\overline{\text{orthogonal}}$ of Y_M. The existence of P_{X_M} is not difficult to prove and is left to the reader.

The choice of the constant C, involved in the definition of θ_{jk}, is directly connected to the time discretization scheme used to approximate (3.3). In this presentation, for sake of simplicity, an Euler time discretization scheme, implicit for the elliptic term and explicit for the other term is used. However it must be emphazised that more sophisticated schemes are currently used for the numerical implementation of the algorithm (see Section 3). Equation (3.3) is then discretized as follows:

$$
\begin{cases}
< (I + \Delta t.L)U_M^{n+1}, \theta_{jk} > = < U_M^n + \Delta t.G_M^n, \theta_{jk} > \\
\\
U_M^0 = U_{0M}
\end{cases}
$$

where V^n stands for the approximation of V at the time $n\Delta t$.

Choosing $C = \Delta t$ one obtains:

$$
\begin{cases}
< U_M^{n+1}, \psi_{jk} > = < U_M^n + G_M^n, \theta_{jk} > \\
\\
U_M^0 = U_{0M}
\end{cases}
\tag{3.4}
$$

At each time step, the computation of U_M^{n+1} is then derived from the knowledge of the wavelet coefficients $< U_M^n + G_M^n, \theta_{jk} >$. According to Sections (2.2) and (2.4), these scalar products can be obtained using fast tree algorithms as soon as G_M^n is known in X_M (see [Liandrat et al. 1990] for details). Specific difficulties are encountered when G_M^n is a nonlinear term (as for example in Section 4). In that case, a classical pseudo-spectral scheme has been used. It consists in performing the nonlinear operations in the physical space (using the values of U_M on collocation points) and then in going back to the wavelet coefficients in X_M.

3.2. ADAPTABILITY

The space of approximation X_M has not yet been described precisely. A naive idea would be to take $X_M = V_{j_M}$ where j_M is fixed once and for all. However, the distribution of scales significantly building a given function in the sense provided by its wavelet decomposition has no reason to be regular on $\mathbb{R}$. The distribution of the energetic wavelet coefficients of a pointwise discontinuous function is known to have a cone shape pointing towards the singularity in the small scales. If we choose, for instance, a given precision α on the L^2 norm of the solution of (4.1) and look at the number of wavelets required to approximate the solution up to this precision, it appears that only a small number of wavelets are needed. For instance, with a L^2 norm precision of 10^{-6} the number of wavelets required to reconstruct the solution of (4) at $t = 1/\pi$ (see Figure 1) is 74. The distribution of these "active" wavelets follows the classical cone shape structure.

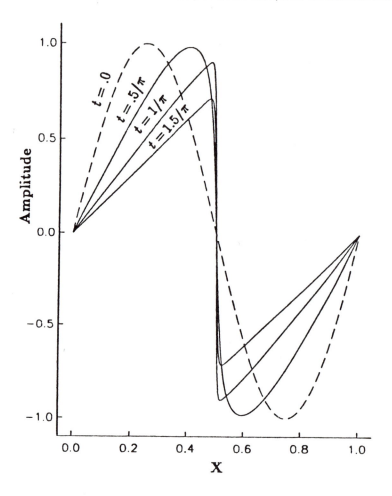

FIGURE 1. Time evolution of the approximated solution. Adaptive algorithm. The maximum number of degrees of freedom during the whole computation is 104

$X_M(n\Delta t)$ will be chosen at each time as close as possible to the set of "active" wavelets. The control of $X_M(n\Delta t)$ (made simple, thanks to orthogonality, by adding or suppressing wavelets) is done by studying the distribution and the evolution in time of the smallest scales of the solution available in $X_M(n\Delta t)$. Figure 2 presents the distribution of the small scales ($j = 6$) for the approximated solution of (4.1) at different times. More precisely, the function $E_S(x, n\Delta t) = |<U_M^n, \psi_{jsk}>|^2$ is plotted where j_S is the smallest scale of $X_M(n\Delta t)$ in the vicinity of x. Clearly, this function is a very precise and sensible indicator for the behaviour of the analyzed function U_M at each point.

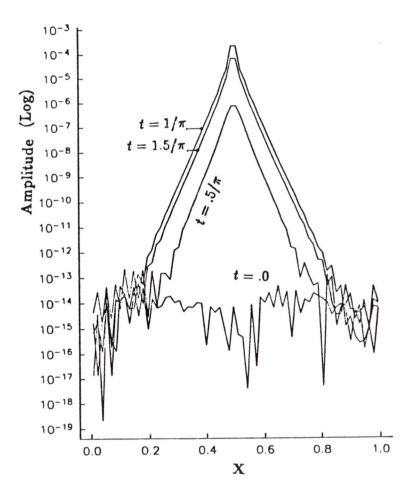

FIGURE 2.   Small scale $(j = 6)$ coefficients of the approximated solution. $(E_S(x, n\Delta t))$

Finally, the wavelet approach provides the two required elements for a flexible adaptable algorithm: the detection of singular regions using the function $E_S(x)$ and the easy management of the quality of approximation of $X_M(n\Delta t)$ due to orthogonality. The time evolution of the adaptable space under the action of differential operators is controled thanks to the localisation properties of wavelets.

4.   NUMERICAL RESULTS

We describe briefly numerical results obtained for Burgers problem with periodic boundary conditions in one dimension. The periodic regularized Burgers equation is:

$$
\begin{cases}
\frac{\partial U}{\partial t} + \frac{U \partial U}{\partial x} = \frac{\nu \partial^2 U}{\partial x^2} \\[2mm]
t \geq 0, x \in [0, 1], \nu = \frac{10^{-2}}{\pi} \\[2mm]
U(0, t) = U(1, t) \\[2mm]
U(x, 0) = \sin(2\pi x).
\end{cases}
\tag{4.1}
$$

The wavelets used for the numerical implementation are the periodic spline wavelets of order $m = 6$ (see [Perrier et al. 1990] and [Liandrat et al. 1990]). All the following results are obtained using an Adams-Bashforth time scheme for the nonlinear convection term and a Crank-Nicholson time scheme for the diffusion term. The integration time step, for every scale is $\Delta t = 10^{-3}$. Figure 1 shows the time evolution of an approximated solution obtained with a computation started with $X_M(0) = V_6$ (that is to say $2^6 = 64$ degrees of freedom) and which was run while using the adaptative procedure. From a time close to $t_a = .5/\pi$ when the local gradients begin to be very large the adaptative procedure adapts progressively the space by adding to $X_M(n\Delta t)$ 20 new wavelets of scale 6 around $x = .5$. For $.66/\pi \leq t \leq 1.4/\pi$, 20 other wavelets of scale 7 are again added around $x = .5$. A maximum of 104 degrees of freedom has then been used for this computation. The corresponding dyadic grid, built from the centers of the basic functions of $X_M(n\Delta t)$, is plotted in Figure 3.

   Table 1 presents a comparison of the numerical results obtained with different algorithms and different options. T_{max} is the time needed to reach the maximum slope at $x = 0.5$, and S_{max} is the value of this slope. Taking into account that no optimization study of the time marching scheme has been made, the general results presented in [Basdevant et al. 1986] indicate the adaptative version of the algorithm can be considered very efficient.

5.   CONCLUSIONS

A new numerical algorithm for resolution of nonlinear and diffusion partial differential equations has been presented. It is based on the approximation properties and the multiscale approach of orthonormal wavelets.

   This algorithm is adaptative in the sense that the space of approxima-

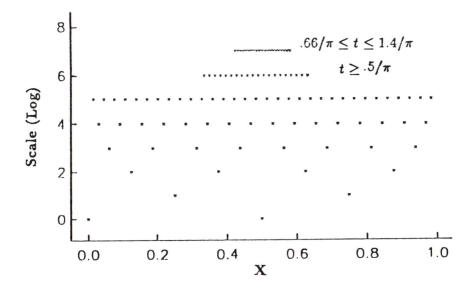

FIGURE 3. Adaptated diadic grid (centers of the wavelets building $X_M(n\Delta t)$)

Algorithm (AB-CN Time scheme)	$2\pi T_{max}$ Exact: 1.6037	$S_{max}/2.$ Exact: 152.005		Degrees of freedom
Present Algorithm. $m = 6$, $X_M(n\Delta t) = V_8$	1.64	150.3	No oscillations	256
Present Algorithm. $m = 6$, $X_M(n\Delta t) = V_7$	1.63	135.0	Localized oscillations	128
Present Algorithm. $m = 6$, $X_M(n\Delta t) \subset V_8$, adaptative	1.64	150.3	No oscillations	≤ 104
Fourier pseudo-spectral $n = 128$	1.62	134.8	Spread oscillations	128

TABLE 1.   Comparison of some different calculations.

tions is fitted at every time with the minimal wavelet space required to decompose the solution. It remains fast thanks to the introduction of the "θ_{jk}" functions that leads to the derivation of a local tree algorithm.

The efficiency of this algorithm has been demonstrated on the 1D periodic Burgers equation. P. Laurencot, [Laurencot, 1990], recently applied this algorithm successfully to the periodic 1D Korteweg-De Vries equation.

The presentation of a 1-D problem with periodic boundary conditions is not connected to limitations of the algorithm or the wavelet approach. This choice has been done for clarity and exposition reasons. Extensions to more realistic problems with different boundary conditions and multidimensions are currently underway.

REFERENCES

[1] C. Basdevant, M. Deville, P. Haldenwang, J. M. Lacroix, J. Ouazzani, and R. Peyret: "Spectral and finite difference solution of the Burgers equation," Computers and Fluids, Vol. 14, No. 1, pp. 23-41 (1986).

[2] C. Canuto, M. Y. Hussaini, A. Quarteroni, and T. A. Zang: "Spectral methods in fluid dynamics", Springer Series in Computational Physics (1987).

[3] I. Daubechies: " Ten Lectures on Wavelets", CBMS-NSF Series in applied mathematics, SIAM (in press 1991).

[4] D. Gottlieb and S. Orszag: "Numerical analysis of spectral methods: Theory and Applications," SIAM-CBMS Philadelphia (1977).

[5] P. Laurencot: "Résolution par ondelettes 1D de l'équation de Burgers avec viscosité et de l'équation de Korteweg-De Vries avec conditions aux limites périodiques", D.E.A report, E.N.S Cachan (1990)

[6] J. Liandrat, Ph. Tchamitchian: " Resolution of the 1D regularized Burgers equation using a spatial wavelet approximation-algorithm and numerical results", ICASE report (1990).

[7] S. Mallat: "Multiresolution representations and wavelets", PhD, MS-CIS-88-68, Grasp Lab Tech. Report 153, University of Pennsylvania, Philadelphia, PA 19104 (1988).

[8] Y. Meyer: "Ondelettes et Opérateurs 1: Ondelettes", Herman (1990).

[9] V. Perrier: "Ondelettes et simulations numériques", PhD Thesis, University of Paris 6 (1991)

IV
OTHER APPLICATIONS

THE OPTICAL
WAVELET
TRANSFORM

A. ARNEODO

F. ARGOUL

J. F. MUZY

B. POULIGNY

Centre de Recherche Paul Pascal,
Avenue Schweitzer 33600 Pessac, France

E. FREYSZ

Centre de Physique Moléculaire
Optique et Hertzienne, Université de
Bordeaux I, 351 Avenue de la Libération
33405 Talence Cedex, France

1. INTRODUCTION

The relevance of *fractals* to physics was stressed by B. Mandelbrot[1] in his seminal book, *The Fractal Geometry of Nature*. Fractals are complex mathematical objects that have no characteristic length scale. It is often the case in the real world that different scales are needed to characterize physical properties. For example, scale invariance is commonly encountered in the context of critical phenomena[2,3] at phase transitions where the divergence of the correlation length leads to universality. Fractals are also observed in systems maintained far from equilibrium[4,5] such as chaos in dissipative systems[6−11], pattern formation[12−18] in viscous fluids, electrochemical systems or porous media, aggregation phenomena in colloidal systems[12−18], and fully developed turbulent flows[19−24] .... A principal challenge to physicists is to characterize the geometry of these nonequilibrium patterns and to understand the dynamics which gives rise to these geometries[25].

In the early eighties, much effort has been devoted to the characterization of fractal objects[4]. Numerical and experimental techniques have been developed in order to measure the fractal dimension D of these structures.

Most of the experimental determinations of D have been based on scattering experiments with X-rays, neutrons or light[26,27]. Generally, D is extracted from the power-law behavior of the scattered intensity $I(k) \sim k^{-D}$, where k is the scattering wavevector modulus. This scaling law indicates that the density-density correlation function behaves as $g(r) \sim r^{D-d}$, where d is the dimension of space. In the pioneering studies[4,12−14,26,27] this scaling analysis was mainly concerned with stochastic structures. More recently it has been generalized to deterministic fractals[28]. But the fractal dimension is a global property that does not provide a deep insight into the geometrical complexity of fractal objects[4,5,25]. A better characterization would, for example, require measuring the scaling exponents D_n of higher order correlation functions. These exponents are inaccessible to classical scattering techniques which are based on a simple Fourier transform[26,27]. The aim of this paper is to convince the reader that the optical measurement of higher order correlation exponents is not an elusive goal.

Very recently, it has become a standard procedure to analyze fractals via the multifractal formalism involving the Renyi dimensions[29−32] D_q. These generalized (to real values of q) fractal dimensions are closely related to the so-called $f(\alpha)$-spectrum of singularities[33−38] by means of a Legendre transform. This connection results from a deep analogy that links the multifractal and the thermodynamical formalisms[36−41]. The variables q and $\tau(q) = (q-1)D_q$ play the same roles as the inverse temperature and the free energy in thermodynamics, while the Legendre transform indicates that in place of energy and entropy we have α and f, i.e., the variables conjugate to q and τ. Therefore, the generalized fractal dimensions D_q and the $f(\alpha)$-spectrum of singularities are thermodynamical functions, i.e., statistical averages that provide macroscopic (global) information about the scaling properties of fractals.

In fact, a full description of fractal geometries requires an efficient multiscale analysis that is capable of extracting microscopic information about the scaling properties of these objects. A known method which comes close to satisfying this requirement is the *Wavelet Transform*[42−46]. This mathematical technique recently introduced in signal analysis[47−49], turns out to be the appropriate tool to collect additional information concerning the hierarchy that governs the spatial distribution of singularities[5,25,50,51]. The wavelet transform can be seen as a "mathematical microscope": when increasing the magnification, one gains insight into the intricate internal structure of fractals[5,25,50−55]. Our purpose here is to describe an experimental set-up which optically performs the wavelet transform using coherent optical spatial filtering. The *Optical Wavelet Transform*[56] (OWT) operates in real time and is readily applicable to experimental situations.

We demonstrate its efficiency to characterize local[56-58] and global[59] scaling properties of fractal aggregates in two dimensions. We point out that the application of the OWT to appropriate physical situations may lead to considerable progress in the understanding of the physical laws that govern the interplay between the spatial and temporal degrees of freedom in nonequilibrium as well as equilibrium critical systems.

The paper is organized as follows. Section 2 is devoted to the instrumental aspects[57] of the OWT. In section 3, we elaborate theoretically on the application of the OWT to characterize local scaling properties of fractals[56-58]. We revisit the multifractal formalism in terms of wavelet components[59] and we show that the generalized fractal dimensions D_q and the $f(\alpha)$-spectrum of singularities are in principle accessible to the OWT. In section 4, we illustrate[56-58] the ability of the OWT to reveal the construction rule of deterministic mathematical fractals and its adequacy to capture local scaling properties through the determination of the local pointwise dimension $\alpha(\vec{x})$. The OWT is shown to be a very efficient experimental tool[59] to measure the D_q and $f(\alpha)$-curves. Its ability to resolve geometrical multifractality is demonstrated. Applications of the OWT to experimental situations[56-59] are reported in section 5. The statistical self-similarity of copper electrodeposition clusters in the limit of small ionic concentration and small current density and of numerical Witten and Sander diffusion-limited aggregates is confirmed. This experimental study brings quantitative evidence indicating that the electrodeposition growth mechanism is likely to be governed by a diffusion-limited aggregation process. In section 6, the results are summarized and the possiblity of applying the OWT to other experimental situations is briefly discussed.

2. THE EXPERIMENTAL SET-UP

2.1. DEFINITIONS

The wavelet transform has been the subject of considerable theoretical developments and practical applications in a wide variety of fields[44]. We will concentrate here on the analysis of fractal objects embedded in two dimensions[53]. We will disregard anisotropic effects in the local scaling behavior of these objects; henceforth we will use radially symmetric analyzing wavelets (for anisotropic wavelets see ref. 60). Thus, let us consider a fractal represented by a real function ρ over $\mathcal{R}^2$, and let $d\mu(\vec{x}) = \rho(\vec{x})d\vec{x}$ be the measure with density $\rho(\vec{x})$ (in practice, a mass density). The function[53,61]

$$T_g(a, \vec{x}) = \left[\rho(\vec{x}') \otimes \frac{1}{a^2} g\left(\frac{\vec{x}'}{a}\right)\right]_{\vec{x}} \tag{2.1}$$

is the wavelet component of ρ for the length scale a (> 0). The symbol $\otimes$ denotes a convolution product, taken at point $\vec{x}$. The analyzing wavelet g is required to be continuous and of zero mean: $\int d\vec{x} g(\vec{x}) = 0$. Whenever g is admissible, no information about ρ is lost since this transformation is likely to be invertible for a large class of functions ρ. From Eq. (2.1) it is clear that the wavelet transform is analogous to a mathematical microscope whose position and magnification are $\vec{x}$ and $1/a$ respectively, and whose optics is given by the choice of the analyzing wavelet g. Of course this microscope is isotropic if g is radially symmetric.

The implementation of the two-dimensional wavelet transform on a computer can be time consuming when a high resolution is required[54]. An alternative method consists in using a coherent optical trick to perform the wavelet transform in real time[56]. By Fourier transforming Eq. (2.1) one gets

$$\hat{T}_g(a, \vec{k}) = \hat{\rho}(\vec{k}) \cdot \hat{g}(a\vec{k}) , \qquad (2.2)$$

where the caret denotes the Fourier transform. $\hat{g}$ has to be zero at $\vec{k} = 0$ as a consequence of the requirement $< g >= 0$. Calculating the wavelet component of scale a of ρ just amounts to filtering its Fourier spectrum with a filter of transparency $\hat{g}(a\vec{k})$. Fraunhofer diffraction is well known to be a very cheap and ultra fast way to perform Fourier transforms of one or two-dimensional objects and has proved to be powerful in various experimental studies[26-28]; it is then readily applicable to wavelet transforms[56].

The experimental set-up[57-59] is sketched in Fig. 1 and photographs of it are shown in Illustration 1 (see color plate). In the OWT, the object ρ is Fourier transformed in a classical diffraction geometry with a lens of focal length f. The electromagnetic field distribution in the rear focal plane of the lens is proportional to $\hat{\rho}(\vec{k})$, with $\vec{k} = 2\pi\vec{u}/\lambda f$, where $\vec{u}$ is the position in the Fourier plane and λ is the light source wavelength. The simplest acceptable shape for the kind of filter to be used in the Fourier plane is a binary approximation of the Fourier tranform of the mexican hat[53,56] (see Fig. 2a). The filter transparency is

$$g(\vec{u}) = \begin{cases} 1 & \text{if} \quad R/\gamma \leq |\vec{u}| \leq R \\ 0 & \text{if} \quad |\vec{u}| > R \text{ or } |\vec{u}| < R/\gamma \end{cases} \qquad (2.3)$$

where R/γ is the radius of the central dark spot and R is the outer radius of the transparent ring. This is just a bandpass filter that admits frequencies between $2\pi R/\lambda f \gamma$ and $2\pi R/\lambda f$. The corresponding analyzing function in

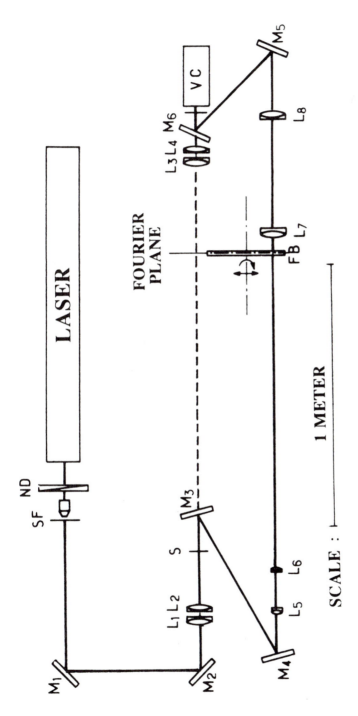

FIGURE 1. Optical Wavelet Transform set-up. $L_1, \cdots, L_8$: doublet lenses (all of them antireflection coated). SF: spatial filter. ND: variable neutral density filter. $M_1, \cdots, M_6$: aluminium coated flat mirrors. S: sample. FB: bandpass filters barrel. VC: video camera.

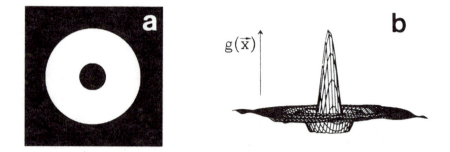

FIGURE 2. (a) Shape of the bandpass filters used in the OWT set-up. The ratio of the largest admissible frequency to the smallest one has been set equal to $\gamma = 3.5$. (Eq. (2.3)). (b) Shape of the corresponding analyzing wavelet in the real space (Eq. (2.4)).

the real space is sketched in Fig. 2b and is simply given by

$$\frac{1}{a^2} \, g\left(\frac{\vec{x}}{a}\right) \propto \frac{1}{a^2} \left[\frac{J_1(|\vec{x}|/a)}{|\vec{x}|/a} - \frac{1}{\gamma^2} \frac{J_1(|\vec{x}|/\gamma a)}{|\vec{x}|/\gamma a} \right] , \qquad (2.4)$$

where J_1 is the first order first kind Bessel function. The tunable parameter $a = \lambda f / 2\pi R$ defines the scale of the OWT. The OWT is carried out by successively using up and down scaled versions of the same filter. The filtered images are sent to a CCD camera; since this detector is quadratic, the OWT actually gives $I(a, \vec{x}) = |T_g(a, \vec{x})|^2$.

2.2. EXPERIMENTAL SET-UP

The optical arrangement[57–59] is shown in Figs. 1 and 2, and in Illustration 1.

SAMPLES

Up to now, we have been working with computer generated objects, namely mathematical sets or growth patterns obtained by numerical simulation. As far as experimental aggregates are concerned, we have used digital image processing of two-dimensional pictures. In any case, our fractal objects were stored into the memory of a computer. A binary image of each fractal was first generated by means of a laser printer. A negative slide of the drawing was then made on a KODAK Kodalith ortho type 3 2256 film (size 9×12 cm^2). This photographic copy was found reliable for details (in fact the pixel size) not much smaller than 40 micrometers. A 1024^2 pixel image could then be stored on the film without any loss of information. Note that these images are far beyond the 512^2 CCD camera resolution.

FILTERS

The bandpass filters were made on a 35 mm AGFA ortho film, starting from laser printer drawings just as for the objects. The ratio γ was chosen equal to 3.5. We found that the value of R/γ (the radius of the central dark spot) had to be at least 100 micrometers for the transparency of the clear part of the slide to keep close to 1 and for the edges of the filter to remain sharp. The maximum value of R compatible with the 35 mm frame is around 10 mm. With this technology, the size of a bandpass filter can be varied by a factor of at most 27. We built a stack of 31 homothetic filters, with a dilation ratio equal to $3^{1/10}$ between two successive filters in the series. The dynamical range covered by the whole stack is $(3^{1/10})^{30} = 27$.

The fundamental limit set by the object is given by the ratio of its largest to smallest sizes, which is at most 1024 and in practice is a little bit less. It is thus clear that a set up[56] with just one instrumental scale factor $2\pi/\lambda f$ for the optical Fourier spectrum does not allow us to take advantage of the whole dynamical range (< 1024) compatible with the object characteristics. To circumvent this difficulty, we designed[57-59] a set up allowing for a change by a factor of about 6.5 of the scale factor. The overall experimental dynamical range is thus raised to about 200. Keeping in mind the fact that the wavelet size a cannot be brought close to the upper and lower cut-off lengths of the objects, we conclude that this number is close to the fundamental limit set by our 1024^2 pixels fractals.

OPTICS

The light source is a COHERENT Innova 90K (Krypton) laser, which is operated at 6471 Å. The beam is first spatially filtered, then sent through a lens made of two identical doublets L_1, L_2 ($f_1 = f_2 = 1300$ mm, $\phi_1 = \phi_2 = 73$ mm). The sample slide (S) is located after $L_1 L_2$. The set-up comprises two paths. Path 1 is in operation when mirrors M_3 and M_6 are removed. The role of the L_3, L_4 lenses ($f_3 = 500$ mm, $\phi_3 = 40$ mm; $f_4 = 250$ mm, $\phi_4 = 40$ mm) is to image the object onto the photosensitive area of a 512^2 pixels CCD camera (I2S IVC 500). The Fourier plane (the plane in which the light beam is focused by $L_1 L_2$) is located about 1100 mm after the sample. The 31 homothetic bandpass filters are located on the periphery of a disk, 300 mm in diameter. Each filter can be brought in operation by moving the whole disk. This is done by two AEROTECH positioning stages, a linear one (ATS 100) and a rotary one (ART 100), driven by a programmable controller (AEROTECH UNIDEX 11). With this technology, it takes about five seconds to move from one filter to the next one, in other words, to "calculate" two successive wavelet components.

Settling back the mirrors M_3 and M_6 brings path 2 into operation. The lens L_5 ($f_5 = 100$ mm, $\phi_5 = 30$ mm) is located near the Fourier plane corresponding to L_1L_2, and images the sample near the plane of lens L_6 ($f_6 = 100$ mm, $\phi_6 = 30$ mm). L_6 forms a magnified image ($\times 6.5$) of the Fourier spectrum in the plane of the above described disk. The filtered image of the sample, i.e., its wavelet component, is formed by lenses L_7 and L_8 ($f_7 = 850$ mm, $\phi_7 = 60$ mm; $f_8 = 1000$ mm, $\phi_8 = 40$ mm) on the photosensitive area of the video camera, with the same magnification as in path 1. Combining the two paths, the wavevectors covered by the filters range from 1.4 cm^{-1} to 900 cm^{-1}. These two optical paths are clearly visible in the color pictures presented in Illustration 1.

IMAGE CAPTURE AND DATA PROCESSING

The video camera is coupled to a system allowing the images to be processed and stored at the video rate (25 frames per second). A synoptic diagram of this image processing system is shown in Fig. 3. The heart of the system is connected to a VME bus. It is made of a video acquisition and display module (CCIR standard), of a storage subunit and of a BUS adapter board. The video acquisition and display module (DATACUBE DIGIMAX 10) digitalizes the video signal at a 10 MHz conversion rate, with a $512 \times 512 \times 8$ bits resolution. The digitized images are displayed in RGB through three programmable look-up tables. The storage subunit comprises a digital video storage module and a large capacity storage unit. The digital video storage module allows for the storage of 3 images and for the data transfer between the video and VME buses. The large capacity storage unit (AVELEM RAMAGE) is made of a video BUS adapter and of two 8 Mb storage boards, and is able to store 64 images (512×512).

A DEC 3200 workstation drives the VME part through a VME-QBUS adapter (PERFORMANCE TECHNOLOGIES), and allows for images transfer and processing. This workstation possesses a 8 Mb memory, a 320 Mb disk, and is connected to the lab central computer (DEC VAX 8600) via an ethernet network.

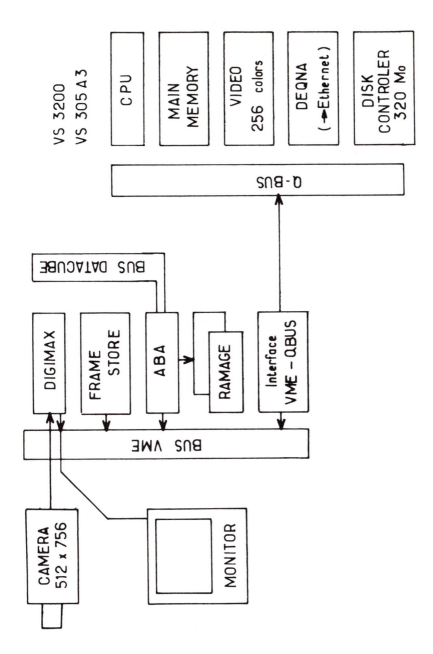

FIGURE 3. Synoptic diagram of the image acquisition and data processing system of the OWT.

3.   OPTICAL WAVELET TRANSFORM AND SCALING PROPERTIES OF FRACTALS

3.1.   DEFINITIONS

A typical property of fractals is that they are asymptotically self-similar at small length scales[1]. Usually the dimension D is introduced to describe the increase of the mass $\mu(\mathcal{B}(\vec{x}_o, \epsilon))$ with the size ϵ

$$\mu(\mathcal{B}(\vec{x}_o, \epsilon)) \;=\; \int_{\mathcal{B}(\vec{x}_o, \epsilon)} \rho(\vec{x})\, d\vec{x} \;\sim\; \epsilon^D \,, \qquad (3.1)$$

where $\mathcal{B}(\vec{x}_o, \epsilon)$ is an ϵ-ball centered at $\vec{x}_o$. In general, however, fractal measures display multifractal properties[19,33−41,62], i.e., the mass scales differently from point to point. Then one is led to consider pointwise dimensions[62,63]

$$\alpha(\vec{x}_o) \;=\; \lim_{\epsilon \to 0^+} \alpha(\vec{x}_o, \epsilon) \;=\; \lim_{\epsilon \to 0^+} \frac{\ln \mu(\mathcal{B}(\vec{x}_o, \epsilon))}{\ln \epsilon} \,. \qquad (3.2)$$

There exist several balls with a given crowding index $\alpha(\vec{x}, \epsilon)$. Their number $N_\alpha(\epsilon)$ scales with ϵ like

$$N_\alpha(\epsilon) \sim \epsilon^{-f(\alpha)} \,. \qquad (3.3)$$

In the limit $\epsilon \to 0^+$, $f(\alpha)$ was shown[37,38] to be the Hausdorff-Besicovitch dimension of the subset on which the measure has a singularity of strength α. The function $f(\alpha)$ is usually called the spectrum of singularities[33−38] of the measure μ. The generalized fractal dimensions[29−32] D_q can be directly computed from this spectrum by means of the following Legendre transform:

$$q \;=\; \frac{df(\alpha)}{d\alpha} \,,$$

$$\qquad (3.4)$$

$$D_q \;=\; \frac{1}{q-1}\, [q\alpha - f(\alpha)] \,,$$

where D_0, D_1 and D_2 are the capacity, information and (two-point) correlation dimensions.

It is noteworthy that uniform (strictly self-similar) fractals[19,29−32,62,63] correspond to the special class of fractals such that all the D_q's are equal. Thus, from Eq. (3.4), the $f(\alpha)$-spectrum of uniform fractals is concentrated on a single point $\alpha = D_0 = D_q$, $\forall q$. In contrast, multifractals[19,33−41,62] are usually characterized by a monotonic decreasing dependence of D_q versus q; hence α is no longer unique but may take on values in a finite

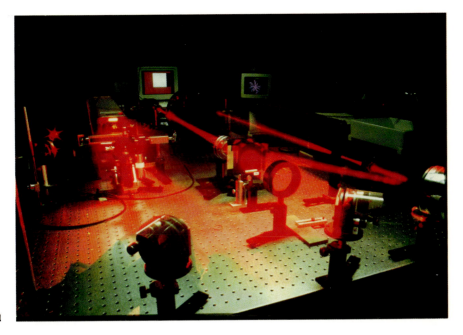

a

b

ILLUSTRATION 1. Optical wavelet transform set-up (see Figures 1, 2, and 3, pages 245–249). The two optical paths 1 and 2 are illustrated in (a). The optical Fourier transform (red) and the optical wavelet transform at a given scale (purple) are shown in (b).

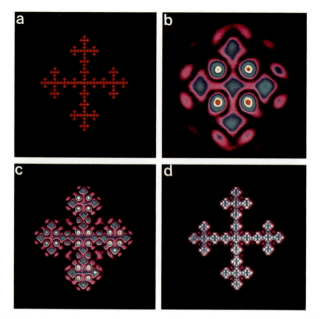

ILLUSTRATION 2. One-scale snowflake fractal (a) and its OWT's, at different scales $a = a^*$ (b), $a^*/3$, (c) and $a^*/3^2$. Pictures show rescaled intensities $S(a, \vec{x}) = a^{-\beta} I(a, \vec{x})$ which are color-coded according to the natural light spectrum from ultra-violet ($S = 0$) to red ($S = 255$); $\beta = 2(\alpha_{SF} - 2)$, $\alpha_{SF} = \log 5/ \log 3$.

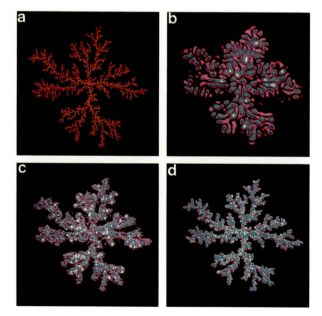

ILLUSTRATION 3. DLA fractal (a) and its OWT's, at different scales: $a = a^*$ (b), $a^*/3^{9/10}$ (c), and $a^*/3^{9/5}$ (d). Pictures show rescaled intensities $S(a, \vec{x}) = a^{-\beta} I(a, \vec{x})$ with $\beta = 2(\alpha_{DLA} - 2)$, $\alpha_{DLA} = 1.66$.

range $[\alpha_{min}, \alpha_{max}]$, while $f(\alpha)$ turns out to be a single humped function with D_0 as its maximum; $\alpha_{min} = \lim_{q \to +\infty} D_q$ and $\alpha_{max} = \lim_{q \to -\infty} D_q$ characterize the scaling properties of the regions where the measure is the most concentrated and the most rarified respectively.

3.2. OWT AND LOCAL SCALING PROPERTIES OF FRACTALS

Although the measurement of the D_q's and $f(\alpha)$-spectrum is undoubtedly very instructive, it only provides global and statistical information about the scaling properties of fractal measures[5,25,64]. As previously mentioned in section 1, in the context of the so-called "thermodynamical description of fractals", the generalized fractal dimensions and the $f(\alpha)$-spectrum of singularities play the role of thermodynamical functions[36-41], i.e., statistical averages, that do not keep memory on the spatial location of the singularities. In order to capture the intricate singularities arrangement, we thus need a tool that gives access to microscopic information. The wavelet transform has been recently shown to be the appropriate mathematical technique to characterize local scaling invariance[5,25,50-55]. Our purpose here is to demonstrate that its optical version[56], the OWT, provides a very efficient experimental tool that can be readily applied to a variety of physical situations[57-59].

Let us assume that μ displays the following scaling behavior near an arbitrary point $\vec{x}_o$:

$$\mu(\mathcal{B}(\vec{x}_o, \lambda \epsilon)) \sim \lambda^{\alpha(\vec{x}_o)} \mu(\mathcal{B}(\vec{x}_o, \epsilon)) \, . \tag{3.5}$$

For an analyzing wavelet that decays sufficiently fast at infinity, this scaling behavior is mirrored by the WT (Eq. (2.1)) which scales in the limit $\lambda \to 0^+$, like[56-58]

$$T_g(\lambda a, \vec{x}_o + \lambda \vec{b}) \sim \lambda^{\alpha(\vec{x}_o)-2} \, T_g(a, \vec{x}_o + \vec{b}) \, . \tag{3.6}$$

Note that for integer exponents, the singularities may be masked by polynomial behavior; in order to overcome this practical difficulty, one generally considers wavelets that operate modulo some polynomials[50-55]. A rigourous relation between the wavelet transform of a measure and its local scaling exponents has been recently derived in ref. 65.

As described in section 2, the OWT provides a direct measurement of the intensity $I(a, \vec{x}) = |T_g(a, \vec{x})|^2$. Thus from Eq. (3.6) one gets the following local scaling behavior:

$$I(\lambda a, \vec{x}_o) = |T_g(\lambda a, \vec{x}_o)|^2 \sim \lambda^{2[\alpha(\vec{x}_o)-2]} \, I(a, \vec{x}_o) \, , \tag{3.7}$$

i.e., the intensity recorded on the CCD camera at the point $\vec{x}_o$ belonging to the fractal behaves as a power-law with exponent $\beta(\vec{x}_o) = 2(\alpha(\vec{x}_o) - 2)$. The OWT can thus be seen as a *singularity scanner*. It allows us[5,25,50−55] (i) to locate the singularities of μ: every local singularity of μ produces a conelike structure in the OWT pointing towards the point $\vec{x}_o$ when the magnification a^{-1} is increased; (ii) to estimate the strength of the singularities of μ: every singularity of μ manifests itself through a power-law behavior of $I(a, \vec{x}_o)$ in the limit $a \to 0^+$, with exponent $\beta(\vec{x}_o) = 2(\alpha(\vec{x}_o) - 2)$. Note that the analyzing wavelet (2.4) implemented in our OWT device displays an oscillatory decreasing tail at large $|\vec{x}|$. When studying fractal aggregates of finite size such as those investigated in the next sections, this oscillatory tail produces oscillations in the log-log representation of $I(a, \vec{x})$ versus a which superpose onto the intrinsic oscillations due to the lacunarity of the aggregates[5,25,50−53]. We have checked[56−59] that with our experimental set-up, the amplitude of these oscillations does not seriously affect the estimate of the local scaling exponents for $\alpha \leq 2$.

3.3.   OWT MEASUREMENT OF THE GENERALIZED FRACTAL DIMENSIONS D_q AND $f(\alpha)$-SPECTRUM OF SINGULARITIES

In the spirit of the multifractal formalism[36−41], let us define a partition function in terms of the wavelet components[50,66]:

$$\mathcal{Z}(a, q) = \int d\mu(\vec{x}) \, |a^2 T_g(a, \vec{x})|^{q-1} \, . \qquad (3.8)$$

One can show that the q-order generalized dimensions can be extracted from the power-law behavior of this partition function in the limit $a \to 0^+$:

$$\mathcal{Z}(a, q) \sim a^{(q-1)D_q} \, . \qquad (3.9)$$

In the OWT, $\mathcal{Z}(a, q)$ is obtained by summing $I(a, \vec{x})^{\frac{q-1}{2}}$ for all points $\vec{x}$ belonging to the fractal set or aggregate. The way in which $\mathcal{Z}(a, q)$ depends on the filter size directly gives the dimensions[59] D_q:.

$$\mathcal{Z}(a, q) = \int d\mu(\vec{x}) \left[a^2 \sqrt{I(a, \vec{x})} \right]^{(q-1)} \sim a^{(q-1)D_q} \, . \qquad (3.10)$$

The $f(\alpha)$-spectrum of singularities is usually deduced[33−41] from the Legendre transform of the curve $\tau(q) = (q-1)D_q$ (Eq. (3.4)). But such a computation requires first a smoothing of the D_q curve and then Legendre transforming. This procedure has a main disadvantage: the smoothing operation prevents the observation of any singularity in the curves $f(\alpha)$

and $\tau(q)$, and the interesting physics of phase transition in the scaling properties of fractal measures can be completely missed[4,62].

Several methods[62,67−69] have been recently proposed for a direct estimate of the $f(\alpha)$-spectrum based on log-log plots of the quantities in Eqs. (3.2) and (3.3). Unfortunately, the application of these methods suffers from neglected logarithmic corrections[69−72] which arise from the scale-dependent prefactors in Eq. (3.3). To overcome these difficulties we will use the following trick[53]. Let us differentiate the partition function $\mathcal{Z}(a, q)$ with respect to q:

$$\frac{d}{dq}\mathcal{Z}(a, q) = \int d\mu(\vec{x}) \left[a^2 \sqrt{I(a, \vec{x})}\right]^{q-1} \ln \left[a^2 \sqrt{I(a, \vec{x})}\right] \cdot \qquad (3.11)$$

From Eqs (3.4), (3.9) and (3.11) one deduces:

$$\alpha(q) = \frac{d}{dq}(q-1)D_q ,$$
$$= \lim_{a \to 0^+} \frac{1}{\ln a} \int d\mu(\vec{x}) \mathcal{I}_q(a, \vec{x}) \ln \left[a^2 \sqrt{I(a, \vec{x})}\right] , \qquad (3.12)$$

where

$$\mathcal{I}_q(a, \vec{x}) = \left[a^2 \sqrt{I(a, \vec{x})}\right]^{q-1} / \mathcal{Z}(a, q) \cdot \qquad (3.13)$$

Then upon using the definition of $f(\alpha)$ one gets:

$$f(\alpha(q)) = \alpha(q)q - (q-1)D_q ,$$
$$= \lim_{a \to 0^+} \int d\mu(\vec{x}) \mathcal{I}_q(a, \vec{x}) \ln \mathcal{I}_q(a, \vec{x}) \cdot \qquad (3.14)$$

Therefore we will directly extract the set of local exponents α and the related $f(\alpha)$-spectrum from log-log plots of the quantities in Eqs. (3.12) and (3.14), without explicitly Legendre transforming and without neglecting logarithmic corrections[59]. As shown in refs 71 and 72, this alternative definition of the $f(\alpha)$-spectrum corresponds to the canonical counterpart of the microcanonical thermodynamic formalism originally introduced in refs 33-38.

4. OPTICAL WAVELET TRANSFORM OF DETERMINISTIC SNOWFLAKE PATTERNS

4.1. ONE-SCALE SNOWFLAKE

Dendritic crystal growth[73] enters the wide class of pattern formation in a diffusion field[12−18]. In general, dendritic crytals have a well organized branched structure, such as seen in snowflake patterns[74,75]. In Fig. 4a, we show a one-scale snowflake fractal which is commonly thought of as a paradigm for self-similar fractal aggregates[76]. Its construction rule can be considered as a deterministic model for aggregation process. The configuration at the nth stage is obtained by adding to the four corners of the $(n-1)^{th}$ stage configuration, the cluster corresponding to the $(n-1)^{th}$ stage. Therefore, from one stage to the next stage of the construction, the cluster length size increases by a factor $\ell^{-1} = 3$, while the number of motifs is multiplied by a factor $m = 5$. A straightforward calculation yields[53] $D_q = \ln 5/\ln 3$ for all q. By Legendre transforming the D_q's one quantifies the self-similarity of the snowflake into a single scaling index $\alpha = \ln 5/\ln 3$ with density $f(\alpha = \ln 5/\ln 3) = \ln 5/\ln 3$.

In Fig. 4 we present an overview of the OWT of the one-scale snowflake fractal[56] (see also color plate for Illustration 2). The OWT is shown for decreasing values of the scale parameter a. This OWT "zooming" provides conspicuous information about the construction rule of the snowflake. At large scales, one observes a single object of length size a^*. In the next step, this object is divided into 9 identical pieces, each of which is a reduced version of the original object with length scale $a^*/3$ (Fig. 4b); 4 among the 9 pieces are removed, while the central piece with the 4 pieces at each corner are retained. Then the same procedure is repeated in the next step for each of the 5 remaining pieces (Fig. 4c). The snowflake is indeed obtained by applying the same rule subsequently (Fig. 4d) ad infinitum. Note that (as compared to the numerical wavelet analysis in ref. 53), at a given step in the construction process, the negative parts of the wavelet transform which correspond to the removed pieces contribute constructively to the recorded intensity $I = |T_g|^2$. But these negative parts can be easily distinguished from the positive parts which delimit the retained pieces because they completely vanish at the next step.

Besides its edge detection ability, the OWT provides clear evidence for the global self-similarity of the one-scale snowflake[56−58]; at each point $\vec{x}$ of the aggregate, the OWT displays a power-law behavior with an exponent α independent of $\vec{x}$. The values of this exponent at three different points of the aggregate were determined from the graphs shown in Fig. 5a, where the rescaled intensity $a^2\sqrt{I}$ was plotted vs a in a $\log_2$-$\log_2$ scale representation. Disregarding finite size effects and despite some oscillations in the log-log procedure partly due to the lacunarity of the snowflake[53], the OWT provides an estimate of the local exponent α in good agreement with the theoretical prediction ($\alpha = \ln 5/\ln 3$), and previous numerical wavelet transform analysis[53]. The overall accuracy in the determination of α depends on both the wavelet shape (the value of γ) and the size of

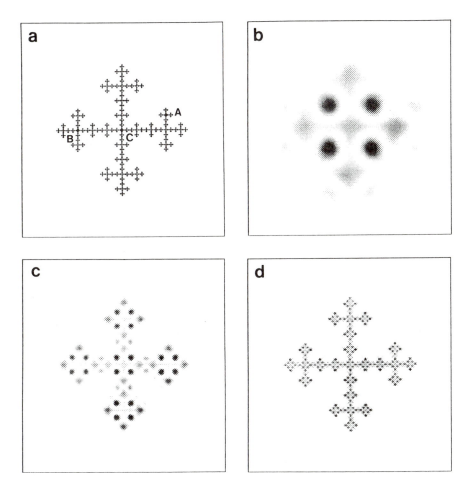

FIGURE 4. One-scale snowflake fractal (a) and its OWT's, at different scales: $a = a^*/3$ (b), $a^*/3^2$ (c) and $a^*/3^3$ (d). Pictures show recaled intensities $S(a, \vec{b}) = a^{-\beta} I(a, \vec{b})$; $\beta = 2(\alpha - 2)$, $\alpha = \ln 5 / \ln 3$.

the aggregate. Extending this optical technique to larger values of α needs more sophisticated filters, whose realization is currently in progress.

The statistical self-similarity of the one-scale snowflake is confirmed by the measurement of the generalized fractal dimensions[59] D_q in Fig. 6. The partition function $\mathcal{Z}(a, q)$ (Eq. (3.9)) is plotted versus the filter size a in a $\log_2$-$\log_2$ representation in Fig. 6b. The slopes D_q obtained from the graphs in Fig. 6b are plotted versus q in Fig. 6c. The error bars correspond to the fluctuations observed around the straight lines in the $\log_2$-$\log_2$ plots in Fig. 6b. Besides finite-size effects and experimental noise, one clearly distin-

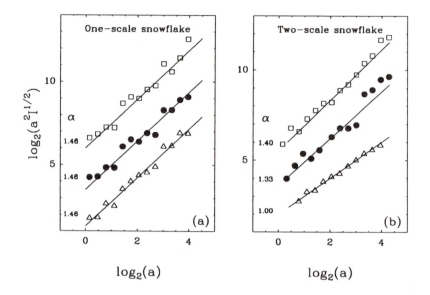

FIGURE 5. Rescaled intensities of the OWT's of the one-scale (a) and two-scale (b) snowflakes vs the filter size in log₂-log₂ representation. The symbols ■ (A), • (B), and △ (C) correspond to different points on the aggregates where the intensity was recorded. Different arbitrary units have been used for the different sets of intensities since only the slopes are relevant. The solid lines correspond to the theoretical predictions.

guishes intrinsic periodic oscillations of period $P = \ln \ell^{-1} = \ln 3$. The amplitude of these oscillations increases with $|q|$. These oscillations are due to the invariance of the one-scale snowflake under dilation of length scales by a factor $\ell^{-1} = 3$. The OWT provides a very accurate estimate of the D_q's: within the experimental uncertainty, the generalized fractal dimensions are found to be all equal to the theoretical prediction $D_q = \ln 5 / \ln 3$. These results confirm the capability of our experimental technique to recognize statistical self-similarity[56−59].

4.2.   *TWO-SCALE SNOWFLAKE*

The main difference between the construction of multi-scale and one-scale fractals is the fact that the starting object is now divided into m parts which are not identical. However, all of these are reduced (by scaling factors that are not identical) versions of the original object. In Fig. 7a we show a generalization of the globally self-similar snowflake fractal studied in the previous section. The configuration at the nth stage is obtained by

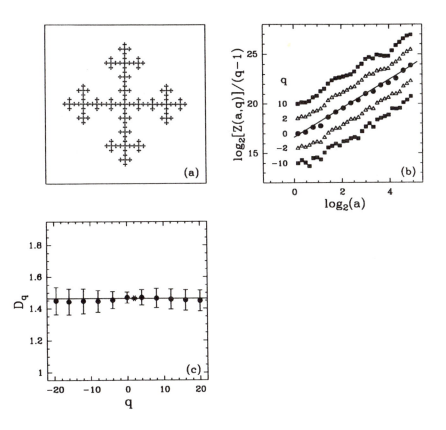

FIGURE 6. OWT determination of the generalized fractal dimensions D_q of the one-scale snowflake fractal. (a) The snowflake. (b) $\mathrm{Log}_2\,\mathcal{Z}(a,q)/(q-1)$ vs $\log_2 a$ (arbitrary scales) for different values of q. (c) D_q vs q. The symbol $*$ indicates the value of D_2, the only dimension accessible to classical scattering experiments. The solid lines correspond to the theoretical prediction $D_q = \ln 5/\ln 3$, $\forall q$.

adding the configuration at the $(n - 1)^{th}$ stage of the growth to the four corners of the twice enlarged version of the cluster corresponding to the $(n - 1)^{th}$ stage configuration. An analogous structure is constructed by dividing a single object of size 1 and measure 1 into $m_1 = 4$ identical pieces with size $\ell_1 = 1/4$ (S for small) and $m_2 = 1$ piece with size $\ell_2 = 1/2$ (L for large). At the next stage of the construction, the same process is applied to each of these $m = m_1 + m_2 = 5$ pieces. The procedure is then repeated again and again. Each square of the cluter obtained after n generations can be labelled by a sequence of n symbols L or S. All squares having the same number s of S's (equivalently $n-s$ of L's) in their symbolic sequences correspond to the same value of the local scaling exponent α. In the limit

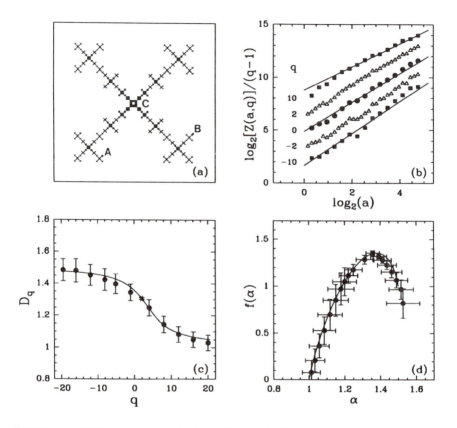

FIGURE 7. OWT measurement of the D_q's and $f(\alpha)$-spectrum of the two-scale snowflake fractal. (a) The multifractal snowflake. (b) $\mathrm{Log}_2 \mathcal{Z}(a,q)/(q-1)$ vs $\log_2 a$ (arbitrary scales) for different values of q. (c) D_q vs q. (d) $f(\alpha)$ vs α. The solid lines in (b), (c) and (d) correspond to the theoretical predictions. The symbol $*$ in (c) indicates the value of the two-point correlation dimension D_2, the only dimension accessible to classical scattering experiments.

of large n, one obtains

$$\alpha(s, n) = \frac{2s + n}{s + n}. \tag{4.1}$$

Since the corresponding measure has an exact recursive structure, one can compute analytically[53] the $f(\alpha)$-spectrum of singularities and the related generalized fractal dimensions D_q. These theoretical predictions are shown in Fig. 7. The scaling exponent α takes on values in a finite range: $\alpha_{min} \leq \alpha \leq \alpha_{max}$. The lower bound $\alpha_{min} = D_{+\infty} = 1$, $f(\alpha_{min}) = 0$, is associated with the symbolic sequence LLL...LLL... which addresses the central square and corresponds to the region of highest mass.

Conversely, $\alpha_{max} = D_{-\infty} = 3/2$ is associated with the set of squares with symbolic sequences SSS...SSS... that corresponds to the regions of lowest mass; these regions have a finite density exponent $f(\alpha_{max}) = 1$. Note that the maximum of the $f(\alpha)$-curve gives the fractal dimension $D_0 = 3 - \ln(\sqrt{17} - 1)/\ln 2 \sim 1.357 \ldots$.

Fig. 5b is a $\log_2$-$\log_2$ plot of the rescaled OWT intensity taken at three different points in a 6-generations two-scale snowflake, versus the filter size a. The data are in remarkable agreement with the predicted slopes[58] α (LLLLLL) = 1 ($\triangle$), α (SSSLLL) = 1.33 ($\bullet$) and α (SSSSLL) = 1.40 ($\blacksquare$). This is the experimental demonstration that the OWT is capable to resolve local scaling behavior of multifractal aggregates. This demonstration is perfected by the determination of the generalized fractal dimensions and $f(\alpha)$-spectrum in Fig. 7. The theoretical D_q and $f(\alpha)$ curves are remarkable fits of the OWT data[59]. The geometrical multifractality of the two-scale snowflake is thus recovered experimentally. These results are compared in Fig. 7 to the information usually extracted in classical scattering experiments, namely the two-point correlation dimension D_2 alone. This comparison is very instructive about the efficiency and the potentiality of the OWT.

5. OPTICAL WAVELET TRANSFORM OF LAPLACIAN FRACTAL AGGREGATES

In recent years, considerable interest has developed in a variety of nonequilibrium growth processes and pattern formation phenomena[4,12-18]. Some of these patterns have been found to have fractal or self-similar structures. Among many nonequilibrium growth models proposed so far, the most intensively studied is undoubtedly the diffusion-limited aggregation (DLA) model introduced by Witten and Sander[77] in the early eighties. This model can be regarded as describing cluster growth in a diffusion field, because particles sticking to the cluster come (one at a time) from far away through pure random walk. The structures generated by the DLA model exhibit branching over a wide range of scales (Fig. 8a); the fractal dimension of these clusters was calculated both theoretically[78-83] and numerically[77,84-88], and found to be strictly less than the Euclidean dimension of the space in which the aggregation process takes place. Since the pioneering simulations of the DLA model, numerous extensions of this model were considered for various purposes e.g. to incorporate surface tension or to mimic anisotropic dendritic growth[12-18]. It is now well admitted that the DLA model and its variants are relevant to account for the pat-

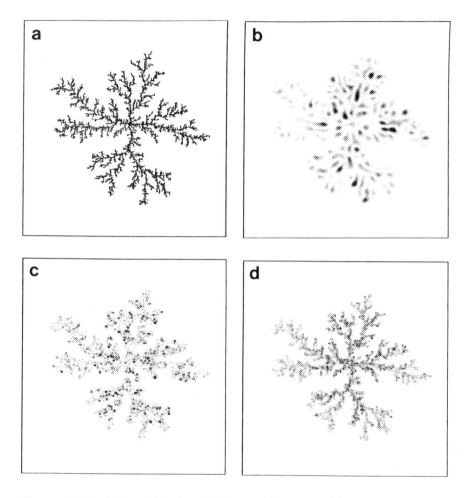

FIGURE 8. On-lattice DLA fractal of mass $M = 13000$ (a) and its OWT's, at different scales: $a = a^*$ (b), $a^*/3^{9/10}$ (c) and $a^*/3^{9/5}$ (d). The intensity of the laser was tuned so as to keep the intensity maxima nearly independent of the filter size.

terns observed in a broad class of experimental situations[4,12−18] including electrodeposition, fluid-fluid displacement in Hele-Shaw cells and porous media, formation of snowflakes and dendritic patterns, colloidal aggregation and dielectric breakdown.

5.1. NUMERICAL DIFFUSION-LIMITED AGGREGATES

Despite the importance and the apparent simplicity of the Witten and Sander model[77], there is still no rigourous theory for diffusion-limited growth processes. In particular, it is still an open question whether the resulting fractal geometry is a product of the randomness in the growth process[89] or the result of a cascade of deterministic tip-splitting instabilities[90−92]. In fact, only a little is known in quantitative terms about the structure of DLA clusters. Most numerical analysis[12−18,77,84,85] have mainly focused on the particularly meaningful dimensions: the fractal dimension or *capacity* dimension D_0, the *information* dimension D_1 and the (two-point) *correlation* dimension D_2. Only very recently, more attention has been paid to the computation of the whole spectrum of generalized dimensions[93]. In a preliminary study[53], we have carried out a quantitative analysis of the D_q's of numerical DLA clusters computed using both the on-square lattice algorithm used by Ball and Brady[87] and the off-lattice algorithm introduced by P. Meakin[94]. The results of box-counting ($q > 0$) and nearest neighbour ($q < 0$) measurements converge to a value $D_q = 1.60 \pm 0.02$ that is independent of q for on-lattice clusters of small mass ($M \leq 50000$ particles). We have carried out a similar analysis for off-lattice DLA clusters which confirms that indeed our numerical estimate of D_q is not affected by the anisotropy of the underlying lattice[95,96].

In order to provide deeper insight into the structural complexity of DLA clusters we have recently examined the geometry of these aggregates with the wavelet transform microscope[53]. We will concentrate here on the experimental results[56,58,59] obtained with the OWT. Fig. 8 shows the OWT of an on-lattice DLA cluster of mass $M = 13000$ particles (Fig. 8a) through different panels corresponding to increasing values of the magnification a^{-1} (see also color plate for Illustration 3). As expected from our previous OWT analysis of snowflake fractals, this method provides insight into thinner internal details in the shape of DLA patterns. Very much like in the numerical analysis of ref. 53, at large scales (Fig. 8b), DLA clusters look very much like viscous fingers observed in Hele-Shaw experiments[97,98]. When looking at smaller scales, the main fingers split into fingers of smaller width (Fig. 8c) and successive generations of fingers show up (Fig. 8d), leading to the arborescent structure of DLA clusters. This structure of the cluster boundary when increasing the magnification of the OWT presents a strong analogy with the tip-splitting and side-branching dynamical instabilities observed in viscous fingering experiments[99−103]. What is remarkable in Fig. 8 is that successive generations appear at the same rate all over the aggregates when the magnification is increased (in logarithmic scale). This observation strongly suggests that DLA clusters are likely to display global scale invariance[56]. Moreover, as previously noticed in the numerical study in ref. 53, this statistical self-similarity is strongly related to the existence

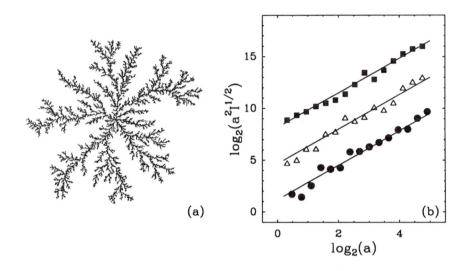

FIGURE 9. OWT determination of the local scaling exponent of a M = 50000 off-lattice DLA cluster. (a) The off-lattice DLA cluster (courtesy of P. Meakin) (b). Rescaled intensities of the OWT vs the filter size in $\log_2$-$\log_2$ representation. The symbols ■, △ and ● correspond to different points selected at random on the aggregates where the intensity was recorded. Different arbitrary units have been used for the different sets of intensities since only the slopes are relevant. The solid lines correspond to the theoretical slope $\alpha = 5/3$ (see text).

of a characteristic screening angle β between branches of two successive generations[104]. Even though our statistical sample is not large enough to allow us to produce any reliable estimate of this angle, our average value is quite compatible with a rough pentagonal symmetry $< \beta > \sim 36°$, already noticed in previous works[79−81,105,106]. Note that the implementation of an anisotropic wavelet transform[60] could provide a firm experimental support to this conjecture.

In Fig. 9 we illustrate the experimental determination of the local scaling properties of a $M = 50000$ off-lattice DLA cluster (Fig. 9a). The rescaled intensity $a^2\sqrt{I}$ of the OWT is plotted versus a in a $\log_2$-$\log_2$ representation (Fig. 9b). To the previously mentioned experimental difficulties (extrinsic oscillations due to the oscillatory tail of the analyzing wavelet g), additional intrinsic aperiodic oscillations, inherent to these chaotic fractals, make the measurement of the local slope $\alpha(\vec{x})$ very demanding[56−58]. The data collected[58] for three points selected at random on the off-lattice DLA boundary strongly suggest the existence of a unique exponent $\alpha \sim 5/3$. Quantitative confirmation of these local measurements is brought in Fig. 10, where the experimental determination of the generalized fractal dimensions does not reveal any significant deviation from a constant D_q curve[59]. These results provide the experimental demonstration that off-lattice DLA clusters are statistically self-similar. The measured dimensions $D_q = 1.65 \pm 0.06$ are found in good agreement with the mean-field prediction[78,82,83] $D_F = 5/3$ and the results of previous numerical analysis[53,107]. Note that similar quantitative estimates[56] are obtained with on-lattice DLA clusters despite the fact that their Fourier spectra display the characteristic four-fold anisotropy of the underlying square lattice.

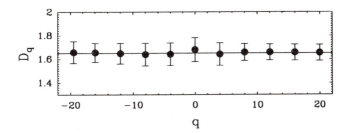

FIGURE 10. OWT measurement of the generalized fractal dimensions D_q of the off-lattice DLA cluster shown in Fig. 9a (Eq. (3.10)). The solid line corresponds to the mean field prediction $D_F = 5/3$.

5.2. EXPERIMENTAL ELECTRODEPOSITION CLUSTERS

Among the various experimental illustrations of fractal pattern forming phenomena, electrochemical metallic deposition[108] is commonly considered as the paradigm for theoretical studies of diffusion-limited aggregation. In fact, by varying the concentration of metal ions and the cathode potential (or current), one can explore different morphologies like dense radial[109], dendritic[110−112], and fractal aggregates[53,110,112−117]. Fractal patterns are usually obtained in the limit of small ionic concentration and small current density. These fractal electrodeposition clusters have been extensively studied and it has been conjectured[115−117] that they are similar to Witten and Sander clusters. In fact, the deep connection between highly ramified electrodeposits and DLA clusters has been explored to a limited extent both theoretically and experimentally. In a preliminary study[113], in collaboration with H. L. Swinney, we have developed a comparative study of zinc electrodeposition and DLA clusters grown in a strip geometry. This study was essentially based on box-counting and fixed-mass computations of the spectrum of generalized fractal dimensions. In a more recent study[53,114], we have confirmed these results in a similar analysis of copper electrode-

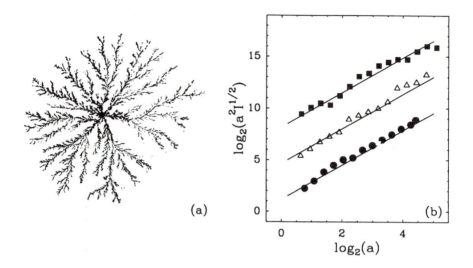

FIGURE 11. OWT determination of the local scaling exponent of a copper elctrodeposit. (a) The experimental fractal aggregate. (b) Rescaled intensities of the OWT vs the filter size in $\log_2$-$\log_2$ representation. The symbols ■, △ and ● correspond to different points chosen at random on the aggregates where the intensity was recorded. Different arbitrary units have been used for the different sets of intensities since only the slopes are relevant. The solid lines correspond to the theoretical prediction $\alpha = 5/3$ for DLA clusters.

position clusters grown in a circular geometry at fixed current intensity. When the migration of electroactive species becomes negligible as compared to diffusion, the highly ramified electrodeposition clusters are found statistically self-similar with dimensions $D_q = 1.63 \pm 0.03$, independently of q, in remarkable quantitative agreement with the numerical values found for DLA clusters. These results strongly suggest that, at least on a statistical level, the geometry of electrodeposition clusters mimics the geometrical complexity of the theoretically proposed diffusion-limited aggregates[53].

In this section we use the OWT to make more accurate the quantitative comparison between the electrodeposition mechanism and diffusion-limited aggregation. The corresponding results of a similar numerical wavelet analysis can be found in ref. 53. In Fig. 11 we concentrate on the experimental determination of the local scaling properties of a copper electrodeposition cluster (Fig. 11a) grown in a circular geometry. The experimental configuration[53] is a thin layer of a 0.05 M $CuSO_4$ solution confined between two glass plates. The gap between the plates is 0.1 mm. A circular copper electrode of diameter 10 cm surrounds the fluid, and a copper cathode of diameter 0.1 mm is introduced vertically through a hole in the upper plate

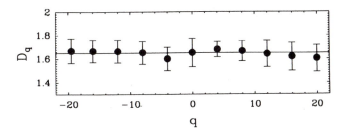

FIGURE 12. OWT measurement of the generalized fractal dimensions D_q of the copper electrodeposition cluster shown in Fig. 11a (Eq. (3.10)). The solid line corresponds to the mean-field prediction $D_F = 5/3$ for DLA clusters.

centered within the ring. The electrodeposition is initiated by applying a constant current $I = 0.1$ mA between the anode and the cathode. After 45-50 min, the ramified copper electrodeposit has grown to a size of a few centimeters. The OWT measurement of the local scaling exponent $\alpha(\vec{x})$ is mainly affected at large scales by finite size effects and by the ultraviolet cut-off due to the "capillary length" of the electrodeposit fingers. The results in Fig. 11b bring strong indication that the exponent α extracted from the local power-law behavior of the rescaled intensity versus the filter size does not depend on the point which has been selected on the cluster[58]. The statistical self-similarity of these experimental aggregates is quantitatively demonstrated in Fig. 12, where the OWT data for the generalized fractal dimensions D_q do not reveal any significant dependence of D_q as a function of q. Within the experimental uncertainty the dimensions $D_q = 1.66 \pm 0.08$ of the copper electrodeposit are the same as the dimensions of numerical DLA clusters[59].

The results reported in this section provide the experimental confirmation that electrodeposition clusters, when diffusion is the dominating transport process of metal ions, and DLA clusters are statistically self-similar and that their fractal properties, as characterized by such quantities as their generalized fractal dimensions, or the related $f(\alpha)$-spectrum are identical.

6.   CONCLUSION

To summarize, we have presented an experimental arrangement which performs the OWT of two-dimensional objects. Its ability to resolve geometrical multifractality has been demonstrated. So far we have mainly analyzed slides representing fractal aggregates. This optical technique can be directly integrated into some experimental arrangements without recourse to the intermediate photographic procedure. Some opening does exist in three dimensions. For its ability to capture local scaling properties and to study the scaling behavior of high-order correlation functions, the OWT is a definite step beyond classical scattering techniques.

In terms of speed, the main limitation of the present OWT set up comes from the delay needed to bring the successive filters in operation. Our hope is to replace the present stack of static filters by a single liquid-crystal television screen. Such a device would allow us to take full technical advantage of the performance offered by the image analysis system and bring the OWT to a video-rate speed. The progress would be fundamental as well, since a variety of anisotropic and grey-level (smooth) filters could be displayed instantaneously on a TV screen; this would make the OWT a very efficient experimental tool for direction or angle detection. The potential of the OWT to operate in real time is likely to pave the way to time-resolved studies of dynamical phenomena. The application of the OWT to a variety of experimental situations, e.g., percolation, colloidal aggregation, growth processes, fracture patterns, nucleation and coalescence phenomena, two-dimensional melting, and turbulent flows, looks very promising. Of particular interest is the use of the OWT to revisit phase transitions in critical phenomena. The study of critical colloidal systems is currently in progress.

ACKNOWLEDGEMENTS

We are very grateful to G. Gabriel and R. Vega for their technical assistance. This work was supported by the Conseil Régional d'Aquitaine, the Ministère de l'Education Nationale under contract (DRED3 - 1322), the Direction des Recherches Etudes et Techniques under contract (DRET N° 89 / 196) and the Centre National des Etudes Spatiales under contract (N°90/CNES/215).

REFERENCES

[1] B.B. Mandelbrot, *The Fractal Geometry of Nature* (Freeman, San Francisco, 1982).

[2] S.K. Ma, *Modern Theory of Critical Phenomena* (Benjamin Reading, Mass., 1976).

[3] D. Amit, *Field Theory, the Renormalization Group and Critical Phenomena* (Mc Graw-Hill, New York, 1978).

[4] Essays in honour of B.B. Mandelbrot, *Fractals in Physics*, edited by A. Aharony and J. Feder, Physica D **38** (1989).

[5] A. Arneodo, F. Argoul, J. Elezgaray and G. Grasseau, in *Nonlinear Dynamics*, edited by G. Turchetti (World Scientific, Singapore, 1988) p. 130.

[6] P. Cvitanovic, ed., *Universality in Chaos* (Hilger, Bristol, 1984).

[7] B.L. Hao, ed., *Chaos* (World Scientific, Singapore, 1984).

[8] H.G. Schuster, *Deterministic Chaos* (Physik-Verlag, Weimheim, 1984).

[9] P. Bergé, Y. Pomeau and C. Vidal, *Order within Chaos* (Wiley, New York, 1986).

[10] H.B. Stewart, *Nonlinear Dynamics and Chaos* (Wiley, New York, 1986).

[11] M.F. Barnsley and S.G. Demko, eds, *Chaotic Dynamics and Fractals* (Academic Press. Inc., Orlando, 1986).

[12] H.E. Stanley and N. Ostrowski, eds, *On Growth and Form: Fractal and Non-Fractal Patterns in Physics* (Martinus Nijhof, Dordrecht, 1986).

[13] L. Pietronero and E. Tosatti, eds, *Fractals in Physics* (North-Holland, Amsterdam, 1986).

[14] H.E. Stanley, ed., *Statphys. 16* (North-Holland, Amsterdam, 1986).

[15] W. Güttinger and D. Dangelmayr, eds, *The Physics of Structure Formation* (Springer-Verlag, Berlin, 1987).

[16] H.E. Stanley and N. Ostrowki, eds, *Random Fluctucations and Pattern Growth* (Kluwer Academic, Dordrecht, 1988).

[17] J. Feder, *Fractals* (Pergamon, New York, 1988).

[18] T. Viscek, *Fractal Growth Phenomena* (World Scientific, Singapore, 1989).

[19] B.B. Mandelbrot, J. Fluid Mech. **62**, 331 (1974).

[20] U. Frisch, *Physica Scripta* Vol. T**9**, 137 (1985).

[21] Y. Gagne, E.J. Hopfinger and U. Frisch, in *New Trends in Nonlinear Dynamics and Pattern Forming Phenomena: The Geometry of Nonequilibrium*, edited by P. Huerre and P. Coullet (Plenum, New York, 1988).

[22] F. Argoul, A. Arneodo, G. Grasseau, Y. Gagne, E.J. Hopfinger and U. Frisch, Nature **338**, 51 (1989).

[23] E. Bacry, A. Arneodo, U. Frisch, Y. Gagne and E. Hopfinger, in *Turbulence and Coherent Structures*, edited by M. Lesieur and O. Metais (Kluwer Academic Publishers, 1990) p. 203.

[24] U. Frisch and S.A. Orszag, Physics Today (1990) p. 24.

[25] A. Arneodo, F. Argoul, E. Bacry, J. Elezgaray, E. Freysz, G. Grasseau, J.F. Muzy and B. Pouligny, "Wavelet transform of fractals: I. From the transition to chaos to fully developed turbulence. II. Optical wavelet transform of fractal growth phenomena", preprint (1990) to appear in the Proceedings of the Conference "Wavelets and some of their applications" (Marseille, Luminy, 1989).

[26] J. Texeira, in ref. 12, p. 145.

[27] S.K. Sinha, in ref. 4, p. 310.

[28] C. Allain and M. Cloitre, Phys. Rev. B **33**, 3566 (1986); in ref. 13, p. 61.

[29] A. Renyi, *Probability Theory* (North Holland, Amsterdam, 1970).

[30] P. Grassberger, Phys. Lett. A **97**, 227 (1983).

[31] H.G. Hentschel and I. Procaccia, Physica D **8**, 435 (1983).

[32] P. Grassberger and I. Procaccia, Physica D **13**, 34 (1984).

[33] E.B. Vul, Ya.G. Sinai and K.M. Khanin, Usp. Mat. Nauk. **39**, 3 (1984); J. Russ. Math. Surv. **39**, 1 (1984).

[34] R. Benzi, G. Paladin, G. Parisi, and A. Vulpiani, J. Phys. A **17**, 3521 (1984).

[35] G. Parisi and U. Frisch, in *Turbulence and Predictability in Geophysical Fluid Dynamics and Climate Dynamics*, edited by M. Ghil, R. Benzi and G. Parisi (North-Holland, Amsterdam, 1985) p. 84.

[36] T.C. Halsey, M.H. Jensen, L.P. Kadanoff, I. Procaccia and B.I. Shraiman, Phys. Rev. A **33**, 1141 (1986).

[37] P. Collet, J. Lebowitz and A. Porzio, J. Stat. Phys. **47**, 609 (1987).

[38] D. Rand, "The singularity spectrum for hyperbolic Cantor sets and attractors", preprint (1986).

[39] B.B. Mandelbrot, *Fractals and Multifractals: Noise, Turbulence and Galaxies Selecta*, (Springer, New York, 1990).

[40] See for example: T. Bohr and T. Tèl, in *Direction in Chaos*, vol. 2, edited by Hao Bai-Lin (World Scientific, Singapore, 1988); T. Tèl, Z. Naturforsch **43a**, 1154 (1988); R. Badii, Thesis, University of Zurich (1987).

[41] M.J. Feigenbaum, J. Stat. Phys. **46**, 919 and 925 (1987).

[42] A. Grossmann and J. Morlet, in *Mathematics and Physics, Lectures on Recent Results*, edited by L. Streit (World Scientific, Singapore, 1985).

[43] I. Daubechies, A. Grossmann and Y. Meyer, J. Math. Phys. **27**, 1271 (1986).

[44] J.M. Combes, A. Grossmann and P. Tchamitchian, eds, *Wavelets* (Springer-Verlag, Berlin, 1988).

[45] Y. Meyer, *Ondelettes* (Herman, Paris, 1990).

[46] P.G. Lemarié, ed., *Les Ondelettes en 1989* (Springer-Verlag, Berlin, 1990).

[47] P. Goupillaud, A. Grossmann and J. Morlet, Geoexploration **23**, 85 (1984).

[48] R. Kronland-Martinet, J. Morlet and A. Grossmann, Int. J. of Pattern Analysis and Artificial Intelligence **1**, n°2, 273 (1987).

[49] A. Grossmann, M. Holschneider, R. Kronland-Martinet and J. Morlet in *Inverse Problem*, edited by P.C. Sabatier, Advances in Electronics and Electron Physics, Supplement **19** (Acad. Press, 1987).

[50] M. Holschneider, J. Stat. Phys. **50**, 963 (1988); Thesis, University of Aix-Marseille II (1988).

[51] A. Arneodo, G. Grasseau and M. Holschneider, Phys. Rev. Lett. **61**, 2281 (1988); in ref. 44, p. 182.

[52] A. Arneodo, F. Argoul and G. Grasseau, in ref. 46, p. 125.

[53] F. Argoul, A. Arneodo, J. Elezgaray, G. Grasseau and R. Murenzi, Phys. Lett. A **135**, 327 (1989); Phys. Rev. A **41**, 5537 (1990).

[54] G. Grasseau, Thesis, University of Bordeaux (1989).

[55] M. Holschneider and P. Tchamitchian, in ref. 46, p. 102.

[56] E. Freysz, B. Pouligny, F. Argoul and A. Arneodo, Phys. Rev. Lett. **64**, 745 (1990).

[57] B. Pouligny, G. Gabriel, J.F. Muzy, A. Arneodo, F. Argoul and E. Freysz, J. Appl. Cryst. (1991) to appear.

[58] J.F. Muzy, B. Pouligny, E. Freysz, F. Argoul and A. Arneodo, "Optical wavelet transform and local scaling properties of fractal aggregates", preprint (1991), submitted to Mod. Phys. Lett. B.

[59] J.F. Muzy, B. Pouligny, E. Freysz, F. Argoul and A. Arneodo, "Optical diffraction measurement of fractal dimensions and $f(\alpha)$-spectrum", preprint (1991), submitted to Phys. Rev. Lett.

[60] J.P. Antoine, P. Carrette, R. Murenzi and B. Piette, "Image analysis with two-dimensional continuous wavelet transform", preprint (1991), submitted to IEEE Trans. Inform. Theory.

[61] R. Murenzi, in ref. 44, p.239; Thesis, University of Louvain la Neuve (1990).

[62] P. Grassberger, R. Badii and A. Politi, J. Stat. Phys. **51**, 135 (1988).

[63] J.D. Farmer, E. Ott and J.A. Yorke, Physica D **7**, 153 (1983).

[64] A.B. Chhabra, R.V. Jensen and K.R. Sreenivasan, Phys. Rev. A **40**, 4593 (1989).

[65] J.M. Ghez and S. Vaienti, J. Stat. Phys. **57**, 415 (1989).

[66] E. Bacry, J.F. Muzy and A. Arneodo, "Fractal dimensions from wavelet analysis: exact results", in preparation.

[67] A. Arneodo, G. Grasseau and E.J. Kostelich, Phys. Lett. A **124**, 426 (1987).

[68] R. Badii and G. Broggi, Phys. Lett. A **131**, 339 (1988).

[69] B.B. Mandelbrot, in ref. 16, p. 279.

[70] W. Van der Water and P. Schram, Phys. Rev. A **37**, 3118 (1988).

[71] A.B. Chhabra and R.V. Jensen, Phys. Rev. Lett. **62**, 1327 (1989).

[72] A.B. Chhabra, C. Meneveau, R.V. Jensen and K.R. Sreenivasan, Phys. Rev. A **40**, 5284 (1989).

[73] J.S. Langer, Rev. Mod. Phys. **52**, 1 (1980); in *Chance and Matter*, edited by J. Souletie, J. Vanimenus and R. Stora (North-Holland, Amsterdam, 1987).

[74] U. Nakaya, *Snow Crystals* (Harvard University Press, Harvard, 1954).

[75] W.A. Bentley and W.T. Humphreys, *Snow Crystals* (Dover, New York, 1962).

[76] T. Vicsek, J. Phys. A **16**, L647 (1983).

[77] T.A. Witten and L.M. Sander, Phys. Rev. Lett. **47**, 1400 (1981); Phys. Rev. B **27**, 5686 (1983).

[78] M. Tokuyama and K. Kawasaki, Phys. Lett. A **100**, 337 (1984).

[79] L.A. Turkevich and H. Scher, Phys. Rev. Lett. **55**, 1026 (1985).

[80] R.C. Ball, Physica A **140**, 62 (1986).

[81] T.C. Halsey, P. Meakin and I. Procaccia, Phys. Rev. Lett. **56**, 854 (1986).

[82] K. Honda, H. Toyoki and M. Matsushita, J. Phys. Soc. Jpn **55**, 707 (1986).

[83] M. Matsushita, K. Honda, H. Toyoki, Y. Hayakawa and H. Kondo, J. Phys. Soc. Jpn. **55**, 2618 (1986).

[84] P. Meakin, Phys. Rev. A **27**, 1495 (1983); Phys. Rev. A **33**, 3371 (1986).

[85] P. Meakin and Z.R. Wasserman, Chem. Phys. **91**, 404 (1984).

[86] P. Meakin and L.M. Sander, Phys. Rev. Lett. **54**, 2053 (1985).

[87] R.C. Ball and R.M. Brady, J. Phys. A **18**, L809 (1985).

[88] P. Meakin, R.C. Ball, P. Ramanlal and L.M.Sander, Phys. Rev. A **35**, 5233 (1987).

[89] J. Nittmann and H.E. Stanley, Nature **321**, 663 (1986); J. Phys. A: Math. Gen. **20**, L1185 (1987).

[90] L.M. Sander, P. Ramanlal and E. Ben-Jacob, Phys. Rev. A **32**, 3160 (1985).

[91] L.M. Sander, in ref. 13, p. 241.

[92] P. Ramanlal and L.M. Sander, J. Phys. A **21**, L995 (1988).

[93]   P. Meakin and S. Havlin, Phys. Rev. A **36**, 4428 (1987).

[94]   P. Meakin, J. Phys. A **18**, L661 (1985).

[95]   G. Li, L.M. Sander and P. Meakin, Phys. Rev. Lett. **63**, 1322 (1989).

[96]   F. Argoul, A. Arneodo, G. Grasseau and H.L. Swinney, Phys. Rev. Lett. **63**, 1323 (1989).

[97]   D. Bensimon, L.P. Kadanoff, S. Liang, B.I. Shraiman and C. Tang, Rev. Mod. Phys. **58**, 977 (1986).

[98]   J. Nittmann, G. Daccord and H.E. Stanley, Nature **314**, 141 (1985).

[99]   L. Paterson, J. Fluid. Mech. **113**, 513 (1981).

[100] S.N. Rauseo, P.D. Barnes and J.V. Maher, Phys. Rev. A **35**, 1245 (1987).

[101] A.R. Kopf-Sill and G.M. Homsy, Phys. Fluids **31**, 242 (1988).

[102] Y. Couder, in ref. 16, p. 75.

[103] H. Thomé, M. Rabaud, V. Hakim and Y. Couder, J. Phys. Fluids A **1**, 224 (1989).

[104] G. Daccord, J. Nittmann and H.E. Stanley, Phys. Rev. Lett. **56**, 336 (1986).

[105] I. Procaccia and R. Zeitak, Phys. Rev. Lett. **60**, 2511 (1988).

[106] G.M. Dimino and J.H. Kaufman, Phys. Rev. Lett. **62**, 2277 (1989).

[107] J. Feder, E.L. Hinrischen, K.J. Maloy and T. Jossang, Physica D **38**, 104 (1989).

[108] L.M. Sander, in ref. 15, p. 257.

[109] D.G. Grier, D.A. Kessler and L.M. Sander, Phys. Rev. Lett. **59**, 2315 (1987).

[110] D.G. Grier, E. Ben-Jacob, R. Clarke and L.M. Sander, Phys. Rev. Lett. **56**, 1264 (1986).

[111] Y. Sawada, A. Dougherty and J.P. Gollub, Phys. Rev. Lett. **56**, 1260 (1986).

[112] F. Argoul and A. Arneodo, J. Phys. France **51**, 2477 (1990).

[113] F. Argoul, A. Arneodo, G. Grasseau and H.L. Swinney, Phys. Rev. Lett. **61**, 2558 (1988).

[114] F. Argoul, A. Arneodo, J. Elezgaray and G. Grasseau, in *Measures of Complexity and Chaos*, edited by N.B. Abraham, A.M. Albano, A. Passamante and P.E. Rapp (Plenum, New York, 1989) p. 433.

[115] R.M. Brady and R.C. Ball, Nature **309**, 225 (1984).

[116] M. Matsushita, M. Sano, Y. Hayakawa, H. Honjo and Y. Sawada, Phys. Rev. Lett. **53**, 286 (1984).

[117] M. Matsushita, Y. Hayakawa and Y. Sawada, Phys. Rev. A **32**, 3814 (1985).

The Continuous Wavelet Transform of Two-Dimensional Turbulent Flows[*]

MARIE FARGE

Laboratoire de Météorologie
Dynamique du C.N.R.S.
École Normale Supérieure
24 rue Lhomond
F-75231 Paris France

'In the last decade we have experienced a conceptual shift in our view of turbulence. For flows with strong velocity shear ... or other organizing characteristics, many now feel that the *spectral description has inhibited fundamental progress*. The next "El Dorado" lies in the *mathematical understanding of coherent structures in weakly dissipative fluids*: the formation, evolution and interaction of metastable vortex-like solutions of nonlinear partial differential equations ... .' Norman Zabusky (1984)

INTRODUCTION

I have chosen to present here a personal point of view concerning the current state of our understanding of fully-developed turbulence. By this I mean the study of dissipative flows in the limit of large Reynolds numbers

[*]Based upon a talk given in Paris on May 5, 1990 on the occasion of the 'Journée Annuelle de la Société Mathématique de France' and originally printed (in French) by the Société Mathématique de France. The translation published here does not include some elementary material on wavelets which appeared in section 2 of the French version; furthermore, this version has been slightly revised and updated to include some recent developments.

(the Reynolds number being a dimensionless number characterizing the ratio of the nonlinear advection to the linear dissipation), that is, the limit where the dissipation becomes negligible so that the dynamics of the flow is essentially dominated by the nonlinear interactions.

After more than a century of turbulence study (Reynolds, 1883), no convincing theoretical explanation has given rise to a consensus among physicists (for a historical review of the various theories of turbulence, see (Von Neumann, 1949), (Monin and Yaglom, 1975), (Farge, 1990). In fact, there exist a large number of *ad hoc* models, called 'phenomenological', that are widely used by fluid mechanicians to interpret experiments and to compute many industrial applications where turbulence plays a role. However, it is still not known whether fully-developed turbulence actually has the universal behavior (independent of initial conditions and boundary conditions) assumed for it in the limit of infinitely large Reynolds numbers and infinitely small scales. Already in 1979, in an unpublished article (Farge, 1979), I expressed reservations about our understanding of turbulence and thought that we did not yet know which are the 'good questions' to ask. Ten years of work on the subject have persuaded me that we have not yet identified the 'good objects,' by which I mean the structures and elementary interactions from which it will be possible to construct a satisfying statistical theory of fully-developed turbulence.

In my opinion, and as Zabusky expresses in the quotation (Zabusky, 1984) I have used as an epigraph to this article, ignorance of the elementary physical mechanisms at work in turbulent flows arises in part from the fact that we reason in Fourier modes (wave vectors), constructed from functions that are not well-localized; this viewpoint ignores the presence of the coherent structures that can be observed in physical space and whose dynamic role seems essential to us. In fact, these coherent structures are observed both in experiments carried out in the laboratory (Jimenez, 1981), (Van Dyke, 1982), (Couder and Basdevant, 1986) and in numerical experiments based on the fundamental equations of fluid mechanics (Kim and Moin, 1979), (Basdevant, Legras, Sadourny and Béland, 1981), (McWilliams, 1984), (Farge and Sadourny, 1989), but the current statistical theory (Monin and Yaglom, 1975) does not take them into account. Thus, the visualization of the evolution of two-dimensional turbulent fields numerically computed (Figure 1) leads us to conjecture that the dynamics of a two-dimensional turbulent flow is essentially dominated by the interactions between the coherent structures that advect the residual flow situated between them; the latter itself seems to play no dynamic role. We think that this point of view can be generalized to three dimensions as well, because the existence of coherent structures has also been observed in the context of three-dimensional flows (Kim and Moin, 1979), (Jimenez, 1981), (Hussain, 1986), but their topology is more complex (Moffatt, 1990). Consequently, the wavelet transform, which decomposes the fields on a set of functions with compact (or quasicompact) support and thus permits an analysis in both space *and* scale, seems to be a good tool, not only for analyzing and interpreting the experimental results obtained in two-dimensional turbulence, but also in the long term for attempting to construct a more satisfactory statistical theory of fully-developed turbulence (see color plate for Figure 1).

1. FULLY-DEVELOPED TURBULENCE

1.1. THE EQUATIONS

The fundamental equation of the dynamics of an incompressible (constant density throughout time) and Newtonian (stress proportional to the velocity gradients) fluid is the Navier-Stokes equation:

$$\begin{cases} \partial_t \vec{V} + (\vec{V} \cdot \nabla)\vec{V} + \frac{1}{\rho}\nabla P = \nu \nabla^2 \vec{V} + \vec{F}, \\ \nabla \cdot \vec{V} = 0, \\ \text{initial conditions,} \\ \text{boundary conditions,} \end{cases} \tag{1}$$

where $\vec{V}$ is the velocity, $\vec{F}$ is the resultant of the external forces per unit of mass, and ν is the kinematic viscosity.

We remark here that the mathematical intracability of the Navier-Stokes equation arises from the fact that the small parameter ν, which tends to zero in the limit of large Reynolds numbers, i.e. for very turbulent flows, appears in the term containing the highest-order derivative, namely the dissipation term $\nu \nabla^2 \vec{V}$. Thus the character of the equation, which is given by the term containing the highest-order derivative, changes as ν tends to zero, since in this limit it is the advection term $(\vec{V} \cdot \nabla)\vec{V}$ that dominates. When $\nu = 0$, or Re $= \infty$, the Navier-Stokes equation is called Euler's equation and the nonlinear advection term is no longer controlled by the linear dissipation term. Moreover, Euler's equation conserves energy whereas the Navier-Stokes equation dissipates it; thus the former is reversible in time whereas the latter is irreversible.

If one takes the curl of equation (1), one can eliminate the pressure term; this gives the equation of the curl of the velocity, also called the vorticity

$\vec{\omega} = \nabla \times \vec{V}$, as:

$$\begin{cases} \partial_t \vec{\omega} + (\vec{V} \cdot \nabla)\vec{\omega} = (\vec{\omega} \cdot \nabla)\vec{V} + \nu \nabla^2 \vec{\omega} + \nabla \times \vec{F} \\ \vec{\omega} = \nabla^2 \psi. \end{cases} \qquad (2)$$

If one considers a regime state, i.e. a state of the flow such that the energy contribution from the external forces is dissipated by the viscous friction, then:

$$\frac{d\vec{\omega}}{dt} = (\vec{\omega} \cdot \nabla)\vec{V}. \qquad (3)$$

Thus, in three dimensions, the Lagrangian variation of vorticity is equal to the product of the vorticity by the velocity gradients, which leads to stretching of the vorticity tubes by the velocity gradients, a mechanism that may explain the transfer of energy towards the smallest scales of the flow in three dimensions (cf. Section 1.2).

In two dimensions, vorticity becomes a pseudo-scalar, for $\vec{\omega} = (0, 0, \omega)$ is then perpendicular to $\nabla \vec{V}$. Therefore in this case, the vorticity stretching by the velocity gradients is no longer possible. Indeed in two dimensions, the vorticity is a Lagrangian invariant of the motion because, in the absence of dissipation, it is conserved throughout time along a fluid trajectory:

$$\frac{d\vec{\omega}}{dt} = 0. \qquad (4)$$

If we now consider the vorticity gradients, we have:

$$\partial_t \overrightarrow{\nabla \omega} + (\vec{V} \cdot \nabla)\overrightarrow{\nabla \omega} = -(\overrightarrow{\nabla \omega} \cdot \nabla)\vec{V}. \qquad (5)$$

Thus, in two dimensions, the Lagrangian variation of the vorticity gradients is equal to the product of the vorticity and velocity gradients, a mechanism that may explain the transfer of enstrophy towards the smallest scales of the flow in two dimensions (cf. Section 1.2).

1.2. THE INVARIANTS

In the absence of external forces ($\vec{F} = 0$) and of dissipation ($\nu = 0$), Euler's equation (i.e. the Navier-Stokes equation for $\nu = 0$) conserves energy in two and three dimensions:

$$\mathcal{E}(t) = \frac{1}{2} \int_{-\infty}^{+\infty} V^2(\vec{x}, t) d^n \vec{x} = \text{constant}, \qquad (6)$$

where n is the dimension of the space. Using Plancherel's identity, we have:

$$\mathcal{E}(t) = \frac{1}{2} \int_{-\infty}^{+\infty} \hat{V}^2(\vec{k}, t) d^n \vec{k} = \int_0^\infty \mathrm{E}(|\vec{k}|, t) d^n |\vec{k}| = \text{constant}, \qquad (7)$$

where

$$\hat{V}(\vec{k}) = \int_{-\infty}^{+\infty} V(\vec{x}) e^{i\vec{k}\cdot\vec{x}} d^n\vec{x}$$

and $E(k)$ is the energy integrated in spectral space over crowns of constant radius $|\vec{k}|$. $E(k)$ characterizes the distribution of energy among the various scales (in the sense of wave numbers) of the motion, a modal distribution predicted by the statistical theory (cf. Section 1.4).

In the special case of two dimensions, which is particularly interesting for studying the large-scale dynamics of geophysical flows for which the two-dimensional approximation is valid, Euler's equation also preserves enstrophy:

$$\Omega(t) = \frac{1}{2} \int_{-\infty}^{+\infty} \omega^2(\vec{x}, t) d^2\vec{x} = \text{constant} \tag{8}$$

or using Plancherel's identity:

$$\Omega(t) = \frac{1}{2} \int_{-\infty}^{+\infty} \hat{\omega}^2(\vec{k}, t) d^2\vec{k} = \int_0^{\infty} Z(|\vec{k}|, t) d^2|\vec{k}| = \text{constant} \tag{9}$$

where

$$\hat{\omega}(\vec{k}) = \int_{-\infty}^{+\infty} \omega(\vec{x}) e^{i\vec{k}\cdot\vec{x}} d^2\vec{x}$$

and $Z(k)$ is the enstrophy integrated in spectral space over crowns of constant radius $|\vec{k}|$. $Z(k)$ characterizes the distribution of enstrophy among the various scales (wave numbers) of the motion and, on imposing hypotheses of homogeneity and of isotropy in the statistical sense (cf. Section 1.4), one can relate it to the modal energy and obtain:

$$Z(k) = k^2 E(k) \tag{10}$$

where $k = |\vec{k}|$.

1.3. UNIQUENESS, REGULARITY AND ANALYTICITY OF THE SOLUTIONS

We shall attempt to summarize briefly the existing theorems on uniqueness, regularity and analyticity of the solutions of the Euler and Navier-Stokes equations in two and three dimensions. In general, these various theorems consider regular initial conditions, i.e. they use C^∞ functions with bounded support in $\mathbf{R}^n$. In this case, in two dimensions the Lagrangian conservation of vorticity (4) implies that:

$$\|\omega\|_{L^\infty} = \sup |\omega(x, y)| = \text{constant}, \tag{11}$$

which leads to the prediction that Euler's solutions remain regular for all times in a bounded domain (Lichtenstein, 1925), (Wolibner, 1933), (Hölder, 1933), (Schaeffer, 1937). Kato (Kato, 1972) has shown that this remains true even if the initial velocity is $C^{1+\epsilon}$ ($\epsilon > 0$). However, we still do not have a regularity theorem for the Euler solutions in an unbounded domain, unless the solutions are constrained to decrease at infinity. An interesting special case, worth noting here, is that of the Kelvin-Helmholtz instability that develops at the interface between two flows of different velocities; if the initial flow presents a discontinuity of the velocity at the interface, it was conjectured by Birkhoff (Birkhoff, 1962), and then proved (Babenko and Petrovich, 1979), (Sulem et al, 1981), that if the interface is initially an analytic curve, then it remains so for a finite time. However, an asymptotic expansion of Moore (Moore, 1979) and numerical results (Meiron and Baker, 1982) suggest that this curve will ultimately always develop a singularity. In two dimensions, the global regularity of the Navier-Stokes equation in an unbounded domain — and this for any viscosity — is a consequence of the regularity of Euler's equation for a bounded domain. Ladyzhenskaya (Ladyzhenskaya, 1963) and Lions (Lions, 1969) have proven the global regularity of the Navier-Stokes equation in two dimensions, provided that the viscosity is sufficiently high. However, for the limit of ν tending to zero, the problem remains open because one does not know how to take into account the boundary layers that develop at the walls.

In three dimensions one shows that, assuming regular initial conditions, one has uniqueness, regularity and analyticity in the following cases:

- for all times, provided that the viscosity is high enough (Reynolds < 1 initially) (Leray, 1933);

- for arbitrary viscosity and arbitrary boundaries, provided that time is sufficiently long (Leray, 1933), or for short times in the absence of boundaries (Kato, 1972);

- for all times, but for dissipation of the form $-\nu'(-\nabla^2)^\alpha$, where $\nu' > 0$ and $\alpha \geq 5/4$ (Ladyzhenskaya, 1963), (Lions, 1969).

For more detailed expositions on this subject, see (Rose and Sulem, 1978), (Frisch, 1983) and (Temam, 1984).

We note here that the Lagrangian conservation of vorticity (4) for Euler's equation in two dimensions implies that, if the initial field of vorticity has N singularities, these will be conserved for all time, and thus there will not be any regularization of the flow in the absence of dissipation (Farge and Holschneider, 1990); we will return to this point in the last part of the article.

1.4. STATISTICAL THEORY AND SPECTRAL SLOPES

It is first of all necessary to note that the statistical theory of fully-developed turbulence, which is attributed to Kolmogorov, was discovered quasi-simultaneously by him (Kolmogorov, 1941a,b,c), and others (Obukhov, 1941), (Onsager, 1945), (Heisenberg, 1948), and (Von Weizsäcker, 1948), each using different methods that we shall not present here. (see (Battimelli and Vulpiani,1982) for a review of the history of this subject.) Kolmogorov studied the way in which the Navier-Stokes equation in three dimensions distributes the energy among the various degrees of freedom of the flow. This type of approach is very classical in statistical mechanics, but the difficulty here arises from the fact that turbulent flow is an open thermodynamical system, that is, not isolated from the exterior, due to the forces acting on the flow either at large scale (external forces) or at small scale (viscous frictional forces). It is therefore necessary to limit oneself to a range of intermediate scales, called the inertial range, where one supposes that energy is transferred between the various degrees of freedom, namely the scales of the flow, and this in a conservative manner. We thus have:

$$\lambda \ll \ell \ll L, \qquad (12)$$

where

- λ denotes the dissipative scales where kinetic energy is transformed into thermal energy under the effect of viscous friction,

- ℓ denotes the scales of turbulent motion dominated by nonlinear advection that transfer kinetic energy between them in a conservative manner, and

- L denotes the integral scales where energy is injected by external forces.

Kolmogorov assumes that for this range of scales ℓ, the flow is statistically homogeneous, that is, invariant under translation, and isotropic, that is, invariant under rotation. ('Statistical' here means in the sense of Gibbs ensemble averages, i.e. in averaging from a set of realizations of the same flow.) He also assumes that the energy is transferred, from the large to the small scales, at a constant rate ε which is independent of scale and equal to the quantity of energy dissipated by the scales smaller than λ. He adds to this the hypothesis that the skewness S , namely the departure from Gaussian behavior of the velocity probability distribution, is constant.

From this he deduces the following scale law for the two-point correlation function of the velocity:

$$\langle |v(r + \ell) - v(r)|^2 \rangle \sim C\varepsilon^{2/3}\ell^{2/3}, \tag{13}$$

where $C = -4\ell/(5S)$ is Kolmogorov's constant. Upon Fourier transforming to the space of wave vectors $\vec{k}$, this yields:

$$\mathcal{E} = \frac{1}{2} \int_{-\infty}^{+\infty} \hat{V}^2(\vec{k}) d^3\vec{k} \sim C\varepsilon^{2/3}k^{-2/3}. \tag{14}$$

Considering now the energy integrated over crowns of constant radius $k = |\vec{k}|$, we obtain the (Kolmogorov) spectrum (Figure 2a):

$$E(k) \sim C\varepsilon^{2/3}k^{-5/3}. \tag{15}$$

Notice here that Kolmogorov never expressed his law in Fourier space, whereas the same result is obtained from the theory of Heisenberg (Heisenberg, 1948) or Von Weizsäcker (Von Weizsäcker, 1948) working directly in spectral space.

Following a remark of Landau (Landau and Lifchitz, 1971) concerning the random character of energy transfers in the inertial zone, Kolmogorov (Kolmogorov, 1961) added to his law (13) a lognormal correction in $(\ln(L/\ell))^\beta$, β being the dispersion constant of the logarithm of ε. Various experiments carried out in wind-tunnels (Batchelor and Townsend, 1949), (Anselmet, Gagne, Hopfinger and Antonia, 1984) have shown that the energy associated with the small scales of a turbulent flow is not densely distributed in space. This observation of a spatial intermittency of the support of the energy transfers has led several authors to conjecture that this support is fractal (Mandelbrot, 1975 and 1976), (Frisch, Sulem and Nelkin, 1978) or multifractal (Parisi and Frisch, 1985), which also gives rise to a correction of the Kolmogorov spectrum (15), of the form $(kL)^{-(3-D/3)}$, D being the Hausdorff dimension of the dissipative structures.

In two dimensions, the conservation of enstrophy, related to energy by the relation (10), leads to a modification of the statistical theory, conjectured by Von Neumann (Von Neumann, 1949), since it prevents the energy from cascading from large to small scales in the limit k tending to infinity. The energy is, on the contrary, transferred to the large scales according to a spectral law similar to that of Kolmogorov (15); this is the inverse energy cascade of two dimensional turbulence (Figure 2b). Kraichnan (Kraichnan, 1967) and Batchelor (Batchelor, 1969) have shown that there is then another cascade, but of enstrophy (9) from large to small scales (Figure 2b), and, assuming the rate of enstrophy transfer η to be

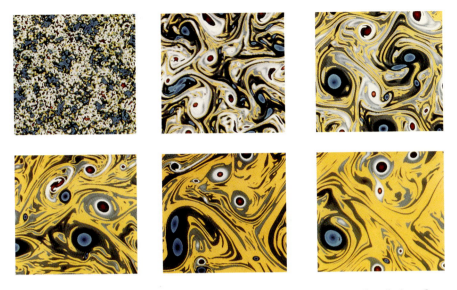

FIGURE 1. Direct numerical simulation of a decaying two-dimensional turbulent flow: time evolution during 10^5 time steps computed from a random initial vorticity field (Farge 1988, Farge and Sadourny 1989).

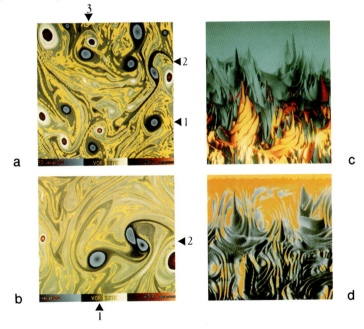

FIGURE 4. Dynamics and topology of coherent structures in two-dimensional turbulence.
a. and b. Elementary interactions: a1) Axisymmetrization, a2) Filamentation,
a3) Binding, b1) Deformation, b2) Merging.
c. and d. Cusp-like shape of the most excited coherent structures (Farge 1988, Farge and Sadourny 1989).

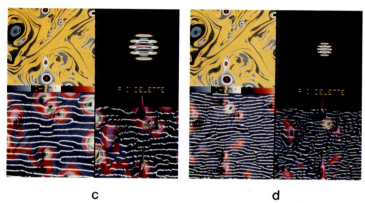

a

b

c

d

FIGURE 7. Two-dimensional wavelet analysis computed in L1-norm, using Morlet wavelet with $k\Psi = 5$ and $\theta = 0$, of the same two-dimensional turbulent flow as Figure 6 (Farge, Holschneider, and Colonna 1990).

On the display we have superposed the vorticity field, in perspective representation, the wavelet coefficients module, color coded in order of increasing luminance (blue, red, magenta, green, cyan, yellow, white), and the zeroes of the wavelet coefficients phase, represented by white isolines.

a. Vorticity field to be analyzed (sampled on 512^2 points).
b. Wavelet coefficients at scale $k = 8$.
c. Wavelet coefficients at scale $k = 16$.
d. Wavelet coefficients at scale $k = 32$.

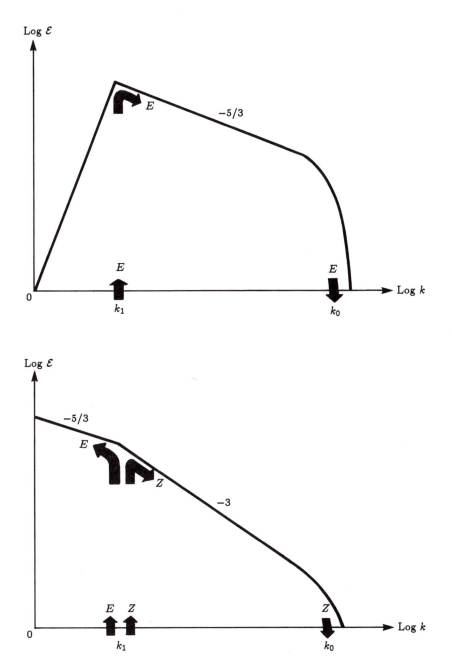

FIGURE 2. Fourier energy spectra predicted by the statistical theory of turbulence: a. Kolmogorov spectrum for three-dimensional turbulence (Kolmogorov 1941a, b, c), b. Kraichnan-Batchelor spectrum for two-dimensional turbulence (Kraichnan 1967, Batchelor 1969).

constant in the inertial range, they predicted the following energy spectrum for two dimensional turbulence:

$$E(k) \sim \eta^{2/3} k^{-3} . \tag{16}$$

Kraichnan (Kraichnan, 1971) added a correction term in $(\ln(kL))^{-1/3}$ to account for the fact that in two dimensional turbulence the transfers are not local in spectral space.

1.5. NUMERICAL SIMULATIONS AND COHERENT STRUCTURES

We shall limit ourselves here to the case of incompressible two-dimensional turbulence. It is very difficult to carry out laboratory experiments under rigorously two dimensional conditions, whereas numerical simulations can attain Reynolds numbers much higher in two than in three dimensions because, for in two dimensions the number of mesh points varies directly with $(Re)^1$, while in three dimensions it varies as $(Re)^{9/4}$. In general, the numerical experiments are carried out with periodic boundary conditions and are initialized with random fields whose energy is distributed over a large spectral band, usually up to the cutoff scale of the mesh (Figure 3a). We then follow the flow evolution both in physical space, by visualizing the vorticity field (Figure 1) which is the most significant quantity since it is a Lagrangian invariant of the flow for $\nu = 0$, as well as in spectral space, by plotting the energy spectrum integrated over the crowns $|\vec{k}| = $ constant (Figure 3b).

In the numerical experiments of two dimensional turbulence, the observed energy spectra usually follow a power law in k^{-4} (Figure 2b), and not in k^{-3} as is predicted by the theory of Kraichnan (Kraichnan, 1967) (Figure 2b). It is thought that this disparity is due to the intermittency of the flow, for which we shall propose a geometric interpretation (cf. Section 2.2) based on the presence of coherent structures we observe in numerical experiments, but that are not addressed by the statistical theory.

What is a coherent structure? We do not presently have any theory to describe them, therefore we must content ourselves with a qualitative description, more similar to the approach of a zoologist than to that of a fluid mechanician. But this taxonomic and descriptive stage is a necessary preliminary to any further theory. Basing ourselves on visualizations of numerically computed two-dimensional turbulent fields (Figure 1), we may characterize coherent structures in the following way:

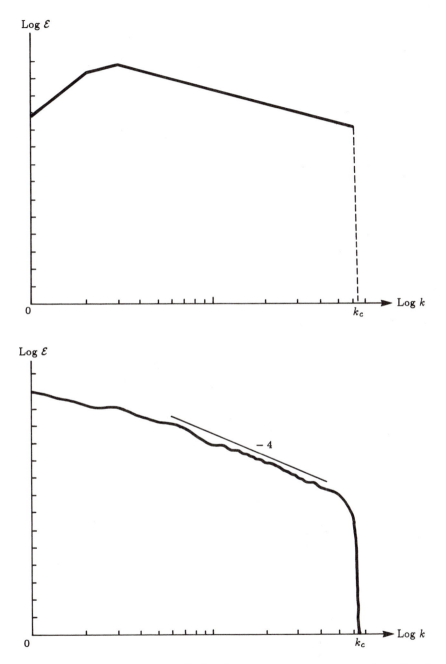

FIGURE 3. Time evolution of the Fourier energy spectrum of a decaying two-dimensional turbulent flow numerically computed (Farge 1988, Farge and Sadourny 1989): a. Initial energy spectrum, b. Energy spectrum after 10^5 time steps.

- they are vortical structures, that is, regions of the flow where vorticity prevails over deformation,

- they contain most of the energy and enstrophy of the flow,

- they form spontaneously by a condensation of the vorticity field, for which we do not have any theory at the present time,

- they are encountered over a large range of scales; in fact throughout the whole inertial range if there is no forcing,

- they survive on time scales much larger than the eddy turnover time $\tau = Z^{1/2}$ selected by the statistical theory as being the characteristic time of the enstrophy transfers.

When one studies the dynamics of coherent structures, one can distinguish the following states and elementary interactions:

1. *relaxed state*, characterized by

    - *aximetrization* (see arrow 1 on Figure 4a) in the absence of interaction with nearby coherent structures,

2. *weakly excited states*, with

    - either *deformation* (see arrow 1 on Figure 4b) under the influence of a nearby coherent structure having the same intensity,

    - or *filamentation* (see arrow 2 on Figure 4a) when the deformation becomes too strong under the influence of a more intense nearby coherent structure,

3. *strongly excited states*, with

    - either *binding* (see arrow 3 on Figure 4a) of two very close coherent structures of opposite sign and comparable intensity,

    - or *merging* (see arrow 2 on Figure 4b) of two or more very close coherent structures having the same sign.

(See color plate for figure 4.)

Although we have defined coherent structures in a purely qualitative manner using observations obtained from numerical simulations, it is equally possible to characterize them in a more quantitative manner by studying the velocity gradient tensor, called the stress tensor:

$$\nabla V = (\nabla V)_{ij} = \partial_j V_i = V_{i,j}. \tag{17}$$

Its symmetric part characterizes the strain undergone by the fluid element, and is written:

$$\frac{1}{2}(\nabla V + \nabla V^t) = \begin{vmatrix} \partial_1 v_1 & \frac{1}{2}(\partial_2 v_1 + \partial_1 v_2) \\ \frac{1}{2}(\partial_1 v_2 + \partial_2 v_1) & \partial_2 v_2 \end{vmatrix} = \frac{1}{2}\begin{vmatrix} s_1 & s_2 \\ s_2 & -s_1 \end{vmatrix}, \quad (18)$$

where ∇V^t is the transposed matrix of ∇V, and $s_1 = 2\partial_1 v_1 = -2\partial_2 v_2$ since the fluid is incompressible.

Its antisymmetric part corresponds to the rotation of the fluid element, and is written:

$$\frac{1}{2}(\nabla V - \nabla V^t) = \begin{vmatrix} 0 & \frac{1}{2}(\partial_2 v_1 - \partial_1 v_2) \\ \frac{1}{2}(\partial_1 v_2 - \partial_2 v_1) & 0 \end{vmatrix} = \frac{1}{2}\begin{vmatrix} 0 & \omega \\ -\omega & 0 \end{vmatrix}. \quad (19)$$

Adding (18) and (19), the equation (17) becomes:

$$\nabla V = \frac{1}{2}\begin{vmatrix} s_1 & s_2 + \omega \\ s_2 - \omega & -s_1 \end{vmatrix}. \quad (20)$$

Calculating the curl of the vorticity equation (2), with neither forcing ($\vec{F} = 0$) nor dissipation ($\nu = 0$), one obtains the equation of the curl of the vorticity, also called the divorticity $\xi = \nabla \times \omega = (\partial_2 \omega - \partial_1 \omega, 0)$:

$$\left\{ \begin{matrix} \partial_1 \xi + (\vec{V} \cdot \nabla)\xi = (\nabla V)\xi \\ \nabla \cdot \xi = 0 \end{matrix} \right\}. \quad (21)$$

Assuming that the spatio-temporal variations of the strain tensor ∇V are slow compared to those of the curl of the vorticity ξ (the hypothesis of (Weiss, 1981)), equation (21) becomes linear in Lagrangian coordinates:

$$\left\{ \begin{matrix} \omega(x_0, y_0, t) = \omega_0 \\ \frac{d\xi}{dt} = (\nabla V)\xi \end{matrix} \right\}. \quad (22)$$

The eigenvalues of the stress tensor are

$$\alpha = \pm\frac{1}{2}\sqrt{-\det(\nabla V)} = \pm\frac{1}{2}[s_1^2 + s_2^2 - \omega^2]^{1/2}. \quad (23)$$

One can then separate the flow into two types of regions where the Lagrangian dynamic is different:

a) **elliptic regions** (Figure 5a) corresponding to the purely imaginary eigenvalues, where rotation dominates strain and for which two initially close fluid particles remain nearby for all time, their distance oscillating only slightly throughout time. These regions are thus associated with the geometrically stable coherent structures,

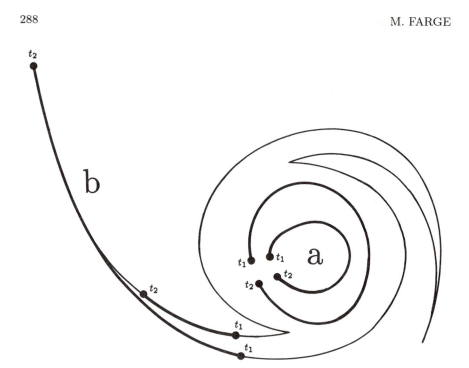

FIGURE 5.   Dynamical characterization of a coherent structure in two-dimensional turbulence. a. Elliptic region corresponding to the coherent structure: two nearby particles at t_1 will remain close at t_2. b. Hyperbolic region corresponding to the vorticity filaments emitted by the coherent structure during a deformation: two nearby particles at t_1 will separate exponentially after t_2.

  b) **hyperbolic regions** (Figure 5b) corresponding to the real eigenvalues of opposite sign, where strain dominates rotation and for which two initially close fluid particles separate exponentially as time goes, their distance being contracted in one direction and dilated in the other. These regions are thus associated with the vorticity filaments stretched by the velocity gradients.

    In conclusion, in view of the numerical simulations we have carried out, we conjecture that the dynamics of two-dimensional turbulent flows is essentially dominated by the interactions between the coherent structures that advect the residual flow. This background flow is formed by the vorticity filaments emitted during the vortex interactions, but it only plays a passive role, contrarily to the coherent structures which are dynamically active. If this conjecture is verified, it then becomes important to find a method capable of separating the coherent structures from the background flow, not by a thresholding technique such as the one we have just presented, but rather by using a filtering technique that respects the local regularity of the flow (cf. Section 2.3).

2. WAVELET ANALYSIS OF TWO-DIMENSIONAL TURBULENT FLOWS

2.1. RESULTS

It is important to realize that the wavelet transform is not being used to study turbulence simply because it is currently fashionable; but rather because we have been searching for a long time for a technique capable of decomposing turbulent flows in both space *and* scale simultaneously. If, under the influence of the statistical theory of turbulence, we had lost in the past the habit of considering the flow evolution in physical space, we have now recovered it thanks to the advent of supercomputers and their associated means of visualization. They have revealed to us a menagerie of turbulent flow patterns, namely, the existence of coherent structures and their elementary interactions (cf. Section 1.5) for which the present statistical theory is not adequate.

During the 1980's, I was very involved with displays of turbulent fields in physical space, and I have proposed a normalization for their representations in order to compare results obtained by numerical simulations and by laboratory experiments which was essentially done in morphological terms (Farge, 1987), (Farge, 1990a). Just before Alex Grossman first spoke to me about wavelets in 1984, I had envisioned making bandpass filters in two-dimensional Fourier space and then reconstructing the filtered field in physical space, spectral band by spectral band, so as to match up certain interactions observed in physical space with the turbulence cascades predicted by the statistical theory in Fourier space. This method, however, ran into problems because of the Gibbs phenomenon, which occurs in physical space when the frequency filtering is too abrupt. The wavelet transform now allows us to analyze two-dimensional turbulence in a much more satisfactory way, and it provides us with new mathematical tools for analyzing the local regularity of a function (Holschneider and Tchamitchian, 1988), (Jaffard, 1989). In an earlier paper (Farge and Rabreau, 1988), we carried out one-dimensional wavelet transforms, using Morlet's wavelet, of sections of several vorticity fields numerically computed (Farge, 1988). This showed how, during the flow evolution, starting from an initial random distribution of vorticity, the smallest scales of the flow become more and more localized and concentrated in the centers of coherent structures (Figure 6). We have subsequently confirmed this result using a two-dimensional Morlet wavelet (Farge, Holschneider and Colonna, 1990) (Figure 7). The smallest scales of the flow are localized in the coherent structures cores and are excited when the latter are deformed by interactions with other nearby structures; this led us to conjecture that, contrary to generally accepted ideas, dissipation also acts in the center of coherent structures. This is confirmed when one visualizes the vorticity Laplacian field which corresponds to the dissipation term; the maxima of the dissipation are well localized in the cores of the coherent structures (Farge, Holschneider and Colonna, 1990). Moreover, some enstrophy dissipation is necessary inside two same-sign coherent structures which are strongly interacting in order that they ultimately merge, which corroborates our hypothesis. (See color plate for Figure 7)

2.2.   *INTERPRETATION*

Now we will present a new model (Farge, 1990c), (Farge and Holschneider, 1990), (Farge, Holschneider and Philipovitch, 1991) in which we relate the energy spectrum of two-dimensional turbulent flows to the presence of cusp-like, or quasi-singular, coherent structures. This model was suggested by our previous wavelet analysis of two-dimensional turbulent fields.

   Consider the immediate neighborhood of a coherent structure which, by definition, is a quasi-stationary solution of Euler's equation

$$\left\{ \begin{array}{c} \partial \omega / \partial t + J(\psi, \omega) = 0 \\ \omega = \nabla^2 \psi \end{array} \right\} \tag{24}$$

where ω denotes the vorticity and ψ the stream function, and suppose that locally, in a domain as small as one wishes, this solution is axially symmetric, i.e.

$$\omega \sim r^\alpha \qquad \text{with} \qquad \alpha \in \mathbf{R} \tag{25}$$

where r denotes the distance to the center of the coherent structure. Then one infers the following scaling laws:

for the circulation

$$\Gamma(r) = 2\pi \int_0^r \omega(r')\, r'\, dr' \sim r^{\alpha+2}, \tag{26}$$

for the energy

$$\varepsilon(r) = 2\pi \int_0^r V(r')^2 r'\, dr' \sim r^{2\alpha+4}, \tag{27}$$

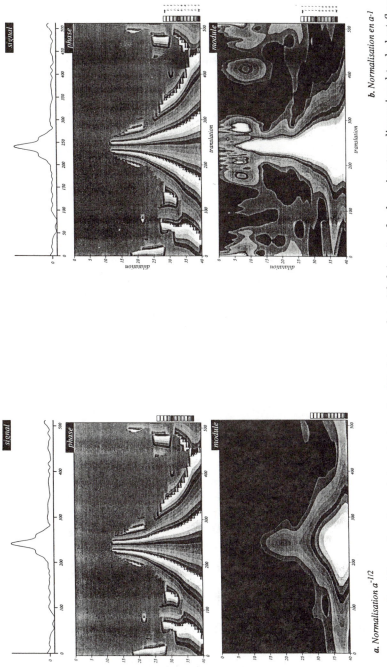
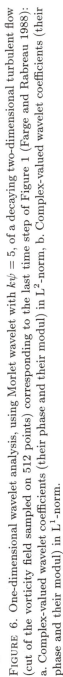

FIGURE 6. One-dimensional wavelet analysis, using Morlet wavelet with $k\psi = 5$, of a decaying two-dimensional turbulent flow (cut of the vorticity field sampled on 512 points) corresponding to the last time step of Figure 1 (Farge and Rabreau 1988): a. Complex-valued wavelet coefficients (their phase and their modul) in L^2-norm, b. Complex-valued wavelet coefficients (their phase and their modul) in L^1-norm.

and for the enstrophy

$$\Omega(r) = 2\pi \int_0^r \omega(r')^2 r' \, dr' \sim r^{2\alpha+2}. \tag{28}$$

In order that these three invariants remain finite, we must have

$$\alpha \geq -1. \tag{29}$$

A coherent structure is characterized by a pointwise relation between the vorticity and the stream function, called the coherent structure function, for which we predict:

$$\omega \sim (\psi - \psi_0)^{\frac{\alpha}{\alpha+2}} \tag{30}$$

where ψ_0 is the value of the stream function at the core of the coherent structure. Finally, calculating the energy in Fourier space and using Plancherel's identity, we obtain the spectral distribution of energy integrated over the crowns $|\vec{k}| = $ constant, which should scale according to a power law of the form

$$E(k) \sim k^{-2\alpha-5}. \tag{31}$$

Relying on the results of numerical experiments (Benzi et al, 1987), we identify the coefficient $\alpha = -1/2$, value which guarantees that the circulation (26), energy (27), and enstrophy (28) all remain finite. In the neighborhood of the coherent structures, we then have the following scaling laws

$$\omega(r) \sim r^{-1/2}, \tag{32}$$

$$\omega(\psi) \sim (\psi - \psi_0)^{-1/3}, \tag{33}$$

$$E(k) \sim k^{-4}. \tag{34}$$

Equation (32) corresponds to a singular axially symmetric vortex distribution in the center of the coherent structures. We have proven (Farge, Holschneider and Philipovitch, 1991) that such cusp-like structures are stable under the flow dynamics, even if they are perturbed by a strong noise. A singular distribution of vorticity in the core of the coherent structures is not a contradiction to the existing theorems (Wolibner, 1933), (Hölder, 1933), (Cottet, 1987) for Euler equation in two dimensions, which predict that in a bounded domain this equation conserves regularity (at least $C^{1+\epsilon}$) and the boundedness (L^∞) of the initial conditions. In fact, we can consider another case where initially one has functions with a finite energy ($L^{4-\epsilon}$) but presenting singularities. Then, due to the Langrangian conservation of vorticity (4), if one has initial singular points ($\omega(x_0, y_0) = \infty$ at $t = 0$), then these initial singularities are advected by the flow, but are conserved for all times ($\omega(x_0, y_0) = \infty$ as $t \to \infty$).

Equation (33) seems confirmed by laboratory experiments of turbulent flow in mercury, where the dynamics is constrained to two dimensions by the presence of a magnetic field (Nguyen Duc and Sommeria, 1988). In such a flow one observes, either a linear coherent structure function, or a cusp-like one very similar in form to (33); this is achieved without modifying the experimental parameters, quite apart from the initial conditions which are never completely reproducible. Our theory permits us to propose the following interpretation of these results: the nature of the coherent structures, characterized by their coherent structure function (30), depends on the initial conditions. If these are *regular*, then the coherence structure function is linear and the distribution of the vorticity is regular with a constant vorticity in the center such that

$$\omega \sim \psi^{\frac{\alpha}{\alpha+2}} \sim \psi \qquad \text{for} \qquad \alpha = \infty, \tag{35}$$

but if the initial distribution has *singularities*, the flow organizes itself around them and one obtains

$$\omega \sim \psi^{-1/3}, \tag{36}$$

corresponding to cusp-like coherent structures of the form

$$\omega \sim r^{-1/2}, \tag{37}$$

with a smoothing of the vorticity in the cores of the vortices caused by dissipation due to viscosity. Contrary to the remark we made previously in Section 1.1, which implied that the nature of the solution to the Navier–Stokes equation may change in the limit $\nu \to 0$ because this singular limit affects the highest order terms of the equation, we are led to the following conjecture: at large scales the singular solutions to Euler's equation dominate, whereas at very small scales, on the order of V/ν, they are locally smoothed by dissipation, and become what we call "quasi-singularities" (Farge and Holschneider, 1990).

Such cusp-like coherent structures are observed by visualizing the vorticity field of two-dimensional turbulent flows, where we see very spiky vortex cores (cf. Figures 4c and 4d). Recently Benzi and Vergassola, by performing a wavelet analysis of numerically computed two-dimensional homogeneous turbulent flows (Benzi and Vergassola, 1990), have confirmed the existance of coherent structures having negative scaling exponents between -0.4 and -0.6, close to the value of -0.5 that we have predicted.

Our theory also suggests a new interpretation of the energy spectrum of two-dimensional turbulent flows in purely geometrical terms, reminiscent of Saffman's interpretation (Saffman, 1971), and quite different from the statistical arguments currently applied to the problem. Thus, according to equation (34), a flow with many coherent structures will have a spectral energy distribution varying as k^{-4} if it has at least one isolated coherent structure with a quasi-singularity in $r^{-1/2}$, up to the dissipative scales where viscosity will smooth the vortex core. Indeed, the spectral slope is determined by the strongest isolated singularity present in the flow. This slope in k^{-4} is steeper than the slope in k^{-3} predicted by the statistical theory of two-dimensional turbulence (Kraichnan, 1967), (Batchelor, 1969), but agrees well with the slopes obtained from most of the numerical experiments of two-dimensional turbulent flows (Basdevant et al, 1981), (McWilliams, 1984), (Farge and Sadourny, 1989). In fact, for these numerical experiments one chooses initial energy spectrum presenting a power-law behavior up to the cut-off scale (Figure 3a), which corresponds to quasi-singular vorticity distributions in physical space (Farge and Holschneider, 1990). The only numerical simulations which obtain k^{-3} energy spectrum (Brachet et al, 1988) are actually initialized with a band-limited spectrum with no energy in the small scales, which corresponds to a smooth vorticity field; in these experiments one does not observe the emergence of isolated coherent structures. If this sensitivity to initial conditions of Euler and Navier-Stokes equations in two dimensions is confirmed, we will have to reconsider the hypothesis of a universal behaviour of turbulence (i.e. independent of the flow initial conditions). Perhaps universality isn't where we think it is; perhaps we should, instead, search for it in the shape of the coherent structures and in their elementary interactions.

2.3. PERSPECTIVES

Applications of the wavelet transform to the theory of turbulence presently follows three directions: analysis, filtering, and numerical integration of turbulent flows. Concerning analysis, the wavelet transform offers the possibility of observing the flow from both sides of the Fourier transform at once (up to the limit of the uncertainty principle); this gives us a method for relating the dynamics of coherent structures in physical space to the redistribution of energy among the various Fourier modes. The wavelet transform is a particularly ideal tool for studying intermittency, one of the major unsolved problems in the theory of turbulence today (Frisch and Orszag, 1990). Indeed, the statistical theory of Kolmogorov (Kolmogorov, 1941a,b,c) assumes that the transfer of energy is distributed densely in physical space, as would be true, for example, for a Gaussian distribution of velocities. But several laboratory results (Batchelor and Townsend, 1949), (Anselmet, Gagne, Hopfinger and Antonia, 1984) seem to violate this hypothesis. Likewise, in two dimensions, we interpret the fact that the numerical results give steeper spectral slopes (of the order of k^{-4}) than the predictions of the statistical theory (k^{-3}) in terms of an intermittency of the enstrophy transfers: the spatial support of enstrophy transfers would not be dense, but would diminish with scale until reaching the dissipative scales, where dissipation would only act on a very small sub-domain of physical space. Many theoretical models, all based on *ad hoc* probabilistic considerations, have been proposed to explain intermittency (Kolmogorov, 1961), (Mandelbrot, 1976), (Frisch, Sulem and Nelkin 1978), (Parisi and Frisch, 1985). The theory we have proposed to interpret our results (cf. Section 2.2), deduced from our wavelet analysis of coherent structures (cf. Section 2.1), allows a purely geometrical interpretation of intermittency: it may be related to the cusp-like shape of some very excited axisymmetric coherent structures whose spatial support would decrease with scale following a power-law behavior until the dissipative scales are reached; this would only represent a very small subdomain in physical space, corresponding to the cores of the coherent structures where vorticity would be locally smoothed by dissipation. Recently, by applying the wavelet transform to analyze three-dimensional flows, we have also found some very strong intermittency that we have related to the presence of coherent structures; this led us to a similar geometrical interpretation of intermittency in three dimensions (Farge, Guezennec, Ho and Méneveau, 1990).

The second direction in which the wavelet transform could play an important role in turbulence is the possibility of extracting coherent sructures from the rest of the flow by filtering the wavelet coefficients. In effect, this would allow us to test our conjecture (cf. Section 1.5) that the dynamics of a turbulent two-dimensional flow may be essentially dominated by nonlinear interactions among the coherent structures, the rest of the flow being only passively advected by them. Thanks to Bruno Torrésani, Pierre Jean Ponenti (Centre de Physique Théorique du CNRS-Luminy) and Richard Kronland-Martinet (Laboratoire de Mécanique et Acoustique du CNRS-Marseille), we have tried a technique (described in the chapter of this book by Tchamitchian and Torrésani) to extract the skeleton of the wavelet coefficients of the vorticity field. This method, which has given interesting results for phase stationary signals (Escudié and Torrésani, 1989), doesn't let us easily separate coherent structures from the rest of the flow. However, the skeleton allows us to delimit the influence cones associated with the coherent structures. Then, after cancelling the wavelet coefficients outside of these influence cones, we may be able to use an inverse wavelet transform to reconstruct new vorticity fields, whose background flow would have been filtered out so that only the coherent structures remain. Another, and quite similar, approach, we are presently trying with Victor Wickerhauser from Yale University, is to perform wavelet packet decomposition of the vorticity field and only retain the wavelet packet coefficients which are attached to coherent structures before reconstructing the filtered vorticity field. We then plan to perform comparative numerical simulations of both filtered and unfiltered flows. If the dynamics of the original and filtered vorticity fields are similar, then our conjecture will have been verified.

Finally, from the point of view of numerical analysis, if our conjecture is verified, one could then hope to reduce the number of degrees of freedom required to numerically integrate the evolution of turbulent flows. Thus, if one defines the degrees of freedom using Fourier modes, their number varies as $(Re)^{9/4}$ in three dimensions or $(Re)^1$ in two dimensions, where Re is the Reynolds number of the flow being calculated. If it could be proved that coherent structures are the dynamically active part of the flow, it would then be sufficient to consider only the degrees of freedom associated with them, that we could express, for example, using orthogonal wavelets or wavelet packets. This would allow us to design new algorithms, variants of the Large Eddy Simulation technique (Ferziger, 1981), based not on the traditional separation between resolved small Fourier wave numbers and parametrized large Fourier wave numbers, but rather on a more physical separation between the wavelet coefficients attached to coherent structures, which would be computed explicitly, and the remaining wavelet coefficients which would be only parametrized globally to take into account their energy and enstrophy but not their phases.

ACKNOWLEDGEMENT

I thank Professor S. Berberian, Prof. L. Tuckerman and Dr. E. Sichel for each translating portions of the French manuscript to English. Part of the work reported here has been done in collaboration with Matthias Holschneider of the Centre de Physique Théorique du CNRS, Marseille, Luminy. The numerical simulations have been computed on the Cray 2 of the Centre de Calcul Vectoriel pour la Recherche, Palaiseau, and the visualization have been done at LACTAMME, Ecole Polytechnique, Palaiseau in collaboration with Jean-François Colonna.

REFERENCES

ANSELMET F., GAGNE Y., HOPFINGER E. J. and ANTONIA R. A. 1984, High-order Velocity Structure Function in Turbulent Shear Flows *J. Fluid Mech.*, **140**, 63–89

BABENKO I. I., and PETROVICH V. V., 1979, *Dokl. Akad. SSSR,* **245**, 551

BASDEVANT C., LEGRAS B., SADOURNY R. and BELAND, 1981, A Study of Barotropic Model Flows: Intermittency, Waves and Predictability *J. Atmos. Sci.,* **38**, 2305–2326

BATCHELOR G. K. and TOWNSEND A. A., 1949, *Proc. Royal Soc.,* **A199**, 238

BATCHELOR G. K., 1969, Computation of the Energy Spectrum in Homogenous Two-dimensional Turbulence *Phys. Fluid, suppl. II,* **12**, 233–239

BATTIMELLI G. e VULPIANI A., 1982, Kolmogorov, Heisenberg, Von Weizsäcker, Onsager: un caso di scoperta simultanea *Terzo Congresso di Storia della Fisica, Palermo*

BENZI R., PATARNELLO G. and SANTANGELO P., 1987, Self-similar Coherent Structures in Two-dimensional Decaying Turbulence *J. Phys. A, Math. Gen.* **21**, 1221–1237

BENZI R. and VERGASSOLA M., 1990, Optimal Wavelet Analysis and its Application to Two-Dimensional Turbulence. *International Workshop on Novel Experiments and Data Processing for Basic Understanding of turbulence*, Tokyo, October 1990

BIRKHOFF G., 1962, Hydrodynamics Instanbility *Proc. Symp. Appl. Math.,* **13**, 55

BRACHET M. E., MENEGUZZI M., POLITANO H. and SULEM P. L., 1988, The Dynamics of Freely Decaying Two-dimensional Turbulence *J. Fluid Mech.,* **194**, 333 *Inverse Problems and Theoretical Imaging*, Springer

COUDER Y., and BASDEVANT C., 1986, Experimental and Numerical Study of Vortex Couples in Two-dimensional Flows *J. Fluid Mech.,* **173**, 225–251

COTTET G. H., 1987, Analyse numérique des méthodes particulaires pour certains problèmes non linéaires *Thèse d'Analyse Numérique, Université Paris VI*

ESCUDIE B. and TORRESANI B., 1989, Wavelet Analysis of Asymptotic Signals *Preprint, Centre de Physique Théorique, CNRS-Luminy*

FARGE M., 1979, Notes sur la turbulence *Rapport pour le cours de J. M. Levy-Leblond, Université Paris VII*

FARGE M., 1987, Normalization of High-Resolution Raster Display Applied to Turbulent Fields *Advances in Turbulence, ed. G. Comte-Bellot and J. Mathieu, Springer*, 111–112

FARGE M., 1988, Vortex Motion in a Rotating Barotropic Fluid Layer *Fluid Dynamics Research*, 3, North-Holland, 282–288

FARGE M., 1990a, Imagerie scientifique: choix des palettes de couleurs pour la visualisation des champs scalaires bidimensionnels *L'Aéronautique et l'Astronomique*, No. 140, 24–33

FARGE M., 1990b, Evolution des idées sur la turbulence: 1870–1970 *Colloque: '100 ans de rapports entre physique et mathématique', Revue du Palais de la Découverte*

FARGE M., 1990c, Transformée en ondelettes et application á la turbulence, *J. Annuelle, Societé Mathématique de France, Mai 1990*

FARGE M. and RABREAU G., 1988, Transformée en ondelettes pour détecter et analyser les structures cohérentes dans les écoulements turbulents bidimensionnels *C. R. Acad. Sci. Paris*, **307**, série II, 1479–1486

FARGE M., and SADOURNY R., 1989, Wave-vortex Dynamics in Rotating Shallow Water *J. Fluid Mech.*, **206**, 433–462

FARGE M., and HOLSCHNEIDER M., 1990, Interpretation of Two-dimensional Turbulence Spectrum in Terms of Singularity in the Vortex Cores *Soumis à Euro. Phys. Lett.*

FARGE M., GUEZENNEC Y., HO C. M., and NINEVEAU, C., 1990, Continuous Wavelet Analysis of Coherent Structures, *Proceedings, Summer Program, Center for Turbulence Research, Stanford University and NASA-Ames.*

FARGE M., HOLSCHNEIDER M. and COLONNA J. F., 1990, Wavelet Analysis o f Coherent Structures in Two-dimensional Turbulent Flows *'Topological Fluid Mechanics', ed. Moffatt*, Cambridge University Press

FARGE M., HOLSCHNEIDER M., and PHILIPOVITCH T., 1991, Formation et stabilité des quasi-singularités $r^{-1/2}$ en turbulence bidimensionnelle, *C. R. Acad. Sci., Paris,* submitted.

FERZIGER J. H., 1981, Higher-level Simulations of Turbulent Flows *Report No. TF-16, NASA-Ames, March 1981*

FRISCH U., 1983, Fully Developed Turbulence and Singularities *'Comportement chaotique des systèmes déterministes', Les Houches,* **36**, North-Holland, 668–701

FRISCH U. and ORSZAG, 1990, Turbulence: Challenge for Theory and Experiment *Physics Today, January 1990,* 24–32

FRISCH U., SULEM P. L., NELKIN M., 1978, A Simple Dynamical Model of Intermittent Fully Developed Turbulence *J. Fluid Mech.,* **87**, 719–736

HEISENBERG W., 1948, Zur statistischen Theorie der Turbulenz *Zeit für Phys.,* **124**, 628–657

HÖLDER E., 1933, *Math. Zeit.,* **37**, 698–726

HOLSCHNEIDER and TCHAMITCHIAN, 1988, On the Wavelet Analysis of the Riemann's Function *Preprint, Centre de Physique Théorique, CNRS-Luminy*

HUSSAIN A. K. M. F., 1986, Coherent Structures and Turbulence *J. Fluid Mech.,* **173**, 303–356

JAFFARDS S., 1989, Construction of Wavelets on Open Sets *Proceedings of the 1st International Conference on Wavelets, Marseille, 14–18 December 1987, ed. Combes et al., Springer,* 247–252

JIMENEZ J., 1981, The Role of Coherent Structures in Modelling Turbulence and Mixing *Proceedings of the International Conference on Coherent Structures, Madrid, Springer*

KATO T., 1972, *J. Funct. Anal.,* **9**, 296

KIM J. and MOIN P., 1979, Large Eddy Simulation of Turbulent Channel Flow *AGARD Symposium on Turbulent Boundary Layers, The Hague*

KOLMOGOROV A. M., 1941a, The Local Structure of Turbulence in Incompressible Viscous Fluid for Very Large Reynolds Numbers *C. R. Acad. Sc. USSR,* **30**, 301–305

KOLMOGOROV A. M., 1941b, On Degeneration of Isotropic Turbulence in an Incompressible Viscous Liquid *C. R. Acad. Sc. URSS,* **31**, 538–540

KOLMOGOROV A. M., 1941c, Dissipation of Energy in the Locally Isotropic Turbulence *C. R. Acad. Sc. URSS,* **32**, 16–18

KOLMOGOROV A. M., 1961, A Refinement of Previous Hypotheses Concerning the Local Structure of Turbulence in Viscous Incompressible Fluid at High Reynolds Number *J. Fluid Mech.,* **13**, 1, 82–85

KRAICHNAN, R. H., 1967, Inertial Ranges in Two-dimensional Turbulence *Phys. Fluid,* **10**, 1417–1423

KRAICHNAN R. H., 1971, Inertial-range Transfer in Two- and Three-dimensional Turbulence *J. Fluid Mech.,* **146**, 21–43

LADYZHENSKAYA O. A., 1963, The Mathematical Theory of Viscous Incompressible Flow *Gordon and Breach*

LANDAU L. D. and LIFCHITZ, 1971, Mécanique des fluides *Edition Mir, Moscou*

LIONS J. L., 1969, Quelques méthodes de résolution des problèmes aux limites non linéaires *Dunod-Gauthier Villars*

LERAY J., 1933, Etude de diverses équations intégrales non-linéaires et de quelques problèmes que pose l'hydrodynamique *J. Mathématiques Pures et Appliquées,* **12**, 1–82

MANDELBROT B., 1975, Intermittent Turbulence in Self-similar Cascades: Divergence of High Moments and Dimension of Carrier *J. Fluid Mech.,* **62**, 331–358

MANDELBROT B., 1976, Turbulence and Navier-Stokes Equation, ed. R. Temam, *Lectures Notes in Mathematics,* Springer, Vol. 565, 121

McWILLIAMS J. C., 1984, The Emergence of Isolated Coherent Vortices in Turbulent Flow *J. Fluid Mech.,* **146**, 21–43

MEIRON D. I. and BAKER G. R., 1982, Analytic Structure of Vortex Sheet Dynamics. Part 1. Kelvin-Helmholtz Instability *J. Fluid Mech,m* **114**, 283

MOORE, D. W., 1979, *Proc. Royal Soc. London,* **A365**, 105

MONIN A. S., and YAGLOM, A. M., 1975, Statistical Fluid Mechanics *The MIT Press*

MOFFATT h. k., 1990, The Topological Approach to Fluid and Plasma Flow Problems *'Topological Fluid Mechanics'*, ed. *Moffett, Cambridge University Press*, 1–10

NGUYEN DUC J. M. and SOMMERIA J., 1988, Experimental Characterization of Steady Two-dimensional Vortex Couples *J. Fluid Mech.*, **192**, 175–192

OBUKHOV A. M., 1941, Energy Distribution in the Spectrum of a Turbulent Flow *Izvestiya AN URSS, No. 4–5*, 453–466

ONSAGER, L., 1945, The Distribution of Energy in Turbulence *Phys. Rev.*, **68**, 286

PARISI G. and FRISCH U., 1985, Turbulence and Predictability in Geophysical Fluid Dynamics and Climate Dynamics, *ed. M. Ghil, R. Benzi and G. Parisi*, North-Holland, 71–88

REYNOLDS O., 1883, An Experimental Investigation of the Circumstances which Determine whether the Motion of Water Shall Be Direct or Sinuous, and the Law of Resistance in Parallel Channels *Phil. Trans. Roy. Soc. London*, **174**, 935–982

ROSE H. and SULEM P. L., 1978, Fully Developed Turbulence and Statistical Mechanics *Le Journal de Physique*, **39**, No. 5

SAFFMAN P. G., 1971, A Note on the Spectrum and Decay of Random Two-dimensional Vorticity Distribution *Stud. Appl. Math.*, **50**, 377–383

SCHAEFFER A. c., 1937, *Transl. American Mathematical Society*, **42**, 497

SULEM C., SULEM P. L., BARDOS C., FRISCH U., 1981, *Comm. Math. Phys.*, **80**, 485

TEMAM R., 1984, Navier-Stokes Equation *North-Holland*

VAN DYKE M., 1982, An Album of Fluid Motion *The Parabolic Press*

VON NEUMANN J., 1949, Recent Theories of Turbulence *Oeuvres Complètes*, **6**, 437–471

VON WEIZSÄCKER C. F., 1948, Das Spectrum der Turbulenz bei grossen Reynoldschen Zahlen *Zeit. für Phys.*, **124**, 628–657

WEISS J., 1981, The Dynamics of Enstrophy Transfer in Two-dimensional Hydrodynamics *report LJI-TN-121, La Jolla Institute, San Diego*

WEISS J., 1991, The Dynamics of Enstrophy Transfer in Two-dimensional Hydrodynamics, *Physica D*, **48**, 273–294

WOLIBNER W., 1933, *Math. Zeit.*, **37**, 698–726

ZABUSKY N., 1984, Computational Synergetics *Physics Today, July 1984,* 2–11

WAVELETS AND QUANTUM MECHANICS

T. PAUL[1]

CEREMADE
Université Paris Dauphine
Place de Lattre de Tassigny
75775 Paris Cedex 16, France

K. SEIP

Division of Mathematical Sciences
University of Trondheim
N-7034 Trondheim. N.T.H., Norway

INTRODUCTION

Coherent states have a long history in quantum mechanics, since their introduction by Schrödinger in 1926 [32]. Coherent states were used in quantum field theory and in optics in the 50's. Spaces of coefficients on coherent states were studied in [3] and used for semi-classical approximation of the Schrödinger equation in [14], [16] and [31]. Propagation of wave packets, namely coherent states in the semi-classical limit, were studied in the setting of Fourier integral operators ([13]), through the Schrödinger equation ([15]) and using path integrals ([10]). The reader can find a list of references and a collection of reprints in [18].

Affine coherent states were introduced in [1] for a one-dimensional model of quantum gravity. They were systematically studied in relation to the Schrödinger equation in [21], [22] and [23].

What is the relation between affine coherent states and wavelets?

Briefly an affine coherent state is the Fourier transform of a continuous wavelet, with positive dilation parameter, where the analyzing wavelet is

---

[1]member of the CNRS

the function

$$\varphi(t) \; = \; \frac{1}{(t+i)^{l+1}} \qquad \text{with} \quad l > 0 \,.$$

Affine coherent states are elements of $L^2(\mathbb{R}^+)$ and play a role in $L^2(\mathbb{R}^+)$ analogous to the role of the Weyl-Heisenberg coherent states in $L^2(\mathbb{R})$. In particular they give rise to a representation for quantum mechanics in which the Schrödinger equation for the hydrogen atom becomes first order and can be solved by an exact W.K.B. expression: this gives a very simple explanation of the fact that the spectrum of the hydrogen atom is given exactly by the semi-classical formula. Let us recall that this formula, namely the Bohr rule, was the starting point of quantum mechanics.

This paper is divided in two sections:

- Section A by T. Paul is devoted to wavelets and quantum mechanics on the half-line, with emphasis on the hydrogen atom.

- Section B by K. Seip is concerned with curves of maximum modulus in the space of wavelet coefficients.

The link between these two somewhat disconnected sections is the following.

The semi-classical approximation of the one-dimensional Schrödinger equation on a space of coherent states coefficients makes clear the fact that, when the Planck constant $\hbar \to 0$, the wave function of a bound state concentrates in phase-space on the classical trajectory corresponding to its eigenvalue (Since we are in one dimension there is only one such trajectory for a given eigenvalue). For the case of the harmonic oscillator this concentration is in a certain sense "sharp": the modulus of the wave function, in this representation, has a curve of local maxima on the classical trajectory. This has been shown in [27] for the harmonic oscillator in a space of entire functions (Bargmann space), (See section 2.3 of the chapter by Feichtinger and Grochenig for further discussion of the Bargmann and Bergman spaces.) and for the hydrogen atom in the space of analytic functions on the upper half-plane described in Section A (Bergman space).

In fact a complete classification of such functions and associated M-curves was given in [27]: in particular the set of such functions having zeros of different orders but at the same point were shown to form an orthonormal basis.

In Section B the case of M-curves on the unit disc corresponding to functions having two distinct zeros is discussed: although the corresponding set of functions does not provide an orthonormal basis, it gives a Riesz basis for the Hilbert space. The quantum mechanical interpretation of such a basis is still open.

This review is by no means exhaustive. The use of wavelets in constructive theory of fields [2], in the study of stability of matter [13] and in propagation through the Schrödinger equation [26] are missing. Each would need an entire article.

A. WAVELETS AND THE SCHRÖDINGER EQUATION ON THE HALF-LINE

Thierry Paul

1. INTRODUCTION

Coherent states are useful in quantum mechanics (and micro-local analysis) because they provide a natural phase-space setting for it.

Fourier analysis consists of analyzing a function by its projection onto the eigenvectors of the operator $P = -i\frac{d}{dx}$; this is particulary useful for solving equations where $-\Delta = P^*P = P^2$ is involved (here the $*$ means adjoint). Weyl-Heisenberg coherent states (windowed Fourier transform, Gabor states...) are eigenvectors of $a = \frac{1}{\sqrt{2}}\left(-i\frac{d}{dx} + ix\right)$ (annihilation operator); they are of great interest in the study of the operator $aa^+ + \frac{1}{2} = -\frac{1}{2}\Delta + \frac{x^2}{2}$ (harmonic oscillator). The main difference between the operators a and P is the fact that a is not a symmetric operator; in particular it has a "spectrum" which is the whole plane and "eigenvectors" which are square integrable and well localized around the points in phase space corresponding to the corresponding eigenvalues. In fact it is easy to see that being an eigenvector of a is equivalent to minimizing the Heisenberg inequalities [17]; Weyl-Heisenberg coherent states are the best localized vectors on phase space.

This section is devoted to the study of affine coherent states and the role they play in $L^2(\mathbb{R}^+)$ compared to the role of the Weyl-Heisenberg coherent states in $L^2(\mathbb{R})$.

There are several reasons to consider the Schrödinger equation on the half-line: first of all it is what remains of the Schrödinger equation in n dimensions after a partial wave decomposition, namely a decomposition into spherical harmonics. Secondly it gives an example of quantization of a non-flat space, namely the Poincaré-Lobatchevsky plane, the simplest example of a manifold with constant negative curvature. Finally, the affine coherent states provide a representation of the state space on which the eigenvectors of the hydrogen atom are given exactly by the W.K.B. form. This explains why the spectrum of the hydrogen is given precisely by the Bohr rule of quantization of the action: in a certain sense the affine coherent states play the role for the hydrogen atom that the Weyl-Heisenberg coherent states play for the harmonic oscillator.

After first presenting properties of the affine coherent states — especially their relation with the Heisenberg inequalities associated with the half-plane — we will revisit in detail the hydrogen atom (see equation (4.1) below), first because the theory becomes transparent and also because this will introduce the tools needed for the more general study of semi-classical approximation of the Schrödinger equation.

2. WAVELETS — AFFINE COHERENT STATES: BASIC PROPERTIES

We want to construct affine coherent states as eigenvectors of an operator, as discussed in the preceding paragraph.

Since the eigenvectors of a are not supported on the half-line, one has to choose another operator. We choose the operator:

$$\overline{Z} = i\frac{d}{dx} - \frac{il}{x} \quad , \quad \text{with} \quad l > 0 , \, l \text{ integer.} \qquad (2.1)$$

There are several reasons for this choice:
- $\overline{Z}$ is going to be of crucial interest for the hydrogen atom;
- $\overline{Z}^+\overline{Z} = -\frac{d^2}{dx^2} + \frac{l(l+1)}{x^2}$ is the reduced Laplacian on $\mathbb{R}^3$ after decomposition into partial waves (note the analogy with Fourier analysis!).

The eigenvectors of $\overline{Z}$ in $L^2(\mathbb{R}^+, dx)$ are the vectors:

$$\psi_{\overline{z}}(x) = x^l e^{-i\overline{z}x} \qquad (2.2)$$

for Im $z > 0$.

We call $\psi_{\overline{z}}$, Im $z > 0$, affine coherent states.

It is easy to see ([23]) that the eigenvectors of $\overline{Z}$ minimize the Heisenberg inequalities associated with $P = -i\frac{d}{dx}$ and $\frac{l}{x}$. By this we mean the following: let A and B be two densely defined operators of a Hilbert space with scalar product $< \cdot, \cdot >$. Let ψ be both in the domains of A and B. We define $\Delta_\psi(A)$ by:

$$\Delta_\psi(A) = \frac{\sqrt{< \psi, A^2\psi > - < \psi, A\psi >^2 / < \psi, \psi >}}{\|\psi\|} . \qquad (2.3)$$

Proposition 1. *Let A, B and ψ be as above :*

$$\Delta_\psi(A) \cdot \Delta_\psi(B) \geq \frac{1}{2}\frac{| < \psi, [A, B]\psi > |}{\|\psi\|} \qquad (2.4)$$

where $[A, B] = AB - BA$.

Proof 1. See [17] or any standard quantum mechanics textbook. (Note that $A = -i\frac{d}{dx}$ and $B = x$ gives the usual Heisenberg inequalities).

In the special case $A = -i\frac{d}{dx}$ and $B = \frac{l}{x}$, (2.4) becomes

$$\Delta_\psi \left(-i\frac{d}{dx}\right) \cdot \Delta_\psi \left(\frac{l}{x}\right) \geq \frac{1}{2} \frac{<\psi, \frac{l}{x^2}\psi>}{\|\psi\|} . \tag{2.5}$$

This becomes an equality if ψ is given by (2.2).

Proposition 2.

$$\Delta_{\psi_{\bar{z}}} \left(-i\frac{d}{dx}\right) \cdot \Delta_{\psi_{\bar{z}}} \left(\frac{l}{x}\right) = \frac{1}{2} \frac{<\psi_{\bar{z}}, \frac{l}{x^2}\psi_{\bar{z}}>}{\|\psi_{\bar{z}}\|} . \tag{2.6}$$

Proof 2. This follows directly by computation or by consequence of the eigenvalue property.

These (affine) Heisenberg inequalities have a nice meaning in terms of phase-space localization on the upper half-plane . Indeed if we decide to define localisation in phase-space by "measuring" the observables

$$B = -i\frac{d}{dx} \quad \text{and} \quad A = \frac{l}{x} ,$$

a given function $\psi(\ \|\psi\| = 1)$ is "localized" around the point $(b = <\psi, B\psi>, a =<\psi, A\psi>)$ in a region $\Delta b \, \Delta a$ satisfying

$$\frac{\Delta b \, \Delta a}{a^2} \geq \frac{1}{2l} . \tag{2.7}$$

This inequality is associated with the volume element of the Poincaré-Lobatchevsky plane, as we will see in the next section, where $\frac{1}{l}$ plays the role of the Planck constant. The link with the usual Heisenberg inequalities is clear if we remark that, in terms of p and q, the symbols of the operators $-i\frac{d}{dx}$ and x, respectively, (2.7) can be written, in the limit with Δb and Δa small, as

$$\Delta p \, \Delta q \geq \frac{1}{2} .$$

The space of wavelet coefficients is the so-called Bergmann space which we want to describe now. We will also need to "transport" some nice operators to the Bergmann space for later use.

Proposition 3.   *Let*

$$(A_l\psi)(z) = \frac{2^l}{\sqrt{2\pi\Gamma(2l)}} \, (\psi_{\bar{z}}, \psi) \qquad (2.8)$$

for $\psi \in L^2(\mathbb{R}^+, dx)$.

Then A_l is a unitary transform between $L^2(\mathbb{R}^+, dx)$ and the (weighted Bergman) space B_{2l+1} defined by:

$$B_{2l+1} = \left\{ f(z) \text{ analytic on } \mathbb{C}^+ \,,\; (\text{Im } z > 0) \,,\; \text{such that} \right.$$

$$\left. \int_{\mathbb{C}^+} |f(z)|^2 \, (\text{Im } z)^{2l+1} \, \frac{dz \, d\bar{z}}{(\text{Im } z)^2} < +\infty \right\}.$$

Moreover

$$A_l \, Z \, A_l^{-1} = A_l \left[i\frac{d}{dx} + i\frac{l}{x} \right] A_l^{-1} = z \qquad (2.9)$$

(multiplication by z)

and

$$A_l \, x \, A_l^{-1} = -i\frac{d}{dz} \,. \qquad (2.10)$$

Proof 3.   - The first part is obvious by noting that A_l is nothing but the l'th derivative of the Laplace transfom.

- (2.9) is an immediate consequence of the eigenvalue property $\overline{Z} \, \psi_{\bar{z}} = \bar{z}\psi$ and the (physicists) convention of left antilinearity of scalar product.

- (2.10) is immediate from $+i\frac{d}{dz} \left(x^l \, e^{-i\bar{z}x} \right) = x \cdot x^l \, e^{-i\bar{z}x}$.

Remarks 1.   - We have not discussed here the problem of different self-adjoint extensions of the operator $-i\frac{d}{dx}$ on the half-line, the identity (2.9) has to be taken on a formal level, which is sufficient for the rest of the paper.

- B_{2l+1} is a space of analytic functions on the upper half-plane [5], phase space of the half-line, which is similar to the Bargmann space on the whole plane.

- Operators are built on L^2-spaces in terms of operators $x \equiv Q$ and $-i\frac{d}{dx} \equiv P$. Operators will be built on B_{2l+1} in terms of $i\frac{d}{dz}$ and z. Since $\left[z, -i\frac{d}{dz}, z \right] = \left[x, -i\frac{d}{dx} \right] = i$, this simply amounts to a change of representation of the canonical commutation relations.

- The variable z, denoting both the eigenvalue and the symbol of (namely the classical observable corresponding to) the operator $\overline{Z}^+ = -i\frac{d}{dz} + i\frac{l}{x} =$

$P + ilQ^{-1}$, has the classical meaning $z = p + i\frac{l}{q}$. We will see in the next section the importance of this classical complex variable.

3. HYDROGEN ATOM: CLASSICAL

A particle in $\mathbb{R}^3$, of unit mass, in a Coulomb potential has the following Hamiltonian $h_c(\vec{p}, \vec{q})$, $(\vec{p}, \vec{q}) \in T^*(\mathbb{R}^3)$,

$$h_c(\vec{p}, \vec{q}) = \frac{\vec{P}^2}{2} - \frac{\lambda}{|\vec{q}|} \quad , \tag{3.1}$$

$\lambda > 0$ and $|\vec{q}| = \sqrt{\vec{q} \cdot \vec{q}}$.

Since the angular momentum $\vec{L} = \vec{q} \wedge \vec{p}$ is a constant of motion, it is better to express h_c in terms of $\vec{L}$, $q = |\vec{q}|$ and p, the conjugate momentum of q (see [17]). The Hamiltonian then becomes :

$$h^L(q, p) = \frac{P^2}{2} + \frac{L^2}{2q^2} - \frac{\lambda}{q} \quad , \tag{3.2}$$

where $L = |\vec{L}|$ is now a parameter.

Everything reduces to the study of the family of Hamiltonians h^L for each value of L. h^L is an Hamiltonian defined on the half-plane, the reduced phase space.

In order to make clear the geometrical structure of the half-plane, we introduce the complex variable:

$$z = p + i\frac{L}{q} \quad . \tag{3.3}$$

The change of variable between (q, p) and $(z, \bar{z})$ is one between euclidean and curved coordinates: namely the volume element $dp\, dq$ becomes $L\frac{dz\, d\bar{z}}{2(\operatorname{Im} z)^2}$ and the symplectic form $dq \wedge dp$ becomes $L\frac{dz \wedge d\bar{z}}{2(\operatorname{Im} z)^2}$. We saw in the preceding sections the relationship between the Lobatchevsky volume element and the localization properties of the affine coherent states.

In terms of the variable z the Hamiltonian h^L becomes

$$\tilde{h}^L(z) = \frac{1}{2}\left|z - i\frac{\lambda}{L}\right|^2 - \frac{\lambda^2}{2L^2} \quad . \tag{3.4}$$

Classical orbits are given by the level surfaces of $\tilde{h}^L(z)$, namely the curves $\gamma(E)$ defined by $\tilde{h}^L(z) = E$; they are (Euclidean) circles, centered at $i\frac{\lambda}{L}$ and radius $\sqrt{2(E + \frac{\lambda^2}{2L^2})}$, for $E > -\frac{\lambda^2}{2L^2}$.

Three cases can be considered :

$-\dfrac{\lambda^2}{2L^2} < E < 0:$ elliptic orbits: circles entirely contained in the upper half-plane

$-\;\;E = 0:$ parabolic orbit: circle touching the origin

$-\;\;E > 0:$ hyperbolic orbits: arc of circle cut by the real line.

Remarks 2.   If we put $\lambda = 0$ (free motion), the trajectories become semi-circles orthogonal to the real axis, i.e., geodesics of the half-plane.

For the semi-classical study presented below we need to discuss the Hamilton-Jacobi equation in the variable z.

As we mentioned before, (q, z) is a pair of conjugate variables with respect to the usual Poisson bracket: here $z = p + i\dfrac{L}{q}$, the Poisson bracket is defined by $\{f, g\} = \dfrac{\partial f}{\partial p}\dfrac{\partial g}{\partial q} - \dfrac{\partial f}{\partial q}\dfrac{\partial g}{\partial p}$ and $\{z, q\} = 1$. Since the system is one dimensional, one can find a canonical transformation mapping (z, q) on action-angle variables (A, φ) having the property that the Hamiltonian expressed in these new variables is φ-independent (integrable system). A canonical transformation Φ can be defined through a generating function $S(A, z)$ which gives Φ via the relations

$$\varphi = \frac{\partial S}{\partial A}(A, z)$$

$$q = \frac{\partial S}{\partial z}(A, z)$$

in order to express $(A, \varphi) = \Phi(z, q)$.

The Hamilton-Jacobi equation is the equation that S has to satisfy. Starting with a general Hamiltonian $h(q, z)$, this equation has the form (since $q = \frac{\partial S}{\partial z}$),

$$h\Big(\frac{\partial S}{\partial z}(A, z), z\Big) = E(A), \quad \text{independent of } \varphi \text{ and } z.$$

The relation between the Hamilton-Jacobi and Schrödinger equations will be made clear in the following section. In the case of the Hamiltonian given by (3.4), $\tilde{h}^L$ expressed as a function of z and q (not $\bar{z}$!) becomes

$$\tilde{h}^L(z, q) = \frac{1}{2}\Big(z - i\frac{\lambda}{L}\Big)\Big(z + i\frac{\lambda}{L} - \frac{2il}{q}\Big) - \frac{\lambda^2}{2L^2}. \tag{3.5}$$

This gives the following Hamilton-Jacobi equation

$$\frac{1}{2}\Big(z - i\frac{\lambda}{L}\Big)\Big(z + i\frac{\lambda}{L} - \frac{2iL}{\frac{\partial S}{\partial z}(A, z)}\Big) - \frac{\lambda^2}{2L^2} = E(A) \tag{3.6}$$

or, equivalently

$$(z^2 - 2E(A)) \frac{\partial S}{\partial z}(A, z) - 2iL\left(z - i\frac{\lambda}{L}\right) = 0. \tag{3.7}$$

$E(A)$ is determined by the fact that, on the orbit of energy $E(A)$, (since $dq \wedge dz = dA \wedge d\varphi$),

$$\int_{\gamma(E(A))} \frac{\partial S}{\partial z}(A, z)\, dz = \int_{\gamma(E(A))} q\, dz$$

$$= \int_{\gamma(E(A))} q(dp + iLd(\frac{1}{q})) = \int_{\gamma(E(A))} q\, dp \tag{3.8}$$

$$= \int_{\gamma(E(A))} A\, d\varphi = 2\pi\, A$$

(since A is a constant on $\gamma(E(A))$).

Since $\frac{\partial S}{\partial z}$, given by (3.7), is a meromorphic function on the upper half-plane the residue formula gives

$$\int_{-\gamma(E(A))} \frac{\partial S}{\partial z}(A, z)\, dz = 2\pi i\, \frac{2iL(i\sqrt{-2E} - i\frac{\lambda}{L})}{2i\sqrt{-2E}} \tag{3.9}$$

Substituting (3.9) into (3.8) one gets

$$E = -\frac{\lambda^2}{2(A+L)^2}$$

which is the correct form of the energy [19].

4. HYDROGEN ATOM : QUANTUM

The (time independent) Schrödinger equation for an electron in a Coulomb potential is :

$$\left(-\frac{1}{2}\Delta - \frac{\lambda}{|\vec{x}|}\right) \psi(\vec{x}) = E\, \psi(\vec{x}), \tag{4.1}$$

where Δ is the Laplacian on $\mathbb{R}^3$. After decomposition into spherical harmonics (4.1) becomes [20]:

$$\left(-\frac{1}{2}\frac{1}{x}\frac{\partial^2}{\partial x^2}x + \frac{l(l+1)}{2x^2} - \frac{\lambda}{x}\right) \psi(x) = E\, \psi(x), \tag{4.2}$$

which can be written as:

$$\left(-\frac{1}{2}\frac{\partial^2}{\partial x^2} + \frac{l(l+1)}{2x^2} - \frac{\lambda}{x}\right) x\, \psi(x) = E\, x\, \psi. \tag{4.3}$$

Since we want to "transport" this equation into the Bergman space, we want to express it in terms of the operators $\overline{Z}^+ \equiv Z$ and $Q = x$. An easy computation gives that (4.3) is equivalent to

$$\left[(Z^2 - 2E)Q - 2il\left(Z - i\frac{\lambda}{l}\right)\right]\psi = 0 . \tag{4.4}$$

Using the unitary transform A_l defined in (2.8) and defining the function $f(z) = (A_l\psi)(z) = \frac{2^l}{\sqrt{2\pi\Gamma(2l)}}\int_0^{+\infty} x^l\, e^{izx}\, \psi(x)\, dx$, the equation (4.2), "transported" to B_{2l+1} is:

$$\left[-i(z^2 - 2E)\frac{d}{dz} - 2il\left(z - i\frac{\lambda}{l}\right)\right] f(z) = 0 , \tag{4.5}$$

since $A_l Z A_l^{-1} = z$ and $A_l Q A_l^{-1} = -i\frac{d}{dz}$, as seen previously.

We first remark that (4.5) is a *first-order equation*. We solve it by looking at a solution of the form $f(z) = e^{+iS(z)}$. The equation for S becomes:

$$(z^2 - 2E)\frac{\partial S}{\partial z}(z) - 2il\left(z - i\frac{\lambda}{l}\right) = 0 , \tag{4.6}$$

which is the same as (3.7).

The meaning of this is the following: in the representation of wavelet coefficients, the solution of the Schrödinger equation for the hydrogen atom is given *exactly* by the W.K.B. ansatz, namely by the exponential of $+i$ times the solution of the Hamilton-Jacobi equation.

The quantization condition is obtained by requiring that the solution $f(z) = e^{+iS(z)}$ belongs to the Hilbert space, that is, is analytic (the L^2 condition is always true). Analyticity of $f(z)$ can be achieved by requiring that

$$\int_\gamma \frac{f'(z)}{f(z)}\, dz = 2\pi in \quad , \quad n \in \mathbb{N} \tag{4.7}$$

for any curve γ avoiding the zeros of f. Then substituting, $f(z) = e^{iS(z)}$ into (4.7), one has

$$\int_\gamma \frac{\partial S}{\partial z}(z)\, dz = 2\pi n . \tag{4.8}$$

One can in particular take for γ the classical orbit of energy E; then (4.8) is equivalent to (3.8) with $A = n$.

The result is the following: the eigenvalues are given exactly by the values of the energy of the classical orbits of integer actions. This is the old Bohr rule whose validity is very transparent in this representation.

5. SEMI-CLASSICAL APPROXIMATIONS

We want briefly to show how the same method can be applied to more general Hamiltonians (see [22] and [23]). In order to consider a semi-classical approximation, we have to put the parameter $\hbar$ in all the definitions. We then define:

$$\psi_{\bar{z}}(x) = x^l e^{\frac{-i\bar{z}x}{\hbar}} . \tag{5.1}$$

We now start with a Schrödinger equation on the half line of the form :

$$\left[-\frac{1}{2}\frac{\hbar^2}{x}\frac{\partial^2}{\partial x^2}x + \frac{\hbar^2 l(l+1)}{2x^2} + V(x) \right] \psi(x) = E\,\psi(x) \tag{5.2}$$

with $V(x)$ real analytic on the half-line. It is easy to see that this equation becomes in B_{2l+1}:

$$\left[-i\hbar(z^2 - 2E)\frac{d}{dz} - 2ilz + 2\left(-i\hbar\frac{d}{dz}\right)V\left(-i\hbar\frac{d}{dz}\right) \right] f(z) = 0 . \tag{5.3}$$

Making the substitution $f(z) = e^{i\frac{S(z)}{\hbar}}$, (5.3) becomes

$$(z^2 - 2E)S'(z) - 2ilz + S'(z)V(S'(z))$$
$$\left[\sum_{h=0}^{\infty} \hbar^k R_k(S^1(z), \dots, S^{(k)}(z)) \right] = 0$$

(see [7] and [9]).

A semi-classical approximation consists of:

- Looking at an expansion of S of the form $S(z) = S_0(z) + \hbar S_1(z) + \hbar^2 S_2(2) + \dots$; and

- Imposing some quantization conditions. Here we will restrict the expansions up to order $\hbar$.

The equation satisfied by $S_0(z)$ is

$$(z^2 - 2E)S_0'(z) - 2ilz + S_0'(z)V(S_0'(z)) = 0, \tag{5.4}$$

which can be viewed as a Hamilton-Jacobi equation.

There is no hope of finding a solution of (5.4) such that $e^{i\frac{S_0(z)}{\hbar}}$ is analytic on the whole half-plane. Nevertheless one can find certain regions of the half-plane where $S_0(z)$ is analytic, for instance a neighbourhood Ω of $\frac{il}{x_0}$, where x_0 is an extremum of $\frac{l(l+1)}{2x^2} + V(x)$ (see [20]). $S_1'(z)$ happens them to be meromorphic in this neighborhood with a simple pole at $\frac{il}{x_0}$. The condition for $f_1(z) = e^{\frac{i}{\hbar}[S_0(z)+\hbar S_1(z)]}$ to be analytic in this neighborhood can be written as:

$$\int_\gamma [S_0'(z) + \hbar S_1'(z)]\,dz = 2\pi n\,\hbar \tag{5.5}$$

with n an integer and γ any path around x_0 in this neighborhood. Since $S_0(z)$ is analytic, $\int S_0'(z)\,dz = 0$ and one must have

$$\int_\gamma S_1'(z)\,dz = 2\pi\,n\,. \tag{5.6}$$

Equation (5.6) fixes the different values of E. This rapid discussion is summarized in the following proposition:

Proposition 4. **([20])**   *Let V be as before. Let x_0 be an extremum of $W(x) = \frac{l(l+1)}{2x^2} + V(x)$ and let $E_0 = W(x_0)$. Then there exist a neighborhood Ω of $i\frac{l}{x_0}$ and a discrete sequence of numbers $E_1(n)$ and functions $f^n(z)$, analytic in Ω such that*

$$\left[-i\hbar(z^2 - 2(E_0 + \hbar E_1(n)))\frac{d}{dz} - 2ilz + 2\left(-i\hbar\frac{d}{dz}\right)V\left(-i\hbar\frac{d}{dz}\right) \right]\cdot$$

$$f^n(z) = 0 + 0(\hbar^2)\,.$$

Moreover,

a) if $W''(x_0) > 0$, namely x_0 is a local minimum of W, then E_1 is real; and

b) if $W''(x_0) < 0$, namely x_0 is a local maximum, then E_1 is purely imaginary.

Remarks 3.   - Case a) corresponds to the semi-classical expansions around the bottom of a well.

- Case b) corresponds to the so-called "top resonances" introduced first in [22] and proved to exist in [5] and [30]. Indeed the complex value of the energy $E_0 + \hbar E_1$ (E_1 purely imaginary) can be interpreted as a resonance associated with the (classical) unstable equilibrium point at the top of the potential. The fact that the imaginary part is of order $\hbar$ (and not $\hbar^\infty$ as in the case of usual "shape resonances") can be viewed as a trace of the classical instability of this fixed point.

B. CURVES OF MAXIMUM MODULUS
Kristian Seip

Coherent state representations are, in general, useful in semiclassical arguments due to the fact that they enable us to study quantum mechanics in a phase space setting. This has just been illustrated, for a special instance, in the preceding section. The motivation for the present section has been the discovery of an additional, at least appealing, feature of this model of quantum mechanics. We have in mind the relation, as pointed out in [6], between the possible curves of maximum modulus and the eigenvectors of some Schrödinger operator expressed in a space of coherent states coefficients. The possible closed curves satisfy a quantization rule; a natural subfamily consists of precisely the quantized classical orbits of the operator in question and carries, in fact, essentially all information about the spectrum and the eigenstates.

Before describing more closely this relation, we discuss some generalities about curves of maximum modulus, henceforth briefly denoted as M-curves. For our purposes it is most convenient to work in the unit disk Δ. Our objects of interest are then functions of the form

$$S(z) = (1 - |z|^2)^\alpha f(z),$$

where f is analytic in Δ and $\alpha > 0$ is a fixed number. We say that a curve Γ in Δ is an M-curve of $S \not\equiv 0$, or equivalently of $f \not\equiv 0$, if $\frac{\partial |S|}{\partial x} = \frac{\partial |S|}{\partial y} = 0$ along Γ $(z = x + iy)$. In [27] we noticed that if S has an M-curve Γ, then $|S|$ is constant and attains local maxima along Γ, and S is determined up to a constant factor by any subset of Γ containing an accumulation point. This uniqueness makes it natural to denote the functions associated to Γ by S_Γ and f_Γ. Also, we found that Γ is an M-curve of S if and only if

$$\frac{f'(z)}{f(z)} = \frac{2\alpha \bar{z}}{1 - |z|^2} \tag{1}$$

holds along Γ. (1) is simply a consequence of the Cauchy-Riemann equations. Locally it states that Γ must be an analytic arc.

We showed in [27] that the set of M-curves is invariant under Möbius self-maps of Δ. If Γ is an M-curve of $f(z)$ then $\Phi^{-1}(\Gamma)$ is an M-curve of $(U_\Phi^\alpha f)(z) = (\Phi'(z))^\alpha f(\Phi z)$ for any Möbius self-map Φ. It is straightforward to check that, for $\alpha > \frac{1}{2}$, U_Φ^α is a unitary operator on the weighted Bergman space A_α, which is the Hilbert space of functions analytic in Δ and square integrable with respect to the weighted measure $(1 - |z|^2)^{2\alpha - 2} dx dy$. We should note that for each α in the range $0 \leq \alpha \leq \frac{1}{2}$, there is a Hilbert space on which U_Φ^α acts unitarily, but we shall whenever necessary tacitly assume that $\alpha > \frac{1}{2}$, since the Bergman spaces are the ones of physical interest.

It is an easy exercise to check that all we are doing, could equivalently have been done in the upper half-plane U, i.e. within the representation discussed in the preceding section. To any f analytic in Δ, we then associate the function

$$\tilde{f}(\zeta) = \left(\frac{2}{\zeta + i}\right)^{2\alpha} f\left(\frac{\zeta - i}{\zeta + i}\right),$$

analytic in U, and observe that

$$|\tilde{S}(\zeta)| = \left|S\left(\frac{\zeta - i}{\zeta + i}\right)\right|,$$

where $(\zeta = \xi + i\eta)$

$$\tilde{S}(\zeta) = \eta^{\alpha} \tilde{f}(\zeta).$$

One can notice that the case $\alpha = l + 1/2$ corresponds to the space introduced in Proposition 3 of Section A. One may consult [27] for details on this as well as a complete discussion of the canonical (Bargmann) case.

We restrict now the attention to closed M-curves. It is easy to see that in this case the corresponding function f_Γ must have at least one zero in the domain bounded by the curve. Indeed, denoting this domain by Ω, we have, by (1) and Green's theorem,

$$\frac{1}{2\pi i} \oint_\Gamma \frac{f'_\Gamma(z)}{f_\Gamma(z)} dz = \frac{\alpha}{\pi} \int \int_\Omega (1 - |z|^2)^{-2} dx\, dy = \frac{\alpha}{\pi} \cdot \text{Area}(\Omega), \qquad (2)$$

which by the argument principle equals the number of zeros, counting multiplicities, of f_Γ in Ω. We have thus a quantization rule, similar to the Bohr-Sommerfeld rule.

We consider first the simple case of one distinct zero enclosed by the curve. By the Möbius invariance this zero can be assumed to be at 0. Since, by (1), the analytic function $z f'_\Gamma(z)/f_\Gamma(z)$ is positive along Γ, it must reduce to a constant, and thus

$$f_\Gamma(z) = z^k$$

for some positive integer k. The corresponding Γ is the circle $|z|^2 = k/(k + 2\alpha)$.

We may now indicate the announced link to the underlying Schrödinger operator. Transformed to U, we have the following. All possible M-circles with a common hyperbolic center are the quantized classical orbits of the radial harmonic oscillator in Paul's representation [23] (see Section A of the present paper). In the case of the hydrogen atom, the same is true if we instead consider all possible circles with a common Euclidean center (we get then only a finite number of closed curves contained in U ; it is easy to check that if $\Gamma \cap U \neq \Gamma$, Γ a circle, then $\Gamma \cap U$ is an M-curve). In either case, the functions of the M-curves are the corresponding eigenstates. As to the spectrum, we can only say that, of course, it is given up to an additive constant by the two equations defining the families of circles; see the preceding section.

Returning to Δ, we note that, trivially, the functions z^k, $k = 0, 1, 2, \ldots$, when suitably normalized, constitute an orthonormal basis for A_α (we have included $k = 0$ for technical reasons only). By the quantization rule (2), the associated circles divide phase space into nonoverlapping annuli, each of area π/α. To the sequence of curves we can therefore associate the phase space density α/π.

It should be noted that this basis constitutes the solution to another phase space localization problem. To see this, let P_α denote orthogonal projection onto A_α in L_α^2 (the notation should be self-explanatory), and Q_r the operator of multiplication by the characteristic function of the disk $|z| < r < 1$. The compact positive operator $P_\alpha Q_r$ is called a concentration operator, which is justified by the simple fact that its eigenfunctions, and eigenvalues, can be found by successively maximizing the concentration,

$$\lambda = \frac{\|Q_r f\|}{\|f\|},$$

over the orthogonal complement in A_α of the linear span of the previously determined eigenfunctions. It is easily seen that the eigenfunctions are the z^k, $k = 0, 1, 2, \ldots$, independently of r, see [4] pp. 14-18, [9], [28].

Our concentration operator is a close relative of the operator associated with the prolate spheroidal wave functions, as well as its analogue in the Bargmann case [8], both conceptually and qualitatively in that the action of the operator consists in restriction in time and frequency, followed by projection onto the space at hand. The asymptotic behavior of the eigenvalues, as discussed in [9], reveals however a notable difference. The behavior reflects, loosely speaking, that in our case the area of the region on which each eigenfunction z^k is essentially concentrated, is proportional to k. It is therefore not correct to think of each eigenfunction as essentially localized around its M-curve, and it does not seem meaningful to interpret the density of the M-curves as the density of the associated basis. This observation can be seen as an instance of the general problem of finding a satisfactory analogue to the very natural and useful concept of phase space density in the Bargmann case, see e.g. [29].

The discussion so far has left us with some rather natural questions that should be mentioned and briefly discussed. We saw above that the location and the order of a zero determined an M-curve uniquely. This leads immediately to the general problem of existence and uniqueness of an M-curve Γ enclosing prescribed points $0 = z_0, z_1, \ldots, z_n$ such that $z_0, z_1, \ldots, z_n$ are the zeroes of f_Γ, and that the orders of the zeros take prescribed values $m_0, m_1, \ldots, m_n$, respectively. The one zero example also leads us to ask if the functions associated to a collection of M-curves "filling up" phase space in a similar way, will constitute a complete set, or even a basis.

We have been able to solve these problems completely for $n = 1$.

Theorem 1. *Let m and n be two arbitrary positive integers. Then if*

$$r \leq \frac{\max(m,n)}{2\alpha + \max(m,n)} ,$$

there exists a unique M-curve, the function of which has zeros at 0 and r of multiplicities m and n, respectively.

The following theorem tells us that a suitable collection of such functions constitutes a Riesz basis for A_α; for technical reasons, we are then allowing $m = n = 0$, in which case we associate a constant function. We remind the reader that a Riesz basis is a basis that can be obtained from an orthonormal basis by a bounded invertible linear operator.

Theorem 2. *If $r < (2\alpha + 1)^{-1}$, the functions of Theorem 1 with $m \geq n \geq m - 1$ constitute, suitably normalized, a Riesz basis for A_α.*

We observe that the functions have been picked in such a way that the areas enclosed by the associated curves correspond precisely to the sequence $k\pi/\alpha$, k a positive integer, in agreement with the one zero case.

When $n > 1$, we know the solution only for certain special symmetric distributions of the zeros. The interested reader is referred to [24] for this and the detailed proofs of the two theorems above. We suspect that we have full analogues of these results for every $n > 1$, when assuming that the given zeros are contained in a sufficiently small disk.

In order to get a feeling for the problem, it is illuminating to make the observation on which the proofs of our two theorems rest. To this end, let us assume that Γ is an M-curve with associated function f_Γ, and that the distinct points $0 = z_0, z_1, \ldots, z_n$, enclosed by Γ, are the zeros of f_Γ. We denote the orders of the zeros $m_0, m_1, \ldots, m_n$, respectively. We know from [6] that such a Γ is the image of the unit circle T under a transformation Ψ_Γ of the form

$$\Psi_\Gamma(\zeta) = C \, \zeta \, \frac{(\zeta - a_1) \cdot (\zeta - a_2) \dots (\zeta - a_n)}{(\zeta - b_1) \cdot (\zeta - b_2) \dots (\zeta - b_n)} \, ,$$

with $|a_1|, \dots, |a_n|, |b_1|, \dots, |b_n| > 1$. Under the assumption that $\{\zeta = |\Psi_\Gamma(\zeta)| < 1\}$ has a connected component containing $\overline{\Delta}$ and that Ψ_Γ is univalent there, so that Ψ_Γ^{-1} can be defined in Δ, we deduce a general expression for the associated function f_Γ.

By (1) we must have

$$\frac{f_\Gamma'(z)}{f_\Gamma(z)} = \frac{2\alpha \overline{\Psi}_\Gamma(1/\Psi_\Gamma^{-1}(x))}{1 - z\overline{\Psi}_\Gamma(1/\psi_\Gamma)}^{-1}(z) \, , \tag{3}$$

where $\overline{\Psi}_\Gamma(z) = \overline{\Psi_\Gamma(\overline{z})}$. We make the substitution $z = \Psi_\Gamma(\zeta)$ so that (3) can be written as

$$\frac{(f_\Gamma \circ \Psi_\Gamma)'(\zeta)}{(f_\Gamma \circ \Psi_\Gamma)(\zeta)} = \frac{2\alpha \overline{\Psi}_\Gamma(1/\zeta)\Psi_\Gamma'(\zeta)}{1 - \Psi_\Gamma(\zeta)\overline{\Psi}_\Gamma(1/\zeta)} \, .$$

An elementary calculation, involving simple observations about the residues of the rational function on the right-hand side, integration, and finally the substitution $\zeta = \Psi^{-1}(z)$, yields

$$f_\Gamma(z) = (\Psi_\Gamma^{-1}(z))^{m_0} \prod_{i=1}^{n} \left(\frac{\Psi_\Gamma^{-1}(z) - b_i}{1 - \overline{\Psi_\Gamma^{-1}(z_i)}\Psi_\Gamma^{-1}(z)} \right)^{2\alpha} \left(\frac{\Psi_\Gamma^{-1}(z) - \Psi_\Gamma^{-1}(z_i)}{1 - \overline{\Psi_\Gamma^{-1}(z_i)}\Psi_\Gamma^{-1}(z)} \right)^{m_i} \, .$$

We may now write down a set of equations that must be satisfied for an M-curve of the desired kind to exist. It is necessary that

$$|C|^2 = \frac{m_0}{2\alpha + m_0} \frac{b_1 \dots b_n}{a_1 \dots a_n} \, , \tag{4}$$

and that for $1 \le i \le n$

$$z_i = \Psi_\Gamma(w_i) \tag{5}$$

$$0 = 1 - \Psi_\Gamma(w_i)\overline{\Psi}_\Gamma(1/w_i) \tag{6}$$

$$m_i = \frac{2\alpha \overline{\Psi}_\Gamma(1/w_i)\Psi_\Gamma'(w_i)}{\Psi_\Gamma(w_i)\overline{\Psi}_\Gamma'(1/w_i)w_i^{-2} - \psi_\Gamma'(w_i)\overline{\Psi}_\Gamma(1/w_i)} \tag{7}$$

for some $w_1, \dots, w_n, a_1^{-1}, \dots a_n^{-1}, b_1^{-1}, \dots, b_n^{-1} \in \Delta \backslash \{0\}$ with Ψ_Γ as defined above. If in addition, there is a connected domain of $\{\zeta : |\Psi_\Gamma(\zeta)| < 1\}$

containing $\overline{\Delta}$ and if Ψ_Γ is univalent there, we know that $\Gamma = \Psi_\Gamma(T)$ is an M-curve.

The set of equations (4)–(7) is not easy to handle in general. For $n = 1$ it is however a simple matter to perform explicit calculations, some of which enable us to prove the two theorems.

REFERENCES

[1]   E.W. Alasken and J.R. Klauder, J. Math. Phys. <u>10</u>, (1969), 2267.

[2]   G. Battle, P. Federbush, Com. Math. Phys. <u>109</u>, 417 (1987).

[3]   V. Bargmann, Comm. Pure and Applied Math <u>14</u>, (1961), 187.

[4]   S. Bergman, *The Kernel Function and Conformal Mapping* , Am. Math. Soc. Math. Surveys V, New York (1950).

[5]   P. Briet, J.M. Combs, P. Duclos, Comm. Part. Diff. Equ., <u>12</u>, 201 (1987).

[6]   R. Coifman, R. Rochberg, M.H. Taibleson, G. Weiss, Astérisque, <u>77</u>, (1980).

[7]   C. Cordoba and C. Fefferman, Comm. Part. Dif. Equ., <u>3</u>, 979 (1978).

[8]   I. Daubechies, *Time-frequency localization operators - A geometric phase space approach*, IEEE Trans. Inform. Theory <u>34</u>, 605-612 (1988).

[9]   I. Daubechies and T. Paul, *Time-frequency localization operators - A geometric phase space approach II. The use of dilations*, Inverse Problems, <u>4</u>, 661-680 (1988).

[10] I. Daubechies and J. Klauder, J. Math. Phys. <u>26</u> (1985), 2239.

[11] I. Daubechies and T. Paul, *Wavelets and application* . Proceedings of the IAMP conference in Marseille, editors M. Mebkhout and R. Seneor, World Scientific, 1986.

[12] I. Daubechies, J.R. Klauder, T. Paul, J. Math. Phys. <u>28</u>, 85 (1987).

[13] C. Feffermann, R. de la Lhave, Rev. Ibero America <u>2</u>, 119 (1986).

[14] S. Graffi, T. Paul, Comm. Math. Phys. <u>108</u>, 25 (1987).

[15] Hagedorn, Comm. Math. Phys., <u>71</u>, 77 (1980).

[16] K. Hepp, Comm. Math. Phys., <u>35</u>, 265 (1974).

[17] J.R. Klauder, E.C.G. Sudershan, *Fundamentals of Quantum Optics*, W.A. Benjamin, 1968.

[18] J.R. Klauder and Skargerstam, *Coherent states*, World Scientific. Singapore 1985.

[19] L. Landau and Lifschitz, *Course in theoretical Physics*, vol.1. Mechanics.

[20] L. Landau and Lifschitz, *Course in theoretical Physics*, vol.3. Quantum mechanics.

[21] T. Paul, J. Math. Phys. 25 (1984), 3252.

[22] T. Paul, Thèse d'Etat, Université d'Aix-Marseille II, 1985.

[23] T. Paul, *Affine coherent states and the radial Schrödinger equation*, Annales de l'Institut Henri Poincaré. Physique théorique.

[24] T. Paul and K. Seip, *Curves of Maximum Modulus in Coherent State Representations II. Closed Curves and Associated Bases in Bergman Spaces* , in preparation.

[25] J. Peetre, Lizhong Peng and Genkai Zhang. Report of University of Stockholm. n.11. 1990.

[26] F. Plantevin, thesis, to appear.

[27] K. Seip, *Curves of maximum modulus in coherent state representations* , Ann. de l'Inst. Henri Poincaré, Phys. Théor. 51, 335-350, (1989).

[28] K. Seip, *Reproducing formulas and double orthogonality in Bargmann and Bergman spaces* , SIAM J. Math. Anal. 22 3 (1991) (to appear).

[29] K. Seip and R. Wallstén, *Sampling and Interpolation in the Bargmann-Fock Space* , Report Institut Mittag-Leffler 1990/91, Djursholm, 1990.

[30] J. Sjostrand, Preprint.

[31] A. Voros, Thèse.

[32] E. Schrödinger, Naturwissenschaft, 14, 664 (1926).

WAVELETS: A RENORMALIZATION GROUP POINT OF VIEW*

GUY BATTLE

Mathematics Department
Texas A&M University
College Station, Texas

1. INTRODUCTION

Ten years ago Gawedzki and Kupiainen [1] derived expansion functions on a lattice with an orthogonality property suited for their renormalization group analysis of a lattice gas model in statistical mechanics. One goal in this exposition is to outline the construction of their functions because they become wavelets in the continuum limit. This is no accident. Another pedagogical goal is to relate a wavelet to the Gaussian fixed point of a renormalization group action. Finally, we explain exactly where wavelets fit in the renormalization group analysis of an interaction.

The wavelet transform has proven to be a flexible tool in phase space localization, and the notion of wavelet pursued here is by no means the most general one. Let $\mathcal{H}_\alpha$ denote the Hilbert space of distributions f on $\mathbb{R}^d$ such that $|\nabla|^\alpha f$ is square-integrable, that is, the massless Sobolev space of degree α.

*Supported in part by the National Science Foundation under Grant No. DMS 8904197 and by the CBMS Conference on Wavelets at the University of Lowell in June, 1990.

Definition. An element $\varphi \in \mathcal{H}_\alpha$ is a *wavelet* if and only if $2^{r(\frac{d}{2}-\alpha)}\varphi(2^r x - n), r \in \mathbb{Z}\backslash\{0\}, n \in \mathbb{Z}^d$, are all orthogonal to $\varphi(x)$ in $\mathcal{H}_\alpha$.

Note that – following the approximation theorists [2,3] – we are not requiring orthogonality between functions that live on the same scale, but a great deal of attention has been devoted to orthonormal bases of wavelets, particularly for $\alpha = 0$. One needs $2^d - 1$ wavelets like φ to generate a basis, but for $\mathcal{H}_0 = L^2$, dimension does not play an important role in the construction of wavelets, so one normally sets $d = 1$ for that function space [2-6,8].

The most obvious – and certainly the oldest – example of a wavelet in the one-dimensional square-integrable case is the Haar function

$$\varphi(x) = \begin{cases} 1, & 0 \le x \le \frac{1}{2}, \\ -1, & \frac{1}{2} < x \le 1, \\ 0, & \text{otherwise.} \end{cases} \tag{1.1}$$

This function actually generates an orthonormal basis for $L^2(\mathbb{R})$, but it is also discontinuous. Localization is very poor in frequency space, and the name of the game is phase space localization. However, one can do much better than the Haar basis.

It was Stromberg [4] who first constructed an L^2-wavelet with exponential fall-off and class C^N regularity for arbitrary N, but his result attracted little attention at the time. The proverbial two-by-four was Meyer's construction [5] of an L^2-wavelet φ for which $\hat{\varphi} \in C_0^\infty$! Since Meyer's wavelet obviously cannot have exponential decay, Stromberg's result is complementary. Lemarié [6] and this author [7] independently constructed a wavelet different from Stromberg's but with the same properties. These results were then overshadowed by Daubechies' impressive construction [8] of an L^2-wavelet in C_0^N for arbitrary N. On the negative side, it has been shown [9] that a wavelet cannot have exponential decay in position space *and* momentum space.

Our method for constructing wavelets differs from the method of Meyer, Lemarié, Daubechies, and their co-workers (although not as much as we believed at the time). The latter method is a very elegant multi-scale resolution analysis due to Mallat and Meyer, while the former is based on renormalization group ideas used in the study of physical systems with infinitely many degrees of freedom.

Wavelets have already been applied to phase cell cluster expansions in quantum field theory, and they have the effect of stream-lining the original phase cell analysis of Glimm and Jaffe [10,11]. So far, applications include the traditional ϕ_3^4 quantum field theory [12], the four-dimensional

Yang-Mills theory [13,14], and three-dimensional spinor quantum electro-dynamics [15]. This phase cell decomposition has received less attention than the "momentum-slicing" of Magnen and Seneor (references given be-low) but it is natural and elegant because of the close relationship between wavelets and the renormalization group. As we shall see, there is a reason to expect renormalization group analysis based on wavelets to be efficient.

The renomalization group is an extremely powerful analytic tool that was originally introduced by Wilson. The literature is so vast that we mention only Kogut and Wilson [16], Polchinski [17], and Balaban [18], in addition to Gawedzki and Kupiainen [1]. The renormalization group iteratively maps a space of *actions* – which are positive functionals of *configurations* – into itself. This remark can mean a lot of things, but in our own context the space of configurations is only the space of real-valued functions on $\mathbb{Z}^d$, while the space of actions we consider will often be a relatively small space of translation-invariant quadratic forms. Only in Section 5 do we look at non-quadratic perturbations.

We define the basic renormalization group mapping of a given action to another in Section 3, but such a mapping is always implemented by an averaging transformation – more or less localized – of configurations followed by a re-scaling of the new, "coarser" configurations to make them as "fine" as the old configurations. There are many transformations of this type, and we mention two classes. The first has nothing to do with lattices, but is quite standard.

1) Let K be a Schwartz function on $\mathbb{R}^d$ with $K(0) = 2^d$ and let the space of configurations be the space of tempered distributions on $\mathbb{R}^d$. Let T be the averaging transformation defined by

$$\widehat{T\Phi}(p) = 2^{-\frac{d}{2}-\alpha} K(p)\hat{\Phi}\left(\frac{1}{2}p\right), \tag{1.2}$$

where α is the degree of the Sobolev space whose inner product is the "free" or "kinetic" part of the action. Thus we always have $\alpha = 1$ for a physical model. This kind of averaging has been used by Polchinski [17]. It is also implicit in the extensive work of Feldman, Magnen, Seneor, and Rivasseau [19-23]. Actually, they use a phase cell cluster expansion based on a momentum decomposition defined by iterations of (1.2), but of course they tackle problems similar to those encountered in the renormalization group formalism. We will say no more about this averaging transformation.

2) Let T be an averaging transformation on lattice configurations defined by a periodicized version of (1.2). In lattice position space,

$$(T\phi)(n) = 2^{-\frac{d}{2}-\alpha} \sum_{m\in\mathbb{Z}^d} c_m\phi(2n - m), \tag{1.3}$$

where $\{c_m\}$ is summable and real-valued,

$$\sum_{m \in \mathbb{Z}^d} c_m = 2^d, \tag{1.4}$$

and the translation-invariant quadratic form $(A\phi, \phi)$ to which the renormalization group analysis is to be applied satisfies the following property. If $A_{mn} = A_{m-n}$ are the matrix elements of such a quadratic form, and

$$A(p) = \sum_{m \in \mathbb{Z}^d} A_m e^{im \cdot p}, \tag{1.5}$$

then

$$\lim_{\varepsilon \to 0} \varepsilon^{-2\alpha} A(\varepsilon p) = |p|^{2\alpha}. \tag{1.6}$$

Remark. (1.3) is a generalization of the well-known "block spin" transformation in physics, which is only the case

$$c_m = 1, \quad m \in \{0, 1\}^d, \tag{1.7}$$

and $c_m = 0$ otherwise. The term "spin" is a literary artifact arising from the physical context in which the renormalization group was originally introduced.

In physics one is naturally interested in a much larger space of actions, including nontrivial perturbations of the quadratic forms. Indeed, in Section 3 we define the renormalization group on an arbitrary space of actions. The aim of the analysis is to extract the long-distance behavior of correlation functions associated with the initial action by iterating the renormalization group mapping a large number of times. The *stationary* actions are the fixed points, and the most important problem is to determine the structure of the flow in a neighborhood of a stationary action. The traditional success of the renormalization group is due to the irrelevance – in the infinite-dimensional space of actions – of all but a finite number of directions in such a neighborhood.

Now from a physicist's point of view, the translation-invariant quadratic forms are trivial, but when a physically interesting action is realized as a perturbation of such a quadratic form, it is reasonable to expand an arbitrary configuration in basis functions that are compatible with the action of the renormalization group on the quadratic form because:

(a) The space of translation-invariant positive quadratic forms is invariant with respect to the renormalization group mapping.

(b)   The image under the mapping of such a quadratic form can be computed explicitly.

These facts will be clarified in Section 3, but the point is that our yet-to-be-defined mapping is really acting on the measure

$$e^{-A(\phi)} \prod_{x \in \mathbb{Z}^d} d\phi(x) \tag{1.8}$$

associated with the action A by integrating out the "fluctuations" relative to the averaging transformation. If $A(\phi)$ is a strictly positive quadratic form, then the corresponding measure is a Gaussian, so we may regard the renormalization group as acting on a space of Gaussians rather than quadratic forms.

Since the lattice basis functions turn out to be wavelets in the continuum limit, our goal is to exhibit the connection between wavelets and fixed points of the renormalization group. In this expository effort we prove no convergence theorems and derive no estimates. We simply sketch relationships that show how wavelets arise from the renormalization group and exhibit the conclusions in the last section. For this reason, we deliberately suppress technical assumptions as much as we dare.

Remark.   Every construction of a wavelet – whether our construction or the Mallat-Meyer construction – is based on comparing a function to discrete translates of a half-scale copy of that function. *The subspace corresponding to the scale of the wavelet one constructs is generated by discrete translates of the wavelet that are double the original translates.* Thus we have two natural choices for the definition of "unit scale". The definition given above reflects the unit-scale choice for the translates of the wavelet (and therefore the half-scale choice for the underlying function and the constructive translates). This is the convention of Meyer and his co-workers, and it is certainly the more popular one. We use the other convention:

Definition'.   An element $\varphi \in \mathcal{H}_\alpha$ is a *wavelet* if and only if $2^{r(\frac{d}{2}-\alpha)}\varphi(2^r x - 2n), r \in \mathbb{Z}\backslash\{0\}, n \in \mathbb{Z}^d$, are all orthogonal to $\varphi(x)$ in $\mathcal{H}_\alpha$.

In other words, we use the unit-scale choice for the underlying function, and so in the case of the Haar basis in $L^2(\mathbb{R})$, the "unit-scale" wavelet is

$$
\varphi(x) = \begin{cases}
\sqrt{\tfrac{1}{2}}, & 0 \leq x \leq 1, \\[2mm]
-\sqrt{\tfrac{1}{2}}, & 1 < x \leq 2, \\[2mm]
0, & \text{otherwise.}
\end{cases}
\tag{1.9}
$$

where the function used to construct it is the characteristic function of $[0,1]$ and its unit-scale translates.

In Section 2 we review the constrained minimization technique [7] used for constructing wavelets. The approach is a generalization of the continuum version of a renormalization group computation that is made in [1], but this point is not made until Section 4. The review simply emphasizes the idea of the construction.

In Section 3 we introduce the definition of the renormalization group transformation based on a given averaging transformation T. We compute its effect on translation-invariant quadratic forms, and then we show how a stationary quadratic form can be expressed in terms of the input function for the construction of some basis of wavelets.

The connection between wavelets and Gaussian fixed points – derived in Section 4 – is as simple as can be. If $A(\phi)$ is an initial translation-invariant quadratic form satisfying (1.6) and $A^\infty(\phi)$ denotes the limiting quadratic form under iterations of the renormalization group mapping based on T, then there is a wavelet – depending on $A^\infty(\phi)$ and T – whose momentum expression includes $|p|^{-2\alpha} A^\infty(p)$ as a factor. Moreover, we show that the wavelets appear as the continuum limit of the lattice basis functions of Kupiainen and Gawedzki.

Finally, in Section 5 we tie these relationships together when we look at an interaction. We focus on the massless lattice ϕ_4^4 model and show that iterations of the renormalization group transformation drive the action – *to first order in the dimensionless coupling constant* – to a lattice action that is equivalent to the massless *continuum* ϕ_4^4 action with a small-length scale cutoff based on a wavelet decomposition of the continuum field. Therefore the effect of basing a renormalization group analysis on a wavelet decomposition is to start with a lattice action that is already fixed to first order in the coupling constant.

2. CONSTRAINED MINIMIZATION

Our starting point is a function of the type used in the multi-scale resolution analysis of Mallat and Meyer. Let ζ be a real-valued square-integrable function such that

$$\zeta(x) = \sum_{m \in \mathbb{Z}^d} c_m \zeta(2x - m) \qquad (2.1)$$

for a summable sequence $\{c_m\}$, which – for the sake of being specific – is the d-fold product of some sequence $\{s_i\}_{i \in \mathbb{Z}}$. Suppose

$$\sum_m |\widehat{\zeta}(p + 2\pi m)|^2 > 0 \qquad (2.2)$$

everywhere. We assume the generated set

$$\{\zeta(2^r x - n): \; r \in \mathbb{Z}, \quad n \in \mathbb{Z}^d\} \qquad (2.3)$$

spans $L^2(\mathbb{R}^d)$ to insure completeness of the set of wavelets to be derived from ζ.

The derivation presented here is based on averaging constraints. Instead of following the Meyer prescription [6,8] we solve the problem of minimizing $\int \varphi^2$ with respect to the constraints

$$\int \varphi(x)\zeta(x - m)dx = (-1)^{\sum_\mu \lambda_\mu^{(\iota)} m_\mu} c_{\lambda^{(\iota)} - m}, \qquad (2.4)$$

where $\{\lambda^{(\iota)}\}$ is the set of non-zero elements of $\{0, 1\}^d$. The point of solving such a problem is that the solution to (2.4) $\varphi_\iota(x)$ is automatically orthogonal to $\varphi_\iota(2^r x - 2n)$ provided $r \neq 0$. This follows from the observation that

$$\int \varphi_\iota(2^r x - 2n)\zeta(x - m)dx$$

$$= 2^{-rd} \int \varphi_\iota(x)\zeta(2^{-r}x + 2^{-r+1}n - m)dx$$

$$= 2^{-rd} \sum_{m^{(1)}, \ldots, m^{(r)}} c_{m^{(1)}} \ldots c_{m^{(r)}} \times$$

$$\times \int \varphi_\iota(x)\zeta(x + 2n - 2^r m - 2^{r-1}m^{(1)} - \cdots - 2m^{(r-1)} - m^{(r)})dx$$

$$= 2^{-rd} \sum_{m^{(1)}, \ldots, m^{(r)}} c_{m^{(1)}} \ldots c_{m^{(r)}} (-1)^{\sum_\mu \lambda_\mu^{(\iota)} m_\mu^{(r)}} \times$$

$$\times c_{\lambda^{(\iota)} + 2n - 2^r m - 2^{r-1}m^{(1)} - \cdots - 2m^{(r-1)} - m^{(r)}} = 0, \quad r > 0, \qquad (2.5)$$

because $\lambda^{(\iota)}$ is non-zero and $c_m = \prod\limits_{\mu=1}^{d} s_{m_\mu}$. The point is that if $\lambda_\nu^{(\iota)} = 1$, then the change of index $m^{(r)} \to \widehat{m}^{(r)}$ with

$$
\widehat{m}_\mu^{(r)} = \begin{cases} m_\mu^{(r)}, & \mu \neq \nu, \\ 1 + 2n_\nu - 2^r m_\nu - 2^{r-1} m_\nu^{(1)} - \cdots - m_\nu^{(r)}, & \mu = \nu, \end{cases} \tag{2.6}
$$

reverses the sign of the multiple sum because of the d-fold product structure of $\{c_m\}$. Why does this annihilation condition guarantee the orthogonality property? Since $\varphi_\iota(2^r x - 2n)$ lies in the kernels of the linear functionals defining the constraints on $\varphi_\iota(x)$ and the latter function minimizes the L^2-norm with respect to those constraints, the orthogonality follows from the simple geometric fact that the shortest vector from 0 to an affine subspace is perpendicular to any vector lying in it.

The solution of a minimization problem like this is a routine application of Poisson summation [7,24]. In momentum space we have

$$
\widehat{\varphi}_\iota(p) = \frac{\widehat{\zeta}(p)\eta_\iota(p)}{\sum\limits_{\ell \in \mathbb{Z}^d} |\widehat{\zeta}(p + 2\pi\ell)|^2}, \tag{2.7}
$$

$$
\eta_\iota(p) = \sum\limits_{m \in \mathbb{Z}^d} (-1)^{\sum\limits_{\mu} \lambda_\mu^{(\iota)} m_\mu} c_{\lambda^{(\iota)} - m} e^{im \cdot p}. \tag{2.8}
$$

$\{\varphi_\iota\}$ generates a basis for $L^2(\mathbb{R}^d)$ that is inter-scale orthogonal but not intra-scale orthogonal. One derives a completely orthonormal basis of wavelets by translation-invariant orthogonalization on each scale. This procedure is more standard because one is simply applying the inverse square root of the overlap matrix, and existence of the former follows from the condition

$$
\sum_\iota \frac{|\eta_\iota(p)|^2}{\sum\limits_{\ell \in \mathbb{Z}^d} |\widehat{\zeta}(p + 2\pi\ell)|^2} + \sum_\iota \frac{|\eta_\iota(p + \pi)|^2}{\sum\limits_{\ell \in \mathbb{Z}^d} |\widehat{\zeta}(p + 2\pi\ell + \pi)|^2} > 0 \tag{2.9}
$$

everywhere.

This concludes our description of how wavelets can be constructed by constrained minimization. We have confined our attention here to the function space $L^2(\mathbb{R}^d)$, but there are no fundamental changes in the construction if our function space is a massless Sobolev space. One would simply minimize the quadratic form $\int (|\nabla|^\alpha \varphi)^2$ instead of $\int \varphi^2$. Naturally the momentum expressions would be different; we intend to display the pattern when we relate this construction to the renormalization group analysis of translation-invariant lattice Gaussians.

Remark.   In momentum space the functional equation (2.1) becomes

$$\widehat{\zeta}(p) = 2^{-d} h(p) \widehat{\zeta}\left(\frac{1}{2}p\right),\tag{2.10}$$

$$h(p) = \sum_m c_m e^{i\frac{1}{2}m\cdot p}.\tag{2.11}$$

Obviously the normalization condition is

$$h(0) = \sum_m c_m = 2^d,\tag{2.12}$$

and the most obvious example is simple addition of discrete translates over a block – i.e., with $\{c_m\}$ the lattice characteristic function of $\{0,1\}^d$. In that case, of course, ζ is just the characteristic function of $[0,1]^d$ and the basis of wavelets is only the basis of Haar functions.

Every wavelet constructed by the Mallat-Meyer machine can be constructed by this machine as well. Naturally the sequence $\{c_n\}$ determines the regularity and decay properties of the wavelet produced, and so – from the renormalization group point of view – the kind of wavelet desired is obtained by a judicious modification of the "block spin" transformation.

3.   THE RENORMALIZATION GROUP

Having introduced a machine for the construction of wavelets, we now describe its relation to the renormalization group. The main point to understand here is that wavelets are related to lattice Gaussian fixed points when the underlying transformation T is an averaging of lattice variables

$$(T\phi)(n) = 2^{-\frac{d}{2}-\alpha} \sum_{m \in \mathbb{Z}^d} c_m \phi(2n - m)\tag{3.1}$$

of the kind already described in the Introduction. *Wavelets appear automatically in the renormalization group analysis of lattice models whether one is interested in these basis functions or not.*

First we show that if f is a function with $|p|^\alpha \hat{f}(p)$ square-integrable and satisfying the condition (2.2) but with (2.1) replaced by

$$f(x) = \sum_{m \in \mathbb{Z}^d} 2^{-\alpha} c_m f(2x - m),\tag{3.2}$$

then a lattice Gaussian fixed point can be constructed from f. Let $A(\phi)$ be a lattice action – i.e., a nonnegative functional of lattice configurations ϕ

– such that the measure

$$e^{-A(\phi \chi_\Lambda)} \prod_{x \in \Lambda} d\phi(x) \tag{3.3}$$

has finite mass Z_Λ for all finite $\Lambda \subset \mathbb{Z}^d$. If T is the averaging transformation we have chosen for lattice configurations, then the corresponding renormalization group transformation $\mathcal{R}_T(A)$ of the action A is defined by

$$\mathcal{R}_T(A)(\psi)$$
$$= \lim_{\Lambda \to \mathbb{Z}^d} \left[\ln Z_\Lambda(T\phi = 0) \tag{3.4} \right.$$
$$\left. - \ln \int \prod_{n \in \Lambda} d\phi(n) e^{-A(\phi \chi_\Lambda)} \prod_{n' \in \Lambda'} \delta\left(\psi\left(\frac{1}{2}n'\right) - (T\phi)\left(\frac{1}{2}n'\right) \right) \right]$$

where Λ' is simply the set of points in Λ having only even coordinates. Technically, the computation of the new action involves an infinite-volume limit constrained by fixed averages. This is usually handled by a cluster expansion, the convergence of which would naturally depend on further properties of the action.

The name of the game is to iterate $\mathcal{R}_T$ as a nonlinear mapping from a space of actions into itself and investigate the flow of this semi-group. In particular, physicists are interested in finding fixed points and understanding the nature of the flow in a neighborhood of a fixed point. We restrict our attention to translation-invariant quadratic forms, which comprise an $\mathcal{R}_T$-invariant cone of physical actions: if $A(\phi)$ is such an action, then so is $\mathcal{R}_T(A)(\phi')$. Although these actions are physically trivial, they play a role in the construction of wavelets.

Technically, we are working with strictly positive quadratic forms $A(\phi)$, and we adopt the standard abuses of notation

$$A(\phi) = \frac{1}{2}(A\phi, \phi)$$
$$= \frac{1}{2} \sum_{m,n \in \mathbb{Z}^d} A_{m-n}\phi(m)\phi(n), \tag{3.5}$$

$$A_{m-n} = \frac{1}{(2\pi)^d} \prod_{\mu=1}^{d} \left(\int_0^{2\pi} dp_\mu \right) e^{-ip \cdot (m-n)} A(p). \tag{3.6}$$

Let $d\mu_A$ denote the probability measure on $\mathbb{R}^{\mathbb{Z}^d}$ realized as the $\Lambda = \mathbb{Z}^d$ limit of

$$Z_\Lambda^{-1} e^{-A(\phi \chi_\Lambda)} \prod_{n \in \Lambda} d\phi(n) \tag{3.7}$$

on cylinder sets. Then $d\mu_A$ is the mean-zero Gaussian measure with co-variance matrix A^{-1} – i.e., if $\langle\ \rangle_A$ denotes the expectation functional

$$\langle F(\phi)\rangle_A = \int F d\mu_A, \tag{3.8}$$

then

$$\langle\phi(m)\phi(n)\rangle_A = (A^{-1})_{m-n}$$

$$= \frac{1}{(2\pi)^d} \prod_{\mu=1}^{d}\left(\int_0^{2\pi} dp_\mu\right) e^{-ip\cdot(m-n)} A(p)^{-1}, \tag{3.9}$$

$$\left\langle\prod_{k=1}^{N}\phi(n_k)\right\rangle_A = 0, \quad N \text{ odd}, \tag{3.10}$$

$$\left\langle\prod_{k=1}^{N}\phi(n_k)\right\rangle_A = \sum_{\substack{\text{disjoint pairings} \\ P \text{ of } \{1,...,N\}}} \prod_{\{i,j\}\in P} \langle\phi(n_j)\phi(n_k)\rangle_A, \quad N \text{ even}, \tag{3.11}$$

$$\left\langle\exp i\sum_m \alpha_m\phi(m)\right\rangle_A = \exp\left(-\frac{1}{2}\sum_{m,n}(A^{-1})_{m-n}\alpha_m\alpha_n\right). \tag{3.12}$$

Now it is easy to see that

$$\langle\psi(m)\psi(n)\rangle_{\mathcal{R}_T(A)} = \lim_{\Lambda\to\mathbb{Z}^d} Z_{\Lambda'}(\mathcal{R}_T(A))^{-1} \int \prod_{n'\in\Lambda'} d\psi\left(\frac{1}{2}n'\right)$$

$$\psi(m)\psi(n) \int \prod_{n\in\Lambda} d\phi(n) e^{-A(\phi\chi_\Lambda)}$$

$$\times \prod_{n'\in\Lambda'} \delta\left(\psi\left(\frac{1}{2}n'\right) - (T\phi)\left(\frac{1}{2}n'\right)\right)$$

$$= \langle(T\phi)(m)(T\phi)(n)\rangle_A \tag{3.13}$$

since

$$Z_{\Lambda'}(\mathcal{R}_T(A)) = Z_\Lambda(A). \tag{3.14}$$

Thus

$$\langle\psi(m)\psi(n)\rangle_{\mathcal{R}_T(A)} = 2^{-d-2\alpha} \sum_{m',n'} c_{m'} c_{n'} \langle\phi(2m-m')\phi(2n-n')\rangle_A, \tag{3.15}$$

and so $\mathcal{R}_T(A)$ is just the inverse of the matrix

$$\sum_{m',n'} (A^{-1})_{2m-2n-m'+n'} c_{m'} c_{n'}$$

$$= \frac{2^{-d-2\alpha}}{(2\pi)^d} \prod_{\mu=1}^{d} \left(\int_0^{2\pi} dp_\mu \right) e^{-ip\cdot(2m-2n)} |h(2p)|^2 A(p)^{-1}$$

$$= \frac{2^{-2d-2\alpha}}{(2\pi)^d} \prod_{\mu=1}^{d} \left(\int_0^{4\pi} dp_\mu \right) e^{-ip\cdot(m-n)} |h(p)|^2 A\left(\frac{1}{2}p\right)^{-1}, \quad (3.16)$$

$$h(p) = \sum_{m\in\mathbb{Z}^d} c_m e^{i\frac{1}{2}m\cdot p}. \quad (3.17)$$

Since $h(p)$ is 4π-periodic, we have

$$\mathcal{R}_T(A)(p)^{-1} = 4^{-d-\alpha} \sum_{\vec{\lambda}\in\{0,1\}^d} |h(p+2\pi\vec{\lambda})|^2 A\left(\frac{1}{2}p+\pi\vec{\lambda}\right)^{-1}, \quad (3.18)$$

so it is now easy to obtain lattice actions which are fixed under $\mathcal{R}_T$. If f is a function satisfying (3.2) then

$$\sum_{m\in\mathbb{Z}^d} |\hat{f}(p+2\pi m)|^2$$

$$= 4^{-d-\alpha} \sum_{\vec{\lambda}\in\{0,1\}^d} |h(p+2\pi\vec{\lambda})|^2 \sum_{m\in\mathbb{Z}^d} \left| \hat{f}\left(\frac{1}{2}p+\vec{\lambda}\pi+2\pi m\right) \right|^2. \quad (3.19)$$

Hence the quadratic form $\frac{1}{2}(A\phi,\phi)$ given by

$$A(p) = \frac{1}{\displaystyle\sum_{m\in\mathbb{Z}^d} |\hat{f}(p+2\pi m)|^2} \quad (3.20)$$

is a fixed point under the renormalization group transformation $\mathcal{R}_T$.

Having established the connection between translation-invariant Gaussian fixed points and functions f satisfying functional equations like (3.2), we turn to the second issue. Since f is not uniquely determined by $\{c_n\}$ unless the function space – i.e., the parameter α – is specified, we are dealing with a family of fixed points. If an initial Gaussian is driven to such a fixed point by iterations of $\mathcal{R}_T$, how is f – and therefore the fixed point – determined? As an example of how it is done, consider the quadratic form

$$(A_0\phi,\phi) = \sum_{m\in\mathbb{Z}^d} \phi(m)^2. \quad (3.21)$$

In this case $A(p) = 1$ and $\alpha = 0$, so

$$\mathcal{R}_T(A_0)(p)^{-1} = 4^{-d} \sum_{\vec{\lambda} \in \{0,1\}^d} |h(p + 2\pi\vec{\lambda})|^2, \tag{3.22.1}$$

$$\mathcal{R}_T^2(A_0)(p)^{-1} = 4^{-2d} \sum_{\vec{\lambda}\vec{\lambda}' \in \{0,1\}^d} |h(p + 2\pi\vec{\lambda}')|^2 \left| h\left(\frac{1}{2}p + 2\pi\vec{\lambda} + \pi\vec{\lambda}'\right)\right|^2$$

$$= 4^{-2d} \sum_{\vec{m} \in \{0,1,2,3\}^d} |h(p + 2\pi\vec{m})|^2 \left| h\left(\frac{1}{2}p + \pi\vec{m}\right)\right|^2, \tag{3.22.2}$$

$$\mathcal{R}_T^N(A_0)(p)^{-1} = 4^{-Nd} \sum_{\vec{\lambda} \in \{0,1\}^d} |h(p + 2\pi\vec{\lambda}|^2 \quad \times$$

$$\times \sum_{\vec{m} \in \{0,\ldots,2^{N-1}-1\}^d} \prod_{k=0}^{N-2} |h(2^{-k-1}p + 2^{-k+1}\vec{m}\pi + 2^{-k}\pi\vec{\lambda})|^2$$

$$= 4^{-Nd} \sum_{\vec{m} \in \{0,\ldots,2^N-1\}^d} \prod_{k=0}^{N-1} |h(2^{-k}p + 2^{-k+1}\vec{m}\pi)|^2 \tag{3.22.N}$$

because h is 4π-periodic. Thus the convergence of the sequence $\{\mathcal{R}_T^N(A_0)\}$ of actions depends on the convergence of the infinite product

$$\prod_{k=0}^{\infty} (2^{-d}h(2^{-k}p)) \equiv H(p), \tag{3.23}$$

and in the case of convergence, the fixed point

$$A_0^{\infty} = \lim_{N \to \infty} \mathcal{R}_T^N(A_0) \tag{3.24}$$

is given by

$$A_0^{\infty}(p)^{-1} = \sum_{m \in \mathbb{Z}^d} |H(p + 2\pi m)|^2, \tag{3.25}$$

in which case f is given by $\hat{f} = H$. The convergence of an infinite product like (3.23) is an important issue around which Daubechies' construction [8] of compactly supported wavelets revolves, and the normalization requirement $h(0) = 2^d$ of Section 2 is obviously consistent with convergence of the product.

As another example, consider the quadratic form defined by difference operators:

$$(A_1\phi, \phi) = \sum_{m \in \mathbb{Z}^d} \sum_{\mu} (\phi(m + \vec{e}_\mu) - \phi(m))^2. \tag{3.26}$$

In this case our momentum expression is

$$A_1(p) = \sum_\mu |e^{ip_\mu} - 1|^2 \tag{3.27}$$

and $\alpha = 1$ so we have

$$\mathcal{R}_T(A_1)(p)^{-1} = 4^{-d-1} \sum_{\vec\lambda \in \{0,1\}^d} |h(p + 2\pi\vec\lambda)|^2 \times$$

$$\times \frac{1}{\sum_\mu |e^{i(\frac{1}{2}p_\mu + \pi\lambda_\mu)} - 1|^2}, \tag{3.28.1}$$

$$\mathcal{R}_T^2(A_1)(p)^{-1} = 4^{-2d-2} \sum_{\vec m \in \{0,1,2,3\}^d} |h(p + 2\pi\vec m)|^2 \left|h\left(\frac{1}{2}p + \pi\vec m\right)\right|^2 \times$$

$$\times \frac{1}{\sum_\mu |e^{i(\frac{1}{4}p_\mu + \frac{1}{2}\pi m_\mu)} - 1|^2}, \tag{3.28.2}$$

$$\mathcal{R}_T^N(A_1)(p)^{-1} = 4^{-Nd-N} \sum_{\vec m \in \{0,\dots,2^N-1\}} \prod_{k=0}^{N-1} |h(2^{-k}p + 2^{-k+1}\vec m\pi)|^2 \times$$

$$\times \frac{1}{\sum_\mu |e^{i(2^{-N}p_\mu + 2^{-N+1}\pi m_\mu)} - 1|^2}. \tag{3.28.N}$$

The convergence of the sequence $\{\mathcal{R}_T^N(A_1)\}$ still depends on the convergence of the infinite product (3.23) because

$$\lim_{N \to \infty} \frac{4^{-N}}{\sum_n |e^{i(2^{-N}p_\mu + 2^{-N+1}\pi m_\mu)} - 1|^2} = \frac{1}{(p + 2\pi m)^2}. \tag{3.29}$$

The fixed point

$$A_1^\infty = \lim_{N \to \infty} \mathcal{R}_T^N(A_1) \tag{3.30}$$

is given by

$$A_1^\infty(p)^{-1} = \sum_{m \in \mathbb{Z}^d} \frac{1}{(p + 2\pi m)^2} |H(p + 2\pi m)|^2. \tag{3.31}$$

The third phase in exposing the connection between the renormalization group and the wavelet construction machine is to describe the role played by the wavelets themselves. To this end, we find it useful to review the expansion functions of Gawedzki and Kupiainen [1] that are used in their renormalization group analysis of a lattice system.

4. GAWEDZKI-KUPIAINEN EXPANSION FUNCTIONS

The expansion functions are based on the quadratic form $(A_1\phi, \phi)$ defined by (3.26), and the averaging transformation T is defined by

$$
c_n = \begin{cases} 1, & n \in \{0,1\}^d, \\ \\ 0, & \text{otherwise,} \end{cases} \tag{4.1}
$$

in this context, but the first part of our description applies to arbitrary $(A\phi, \phi)$ and $\{c_n\}$.

Let M be the right-inverse of the transformation T such that $M(\psi)$ minimizes $(A\phi, \phi)$ with respect to the constraints

$$
T(\phi) = \psi. \tag{4.2}
$$

This linear operator M is the *minimizer*, and the idea is to decompose an arbitrary lattice configuration ϕ in the following way:

$$
\phi = MT\phi + \tilde{\phi} \equiv \phi' + \tilde{\phi}. \tag{4.3}
$$

Since M is a right inverse of T, $\tilde{\phi}$ has the property

$$
T\tilde{\phi} = 0. \tag{4.4}
$$

This decomposition has the orthogonality property

$$
(A\phi, \phi) = (A\phi', \phi') + (A\tilde{\phi}, \tilde{\phi}), \tag{4.5}
$$

so the effect of $\mathcal{R}_T$ on A is to integrate out $\tilde{\phi}$ very neatly, as $\tilde{\phi}$ does not contribute to the delta functions. Thus

$$
\mathcal{R}_T(A)(\psi) =
$$
$$
= \lim_{\Lambda \to \mathbb{Z}^d} \Big[\ln Z_\Lambda(\phi = \tilde{\phi})
$$
$$
- \ln \int d\phi' d\tilde{\phi} e^{-A(\phi' \chi_\Lambda)} e^{-A(\tilde{\phi} \chi_\Lambda)} \prod_{n' \in \Lambda'} \delta\left(\psi\left(\frac{1}{2}n'\right) - (T\phi')\left(\frac{1}{2}n'\right) \right) \Big]
$$
$$
= - \lim_{\Lambda \to \mathbb{Z}^d} \ln \int d\phi' e^{-A(MT(\phi') \chi_\Lambda)} \prod_{n' \in \Lambda'} \delta\left(\psi\left(\frac{1}{2}n'\right) - (T\phi')\left(\frac{1}{2}n'\right) \right) \tag{4.6}
$$

since $\phi' = MT\phi'$. But the functional integral now obviously collapses to $e^{-A(M(\psi) \chi_\Lambda)}$, so we have

$$
\mathcal{R}_T(A)(\psi) = A(M\psi). \tag{4.7}
$$

Remark. In the decomposition, ϕ' is the "minimum energy" contribution to the averaged configuration, while $\tilde{\phi}$ is the "fluctuating" part of the configuration.

The iteration of (4.7) is related to the iteration of (4.3) in the following way. We define the configurations $\tilde{\phi}', \tilde{\phi}'', \ldots, \tilde{\phi}^{(N-1)}$ by the successive decompositions

$$\phi' = M^2 T^2 \phi' + \tilde{\phi}' \equiv \phi'' + \tilde{\phi}', \tag{4.8.1}$$

$$\phi'' = M^3 T^3 \phi'' + \tilde{\phi}'' \equiv \phi''' + \tilde{\phi}'', \tag{4.8.2}$$

$$\vdots$$

$$\phi^{(N-1)} = M^N T^N \phi^{(N-1)} + \tilde{\phi}^{(N-1)} \equiv \phi^{(N)} + \tilde{\phi}^{(N-1)}. \tag{4.8.N-1}$$

Thus

$$\phi = \phi^{(N)} + \sum_{k=0}^{N-1} \tilde{\phi}^{(k)}, \tag{4.9}$$

$$T^k \tilde{\phi}^{(k-1)} = 0, \tag{4.10}$$

and since $M^k(\psi)$ minimizes $(A\phi, \phi)$ with respect to the constraints

$$T^k(\phi) = \psi, \tag{4.11}$$

we also have the orthogonality property

$$(A\phi, \phi) = (A\phi^{(N)}, \phi^{(N)}) + \sum_{k=0}^{N-1} (A\tilde{\phi}^{(k)}, \tilde{\phi}^{(k)}). \tag{4.12}$$

On the other hand the identity

$$(M^k T^k)(M^{k-1} T^{k-1}) \ldots (MT) = M^k (T^k M^{k-1})(T^{k-1} M^{k-2}) \ldots (T^2 M) T$$
$$= M^k T^k \tag{4.13}$$

together with the iteration of

$$\phi^{(k)} = M^k T^k \phi^{(k-1)} \tag{4.14}$$

yields

$$\phi^{(k)} = M^k T^k \phi. \tag{4.15}$$

Hence

$$(A\phi^{(N)}, \phi^{(N)}) = (\mathcal{R}_T^N(A) T^N \phi, T^N \phi), \tag{4.16}$$

so the residual part of the decomposition of A is the operator

$$T^{\dagger N}\mathcal{R}^N_T(A)T^N, \tag{4.17}$$

where $\dagger$ denotes operator adjunction on $\ell^2(\mathbb{Z}^d)$. The multi-scale fluctuation part of the decomposition is

$$\sum_{k=0}^{N-1} AP_k, \tag{4.18}$$

where the idempotent operator

$$P_k\phi = \tilde{\phi}_k \tag{4.19}$$

has the property $P_k^{\dagger}A = AP_k$. The point is that P_k is an $(A\phi, \phi)$-orthogonal projection.

Now we turn to the special case where $A = A_1$ and T is given by (4.1) and introduce expansion functions for the fluctuations. Let $\{\psi^{(\iota)}\}$ be a basis for the $(2^d - 1)$-dimensional space of functions on $\mathbb{Z}^d$ supported on $\{0, 1\}^d$ and annihilated by T. Clearly, the fluctuations can be expanded as

$$\tilde{\phi}^{(k)} = \sum_{\iota}\sum_{m} \alpha^{(k)}_{\iota m} M^k \psi^{(\iota)}_m, \tag{4.20}$$

$$\psi^{(\iota)}_m(n) = \psi^{(\iota)}(n - 2m), \tag{4.21}$$

and this expansion is a "localization" of the fluctuation $\tilde{\phi}^{(k)}$ associated with the 2^k-scale. What does this have to do with wavelets? As we shall see, the configurations $M^k\psi^{(\iota)}_m$ are lattice approximations of wavelets. These functions – introduced by Gawedzki and Kupiainen – are easy to compute in momentum space. $M^k\psi^{(\iota)}$ is given by

$$\widehat{M^k\psi^{(\iota)}}(p) = \frac{2^{k(\frac{d}{2}-1)}}{A_1(p)} \frac{1}{\displaystyle\sum_{m\in\{0,1,\ldots,2^k-1\}^d} \frac{|\hat{\chi}_k(p+2^{-k+1}\pi m)|^2}{A_1(p+2^{-k+1}\pi m)}}$$

$$\times \hat{\chi}_k(p) \sum_{m\in\mathbb{Z}^d} \psi^{(\iota)}(m)e^{i2^k m\cdot p}, \tag{4.22}$$

where χ_k is the lattice characteristic function of $\{0, \ldots, 2^k - 1\}^d$. The momentum expression for χ_k is clearly

$$\hat{\chi}_k(p) = \prod_{\mu=1}^{d} \frac{1 - e^{i2^k p_\mu}}{1 - e^{ip_\mu}}. \tag{4.23}$$

The functions $M^k\psi^{(\iota)}_m$ were used by Gawedzki and Kupiainen in their renormalization group analysis [1] of an infrared problem.

Naturally these expansion functions cannot be generated by scaling the functions associated with a single scale, since a minimum-scale cutoff is built into these functions by the unit lattice. The orthogonality property is there, however, and the scaling property appears when we take the continuum limit

$$\widehat{\varphi_{k\iota}}(p) = \lim_{N \to \infty} 2^{-N(\frac{d}{2}-1)} M^{\widehat{N+k}\psi^{(\iota)}}(2^{-N}p), \qquad (4.24)$$

which is obviously defined for negative integers k as well. We get the class C^1 functions given by

$$\widehat{\varphi_{k\iota}}(p) = 2^{k(\frac{d}{2}-1)}\widehat{\varphi_{0\iota}}(2^k p), \qquad (4.25)$$

$$\widehat{\varphi_{0\iota}}(p) = \frac{c}{p^2} \frac{\widehat{\chi}(p)}{\sum_m \frac{|\widehat{\chi}(p+2\pi m)|^2}{(p+2\pi m)^2}} \sum_m \psi^{(\iota)}(m) e^{im \cdot p}, \qquad (4.26)$$

where χ is the continuum characteristic function of $[0,1]^d$. We now have wavelets for which

$$(\nabla \varphi_{k\iota m}, \nabla \varphi_{k'\iota'm'}) = 0, \quad k \neq k', \qquad (4.27)$$

and it is not hard to show [25] that $\varphi_{k\iota}$ has exponential decay. Moreover, we observe that $\varphi_{0\iota}$ minimizes $\int (\nabla \varphi)^2$ with respect to the constraints

$$\int_{S_n} \varphi = \psi^{(\iota)}(n), \qquad (4.28)$$

where S_n is the d-dimensional unit cube with n the vertex of minimum coordinates.

Returning to the general quadratic form $(A\phi, \phi)$ and general transformation T, we now let $\{\psi^{(\iota)}\}$ be $2^d - 1$ functions on $\mathbb{Z}^d$ such that $\{\psi_m^{(\iota)}\}$ is a basis for the kernel of T – the essential property of the specific $\psi^{(\iota)}$ defined above. If one wishes to be a little more concrete in this general setting, one can always consider the case $c_m = \prod_{\mu=1}^{d} s_{m_\mu}$, as we did in Section 2, and define

$$\psi^{(\iota)}(n) = (-1)^{\sum_\mu \lambda_\mu^{(\iota)} n_\mu} c_{\lambda^{(\iota)}-n}, \qquad (4.29)$$

with $\{\lambda^{(\iota)}\}$ the set of all non-zero elements of $\{0,1\}^d$ as in Section 2. The momentum expression for $M^k \psi^{(\iota)}$ can still be computed explicitly. We get

$$\widehat{M^k \psi^{(\iota)}}(p) = \frac{2^{k(\frac{d}{2}-1)}}{A(p) \displaystyle\sum_{m \in \{0,1,\ldots,2^k-1\}^d} \frac{|H_k(p+2^{-k+1}\pi m)|^2}{A(p+2^{-k+1}\pi m)}}$$

$$H_k(p) \sum_{m \in \mathbb{Z}^d} \psi^{(\iota)}(m) e^{i 2^k m \cdot p}, \tag{4.30}$$

where

$$H_k(p) = \prod_{\ell=0}^{k-1} (2^{-\frac{d}{2}} h(2^{k-\ell} p)) \tag{4.31}$$

Since

$$\lim_{N \to \infty} H_{k+N}(2^{-N} p) = \prod_{\ell=0}^{\infty} (2^{-\frac{d}{2}} h(2^{k-\ell} p))$$

$$= H(2^k p), \tag{4.32}$$

it follows that the wavelet $\varphi_{k\iota}$ obtained from the continuum limit (4.24) is given by

$$\widehat{\varphi_{k\iota}}(p) = 2^{k(\frac{d}{2}-\alpha)} \widehat{\varphi_{0\iota}}(2^k p), \tag{4.33}$$

$$\widehat{\varphi_{0\iota}}(p) = c|p|^{-2\alpha} \frac{H(p)}{\sum_m |\hat{f}(p+2\pi m)|^2} \sum_m \psi^{(\iota)}(m) e^{im \cdot p}, \tag{4.34}$$

$$|p|^{2\alpha} = c \lim_{N \to \infty} (4^{\alpha N} A(2^{-N} p)), \tag{4.35}$$

where f denotes

$$\hat{f}(p) = |p|^{-\alpha} H(p). \tag{4.36}$$

f is a solution of (3.2), and $\varphi_{0\iota}$ minimizes the quadratic form $(|\nabla|^{2\alpha}\varphi, \varphi)$ with respect to the constraints

$$\int \varphi(x)(|\nabla|^\alpha f)(x-n)dx = \psi^{(\iota)}(n). \tag{4.37}$$

On the other hand, the calculation in Section 3 shows that if A^∞ is the fixed point to which A is driven by iterations of $\mathcal{R}_T$, then

$$\sum_m |\hat{f}(p+2\pi m)|^2 = A^\infty(p)^{-1}. \tag{4.38}$$

Hence the wavelet is related to the Gaussian fixed point by the formula

$$\widehat{\varphi_{0\iota}}(p) = c|p|^{-2\alpha} A^\infty(p) H(p) \sum_m \psi^{(\iota)}(m) e^{im \cdot p}. \tag{4.39}$$

5. THE ROLE OF WAVELETS IN THE INTERACTING CASE

In this section we show how wavelets are relevant to the renormalization group analysis of an interaction. In the interests of concreteness we concentrate on the massless lattice ϕ_4^4 model, whose action functional is

$$L(\phi) = \frac{1}{2}(A_1\phi, \phi) + \lambda \sum_{m \in \mathbb{Z}^4} : \phi(m)^4 :_{A_1}, \qquad (5.1)$$

where A_1 is the finite-difference Laplacian defined by (3.26), $\lambda > 0$ is the *coupling constant*, and $: :_{A_1}$ denotes Wick ordering with respect to the covariance $\langle \phi(m)\phi(n) \rangle_{A_1}$. For definitions of Wick ordering from various points of view, we refer to [26]. We exploit its properties without comment. It can be regarded as a first-order renormalization of the interaction in the sense of quantum field theory, but of course there are no ultraviolet singularities to cancel in our context. What we propose to do here is iterate the renormalization group transformation to first order in λ to obtain an $O(\lambda)$-fixed point. We show that the new lattice action is equivalent to the action of the massless *continuum* Φ_4^4 model regularized by a *wavelet cutoff* on the field Φ at the unit-scale level.

Our first step in this direction is to re-examine the fixed point to which the free part of L is driven by the renormalization group. For the moment we consider the more general quadratic form $(A\phi, \phi)$ satisfying (1.6). Let $\{c_n\}$ be the summable sequence involved in the averaging transformation T that we are using. If f is the solution of (3.2), then we have

$$A^\infty(p)^{-1} = \sum_{\ell \in \mathbb{Z}^d} |\hat{f}(p + 2\pi\ell)|^2, \qquad (5.2)$$

from which it follows that

$$\begin{aligned}
\langle \phi(m)\phi(n) \rangle_{A^\infty} &= \frac{1}{(2\pi)^d} \sum_{\ell \in \mathbb{Z}^d} \prod_{\mu=1}^{d} \left(\int_0^{2\pi} dp_\mu \right) e^{-ip \cdot (m-n)} |\hat{f}(p + 2\pi\ell)|^2 \\
&= \frac{1}{(2\pi)^d} \prod_{\mu=1}^{d} \left(\int_{-\infty}^{\infty} dp_\mu \right) e^{-ip \cdot (m-n)} |\hat{f}(p)|^2 \\
&= \prod_{\mu=1}^{d} \left(\int_{-\infty}^{\infty} dx_\mu \right) f(x - m) f(x - n). \qquad (5.3)
\end{aligned}$$

On the other hand, consider the Gaussian random field Φ on $\mathbb{R}^d$ with mean zero and covariance operator

$$C = |\nabla|^{-2\alpha} \qquad (5.4)$$

and expand $\Phi(x)$ in the wavelets $2^{r(\frac{d}{2}-\alpha)}\varphi_\iota(2^r x - 2n)$ generated by (4.39). Thus

$$\Phi_1(x) = \sum_{\iota,r}\sum_{n\in\mathbb{Z}^d} \alpha_{\iota rn} 2^{r(\frac{d}{2}-\alpha)}\varphi_\iota(2^r x - 2n) \tag{5.5}$$

and we define the unit-scale wavelet cutoff $\Phi_1(x)$ by

$$\Phi_1(x) = \sum_{\substack{\iota,r\\r\leq 0}}\sum_{n\in\mathbb{Z}^d} \alpha_{\iota rn} 2^{r(\frac{d}{2}-\alpha)}\varphi_\iota(2^r x - 2n). \tag{5.6}$$

Only wavelets with large length scale are included.

Now consider the functions $g_n(x) = g(x - n)$, where g is given by

$$\begin{aligned}\hat{g}(p) &= c|p|^{-2\alpha} A^\infty(p) H(p)\\ &= c|p|^{-\alpha} A^\infty(p)\hat{f}(p).\end{aligned} \tag{5.7}$$

The subspace of the Sobolev space $\mathcal{H}_\alpha$ spanned by $\{g_n\}$ is identical to the subspace spanned by the set of wavelets in the cutoff (5.6). This is a fundamental property of the multi-scale resolution analysis of Mallat and Meyer. To be sure, the wavelets used here are only inter-scale orthogonal in $\mathcal{H}_\alpha$ and are not intra-scale orthogonal, but the analysis of Chui and Wang [2,3] highlights the fact that this property remains a part of the wavelet construction even when orthogonality on the same scale is sacrificed. In any case, the property can be derived from the momentum expressions for our particular wavelets. Thus we have the decomposition

$$\Phi_1(x) = \sum_{n\in\mathbb{Z}^d} \beta_n g_n(x), \tag{5.8}$$

so if $u_m(x) = u(x-m)$ is the dual basis of $\{g_n\}$ in the subspace, the random variables β_m are given by

$$\begin{aligned}\beta_m &= (\Phi_1, \mu_m)_\alpha\\ &= \Phi_1(|\nabla|^{2\alpha}u_m)\\ &= \Phi(|\nabla|^{2\alpha}u_m),\end{aligned} \tag{5.9}$$

where the last equation follows from the orthogonality in $\mathcal{H}_\alpha$ of $u_m(x)$ to the wavelets

$$2^{r(\frac{d}{2}-\alpha)}\varphi_\iota(2^r x - n), \quad r \geq 0, \tag{5.10}$$

excluded from the cutoff Φ_1.

Now actually the dual basis is quite easy to compute. Given (5.7), one can see that

$$\hat{u}(p) = \frac{\hat{g}(p)}{\sum\limits_{\ell \in \mathbb{Z}^d} |p + 2\pi\ell|^{2\alpha} |\hat{g}(p + 2\pi\ell)|^2} = c \frac{|p|^{-\alpha} A^\infty(p) \hat{f}(p)}{A^\infty(p)^2 \sum\limits_{\ell \in \mathbb{Z}^d} |\hat{f}(p + 2\pi\ell)|^2}$$

$$= c|p|^{-\alpha} \hat{f}(p), \tag{5.11}$$

so we have

$$\beta_m = c\Phi(|\nabla|^\alpha f_m). \tag{5.12}$$

Thus the Gaussian expectation $\langle \beta_m \beta_n \rangle_\alpha$ is given by

$$\langle \beta_m \beta_n \rangle_\alpha = \prod_{\mu=1}^d \left(\int_{-\infty}^{\infty} dx_\mu \right) f_m(x) f_n(x), \tag{5.13}$$

and so the lattice action $\frac{1}{2}(A^\infty \phi, \phi)$ is equivalent to the continuum action $\frac{1}{2}\int (|\nabla|^\alpha \Phi)^2$ with expectations restricted to those generated by the random variables β_n. By inner product symmetry, we actually have a choice between the correspondences $\beta_n \leftrightarrow \phi(n)$ and $\beta_n \leftrightarrow \phi(-n)$. The latter choice is more natural, as we see below.

We now return to the massless lattice ϕ_d^4 model and consider the first-order effect $\widehat{\mathcal{R}}_T^N(L)$ of the iterations of the renormalization group transformation applied to L. The $O(\lambda)$ contribution to the defining equation (3.4) for $\mathcal{R}_T(L)$ is straightforward enough to extract:

$$\widehat{\mathcal{R}}_T(L)(\psi) = \frac{1}{2}(\mathcal{R}_T(A_1)\psi, \psi) + \lambda I'(\psi), \tag{5.14}$$

$$I'(\psi) = \lim_{\Lambda \to \mathbb{Z}^d} \left[\frac{d}{d\lambda} \ln Z_\Lambda(T\phi = 0) \Big|_{\lambda=0} - \frac{d}{d\lambda} \ln \int \prod_{n \in \Lambda} d\phi(n) e^{-A_1(\phi \chi_\Lambda)} e^{-\lambda I(\phi \chi_\Lambda)} \prod_{n' \in \Lambda'} \delta\left(\psi\left(\frac{1}{2}n'\right) \right. \right.$$

$$\left. \left. -(T\phi)\left(\frac{1}{2}n'\right) \right) \Big|_{\lambda=0} \right]$$

$$= \lim_{\Lambda \to \mathbb{Z}^d} \left[\left\langle I(\phi \chi_\Lambda) \prod_{n' \in \Lambda'} \delta\left(\psi\left(\frac{1}{2}n'\right) - (T\phi)\left(\frac{1}{2}n'\right) \right) \right\rangle_{A_1} \right.$$

$$\left. -\langle I(\phi \chi_\Lambda) \rangle_{A_1}(T\phi = 0) \right]. \tag{5.15}$$

But by the same $\phi = \phi' + \tilde{\phi}$ calculation that led to (4.7), this reduces to

$$I'(\psi) = \langle I(M\psi + \tilde{\phi}) \rangle_{A_1}(T\phi = \psi) - \langle I(\tilde{\phi}) \rangle_{A_1}(T\phi = \psi), \tag{5.16}$$

where M is the minimizer associated with A_1, the contribution ϕ' is given by $\phi' = MT\phi$, and the contribution $\tilde{\phi}$ satisfies the constraint $T\tilde{\phi} = 0$. Now since

$$(A_1 M\psi, \tilde{\phi}) = 0, \tag{5.17}$$

we know that $M\psi$ and $\tilde{\phi}$ are independent random variables with respect to the A_1-expectation. They also have mean zero, so we can easily verify

$$\langle : \phi(m)^4 :_{A_1} \rangle_{A_1} (T\phi = \psi) =: (M\psi)(m)^4 :_{A_1}, \tag{5.18}$$

and in particular,

$$\langle : \tilde{\phi}(m)^4 :_{A_1} \rangle_{A_1} (T\phi = \psi) = 0. \tag{5.19}$$

Combining all this with (5.14) we obtain

$$\widehat{\mathcal{R}}_T(L)(\psi) = \frac{1}{2}(\mathcal{R}_T(A_1)\psi, \psi) + \lambda I(M\psi)$$
$$= \frac{1}{2}(A_1 M\psi, M\psi) + \lambda I(M\psi), \tag{5.20}$$

where the alternate expression follows from (4.7). Thus we may write

$$\widehat{\mathcal{R}}_T(L)(\psi) = L(M\psi). \tag{5.21}$$

Naturally, the full-blown renormalization group transformation $\mathcal{R}_T(L)$ has far more complex behavior.

Iterating the application of $\widehat{\mathcal{R}}_T$ indefinitely, we drive the action to the $\widehat{\mathcal{R}}_T$-fixed point

$$\lim_{N\to\infty} \widehat{\mathcal{R}}_T^N(L)(\psi) = \frac{1}{2}(A_1^\infty \psi, \psi) + \lambda I^\infty(\psi), \tag{5.22}$$

$$I^\infty(\psi) = \lim_{N\to\infty} I^{(N)}(\psi)$$
$$= \lim_{N\to\infty} I(M^N \psi)$$
$$= \lim_{N\to\infty} \sum_{m\in\mathbb{Z}^d} : (M^N\psi)(m)^4 :_{A_1}. \tag{5.23}$$

Now we know from the calculations in Section 4 that

$$\lim_{N\to\infty} 2^{-N(\frac{d}{2}-1)} \widehat{M^N\psi}(2^{-N}p) = cp^{-2} A_1^\infty(p) H(p) \sum_{n\in\mathbb{Z}^d} \psi(n)e^{in\cdot p}, \tag{5.24}$$

and in position space this becomes

$$\lim_{N\to\infty} 2^{N(\frac{d}{2}-1)}(M^N\psi)(2^N x) = \sum_{n\in\mathbb{Z}^d} \psi(n)g(x-n), \tag{5.25}$$

since we know

$$g(x) = \frac{1}{(2\pi)^d} \int e^{-ix\cdot p} p^{-2} A_1^\infty(p) H(p) dp. \tag{5.26}$$

The point is that the sum in (5.23) may be written as a Riemann sum in the scaling of $M^N \psi$.

$$\sum_{m \in \mathbb{Z}^d} : (M_1^N \psi)(m)^4 : = 2^{N(4-d)} \sum_{x \in 2^{-N} \mathbb{Z}^d} 2^{-Nd} \times$$

$$\times : 2^{N(2d-4)} (M^N \psi)(2^N x)^4 :_{4^N A_1(2^{-N} p)}, \tag{5.27}$$

so in the case $d = 4$ we have existence of a non-zero limit:

$$I^\infty(\psi) = \int : \left(\sum_{n \in \mathbb{Z}^4} \psi(n) g(x - n) \right)^4 :_{-\Delta} dx. \tag{5.28}$$

This completes our derivation of the equivalence.

To summarize what we have shown, we note that if we impose the wavelet cutoff $\Phi_1(x)$ on the massless continuum Φ_4^4 model (which is not asymptotically free at small length scales in any case), the resulting action

$$\frac{1}{2} \int (\nabla \Phi)^2 + \lambda \int : \Phi_1^4 :_{-\Delta} \tag{5.29}$$

is equivalent to

$$\frac{1}{2}(A_1^\infty \phi, \phi) + \lambda I^\infty(\phi), \tag{5.30}$$

where the correspondence of random variables can be $\phi(n) \leftrightarrow \beta_n$ or $\phi(-n) \leftrightarrow \beta_n$ as it was in the free case. Thus the massless Φ_4^4 continuum action with the unit-scale wavelet cutoff is the $O(\lambda)$-fixed point to which the massless ϕ_4^4 lattice action is driven by iterations of $\widehat{\mathcal{R}}_T$.

Finally, if we choose the correspondence $\beta_n \leftrightarrow \phi(-n)$, the coefficients $\alpha_{\iota rn}$ in the decomposition (5.6) of Φ_1 are the variables that are integrated out in the iteration of the full renormalization group transformation $\mathcal{R}_T$ applied to (5.30). If we successively split ϕ into "minimum energy" and "fluctuating" parts as we did in Section 4 – except for A_1^∞ instead of A_1 – we get

$$\phi = \phi^{(N)} + \sum_{k=0}^{N-1} \tilde{\phi}^{(k)}, \tag{5.31}$$

$$\tilde{\phi}^{(k)}(m) = \sum_\iota \sum_{n \in \mathbb{Z}^4} \alpha_{\iota, -k, n} 2^{-k(\frac{d}{2}-1)}$$

$$\int |\nabla| f(x + m) \varphi_\iota(2^{-k} x - 2n) dx. \tag{5.32}$$

This follows from the decomposition of Φ_1 associated with our equivalence:

$$\Phi_1(x) = \sum_{n \in \mathbb{Z}^4} \phi(n) g(x+n). \tag{5.33}$$

Indeed, since $\{u_n\}$ is dual to $\{g_n\}$ and $u = |\nabla|^{-1} f$, we have

$$\phi(m) = \int |\nabla| f(x+m) \Phi_1(x) dx$$

$$= \sum_{\substack{\iota, r \\ r \leq 0}} \sum_{n \in \mathbb{Z}^4} \alpha_{\iota r n} 2^{r(\frac{d}{2}-1)} \int |\nabla| f(x+m) \varphi_\iota(2^r x - 2n) dx, \tag{5.34}$$

and the point is that $\phi^{(k)} = M_\infty^k T^k \phi$ and that

$$\left(T^k \int (|\nabla| f)(x + \cdot) \varphi_\iota(2^r x - 2n) dx \right)(m) =$$

$$2^{-k(\frac{d}{2}+1)} \sum_{m', \ldots, m^{(k)}} c_{m'} \cdots c_{m^{(k)}} \int (|\nabla| f)(2^k x + 2^k m^{(k)}$$

$$-2^{k-1} m^{(k-1)} - \cdots - 2m' - m) \varphi_\iota(2^r x - 2n) dx \tag{5.35}$$

vanishes for $r > -k$ by the same reasoning that was used in Section 2 when we constructed $\alpha = 0$ wavelets with continuum averaging constraints – except $\alpha = 1$ here. On the other hand, if $r \leq -k$, it is equally important to observe that the application of M_∞^k to (5.35) recovers

$$m \mapsto \int (|\nabla| f)(x+m) \varphi_\iota(2^r x - 2n) dx. \tag{5.36}$$

This can easily be verified in momentum space if one recalls that the iteration of the minimizer is given by

$$\widehat{M_\infty^k \psi}(p) = \frac{2^{k(\frac{d}{2}-1)} H_k(p)}{A_1^\infty(p) \sum_{\ell \in \{0,1,\ldots,2^k-1\}^4} \frac{|H_k(p+2^{-k+1}\pi\ell)|^2}{A_1^\infty(p+2^{-k+1}\pi\ell)}} \sum_{m \in \mathbb{Z}^4} \psi(m) e^{i 2^k m \cdot p} \tag{5.37}$$

in this case.

REFERENCES

[1]   K. Gawedzki and A. Kupiainen, Commun. Math. Phys. **77** (1980), 31.

[2]   C. Chui and J. Wang, "A Cardinal Spline Approach to Wavelets", to appear in Proc. Amer. Math. Soc.

[3]   C. Chui and J. Wang, "On Compactly Supported Spline Wavelets and a Duality Principle", to appear in Trans. Amer. Math. Soc.

[4]   J. Stromberg, Conference in Harmonic Analysis in Honor of A. Zygmund, Vol. 2, Wadsworth Inc., Belmont, CA. (1983), 475.

[5]   Y. Meyer, Séminaire Bourbaki **38** (1985-86), 662.

[6]   P. Lemarié, J. Math. Pures et Appl. **67** (1988), 227.

[7]   G. Battle, Commun. Math. Phys. **110** (1987), 601.

[8]   I. Daubechies, Comm. Pure and Appl. Math. **41** (1988), 906.

[9]   G. Battle, J. Math. Phys. **30**, No. 10 (1989), 2195.

[10] J. Glimm and A. Jaffe, Fortschr. Phys. **21** (1973), 327.

[11] J. Glimm and A. Jaffe, *Quantum Physics: A Functional Integral Point of View*, Springer-Verlag, New York, 1987.

[12] G. Battle and P. Federbush, Commun. Math. Phys. **88** (1983), 263.

[13] P. Federbush, Commun. Math. Phys. **107** (1987), 319.

[14] P. Federbush, Commun. Math. Phys. **114** (1988), 317.

[15] G. Battle, Ann. Phys. **201**, No. 1 (1990), 117.

[16] J. Kogut and K. Wilson, Physics Reports **12** (1974), 75.

[17] J. Polchinski, Nuclear Phys. **B231** (1984), 269.

[18] T. Balaban, Commun. Math. Phys. **102** (1985), 255.

[19] J. Feldman, J. Magnen, V. Rivasseau, and R. Seneor, "Construction and Borel Summability of Infrared ϕ_4^4 by a Phase Space Expansion", to appear in Commun. Math. Phys.

[20] J. Feldman, J. Magnen, V. Rivasseau, and R. Seneor, Commun. Math. Phys. **103** (1986), 67.

[21] V. Rivasseau, Commun. Math. Phys. **95** (1984), 445.

[22] V. Rivasseau, "Constructive Renormalization", Ecole Polytechnique preprint (1986).

[23] J. Feldman, J. Magnen, V. Rivasseau, and R. Seneor, Commun. Math. Physics **100** (1985), 23.

[24] P. Federbush and C. Williamson, J. Math. Phys. **28** (1987), 1416.

[25] G. Battle, Commun. Math. Phys. **114** (1988), 93.

[26]   B. Simon, *The $P(\phi)_2$ Euclidean (Quantum) Field Theory*, Princeton University Press, Princeton, N.J., 1974.

[27]   I. Daubechies, "Ten Lectures on Wavelets", CBMS-NSF Series in Applied Mathematics, SIAM (in press 1991).

V
THEORETICAL DEVELOPMENTS

Non-Orthogonal Wavelet and Gabor Expansions, and Group Representations

H. G. FEICHTINGER

Address: Institut für Mathematik,
Universität Wien, Strudlhofg. 4, A-1090
Wien, Austria

K. GRÖCHENIG*

Department of Mathematics U-9,
University of Connecticut,
Storrs, Connecticut

In this article we explain the construction of nonorthogonal series expansions with respect to basis functions which are simple transforms of a single function. In mathematics such expansions are often called atomic decompositions, while in physics they are series expansions with respect to discrete sets of coherent states [22]. This means that all basis functions are derived from a single function by elementary operations such as shifts, modulations, scaling or rotations. An expansion of this type had been proposed very early for applications in the theory of communication and signal analysis [14]. In the last decade, after such series expansions had been obtained for spaces of analytic functions [1] and for Besov spaces [11], it became clear that these atomic decompositions must be part of a more general phenomenon. Inspired by the work of A. Grossmann and the Marseille group [17], such a general theory was found in the series of papers [8, 9, 10, 16] by the authors.

This article is intended to be a counterpart to the abstract theory of [9, 10], which for reasons of length do not contain any examples. Here we will focus on the applications of the general theory, in particular on the

---

*This article is an extended version of lecture notes by the second author. He acknowledges partial funding by the grant AFOSR-90-0311

construction of non-orthogonal wavelet and Gabor-type expansions.

In the excitement and enthusiasm about orthogonal wavelets other representations of functions have been somewhat neglected, but nonorthogonal expansions and frames also have their merits:

(1) Orthogonal wavelets are rather complicated functions, whereas any "nice" function can serve as a basic wavelet for an non-orthogonal expansion.

(2) Instead of translations and dilations other operations can be used to obtain the expanding functions from a "basic wavelet". In some cases of interest, appropriate orthogonal bases of coherent states do not even exist; and then it is natural to look at non-orthogonal expansions of the same structure.

(3) Since no lattice structure is required, one obtains greater stability. As an additional feature one can prove irregular sampling theorems for the continuous wavelet transform and the short time Fourier transform.

The article is organized as follows. Section 1 contains some generalities on non-orthogonal expansions and frames. Section 2 – the main body of the article – presents various examples of non-orthogonal expansions and some applications. All expansions mentioned, in particular wavelet and Gabor-type expansions, are special cases of the general theorem 6.1 in [9]. In Section 3 – 5 we explain the ideas underlying the construction of non-orthogonal expansions with respect to coherent states and give a partial proof for the statements in Section 2. In this part we follow [16], which is devoted to frames for Banach spaces and simplifies the original construction of non-orthogonal expansions. Section 6 treats the computational aspects and states the general theorem.

1. GENERALITIES

1.1. A sequence of functions $\{e_i, i \in I\}$ in a Hilbert space $\mathcal{H}$ gives rise to nonorthogonal expansions, if every function $f \in \mathcal{H}$ has a series expansion

$$f = \sum_{i \in I} c_i e_i , \qquad (1)$$

such that the coefficient sequence c depends linearly on f and satisfies the norm equivalence

$$A_1 \|f\|_{\mathcal{H}}^2 \leq \sum_i |c_i|^2 \leq A_2 \|f\|_{\mathcal{H}}^2 \qquad (2)$$

for some constants $0 < A_1 \leq A_2$ (or in short $\|f\|_{\mathcal{H}} \cong \|c\|_2$) .

Convergence of a sum $\sum_{i \in I} \cdots$ over a general index set I is understood as unconditional convergence, i.e., any sequence of partial sums converges and the limit is independent of the summation procedure.

In general the basis functions are neither orthogonal nor linearly independent, therefore f may be representable with other coefficient sequences. This non-uniqueness of the coefficients in the absence of linear independence is not a serious problem, because from (1) and (2) there is a canonical way to construct the coefficients c_i from the inner products $\langle e_i, f \rangle$, cf. 1.3.

1.2. If $g_i \in \mathcal{H}$ denotes the coefficient functional $f \mapsto c_i = \langle g_i, f \rangle$, then from the equality $\langle h, f \rangle = \langle (\sum_i \langle g_i, h \rangle e_i), f \rangle = \langle h, \sum_i \langle e_i, f \rangle g_i \rangle$ for all $h \in \mathcal{H}$, one obtains $f = \sum_i \langle e_i, f \rangle g_i$ with the norm equivalence

$$A_2^{-1} \|f\|_{\mathcal{H}}^2 \leq \sum_i |\langle e_i, f \rangle|^2 \leq A_1^{-1} \|f\|_{\mathcal{H}}^2 \tag{3}$$

If (3) holds, then the set of function $\{e_i, i \in I\}$ is called a *frame* for $\mathcal{H}$. By (2) the coefficient functionals g_i are also a frame.

CONSTRUCTION OF THE COEFFICIENTS c_i OR RECONSTRUCTION OF f FROM $\langle e_i, f \rangle$

Clearly, f is uniquely determined by the frame coefficients $\langle e_i, f \rangle$. It is a remarkable fact that from the norm equivalence (3), one can derive a method for the reconstruction of f from the frame coefficients and also of the coefficients in the series expansion (1), cf. [5]. For this one observes that by (3), the operator

$$Sf = \sum_i \langle e_i, f \rangle e_i, \tag{4}$$

the so-called *frame operator*, is positive, bounded below and above by A_1^{-1} and A_2^{-1}. Consequently, S is invertible on $\mathcal{H}$ and

$$f = S^{-1} Sf = \sum_i \langle e_i, f \rangle S^{-1} e_i = SS^{-1} f = \sum_i \langle S^{-1} e_i, f \rangle e_i. \tag{5}$$

Since S^{-1} has a Neumann series $S^{-1} = \alpha \sum_{n=0}^{\infty} (1 - \alpha S)^n, \alpha = 2/(A_1^{-1} + A_2^{-1})$, this can easily be written as an iterative algorithm for the construction of the coefficients c_i or the reconstruction of f, cf. Section 6 or [5, 3].

1.3. In our context $\mathcal{H}$ is a Hilbert space of square-integrable functions $L^2(S, \mu)$ and the basis functions e_i are coherent states, i.e., they are transforms of one function g by simple operations. In many applications it is desirable to recognize finer details of functions from their series expansions. This leads to the questions, which Banach spaces can be characterized by means of their expansions with respect to coherent states or by the size of their frame coefficients. If such a description is possible, how can the coefficients be computed or how can the function be reconstructed from its frame coefficients. The question of frames for Banach spaces is more complicated since the argument of 1.3 breaks down in this case. The general answer to these questions has been obtained in [9, 10, 16].

As a practical consequence of the extension to Banach spaces one obtains statements about the quality of the convergence of the reconstruction in 1.3.

2. SOME EXAMPLES OF NONORTHOGONAL EXPANSIONS

This section is a collection of results on non-orthogonal series expansions from different branches of mathematics. All decomposition theorems can be obtained as special cases of a single theorem. Most statements also have direct, "hard analysis" proofs, but in most examples the main theorem in Section 6 provides more detailed information on the class of admissible wavelets, irregular expansions or stability.

NONORTHOGONAL WAVELET EXPANSIONS

Recall first that the moments of a multivariate function f are defined by $\int_{R^n} x_1^{k_1} x_2^{k_2} \ldots x_n^{k_n} f(x)\, dx$, where $x = (x_1, x_2, \ldots, x_n) \in R^n$ and the $k_i \geq 0$ are integers.

Let $g \in L^2(R^n)$ be a function (i) with enough smoothness and a sufficient number of vanishing moments, and (ii) such that $|\hat{g}(\xi)| \geq c > 0$ for $a \leq |\xi| \leq b$, some $a, b, c > 0$, e.g. g is a Schwartz function with some vanishing moments , or supp $\hat{g} \subseteq \{0 < r_0 \leq |x| \leq R_0 < \infty\}$, or g has compact support and vanishing moments, etc. (cf. Section 6). Then there exist $\alpha_0 > 0, \beta_0 > 1$, the size depending only on g, with the property that for $0 < \alpha \leq \alpha_0, 1 < \beta \leq \beta_0$, every $f \in L^2(R^n)$ has a nonorthogonal wavelet expansion

$$f = \sum_{j \in Z, k \in Z^n} c_{jk} \beta^{jn/2} g(\beta^j x - \alpha k) \tag{6}$$

with convergence in $L^2(R^n)$. The coefficients c_{jk} can be constructed as in Section 1.3 and satisfy $\|f\|_2 \cong (\sum_{j \in Z, k \in Z^n} |c_{jk}|^2)^{1/2}$.

Condition (i) guarantees the convergence of the wavelet expansion (6), while condition (ii) implies that the basis functions span $L^2(R^n)$.

WAVELET FRAMES

By duality, the functions $g_{jk}(x) = \beta^{jn/2}g(\beta^j x - \alpha k), j \in Z, k \in Z^n$ constitute a frame for $L^2(R^n)$: thus $f \in L^2(R^n)$ can be completely reconstructed from the frame coefficients $\langle g_{jk}, f \rangle = \beta^{jn/2} \int \bar{g}(\beta^j y - \alpha k)f(y)dy$ and

$$A_1\|f\|_2 \le \left(\sum_{j \in Z, k \in Z^n} |\langle g_{jk}, f \rangle|^2 \right)^{1/2} \le A_2\|f\|_2 \quad . \tag{7}$$

The frame coefficients $\langle g_{jk}, f \rangle$ are a regular sampling of the *continuous wavelet transform*

$$W_g(f)(x, t) = \int_{R^n} t^{-n/2} \bar{g}\left(\frac{y - x}{t}\right)f(y)dy \tag{8}$$

at the points $(x, t) = (\alpha k \beta^{-j}, \beta^{-j})$. Thus (7) can also be interpreted as a sampling theorem for the wavelet transform analogous to the sampling theorem for band-limited functions.

IRREGULAR SAMPLING AND EXPANSIONS

Similar results are true for irregular sampling sequences $X = (x_i, y_i)_{i \in I}$ in $R^n \times R^+$. Let $Q_\alpha(x)$ denote the cube of side length α with center at $x \in R^n$. If g, α, β are given as in (6) and if X is any set in $R^n \times R^+$ that satisfies (i) the density condition $\bigcup_i(Q_{y_i\alpha}(x_i) \times [y_i\beta^{-1/2}, y_i\beta^{1/2}]) = R^n \times R^+$, and (ii) such that this covering is of finite height, then the collection of functions

$$y_i^{-n/2}g(y_i^{-1}(x - x_i)), i \in I, \text{ is a frame for } L^2(R^n)$$

Every $f \in L^2(R^n)$ has an irregular non-orthogonal wavelet expansion

$$f(x) = \sum_{i \in I} c_i y_i^{-n/2} g(y_i^{-1}(x - x_i)) \tag{9}$$

with $\|f\|_2 \cong \|c\|_2$.

DIRECTIONAL SENSITIVITY

In multidimensional image processing, it is desirable to obtain information about the directional behavior of the frequencies of an image. This can be achieved by a variation of (6), which also includes rotations.

Given any $g \in L^2(R^n)$ with sufficient smoothness and a sufficient number of vanishing moments, there exist $\alpha > 0, \beta > 1$ and a finite number of orthogonal matrices $O_l \in \mathcal{O}(n), l = 1, \ldots, r$ such that the collection

$$g_{jkl}(x) = \beta^{-jn/2}g(O_l^{-1}(\beta^{-j}x - \alpha k)), j \in Z, k \in Z^n, l = 1, \ldots, r$$

is a frame for $L^2(R^n)$. By duality every $f \in L^2(R^n)$ admits a non-orthogonal expansion

$$f = \sum_{j,k,l} c_{jkl} \beta^{-jn/2} g(O_l^{-1}(\beta^{-j}x - \alpha k)) \tag{10}$$

with $\|f\|_2 \cong \|c\|_2$.

In order to obtain directional sensitivity, one chooses a basic wavelet g with supp $\hat{g} \subseteq C_{v,\alpha}$, where $C_{v,\alpha} = \{x \in R^n : |x|^{-1}\langle x, v \rangle > \cos \alpha\}$ is the cone with its axis along the unit vector $v \in R^n, |v| = 1$, and angle α. Then for each l the function $f_l = \sum_{j \in Z, k \in Z^n} c_{jkl} g_{jkl}$ has spectrum supp$\hat{f}_l \subseteq C_{O_l^{-1}v,\alpha}$, and f_l contains the essential information about the frequency content of f in the direction of $O_l^{-1}v$.

CHARACTERIZATIONS OF THE CLASSICAL FUNCTION SPACES

Under some additional conditions on g, $g \in S_0$, say, functions and distributions outside $L^2(R^n)$ also have nonorthogonal wavelet expansions of the form (6) and (9). As an illustration we consider the homogeneous Besov spaces

$$\dot{B}_{pq}^s(R^n) = \left\{ f \in S_0' : \|f\|_{\dot{B}_{pq}^s}^q \right.$$

$$= \int \left(\int_{R^n} |W_g(f)(x,t)|^p dx \right)^{q/p} t^{-qs-nq/2} dt/t < \infty \right\} \tag{11}$$

for $1 \leq p, q < \infty, s \in R$. For this family of Banach spaces the decay of the wavelet transform serves as a measure for the smoothness of a function or distribution. In particular, if $p = q = 2$ and if $s \geq 0$ is an integer, then the norm $\|f\|_{\dot{B}_{pq}^s}$ is equivalent to the Sobolev norm $\sum_{s_1+s_2+\ldots+s_n=s} \|\partial^s f/\partial x_1^{s_1} \partial x_2^{s_2} \partial x_n^{s_n}\|_2$. In other words, the Besov spaces fill in the gaps between the ordinary Sobolev spaces.

Functions in Besov spaces admit wavelet expansions (6) and (9) and are characterized by expansions

$$f \in \dot{B}_{pq}^s \iff \left(\sum_{j \in Z} \left(\sum_{k \in Z^n} |c_{jk}|^p \right)^{q/p} \beta^{-jq(s+n/2-n/p)} \right)^{1/q} < \infty \tag{12}$$

Both norms in (12) are equivalent and (6) converges in $\dot{B}_{pq}^s$.

Similarly $f \in \dot{B}_{pq}^s$ can be recognized by the size of the frame coefficients

$\langle g_{jk}, f \rangle$ as follows: there are two constants $A_1, A_2 > 0$ such that

$$A_1 \|f\|_{B^s_{pq}} \leq \left(\sum_{k \in Z^n} |\langle g_{jk}, f \rangle|^p)^{q/p} \beta^{-jq(s+n/2-n/p)} \right)^{1/q} \leq A_2 \|f\|_{B^s_{pq}}$$

Writing $Q_{jk} = \prod_{r=1}^n [\beta^j \alpha k_r, \beta^j \alpha(k_r + 1)], j \in Z, k = (k_1, \ldots, k_n) \in Z^n$,

$$f \in L^p(R^n) \Longleftrightarrow (\int_{R^n} (\sum_{j,k} |c_{jk}|^2 \beta^{-jq(s+n/2)} \chi_{Q_{jk}}(x))^{p/2} dx)^{1/p} < \infty$$

and (6) converges in L^p.

Likewise all Besov-Triebel-Lizorkin (= BTL-) spaces can be character-ized by the size of the coefficients in the nonorthogonal wavelet expansions (compare [13, 16]).

All characterizations are of course true for orthogonal wavelet bases as well. These are even unconditional bases for the BTL-spaces [20, 15]. De-spite the power and elegance of orthogonal wavelet expansions, in partic-ular for fast numerical computations, non-orthogonal expansions might be preferable in certain applications, because *any* reasonable function may serve as a basic wavelet. On the other hand orthogonal wavelets are rather complicated functions and require a delicate construction, see e.g. [21].

REFERENCES

All results mentioned occur as special cases of [9], Theorem 6.1. and [16], Theorem T. An extensive study of non-orthogonal wavelets and some ap-plications has recently appeared in the impressive paper [13], see also the preceding versions in [11, 12]. The case $L^2(R^n)$ is treated in [2, 4].

GABOR TYPE EXPANSIONS

If $g \in L^2(R^n)$ satisfies $\int_{R^n} |g(x)|^2 (1 + |x|)^{2+\epsilon} dx < \infty$ and $\int_{R^n} |\hat{g}(\xi)|^2 (1 + |\xi|)^{2+\epsilon} d\xi < \infty$ for some $\epsilon > 0$, then there are $\alpha_0 > 0, \beta_0 > 0$ with the property that for any $0 < \alpha \leq \alpha_0, 0 < \beta \leq \beta_0$ every function $f \in L^2(R^n)$ has a Gabor-type expansion

$$f = \sum_{k,m \in Z^n} c_{km} e^{2\pi i \alpha k x} g(x - \beta m) \tag{13}$$

The series converges in $L^2(R^n)$ and coefficients satisfy
$\|f\|^2 \cong (\sum_{k,m} |c_{km}|^2)^{1/2}$.

By duality, the set of functions $\{g_{km} = e^{2\pi i \alpha k x} g(x - \beta m), k, m \in Z^n\}$ is a frame for $L^2(R^n)$, i.e.

$$A_1\|f\|^2 \le \left(\sum_{k,m} |\langle g_{km}, f\rangle|^2\right)^{1/2} \le A_2\|f\|_2 \qquad (14)$$

In this case the frame coefficients $\langle g_{km}, f\rangle$ are a regular sampling of the *short time Fourier transform* with window function g:

$$S_g(f)(x,y) = \int_{R^n} e^{-2\pi iy\xi} \bar{g}(\xi - x)f(\xi)d\xi \qquad (15)$$

at the points $(\beta m, \alpha k)$. The equivalence of norms (14) implies that the short time Fourier transform $S_g(f)$ and f can be completely reconstructed from these sampled values by the method of 1.3.

D. Gabor [14] has proposed such an expansion for $g(x) = e^{-\pi x^2}, \alpha\beta = 1$ in order to decompose a signal into components with optimal concentration in the time-frequency plane. In this case, however, (14) is violated, and the expansions (13) are extremely unstable. It can be shown that stable expansions require "oversampling" $\alpha\beta < 1$ [2], but sharp bounds on the required density are still unknown.

IRREGULAR SAMPLINGS AND EXPANSIONS

Let $(x_i, y_i)_{i\in I}$ be any discrete set in $R^n \times R^n$ such that $\bigcup_i (Q_\alpha(x_i) \times Q_\beta(y_i)) = R^n \times R^n$, where $Q_\alpha(x)$ is the cube of side length α centered at $x \in R^n$, and g, α, β are as above. Then the set of functions

$$e^{2\pi iy_i\xi}g(\xi - x_i), i \in I, \quad \text{constitutes a frame for } L^2(R^n)$$

Any $f \in L^2(R^n)$ has a jittered Gabor-type expansion

$$f = \sum_i c_i e^{2\pi iy_i\xi}g(\xi - x_i) \qquad (16)$$

with convergence in $L^2(R^n)$ and $\|f\|_2 \cong \|c\|_2$.

GABOR EXPANSIONS AND THE BEHAVIOR OF FUNCTIONS IN THE TIME-FREQUENCY PLANE

The short time Fourier transform can be viewed as a representation of a signal in the time-frequency plane. The behavior of signals simultaneously in time and frequency is best studied in the frame-work of the *modulation spaces* $M_{pq}^s(R^n)$. Let $g \in \mathcal{S}$ be fixed, $1 \le p, q \le \infty, s \in R$, then $M_{pq}^s(R^n)$ is defined as

$$M_{pq}^s(R^n) = \left\{ f \in \mathcal{S}' : \|f\|_{pqs}^q \right.$$

$$:= \int \left(\int |S_g(f)(x,y)|^p dx \right)^{q/p} (1+|y|)^{sq} dy < \infty \Big\}$$

with the obvious modifications for $p, q = \infty$. In particular $L^2(R^n) = M_{22}^0$ and the Bessel potential spaces $H^s = \{f \in \mathcal{S}' : \|f\|_{H^s}^2 = \int |\hat{f}(\xi)|^2(1 + |\xi|^{2s})d\xi < \infty\} = M_{22}^s$.

The space $S_0(R^n) := M_{11}^0$ has some interesting properties: it is an algebra with respect to pointwise multiplication and convolution, invariant under modulation and under Fourier transform. It is a natural domain for Poisson's formula and can often be used as a substitute for the more complicated Schwartz space, e.g. in generalized Fourier analysis [6].

For appropriate g, say $g \in \mathcal{S}$ and $\alpha, \beta > 0$ small enough, depending only on g and $|s|$, every function $f \in M_{pq}^s$ has a Gabor expansion (13) or (16), which converges in the norm of M_{pq}^s for $p, q < \infty$, and

$$f \in M_{pq}^s \iff \left(\sum_{k \in Z^n} \left(\sum_{m \in Z^n} |c_{km}|^p \right)^{q/p} (1+|k|)^{sq} \right)^{1/q} < \infty \qquad (18)$$

$$\iff \left(\sum_{k \in Z^n} \left(\sum_{m \in Z^n} |\langle g_{km}, f \rangle|^p \right)^{q/p} (1+|k|)^{sq} \right)^{1/q} < \infty \qquad (19)$$

Both expressions are equivalent norms for $f \in M_{pq}^s$.

SYSTEM IDENTIFICATION

A linear time-invariant system T on $L^p(R^n)$ can be modeled by a convolution operator $T = T_\sigma$, where $T_\sigma f = \sigma * f$ is the response of T to the input signal f. It can be shown that σ is always a tempered distribution in $S_0'(R^n) = M_{\infty\infty}^0$. A complete identification would require testing T against all "Dirac pulses" δ_x or all pure modulations $e^{iy\xi}$ and is often impossible. Since the modulation of a signal and its sampling are often technically easier to realize than sharp "Dirac pulses", a more realistic approach to the identification of T proceeds as follows:

1. Test the system T against a collection of input signals which are complex modulates $M_{\alpha k}g(t) = e^{2\pi i \alpha k t}g(t)$ of a given smooth envelope g.

2. Then sample the output of T in each case. (The sampling does not have to be equally spaced, but may depend on the modulation αk, if necessary.)

$$M_{k\alpha}g \xrightarrow{T} \sigma * M_{k\alpha}g \xrightarrow{sampling} \sigma * M_{k\alpha}g(m\beta)$$

In other words, the data obtained from this procedure are the numbers $\sigma * M_{k\alpha} g(m\beta) = e^{2\pi i \alpha \beta km} S_g\, (f)(m\beta, k\alpha) = e^{2\pi i \alpha \beta km} \langle g_{km}, f \rangle$, i.e., nothing but the frame coefficients of σ (here $g\,(x) = \overline{g(-x)}$). Therefore, if $\alpha, \beta > 0$ are small enough, σ is completely determined by these data and can be completely reconstructed by a variation of the frame method.

REFERENCES

Modulation spaces and their atomic decompositions are studied in [7], the above results are a special case of [9], Theorem 6.1 and [16]. The L^2-case is treated in [2, 4], [23] contains a Zak transform approach to Gabor-type expansions of modulation spaces.

OTHER NONORTHOGONAL EXPANSIONS

A) THE BARGMANN-FOCK SPACES [9, 16, 19]

Let $\mathcal{F}^p = \{f \text{ entire on } C^n : \|f\|_{\mathcal{F}^p}^p = \int_{C^n} |f(z)|^p e^{-\pi p|z|^2/2} dz < \infty\}$ for $1 \leq p < \infty$. $\mathcal{F}^2$ is a reproducing kernel Hilbert space with kernel $e_w(z) = e^{\pi \bar{w}z}$, i.e., $f(w) = \langle e_w, f \rangle = \int_{C^n} e^{\pi wz} f(z) e^{-\pi |z|^2} dz$.

These spaces are of interest in signal analysis because they arise essentially as spaces of short time Fourier transforms $S_g(f)$ with the fixed window $g(x) = e^{-\pi x^2}$. They also occur in the study of the canonical commutation relations in quantum mechanics.

For arbitrary $g \in \mathcal{F}^1$, there is a lattice density $\alpha, \beta > 0$ depending only on g, such that

$$f(z) = \sum_{k,m \in Z^n} c_{km} e^{\pi w_{km} z - \pi |w_{km}|^2/2} g(z - \bar{w}_{km}) \tag{20}$$

where $w_{km} = \beta m + i\alpha k \in C^n$ and the series converges in $\mathcal{F}^p$. Moreover, $\|f\|_{\mathcal{F}^p} \cong (\sum_{k,m \in Z^n} |c_{km}|^p)^{1/p}$. In particular, if $g(z) \equiv 1 \in \mathcal{F}^1$ is chosen, then (20) yields a stable expansion with respect to the reproducing kernel functions e_w.

The corresponding result for frames states that every $f \in \mathcal{F}^p$ can be completely reconstructed from the coefficients $\lambda_{km} = \int_{C^n} e^{\pi w_{km} \bar{z} - \pi |w_{km}|^2/2} \bar{g}(z - \bar{w}_{km}) f(z) e^{-\pi |z|^2} dz$ and $\|f\|_{\mathcal{F}^p} \cong \|\lambda\|_p$. If in particular $g \equiv 1$, then $\lambda_{km} = f(w_{km}) e^{-\pi |w_{km}|^2/2}$ and one obtains a well-known sampling theorem for entire functions $\|f\|_{\mathcal{F}^p} \cong (\sum_{k,m} |f(w_{km})|^p e^{-\pi p|w_{km}|^2/2})^{1/p}$.

Instead of the w_{km}, irregularly distributed sets as in (16) can be used in (20) and the sampling theorem.

B) GABOR-TYPE EXPANSIONS IN ABELIAN GROUPS

Let $\mathcal{G}$ be a separable locally compact Abelian group, $\hat{\mathcal{G}}$ its dual group, and g be a "nice" function on $\mathcal{G}$. For instance any $g \in L^1(\mathcal{G})$ with a decomposition $g = \sum_n g_n(x - x_n)$, where $x_n \in \mathcal{G}$, supp $g_n \subseteq Q$, a fixed compact set in $\mathcal{G}$, and $\sum_n \|\hat{g}_n\|_1 < \infty$, qualifies. If the neighborhood $U \times V \subseteq \mathcal{G} \times \hat{\mathcal{G}}$ of the identity is small enough (depending only on g) and if $(x_m, \chi_k)_{k,m \in N} \subseteq \mathcal{G} \times \hat{\mathcal{G}}$ is a discrete set, such that $\bigcup_{k,m \in N}(x_n + U) \times (\chi_m \cdot V) = \mathcal{G} \times \hat{\mathcal{G}}$ then every $f \in L^2(\mathcal{G})$ has a nonorthogonal expansion

$$f(x) = \sum_{k,m \in Z} c_{km}\chi_m(x)g(x - x_k) \tag{21}$$

and $\|f\|_2 \cong (\sum_{k,m} |c_{km}|^2)^{1/2}$.

Also the collection $g_{km}(x) = \chi_m(x)g(x - x_k), k, m \in Z$ is a frame for $L^2(R^n)$.

For $\mathcal{G} = R^n$ one recovers the Gabor-type expansions of Section 2.2. It is easy to extend the theory of modulation spaces and their Gabor expansions to general locally compact Abelian groups.

In particular, for $\mathcal{G} = Z^n$ or $\mathcal{G} = Z_N$ (the cyclic group of order N) we obtain discrete models of Gabor expansions. They furnish a systematic procedure for the discretization and numerical implementation of the Gabor-type expansions (13).

C) ANISOTROPIC WAVELET EXPANSIONS

Let D be a $n \times n$ matrix such that each eigenvalue has strictly positive real part and set $\Delta(t) = |\det e^{-tD/2}|$. If $g \in L^2(R^n)$ satisfies $|\hat{g}(\xi)| \geq c > 0$ for $a \leq |\xi| \leq b$ and mild decay and moment conditions, then for $\alpha, \beta > 0$ small enough, every $f \in L^2(R^n)$ has an anisotropic nonorthogonal wavelet expansion

$$f(x) = \sum_{j \in Z, k \in Z^n} c_{jk}\Delta(\beta j)g(e^{-\beta jD}x - \alpha k) \tag{22}$$

with convergence in $L^2(R^n)$ and $\|f\|_2 \cong \|c\|_2$.

The class of anisotropic Besov-Triebel-Lizorkin spaces can be characterized by this type of nonorthogonal expansions.

If the matrix $A = e^D$ leaves Z^n or some other lattice invariant, it is known that orthogonal wavelet expansions (22) exist. In the other cases non-orthogonal expansions are the only tool available.

D) BERGMAN SPACES ON THE UPPER HALF-PLANE [1]

Let $U = \{z = x + iy, y > 0\} \subseteq C$ by the upper half plane and $A^{p,m}, 1 \leq p < \infty, m \geq 3, m \in N$ be the Bergman space

$$A^{p,m} = \{f \text{ analytic on } U : \|f\|_{p,m}^p = \int \int |f(z)|^p y^{pm/2-2} dx dy < \infty\}$$

(23)

Then for $\alpha > 0$ and $\beta > 1$ small enough, every $f \in A^{p,m}$ has an expansion

$$f(z) = \sum_{j,k \in Z} c_{jk} \beta^{-2jm} (z - \alpha k + i\beta^{2j})^{-m}$$

(24)

with convergence in $A^{p,m}$ and coefficients satisfying $\|c\|_p \cong \|f\|_{p,m}^p$.

3. THE GENERAL STRUCTURE OF THESE EXPANSIONS

Despite the diversity of these expansions, they are but special cases of a single theorem. To gain some insight into the structure of these expansions, we observe their common structure.

3.1. *The basis functions are derived from a single function g by the action of a (square-)integrable unitary representation π of a (Lie) group $\mathcal{G}$ on a Hilbert space $\mathcal{H}$.*

3.2. *All occurring spaces are defined by the magnitude of representation coefficients $V_g(f)(x) = \langle \pi(x)g, f \rangle$, where g is a fixed test function.*

To any appropriate Banach space Y of functions on $\mathcal{G}$ is associated a Banach space of functions of distributions on the level of $\mathcal{H}$, the *coorbit of Y under the representation π* [9]

$$Co_\pi Y = \{f : \langle \pi(x)g, f \rangle \in Y\} \text{ with norm } \|f\|_{Co\,Y} = \|\langle \pi(x)g, f \rangle\|_Y \quad (25)$$

in particular, $Co_\pi L^2(\mathcal{G}) = \mathcal{H}$.

3.3. The nonorthogonal series expansions are of the form

$$f = \sum_i \lambda_i \pi(x_i)g$$

(26)

with convergence in $Co_\pi Y$, where x_i is a discrete, sufficiently dense set in $\mathcal{G}$. A $f \in Co_\pi Y$ can be completely characterized by the size of the set of coefficients λ_i, e.g. $f \in Co_\pi L^p(\mathcal{G}) \iff (c_i) \in l^p$.

3.4. *Frames for $Co_\pi Y$*: $\{\pi(x_i)g, i \in I\}$ being a frame means essentially that f is uniquely determined by the sequence $\langle \pi(x_i)g, f \rangle_{i \in I}$. In other words, the representation coefficient $x \to \langle \pi(x)g, f \rangle$ is completely determined by and can be reconstructed from its sampled values on the discrete set $\{x_i, i \in I\}$ in $\mathcal{G}$. Thus the construction of coherent frames amounts to proving a sampling theorem for representation coefficients. The analogy with the famous Shannon-Whittacker-Kotel'nikov sampling theorem for band-limited functions and its extensions has been very fruitful for understanding the theory of frames.

The examples of Section 2 are attached to the following groups and representations.

i) The group for wavelet expansions is $\mathcal{G} = R^n \times R^+$ ($ax + b$-group) with multiplication $(x, t) \cdot (y, u) = (x + ty, tu), x, y \in R^n, u, v > 0$, a group of affine transformations on R^n. The representation is $\pi(x, t)f(y) = L_x D_t f(y) = t^{-n/2} f(y/t - x/t)$ on $L^2(R^n)$ [17]. The continuous wavelet transformation $W_g(f)(x, t)$ is identical to the representation coefficient $\langle \pi(x, t)g, f \rangle$.

A comparison of (11) and (25) shows that $Co_\pi L^p = \dot{B}_{pp}^{n/p - n/2}(R^n)$.

The expansions (10) use the larger group $\mathcal{G}' = R^n \times (R^+ \times O(n)$ with multiplication $(x, t, O_1) \cdot (y, u, O_2) = (x + tO_1 y, tu, O_1 O_2), x, y \in R^n, u, v > 0, O_i$ being orthogonal matrices. The representation is $\pi'(x, t, O)f(y) = t^{-n/2} f(O^{-1}(y - x)/t)$ on $L^2(R^n)$. In contrast to π, this representation is irreducible and allows us to pick the wavelets in a larger class of functions.

ii) The group for Gabor expansions is the Heisenberg group $\mathcal{G} = R^n \times R^n \times T$ with multiplication $(x, y, \sigma) \cdot (u, v, \tau) = (x + u, y + v, \sigma \tau e^{2\pi i y u})$, $x, y, u, v \in R^n, \sigma, \tau \in C, |\sigma| = |\tau| = 1$. The occurring representation π is the Schrödinger representation $(x, y, \tau) \mapsto \tau L_x M_y f(\xi) = \tau e^{2\pi i y(\xi - x)} f(\xi - x)$ on $L^2(R^n)$. Here L_x is the shift operator and M_y denotes the modulation operator. The short time Fourier transform is identified as $S_g(f)(x, y) = \tau e^{-2\pi i x y} \langle \pi(s, y, \tau)g, f \rangle$. Since the torus acts trivially by scalar multiplication, this additional phase factor is of no consequence and does not occur in the results.

In this case the coorbits of $L^p(\mathcal{G})$ are the modulation spaces M_{pp}^0.

iii) In Example 2.3a) we find again the Heisenberg group, but acting by the Bargmann-Fock representation $\pi(x, y, \tau)f(z) = \tau e^{-i\pi x y} e^{-\pi |w|^2/2} e^{\pi w z} \times f(z - \bar{w})$ on $\mathcal{F}^2(C^n)$, where $w = x + iy \in C^n$. To see that the Bargmann-Fock spaces can also be defined by the size of a representation coefficient, choose $g(z) \equiv 1$, then $|\langle \pi(x, y, \tau)g, f \rangle| = |f(\bar{w})|e^{-\pi |w|^2/2}$ and the representation coefficient is in $L^p(\mathcal{G})$ if and only if $f \in \mathcal{F}^p$, in other words $Co_\pi L^p = \mathcal{F}^p$.

iv) Example 2.3b) is also based on a group of Heisenberg type, namely $\mathcal{G} \times \hat{\mathcal{G}} \times T$ with multiplication $(x, \chi, \sigma) \cdot (y, \psi, \tau) = (x + y, \chi\psi, \sigma\tau\chi(y))$ for $x, y \in \mathcal{G}, \chi, \psi \in \hat{\mathcal{G}}, |\sigma| = |\tau| = 1$. The representation is $\pi(x, \chi, \tau)f(y) = \tau\chi(y - x)f(y - x)$ acting on $f \in L^2(\mathcal{G})$.

v) For the anisotropic wavelet expansions, the "anisotropic" $ax + b$-group is used: $\mathcal{G} = R^n \times R$ with multiplication $(x, t) \cdot (y, v) = (x + e^{tD}y, t + v)$ for $x, y \in R^n, t, v \in R$. In this case the representation acts on $f \in L^2(R^n)$ by $\pi(x, t)f(y) = \Delta(t)f(e^{-tD}(y - x))$.

vi) The last example uses $\mathcal{G} = SL(2, R)$ and its discrete series representations, see [8] for more details.

4. THE MAIN TOOL: THE ORTHOGONALITY RELATIONS

4.1.   Let $\mathcal{G}$ be a locally compact group with (left) Haar measure dx and π be an irreducible, unitary representation of $\mathcal{G}$ on a Hilbert space $\mathcal{H}$ that is square-integrable, i.e. there exists $g \in \mathcal{H}, g \neq 0$, such that

$$\int_{\mathcal{G}} |\langle \pi(x)g, g \rangle|^2 dx < \infty \tag{27}$$

Then there exists a positive, self-adjoint, densely defined operator A on $\mathcal{H}$, such that

$$\int_{\mathcal{G}} \overline{\langle \pi(x)g_1, f_1 \rangle} \langle \pi(x)g_2, f_2 \rangle dx = \langle Ag_2, Ag_1 \rangle \langle f_1, f_2 \rangle \tag{28}$$

for all $g_1, g_2 \in \mathrm{dom} A, f_1, f_2 \in \mathcal{H}$. If $\mathcal{G}$ is unimodular, e.g. for the Heisenberg group, then A is a multiple of the identity operator. See [17, 22, 3] for proofs and details.

Special Case: $g_1 = g_2 = g \in \mathrm{dom}\, A, \|Ag\| = 1, f_1 = \pi(y)g, f_2 = f \in \mathcal{H}$. Set $V_g(f)(x) = \langle \pi(x)g, f \rangle$, then

$$V_g(f) * V_g(g)(y) = \int \langle \pi(x)g, f \rangle \overline{\langle \pi(x)g, \pi(y)g \rangle} dx = \langle \pi(y)g, f \rangle = V_g(f)(y) , \tag{29}$$

where $*$ stands for the usual convolution $F * G(x) = \int_{\mathcal{G}} F(y)G(y^{-1}x)dy$ between two functions F and G on $\mathcal{G}$. Thus the orthogonality relations imply the reproducing formula $V_g(f) * V_g(g) = V_g(f)$ for representation coefficients. In the sequel we assume without mentioning that $g \in \mathcal{H}$ is normalized $\|Ag\| = 1$, and thus the reproducing formula (29) holds true.

In some applications the following modification of (29) is very useful, cf.

e.g. [13]: if $\langle Ah, Ag \rangle = 1$, then by (28)

$$V_g(f) * V_g(h) = \langle Ah, Ag \rangle V_g(f) = V_g(f) \tag{30}$$

holds for all $f \in \mathcal{H}$. This is a general statement of the fact that the analyzing wavelet g and the synthesizing wavelet h can be chosen (almost) independent of each other.

For the $ax + b$-group, (29) is equivalent to Calderòn's reproducing formula. For the Heisenberg group the orthogonality relations are known as Moyal's formulas.

The irreducibility of the representation is only a technical assumption; the reproducing formula holds for a larger class of representations, e.g. for groups of $ax + b$-type. For instance, given the representation $\pi(x, t)f(y) = \Delta(t)f(e^{-tD}(y - x))$ of the anisotropic $ax + b$-group (Example 2.3c) and $g, h \in L^2(R^n)$, the reproducing formula (30) for the wavelet transform $\langle \pi(x, t)g, f \rangle$ holds under the condition

$$\int \overline{\hat{g}(e^{tD}\xi)}\hat{h}(e^{tD}\xi)dt = 1 \tag{31}$$

for all $\xi \in R^n, |\xi| = 1$. This explains the conditions on the support of $\hat{g}$ and the vanishing moments in Examples 2.1 and 2.3c).

EXTENSION TO OTHER SPACES

For the decomposition of general $Co_\pi Y$-spaces, the stronger condition

$$\int_{\mathcal{G}} |\langle \pi(x)g, g \rangle| w(x)\, dx < \infty \tag{32}$$

for some $g \in \mathcal{H}, g \neq 0$ is needed. Given a Banach space Y of functions on $\mathcal{G}$, the weight has to be chosen such that $L^1_w(\mathcal{G}) * Y \subseteq Y$ and $Y * L^1_w(\mathcal{G}) \subseteq Y$, where $L^1_w(\mathcal{G}) = \{f$ measurable on $\mathcal{G} : \int_{\mathcal{G}} |f(x)|w(x)\, dx < \infty\}$.

With this assumption:

a) It is possible to define a suitable space of test functions embedded in $\mathcal{H}$, namely $Co\, L^1_w$, and distributions (the dual of $Co\, L^1_w$) from which to select elements. Then definition (25) makes sense even for functions not contained in $\mathcal{H}$.

b) The reproducing formulas (29) and (30) carry over to general coorbit spaces $Co_\pi Y$. Moreover, it follows from the orthogonality relations that the definition of $Co_\pi Y$ is independent of the choice of g, such that different g's yield equivalent norms for $Co_\pi Y$. A direct proof of this statement for the examples of Section 2 is usually more involved.

c) The convolution $V_g(f) * V_g(g) = V_g(f)$ in $L^p(\mathcal{G}) * L^1(\mathcal{G}) \subseteq L^p(\mathcal{G})$ or in $Y * L^1_w(\mathcal{G}) \subseteq Y$ in general, is easy to analyze.

4.2. If $f = g$ in (29), then $V_g(g) * V_g(g) = V_g(g)$, and the convolution operator $F \mapsto F * V_g(g)$ is the orthogonal projection from $L^2(\mathcal{G})$ onto the closed subspace of representation coefficients $\{\langle \pi(x)g, f \rangle, f \in \mathcal{H}\}$. If $\langle \pi(x)g, g \rangle \in L^1_w(\mathcal{G})$, then this statement extends to $Co_\pi Y$ and yields the fundamental Correspondence Principle.

Correspondence Principle [9], Prop. 4.3. *A function $F \in Y$ on $\mathcal{G}$ is a generalized representation coefficient $V_g(f)(x) = \langle \pi(x)g, f \rangle$ for a unique $f \in Co_\pi Y$, if and only if $F * V_g(g) = F$. The translate $L_x F \in Y$ corresponds to $\pi(x)f \in Co_\pi Y$, where $L_x F(y) = F(x^{-1}y), x, y \in \mathcal{G}$.*

The Correspondence Principle is the very reason why all the different examples of Section 2 can be treated by a single method. As a result of this *unification* by representation theory the objects are now nice, smooth functions $V_g(f)(x)$ on the group $\mathcal{G}$, instead of rough functions, measures, or distributions on R^n, no matter what the original space looked like! Instead of a possibly complicated integral formula these representation coefficients satisfy a simple convolution equation, which can be analyzed with methods of abstract harmonic analysis.

5. THE CONSTRUCTION OF NONORTHOGONAL EXPANSIONS

We present the main ideas for the series expansions of $Co_\pi L^p$, which for the various groups contains the Besov spaces and the modulation spaces with indices p, p, the Bargmann-Fock spaces etc. The rigorous proof for the decomposition of general coorbit spaces $Co_\pi Y$ differs only in technical details.

Our goal is to approximate the reproducing formula (29) by a sum of translates of $V_g(g)$ and then iterate on the remainder (see [1], where a similar idea is first used to construct atomic decompositions.)

In the following we abbreviate $F(x) = \langle \pi(x)g, f \rangle, G(x) = \langle \pi(x)g, g \rangle$.

APPROXIMATION OPERATORS

Let $U \subseteq \mathcal{G}$ be a neighborhood of the identity e in $\mathcal{G}$. A subset $(x_i)_{i \in I}$ in $\mathcal{G}$ is called U-dense, if $\cup_{i \in I} x_i \cdot U = \mathcal{G}$. Given (x_i), choose a partition of unity $\Psi = (\psi_i)_{i \in I}$ in $\mathcal{G}$, such that (1) $\sum_i \psi_i \equiv 1$ almost everywhere, (2) $0 \leq \psi_i \leq 1$, (3) supp $\psi_i \subseteq x_i \cdot U$.

Example. In the n-dimensional $ax+b$-group $R^n \times R^+$ the set $(\alpha k \beta^j, \beta^j)$, $j \in Z, k \in Z^n$ is dense with respect to the neighborhood $U = [-\alpha/2, \alpha/2]^n \times [1/\sqrt{\beta}, \sqrt{\beta}]$ and the characteristic functions of $[\beta^j \alpha(k-1/2), \beta^j \alpha(k+1/2)] \times [\beta^{j-1/2}, \beta^{j+1/2}]$ form a partition of unity. The U-density for an irregular sampling set (x_i, y_i) in $R^n \times R^+$ has been explicitly stated in Section 2.1.

Definition. Given a partition of unity Ψ, the operator T_Ψ, acting on functions on $\mathcal{G}$, is defined by

$$T_\Psi F = \sum_i \langle \psi_i, F \rangle L_{x_i} G \qquad (33)$$

where $\langle \psi_i, F \rangle = \int_{\mathcal{G}} \bar{\psi}_i F(x)\, dx$ is essentially a local average of F near x_i. Given a neighborhood $U \subseteq \mathcal{G}$ of e, the *U-oscillation* of a function G is

$$\mathrm{osc}_U G(x) = \sup_{u \in U} |G(ux) - G(x)| \qquad (34)$$

Facts [9]. Assume that $g \in \mathcal{H}$, such that $G(x) = \langle \pi(x)g, g \rangle \in L^1(\mathcal{G})$ and also $\mathrm{osc}_U G \in L^1(\mathcal{G})$ for some, hence for all, neighborhoods U of e. Then:

1. T_Ψ is bounded from $L^p(\mathcal{G})$ into the closed subspace $L^p * G$ with a bound independent of Ψ.

2. If $f \in Co_\pi L^p$, then $\sum_i \langle \pi(x_i)g, f \rangle \psi_i \in L^p(\mathcal{G})$. In particular, if $(x_i)_{i \in I}$ is a discrete subset in $\mathcal{G}$, then $\langle \pi(x_i)g, f \rangle \in l^p(I)$.

3. If $(x_i)_{i \in I}$ is a discrete subset in $\mathcal{G}$ and $(c_i)_{i \in I} \in l^p(I)$, then $F = \sum_i c_i L_{x_i} G \in L^p(\mathcal{G})$ and thus by the Correspondence Principle $f = \sum_i c_i \pi(x_i)g \in Co_\pi L^p$.

THE MAIN ESTIMATE

We approximate the convolution $F * G = F$ (29) by a Riemann type sum, i.e. by T_Ψ, as follows:

$$F(x) = F * G(x) = \int_{\mathcal{G}} F(y)G(y^{-1}x)dy = \sum_i \int F(y)\psi_i(y)G(y^{-1}x)dy \approx$$

$$\approx \sum_i (\int F(y)\psi_i(y)dy)G(x_i^{-1}x) = T_\Psi F$$

Since $y \in x_i \cdot U \Leftrightarrow y = x_i u \Leftrightarrow x_i^{-1} = uy^{-1}$ for some $u \in U$, one obtains a pointwise estimate

$$|F(x) - T_\Psi F(x)| = |\int_{\mathcal{G}} F(y) \left(\sum_i \psi_i(y)\right) (G(y^{-1}x) - G(x_i^{-1}x))\,dy| \le$$

$$\le \sum_{i \in I} \int |F(y)| \psi_i(y) \sup_{u \in U} |G(y^{-1}x) - G(uy^{-1}x))|dy = |F| * \mathrm{osc}_U G(x) \quad (35)$$

Upon taking norms

$$\|F - T_\Psi F\|_p \le \| |F| * \mathrm{osc}_U G\|_p \le \|F\|_p \| \mathrm{osc}_U G\|_1 \quad (36)$$

holds for all $F \in L^p * G$. If the neighborhood U is chosen so small that $\|\mathrm{osc}_U G\|_1 < 1$, then $\mathbf{1} - T_\Psi$ is a contraction on the closed subspace $L^p * G$ and the operator norm satisfies

$$|\|\mathbf{1} - T_\Psi\|| < 1 \quad (37)$$

THE ITERATION

By (37) T_Ψ is invertible on $L^p * G$ and

$$T_\Psi^{-1} = \sum_{n=0}^\infty (\mathbf{1} - T_\Psi)^n \quad \text{on } L^p * G \quad (38)$$

Consequently $F = T_\Psi T_\Psi^{-1} F = \sum_{i \in I} \langle \psi_i, T_\Psi^{-1} F \rangle L_{x_i} G$ and the Correspondence Principle implies

$$f = \sum_{i \in I} c_i \pi(x_i) g \quad (39)$$

with coefficients $c_i = \langle \psi_i, T_\Psi^{-1} F \rangle \in l^p(I)$.

FRAMES

From (39) it is clear that the coefficients c_i are given by functionals e_i in the dual of $Co_\pi L^p$: $c_i = \langle e_i, f \rangle$. Precisely e_i is given by $V_g(e_i) = T_\Psi^{*-1} \psi_i$. Using the duality theory for coorbit spaces [9, 10], the argument of Section 1 carries over and yields that $\{\pi(x_i)g, i \in I\}$ is a (Banach) frame for $Co_\pi L^p, 1 \le p < \infty$, i.e. for some constants $0 < A_1 \le A_2$

$$A_1 \|f\|_{Co_\pi L^p} \le \|\langle \pi(x_i)g, f \rangle_{i \in I}\|_p \le A_2 \|f\|_{Co_\pi L^p}, \quad (40)$$

and the reconstruction of f from the frame coefficients is

$$f = \sum_i \langle \pi(x_i)g, f \rangle e_i .$$

Note, however, that for Banach spaces the norm equivalence (40) by itself does not imply a reconstruction method.

6.   CONCLUSION

NUMERICAL ASPECTS

For numerical calculations it is better to work with the modified approximation operator $S_\Psi F = \sum_i F(x_i) \gamma_i L_{x_i} G$ on $L^p * G$, where the partition of unity Ψ comes in only through the weight factors $\gamma_i = \int \psi_i(x) dx$.

If $(x_i)_{i \in I}$ is dense enough in $\mathcal{G}$, then S_Ψ is invertible on $L^p * G$ [16] and

$$F = S_\Psi^{-1} S_\Psi F = \sum_i F(x_i) \gamma_i S_\Psi^{-1} L_{x_i} G = \sum_{n=0}^{\infty} (1 - S_\Psi)^n S_\Psi F \qquad (41)$$

provides a complete reconstruction of F from the samples $F(x_i) = \langle \pi(x_i) g, f \rangle$.

With the Correspondence Principle the latter expression translates into the following reconstruction scheme on $Co_\pi L^p$:

$$\phi_0 = \sum_{i \in I} \langle \pi(x_i) g, f \rangle \gamma_i \pi(x_i) g \qquad (42)$$

$$\phi_{n+1} = \phi_n - \sum_{i \in I} \langle \pi(x_i) g, \phi_n \rangle \gamma_i \pi(x_i) g \qquad (43)$$

$$f = \sum_{n=0}^{\infty} \phi_n \qquad (44)$$

COMPUTATION OF THE COEFFICIENTS c

The identity (41) also implies the series expansion $f = \sum_i c_i \pi(x_i) g$. The reconstruction scheme (42) can be turned into an algorithm to compute the coefficients c_i as follows. Set $\phi_n = \sum_{i \in I} c_i^{(n)} \pi(x_i) g$, where the initial data $c_i^{(0)} = \gamma_i \langle \pi(x_i) g, f \rangle$ is the sampling of the representation coefficient. Let $A = (A_{ij})_{i,j \in I}$ be the matrix with elements $A_{ij} = \gamma_i \langle \pi(x_i) g, \pi(x_j) g \rangle$. Then

$$c^{(n+1)} = c^{(n)} - Ac^{(n)} \quad \text{and} \quad c_i = \sum_{n=0}^{\infty} c_i^{(n)} , \qquad (45)$$

as follows readily from

$$\phi_{n+1} = \sum_i c_i^{(n+1)} \pi(x_i) g = \phi_n - \sum_{i \in I} \langle \pi(x_i) g, \phi_n \rangle \gamma_i \pi(x_i) g =$$

$$= \sum_i \left(c_i^{(n)} - \sum_j c_j^{(n)} \gamma_i \langle \pi(x_i)g, \pi(x_j)g \rangle \right) \pi(x_i)g \, .$$

This is essentially the same method as in 1.3. At this point the group theoretic background of the reconstruction method disappears and the implementation becomes a problem in linear algebra.

The following theorem summarizes the conclusions of Sections 3–5 and contains the examples discussed in Section 2 as special cases.

Theorem 1. *Given* $\mathcal{G}, \pi, \mathcal{H}, L^p(\mathcal{G}), 1 \le p < \infty$. *Assume that* $g \in \mathcal{H}$ *is normalized* $\|Ag\| = 1$ *and satisfies* $G(x) = \langle \pi(x)g, g \rangle \in L^1(\mathcal{G})$ *and* $osc_{U_0}G \in L^1(\mathcal{G})$ *for some neighborhood* U_0 *of* e. *Then choose* U *so small that*

$$\|osc_U G\|_1 < 1$$

a) If $(x_i)_{i \in I}$ *in* $\mathcal{G}$ *is* U-*dense and discrete, then any* $f \in Co_\pi L^p$ *has a (nonorthogonal) expansion*

$$f = \sum_{i \in I} c_i \pi(x_i)g$$

The series converges unconditionally in $Co_\pi L^p$ *and the coefficients* c *satisfy*

$$A_1 \|f\|_{Co_\pi L^p} \le \|(c_i)_{i \in I}\|_p \le A_2 \|f\|_{Co_\pi L^p}$$

b) On the other hand, $\{\pi(x_i)g, i \in I\}$ *is a frame for* $Co_\pi L^p(\mathcal{G})$, *i.e.,* $f \in Co_\pi L^p$ *is uniquely determined by the coefficients* $\langle \pi(x_i)g, f \rangle_{i \in I}$,

$$\|f\|_{Co_\pi L^p} \cong \left(\sum_i |\langle \pi(x_i)g, f \rangle|^p \right)^{1/p}$$

and $f = \sum_i \langle \pi(x_i)g, f \rangle e_i$ *for a fixed set* $e_i \in Co_\pi L^1$.

Final Remarks. a) The Theorem gives a universal condition on g to be a wavelet and on the sampling density U required for a nonorthogonal expansion. The conditions on the wavelets in the concrete examples are sufficient to ensure the required properties of the representation coefficient $\langle \pi(x)g, g \rangle$. The explicit determination of the size of U is more tricky and is being investigated intensively.

b) It contains an extension of the theory of frames to Banach spaces [16].

c) Since the construction works for *any* U-dense subset (x_i), the Theorem allows for irregular sampling of representation coefficients (which in turn yields sampling theorems for wavelet transforms and short time Fourier transforms) and for "jittered" expansions.

d) The missing technical details and the exact proof for general coorbit spaces can be found in [9, 16].

e) The original papers [9, 16] also contain a stability theory for non-orthogonal expansions and frames of Banach spaces. It is shown that the reconstruction is robust under small changes of the wavelet g or the sampling set $x_i, i \in I$.

f) With Theorem 1 and the identification of the underlying groups and representations it is now a routine task to obtain the explicit expansions which were discussed in Section 2.

Since each of the examples of Section 2 possesses a rich background, it is impossible to include all substantial contributions. The following list of references can therefore serve only as a guide to more detailed collections of references. Especially useful should be I. Daubechies' lecture notes [3], the review article [18], Y. Meyer's book [21] and the original paper [8] on the unified theory.

REFERENCES

[1] R. R. Coifman and R. Rochberg, Representation theorems for holomorphic and harmonic functions in L^p. Asterisque 77 (1980), 11–66.

[2] I. Daubechies. The wavelet transform, time-frequency localization and signal analysis. IEEE Trans. Inf. Theory, Vol. 36 (1990), 961–1005.

[3] I. Daubechies. Wavelets and Applications. CBMS-Lecture Notes, to appear.

[4] I. Daubechies, A Grossman, and Y. Meyer. Painless nonorthogonal expansions. J. Math. Phys. 27 (1986), 1271-1283.

[5] R. Duffin, A. Schaeffer. A class of nonharmonic Fourier series. Trans. Amer. Math. Soc. 72 (1952), 341–366.

[6] H. G. Feichtinger. On a new Segal algebra. Monatsh. Math. 92 (1981), 269–289.

[7] H. G. Feichtinger. Atomic characterizations of modulation spaces through Gabor-type representations. Proc. Conf "Constructive Function Theory", Edmonton, July 1986, Rocky Mount. J. Math. 19 (1989), 113–126.

[8] H.G.Feichtinger, K. Gröchenig. A unified approach to atomic decompositions via integrable group representations. Proc. Conf. Lund 1986, "Function Spaces and Applications", Lecture Notes in Math. 1302 (1988), 52–73.

[9] H. G. Feichtinger, K Gröchenig. Banach spaces related to integrable group representations and their atomic decompositions I. J. Funct. Anal. 86 (1989), 307–340.

[10] H. G. Feichtinger, K. Gröchenig. Banach spaces related to integrable group representations and their atomic decompositions II. Monatsh. f. Math. 108 (1989), 129–148.

[11] M. Frazier, B. Jawerth. Decomposition of Besov spaces. Indiana Univ. Math. J. 34 (1985), 777–799.

[12] M. Frazier, B. Jawerth. The ϕ-transform and applications to distribution spaces. Proc. Conf. Lund 1986, "Function Spaces and Applications", Lecture Notes in Math. 1302 (1988), 223–246.

[13] M. Frazier, B. Jawerth. A discrete transform and decomposition of distribution spaces. J. Functional Anal. 93 (1990), 34–170.

[14] D. Gabor. Theory of Communication. J. Inst. Elect.Eng. 93, p. 429–457, 1946.

[15] K. Gröchenig. Unconditional bases in translation and dilation invariant function spaces on R^n. In "Constructive Theory of Functions". Conf. Varna 1987, B. Sendov et al., eds., p. 174–183. Bulgarian Acad. Sci. 1988.

[16] K. Gröchenig. Describing functions: atomic decompositions versus frames. Submitted.

[17] A. Grossmann, J. Morlet, and T. Paul. Transforms associated to square integrable group representations I: general results. J. Math. Phys. 26 (1985), 2473–2479.

[18] C.E. Heil, D.F. Walnut. Continuous and discrete wavelet transforms. SIAM Rev. 31 (1989), 628–666.

[19] S. Janson, J. Peetre, R. Rochberg. Hankel forms and the Fock space. Revista Mat. Iberoam. 3 (1987), 61–138.

[20] P. G. Lemarie and Y. Meyer. Ondelettes et bases hilbertiennes. Revista Mat. Iberoam. 2 (1986), 1–18.

[21] Y. Meyer. Ondelettes. Vol. I. Hermann, Paris. 1990.

[22] A. Perelomov. Generalized Coherent States and Their Applications. Texts and Monographs in Physics. Berlin-Heidelberg, Springer. 1986.

[23] D. F. Walnut. Lattice size estimates for Gabor decompositions. Preprint.

APPLICATIONS OF THE φ AND WAVELET TRANSFORMS TO THE THEORY OF FUNCTION SPACES

MICHAEL FRAZIER

Department of Mathematics
Michigan State University
East Lansing, Michigan

BJÖRN JAWERTH[1]

Department of Mathematics
University of South Carolina
Columbia, South Carolina

1. INTRODUCTION

In this exposition, we discuss the φ and wavelet transforms in the context of mathematical harmonic analysis. These transforms are presented as part of a natural progression within what is known as "Littlewood-Paley theory" (which will be described in an introductory fashion in Section 2 — for a more detailed account, see for example [FJW]). We will see that these transforms can be used to clarify and extend some important recent developments in harmonic analysis related to the study of function spaces and Calderón-Zygmund operators.

Littlewood-Paley theory originated on the unit disc, in the 1930's, in the context of using "complex methods" to study Fourier series (see e.g. [LP], and [Z], Ch. 14–15). The basic feature of the theory is that various function space norms can be characterized by certain auxiliary "Littlewood-Paley" expressions. The corresponding theory for $\mathbb{R}^n$ was developed in the 1950's and 1960's by Stein and his school (see [S]); among other things, it led to the

[1]Partially supported by NSF grant DMS-8803585, AFOSR grant 89-0455, and ONR grant N00014-90-J-1343

modern theory of H^p-spaces on $\mathbb{R}^n$ (see e.g. [SW], [BGS], [FS2], and [Co]). Meanwhile, an early version of the "Calderón formula" appeared in 1964 in Calderón's paper [Cal1] introducing the complex interpolation method. This formula was applied in [Cal2] to obtain a proof of Coifman and Latter's famous "atomic decomposition" of the H^p-spaces (see [Co], [L], [W], and [Ch-F]). Uchiyama [U] used the Calderón formula to characterize L^2 and BMO in terms of size conditions on the coefficients $\{s_Q\}_{Q\in\mathcal{Q}}$ of an expansion $f(x) = \sum_{Q\in\mathcal{Q}} s_Q a_Q(x)$, where $\mathcal{Q}$ is the collection of dyadic cubes in $\mathbb{R}^n$, $a_Q \in C^\infty(\mathbb{R}^n)$, $\int a_Q = 0$, and supp $a_Q \subseteq 3Q$, the triple of Q. We call such a representation formula a "smooth atomic decomposition" of f.

In [FJ1] we introduced the "φ-transform", which yields an identity closely related to the smooth atomic decomposition (see antecedents in [J], [P2]). (These topics will be discussed in more detail in Section 3.) For the φ-transform, for each $Q \in \mathcal{Q}$ there are functions φ_Q and ψ_Q, obtained from certain original $\varphi, \psi \in \mathcal{S}(\mathbb{R}^n)$ by translations and dilations. The φ-transform identity is

$$f = \sum_{Q\in\mathcal{Q}} \langle f, \varphi_Q\rangle \psi_Q. \tag{1.1}$$

For example, in one dimension, (1.1) takes the form

$$f = \sum_{v\in\mathbb{Z}} \sum_{k\in\mathbb{Z}} \langle f, \varphi_{vk}\rangle \psi_{vk}, \tag{1.1'}$$

where $\varphi_{vk}(x) = 2^{v/2}\varphi(2^v x - k)$ and $\psi_{vk}(x) = 2^{v/2}\psi(2^v x - k)$. (See Section 3 for the definitions in the general case.)

The advantages of (1.1) over the smooth atomic decomposition are that the representing functions $\{\psi_Q\}_{Q\in\mathcal{Q}}$ are "canonical" (that is, independent of f), the ψ_Q's have a very particular form (e.g. supp $\hat{\psi}_Q$ is compact), and the coefficients $\langle f, \varphi_Q\rangle$ depend linearly on f. We proved (in [FJ1], [FJ2], see also [FJ3]) that all of the function spaces characterized by Littlewood-Paley expressions could also be characterized by size conditions on the coefficients in either the smooth atomic decomposition or the φ-transform identity. This includes the L^p spaces, $1 < p < +\infty$, the H^p spaces, $0 < p \leq 1$, the space BMO, the Besov spaces, and the Bessel or Riesz potentials of these spaces (in particular, the Sobolev spaces, $1 < p < +\infty$). Our purpose, initiated in [FJ1], and carried out in [FJ3], was to use Littlewood-Paley theory, the Calderón formula, and the φ-transform, to obtain a systematic analysis of the commonly occurring function spaces (see also [CMS] in this regard).

Independently of these developments in harmonic analysis, Grossman, Morlet, and Paul ([GM1-2] and [GMP]) constructed the wavelet transform in its continuous form (which is essentially the Calderón formula). The discretization of the Calderón formula occurred first in the smooth atomic decomposition, noted earlier, and then in the development of (1.1) described above. Shortly thereafter, the theory of "frames" ([DGM], [Dau2]) provided a general context for the study of discrete expansions. Some, but not all, frames can be viewed as examples of φ-transforms. A synthesis between this wavelet development and harmonic analysis was effected by Meyer (similar to the point of view in [FJ1-3]). This resulted in the construction by Lemarié and Meyer [LM] of orthonormal wavelet bases $\{\psi_Q\}_{Q\in\mathcal{Q}}$ for $L^2(\mathbb{R})$ (or $L^2(\mathbb{R}^n)$, similarly, with $2^n - 1$ functions $\psi_Q^{(i)}$ for each $Q \in \mathcal{Q}$). In this construction, carried out shortly after [FJ1], the identity (1.1') holds with $\varphi = \psi$. A similar basis had been discovered earlier by Strömberg [Str] in studying the Hardy spaces. As for the φ-transform, all of the classical Littlewood-Paley spaces on $\mathbb{R}^n$ are characterized by the size of the coefficients in the expansion with respect to the orthonormal wavelets (see [Me3] or [FJW]). The wavelet construction in [LM] is clever but somewhat mysterious; since then the understanding of orthonormal wavelet bases has developed considerably. Now there are many constructions of orthonormal wavelet bases (see e.g. [Dau1]), which are naturally described by the theory of "multi-resolution analysis" (see [Ma]).

An exposition of the theory of function spaces along the lines we have just presented can be found in [FJW]. More extensive information about wavelets can be found in Meyer's book [Me3]. Other references containing considerable historical background are [Dau3] and [HW].

Given the identity (1.1), the statement that a function space X on $\mathbb{R}^n$ is characterized by (1.1) means that associated to X is a space Y of sequences $s = \{s_Q\}_{Q\in\mathcal{Q}}$, such that

$$\| \sum_{Q\in\mathcal{Q}} s_Q \psi_Q \|_X \leq c \| s \|_Y \quad \text{for all } s \in Y, \tag{1.2}$$

and

$$\| \{\langle f, \varphi_Q\rangle\}_{Q\in\mathcal{Q}} \|_Y \leq c \| f \|_X . \tag{1.3}$$

(We follow the analyst's convention that "c" may designate a different constant, perhaps depending on the dimension n, or other fixed parameters, at each occurrence.) Note that (1.1-3) implies

$$\| \{\langle f, \varphi_Q\rangle\}_{Q\in\mathcal{Q}} \|_Y \approx \| f \|_X, \tag{1.4}$$

where "$\approx$" means the norms are equivalent. For the φ-transform, it is not necessarily true, however, that $\| s \|_Y \approx \| \sum_{Q\in\mathcal{Q}} s_Q \psi_Q \|_X$; the non-orthogonality of the decomposition allows the 0 function to be non-trivially represented in the form $\sum_{Q \in \mathcal{Q}} s_Q \psi_Q$.

Much work has been done recently to obtain results like (1.1-4) with more and more specific choices of φ and ψ. Here we consider questions in the opposite direction: what are the most general families $\{m_Q\}_{Q \in \mathcal{Q}}$ for which analogues of (1.2) or (1.3) hold? We say that $\{m_Q\}_{Q \in \mathcal{Q}}$ is a family of molecules for X if there exists $c > 0$ such that

$$\| \sum_{Q \in \mathcal{Q}} s_Q m_Q \|_X \leq c \| s \|_Y, \text{ for all } s = \{s_Q\}_{Q \in \mathcal{Q}} \in Y, \qquad (1.5)$$

and we say that $\{m_Q\}_{Q \in \mathcal{Q}}$ is a norming family for X if there is a constant $c > 0$ such that

$$\| \{\langle f, m_Q \rangle\}_{Q \in \mathcal{Q}} \|_Y \leq c \| f \|_X, \text{ for all } f \in X. \qquad (1.6)$$

As an example of the interest in such families, note that if T is a linear operator (continuous from $\mathcal{S}$ to $\mathcal{S}'$), then T is (or extends to be) a bounded operator on X if and only if $\{T\psi_Q\}_{Q \in \mathcal{Q}}$ is a family of molecules for X, where $\{\psi_Q\}_{Q \in \mathcal{Q}}$ is the family in (1.1). ("If" follows from (1.1) and (1.4) easily; "only if" is immediate from (1.2).) Thus, conditions guaranteeing that (1.5) holds result in a way of analyzing T in terms of the action of T on the canonical functions $\{\psi_Q\}_{Q \in \mathcal{Q}}$. We will see that conditions (1.5) and (1.6) are dual to each other.

For the class of Littlewood-Paley spaces, families of molecules and norming families are precisely characterized in terms of infinite matrices $A = \{a_{QP}\}_{Q, P \in \mathcal{Q}}$. Such a matrix acts on a sequence $s = \{s_Q\}_{Q \in \mathcal{Q}}$ in the usual way: $As = \{(As)_Q\}_{Q \in \mathcal{Q}}$ where

$$(As)_Q = \sum_{P \in \mathcal{Q}} a_{QP} s_P.$$

Proposition 1.1. *Suppose X is a function space, $\mathcal{Q}$ some index set, and $\{\varphi_Q\}_{Q \in \mathcal{Q}}$ and $\{\psi_Q\}_{Q \in \mathcal{Q}}$ are families of functions such that (1.1) holds for all $f \in X$. Suppose Y is a space of sequences $s = \{s_Q\}_{Q \in \mathcal{Q}}$ such that (1.2–3) hold. Let $\{m_Q\}_{Q \in \mathcal{Q}}$ be a given family of functions.*

A.) Define a matrix $A = \{a_{QP}\}_{Q, P \in \mathcal{Q}}$ by

$$a_{QP} = \langle m_P, \varphi_Q \rangle, \text{ for } P, Q \in \mathcal{Q}.$$

Then $\{m_Q\}_{Q \in \mathcal{Q}}$ is a family of molecules for X if and only if A is bounded on Y.

B.) Define a matrix $B = \{b_{QP}\}_{Q,P \in \mathcal{Q}}$ by

$$b_{QP} = \langle \psi_P, m_Q \rangle, \quad \text{for } P, Q \in \mathcal{Q}.$$

Then $\{m_Q\}_{Q \in \mathcal{Q}}$ is a norming family for X if and only if B is bounded on Y.

Proof. **A.)** Let $f = \sum_{Q \in \mathcal{Q}} s_Q m_Q$. Then for $Q \in \mathcal{Q}$,

$$(As)_Q = \sum_{P \in \mathcal{Q}} a_{QP} s_P = \sum_{P \in \mathcal{Q}} \langle m_P, \varphi_Q \rangle s_P = \langle \sum_{P \in \mathcal{Q}} s_P m_P, \varphi_Q \rangle = \langle f, \varphi_Q \rangle.$$

Hence

$$\| As \|_Y = \| \{ \langle f, \varphi_Q \rangle \}_{Q \in \mathcal{Q}} \|_Y \approx \| f \|_X,$$

by (1.4). Thus (1.5) is equivalent to the boundedness of A on Y.

B.) Suppose $f = \sum_{Q \in \mathcal{Q}} s_Q \psi_Q$. Then for $Q \in \mathcal{Q}$,

$$\langle f, m_Q \rangle = \langle \sum_{P \in \mathcal{Q}} s_P \psi_P, m_Q \rangle = \sum_{P \in \mathcal{Q}} s_P \langle \psi_P, m_Q \rangle = \sum_{P \in \mathcal{Q}} b_{QP} s_P = (Bs)_Q.$$

$$(1.7)$$

Thus $\| Bs \|_Y = \| \{ \langle f, m_Q \rangle \}_{Q \in \mathcal{Q}} \|_Y$. If we assume (1.6), this observation and (1.2) imply that B is bounded. If we assume B is bounded, and let $s_Q = \langle f, \varphi_Q \rangle$ for $Q \in \mathcal{Q}$, then by (1.1), $f = \sum_{Q \in \mathcal{Q}} s_Q \psi_Q$, and $\| s \|_Y \approx \| f \|_X$ by (1.4). This, (1.7) and the boundedness of B show that (1.6) holds. $\square$

If we combine Proposition 1.1A with the simple observation above regarding a linear operator T on X, we see that T is bounded on X if and only if the matrix $A = \{a_{QP}\}_{Q,P \in \mathcal{Q}}$, defined by $a_{QP} = \langle T\psi_P, \varphi_Q \rangle$, is bounded on Y (this is easy to see directly as well). By this observation and Proposition 1.1, we see that to every class of bounded matrices on Y there corresponds a class of bounded operators on X, a class of families of molecules for X, and a class of norming families for X.

Our purpose in this expository article is to develop certain aspects of the theory of function spaces on $\mathbb{R}^n$, using either the φ-transform decomposition (as in [FJ3]) or wavelets. Most of the material discussed is from [FJ3]; it is, however, somewhat reorganized here around the central notion of matrix boundedness. We will discuss several conditions on a matrix $\{a_{QP}\}_{Q,P \in \mathcal{Q}}$ that imply boundedness. As noted above, such a condition yields a corresponding condition for families of molecules, for norming families, and for operator boundedness on function spaces. In some cases the corresponding conditions are explicit and well-understood, as we will describe. There are some interesting open questions for the cases that are not so well understood.

In Section 2 we describe the full class of Littlewood-Paley spaces X and their corresponding sequence spaces Y for which we will be able to obtain (1.1–3); these are the spaces for which the previous discussion is meaningful. We discuss the relevant decompositions in Section 3: the smooth atomic decomposition, the φ-transform decomposition, and the wavelet decomposition. The last two are of the type for which Proposition 1.1 can be applied, for the entire range of function spaces described in Section 2. We begin our discussion of matrix boundedness in Section 4, by describing the few cases for which there is a simple, complete characterization of the algebra of bounded matrices. A general condition, called "almost diagonality," which gives a boundedness condition for each space discussed in Section 2, is presented in Section 5. In that section we also discuss the families of molecules and norming families corresponding to the almost diagonal matrices; they consist of "smooth molecules". For the L^p spaces, $1 < p < +\infty$, the class of almost diagonal matrices corresponds, in the manner described above, to the reduced David-Journé class of bounded, generalized Calderón-Zygmund operators T satisfying $T1 = T^*1 = 0$. This correspondence is presented in Section 6. We apply the smooth atomic decomposition in Section 7 to obtain a general version of the H^p atomic decomposition; this in turn is applied in Section 8 to obtain a matrix boundedness criterion sharper than almost diagonality. Finally, in Section 9 we discuss the full David-Journé class and some questions it raises regarding matrix conditions, families of molecules, and norming families. We will refer to the existing literature for most of the proofs.

2.   THE LITTLEWOOD-PALEY SPACES

Let $P(x,t) = c_n t(t^2 + |x|^2)^{-(n+1)/2}$ be the Poisson kernel for the upper half space $\mathbb{R}^{n+1}_+ = \{(x,t) : x \in \mathbb{R}^n, t > 0\}$. Let $\varphi(x,t) = t\frac{\partial}{\partial t}P(x,t)$; note that $\varphi(x,t) = t^{-n}\varphi(x/t, 1)$. Define the Littlewood-Paley g-function of $f : \mathbb{R}^n \to \mathbb{C}$ by

$$g(f)(x) = \left(\int_0^\infty |\varphi(\cdot, t) * f(x)|^2 dt/t \right)^{1/2}.$$

Then Littlewood-Paley theory gives the following equivalences ([S], [FS2]):

$$\| f \|_{L^p} \approx \| g(f) \|_{L^p}, \quad \text{for } 1 < p < +\infty, \text{ if } f \in L^p = L^p(\mathbb{R}^n),$$

and

$$\| f \|_{H^p} \approx \| g(f) \|_{L^p}, \text{ for } 0 < p \le 1, \text{ if } f \in H^p = H^p(\mathbb{R}^n).$$

Notice that the expression $\| g(f) \|_{L^p}$ is a mixed norm expression: it involves first an L^2 integration in t (vertically), followed by an L^p integration in x (horizontally).

Lipschitz spaces are also characterized by an expression involving $\varphi(\cdot, t) * f$ (see [Ta] or [S]): for $0 < \alpha < 1$,

$$\| f \|_{\Lambda_\alpha} \approx \| f \|_{L^\infty} + \sup_{(x,t) \in \mathbb{R}_+^{n+1}} t^{-\alpha} |\varphi(\cdot, t) * f(x)|.$$

The same equivalence holds for $\alpha \ge 1$ if we let $\varphi(x, t) = t^k \frac{\partial^k}{\partial t^k} P(x, t)$ for some $k > \alpha$; what is essential here is that $\varphi(x, t)$ should satisfy $\int x^\gamma \varphi(x, t) dx = 0$ whenever $x^\gamma \equiv x_1^{\gamma_1} \cdots x_n^{\gamma_n}$ is a monomial with $|\gamma| \equiv \gamma_1 + \ldots + \gamma_n \le \alpha$. With the same restriction on φ, the generalized Lipschitz spaces $\Lambda_\alpha^{p,q}$ (see [Ta] or [S]) are similarly characterized:

$$\| f \|_{\Lambda_\alpha^{p,q}} \approx \| f \|_{L^p} + \left(\int_0^\infty (t^{-\alpha} \| \varphi(\cdot, t) * f \|_{L^p})^q dt/t \right)^{1/q}.$$

Notice that the last expression is a mixed norm with the horizontal L^p integration first, followed by the vertical L^q-integration, i.e. the reverse of the expression $\| g(f) \|_{L^p}$ above. The remarkable aspect of Littlewood-Paley theory is its unifying perspective on function spaces: the seemingly unrelated scales of H^p/L^p spaces and generalized Lipschitz spaces are both characterized by closely related mixed norm expressions.

Peetre and Triebel (see e.g. [P2] and [Tr]) made a systematic study of function spaces from this point of view. For simplicity, they replaced the integrals in t above by discrete expressions and, to avoid changing φ with the index α, selected $\varphi \in \mathcal{S}(\mathbb{R}^n)$, all of whose moments vanish. Let $\mathcal{A}$ be the class of all $\varphi \in \mathcal{S}(\mathbb{R}^n)$ such that supp $\hat{\varphi} \subseteq \{\xi : 1/2 \le |\xi| \le 2\}$ and $|\hat{\varphi}(\xi)| \ge c > 0$ if $3/5 \le |\xi| \le 5/3$ ($\hat{\varphi}$ is the Fourier transform of φ). Let $\mathcal{B}$ be the class of $\Phi \in \mathcal{S}(\mathbb{R}^n)$ such that supp $\hat{\Phi} \subseteq \{\xi : |\xi| \le 2\}$ and $|\hat{\Phi}(\xi)| \ge c > 0$ if $|\xi| \le 5/3$. For $v \in \mathbb{Z}$, let $\varphi_v(x) = 2^{vn} \varphi(2^v x)$. For $\alpha \in \mathbb{R}$, $0 < p, q \le +\infty$, $\varphi \in \mathcal{A}$, $\Phi \in \mathcal{B}$, and $f \in \mathcal{S}'(\mathbb{R}^n)$, let

$$\| f \|_{\dot{B}_p^{\alpha q}} = \left(\sum_{v \in \mathbb{Z}} (2^{v\alpha} \| \varphi_v * f \|_{L^p})^q \right)^{1/q},$$

$$\| f \|_{B_p^{\alpha q}} = \| \Phi * f \|_{L^p} + \left(\sum_{v=1}^\infty (2^{v\alpha} \| \varphi_v * f \|_{L^p})^q \right)^{1/q},$$

and if $p < +\infty$, let

$$\| f \|_{\dot{F}_p^{\alpha q}} = \| \left(\sum_{v \in \mathbb{Z}} (2^{v\alpha} |\varphi_v * f|)^q \right)^{1/q} \|_{L^p},$$

and

$$\| f \|_{F_p^{\alpha q}} = \| \Phi * f \|_{L^p} + \| \left(\sum_{v=1}^{\infty} (2^{v\alpha} |\varphi_v * f|)^q \right)^{1/q} \|_{L^p}.$$

The spaces of $f \in \mathcal{S}'$ such that these quasi-norms are finite are known as homogeneous and inhomogeneous Besov spaces ($\dot{B}_p^{\alpha q}$ and $B_p^{\alpha q}$, respectively), and homogeneous and inhomogeneous Triebel-Lizorkin spaces ($\dot{F}_p^{\alpha q}$ and $F_p^{\alpha q}$, respectively). Up to equivalence of quasi-norms, these spaces are independent of the choices of $\varphi \in \mathcal{A}$ and $\Phi \in \mathcal{B}$.

Selecting $\varphi \in \mathcal{S}$ with supp $\hat{\varphi} \subseteq \{\xi : \frac{1}{2} \leq |\xi| \leq 2\}$ guarantees that all moments of φ vanish. More importantly, since $\hat{\varphi}_v(\xi) = \hat{\varphi}(2^{-v}\xi)$, $\varphi_v * f$ carries the frequency content of f for $|\xi| \approx 2^v$. Thus the expressions above are natural for Fourier analysis. Notice that the $\dot{B}$ and $\dot{F}$-spaces are defined via discrete mixed norm expressions on $\mathbb{R}^n \times \mathbb{Z}$ similar to the continuous expressions on $\mathbb{R}^n \times (0, \infty)$ described earlier, where we think of $v \in \mathbb{Z}$ as corresponding to $t = 2^{-v}$. Thus the B and F spaces constitute, in a natural, albeit notationally clumsy way, the full range of distribution spaces of $\mathbb{R}^n$ characterized by mixed norm Littlewood-Paley expressions.

Adapting Littlewood-Paley theory to this discrete context yields characterizations of most of the classical spaces in analysis: Lebesgue spaces L^p, $1 < p < +\infty$, Hardy spaces H^p, $0 < p \leq 1$, Lipschitz and generalized Lipschitz spaces (Λ_α and $\Lambda_\alpha^{p,q}$, respectively), Bessel potential spaces L_α^p (including the Sobolev spaces L_k^p), and the Riesz potential spaces $\dot{L}_\alpha^p$. Specifically, we have

$$L^p \approx \dot{F}_p^{02} \approx F_p^{02}, \qquad 1 < p < +\infty$$

$$H^p \approx \dot{F}_p^{02}, \qquad 0 < p \leq 1$$

$$L_\alpha^p \approx F_p^{\alpha 2}, \qquad \alpha > 0, \ 1 < p < +\infty$$

$$\dot{L}_\alpha^p \approx \dot{F}_p^{\alpha 2}, \qquad \alpha > 0, \ 1 < p < +\infty$$

$$\Lambda_\alpha \approx B_\infty^{\alpha \infty}, \qquad \alpha > 0$$

and

$$\Lambda_\alpha^{p,q} \approx B_p^{\alpha q}, \qquad \alpha > 0, \ 1 \leq p, q \leq +\infty.$$

(See e.g. [Tr] for these equivalences and more background.) We can also include the space BMO in this list. By definition,

$$\| f \|_{BMO} = \sup_Q \frac{1}{|Q|} \int_Q |f - f_Q|, \text{ for } f_Q = \frac{1}{|Q|} \int_Q f,$$

where the sup is taken over all cubes Q in $\mathbb{R}^n$ with sides parallel to the axes. For $\alpha \in \mathbb{R}$ and $0 < q \le +\infty$, let

$$\| f \|_{\dot{F}_\infty^{\alpha q}} = \sup_{Q \in \mathcal{Q}} \left[\frac{1}{|Q|} \int_Q \sum_{v=-\log_2 \ell(Q)}^{\infty} (2^{v\alpha} |\varphi_v * f|)^q \right]^{1/q},$$

where $\ell(Q)$ is the side length of Q, and, as above, $\mathcal{Q}$ is the collection of all dyadic cubes in $\mathbb{R}^n$. The definition was introduced in [FJ3], where we showed that $\dot{F}_\infty^{\alpha q}$ is the natural extension of the $\dot{F}$ spaces for $p < +\infty$. The definition is analogous to the Carleson measure Littlewood-Paley characterization of BMO in [FS2], and in fact we have

$$BMO \approx \dot{F}_\infty^{02}.$$

The sequence spaces that correspond to the B and F spaces via the decompositions to be described in Section 3 are discrete Littlewood-Paley spaces. For $Q \in \mathcal{Q}$, let $\tilde{\chi}_Q$ be the L^2 normalized characteristic function of $Q : \tilde{\chi}_Q = |Q|^{-1/2}\chi_Q$. For $\alpha \in \mathbb{R}$, $0 < q \le +\infty$, and a sequence $s = \{s_Q\}_{Q \in \mathcal{Q}}$, let

$$g^{\alpha q}(s)(x) = (\sum_{Q \in \mathcal{Q}} (|Q|^{-\alpha/n} |s_Q| \tilde{\chi}_Q(x))^q)^{1/q}.$$

For $v \in \mathbb{Z}$, let $\mathcal{Q}_v = \{Q \in \mathcal{Q} : \ell(Q) = 2^{-v}\}$. Notice that for each $v \in \mathbb{Z}$, there is only one $Q \in \mathcal{Q}_v$ with $\tilde{\chi}_Q(x) \ne 0$ (for a.e. x), so we could write

$$g^{\alpha q}(s) = (\sum_{v \in \mathbb{Z}} (\sum_{Q \in \mathcal{Q}_v} |Q|^{-\alpha/n} |s_Q| \tilde{\chi}_Q)^q)^{1/q}.$$

If we define f on $\mathbb{R}^n \times \{2^{-v} : v \in \mathbb{Z}\} \subseteq \mathbb{R}_+^{n+1}$ by $f(x, 2^{-v}) = \sum_{Q \in \mathcal{Q}_v} s_Q \tilde{\chi}_Q$, we see that $g^{\alpha q}(s)$ is a mixed norm expression of Littlewood-Paley type associated with the discrete function $f(x, 2^{-v})$. For $\alpha \in \mathbb{R}$, $0 < p < +\infty$, and $0 < q \le +\infty$, let

$$\| s \|_{\dot{f}_p^{\alpha q}} = \| g^{\alpha q}(s) \|_{L^p} .$$

As in the case of $\dot{F}_p^{\alpha q}$ and $\dot{B}_p^{\alpha q}$, we reverse the order of the mixed norm to define $\dot{b}_p^{\alpha q}$: for $\alpha \in \mathbb{R}$, $0 < p, q \leq +\infty$, and $s = \{s_Q\}_{Q \in \mathcal{Q}}$, let

$$\| s \|_{\dot{b}_p^{\alpha q}} = (\sum_{v \in \mathbb{Z}} \| \sum_{Q \in \mathcal{Q}_v} |Q|^{-\alpha/n} |s_Q| \tilde{\chi}_Q \|_{L^p}^q)^{1/q}$$

$$= (\sum_{v \in \mathbb{Z}} (\sum_{Q \in \mathcal{Q}_v} (|Q|^{-\alpha/n - 1/2 + 1/p} |s_Q|)^p)^{q/p})^{1/q},$$

by a trivial computation.

We define $\dot{f}_\infty^{\alpha q}$ for $\alpha \in \mathbb{R}$, $0 < q \leq +\infty$, by

$$\| s \|_{\dot{f}_\infty^{\alpha q}} = \sup_{P \in \mathcal{Q}} (\frac{1}{|P|} \int_P \sum_{Q \subseteq P, Q \in \mathcal{Q}} (|Q|^{-\alpha/n} |s_Q| \tilde{\chi}_Q)^q)^{1/q}$$

$$= \sup_{P \in \mathcal{Q}} (\frac{1}{|P|} \sum_{Q \subseteq P, Q \in \mathcal{Q}} (|Q|^{-\alpha/n - 1/2} |s_Q|)^q |Q|)^{1/q}.$$

There are also inhomogeneous versions, for which we consider sequences $s = \{s_Q\}_{Q \in \mathcal{Q}'}$, where $\mathcal{Q}' = \{Q \in \mathcal{Q} : \ell(Q) \leq 1\}$. Then $\| s \|_{b_p^{\alpha q}}$ and $\| s \|_{f_p^{\alpha q}}$ are defined exactly as above except that the sum on Q runs only over $\mathcal{Q}'$.

In the remainder of this exposition we will, for simplicity, only state our results for $\dot{F}_p^{\alpha q}$ and $\dot{f}_p^{\alpha q}$. In every case there are similar results for $\dot{B}_p^{\alpha q}$, $F_p^{\alpha q}$, $B_p^{\alpha q}$, and their sequence space analogues. For details on these, see [FJ1] and [FJ3]. The reader who is put off by the three-index notation should simply understand that each result, when properly interpreted, holds for his or her favorite traditional space, e.g. $L^p, H^p, \Lambda_\alpha, L_\alpha^p$, or BMO.

3. DECOMPOSITIONS OF FUNCTION SPACES

We will discuss three decompositions that characterize the function spaces discussed in Section 2. The first is the smooth atomic decomposition, derived from Calderón's formula. The other two are the φ-transform identity and the wavelet representation.

There are many variants of the Calderón formula. For one involving the Poisson kernel, see § 0 in [FJW]. Uchiyama [U] used a C^∞ function of compact support (cf. [Cal2] and [Ch-F]). The version we discuss here is adapted to the discrete notation introduced in Section 2.

It is not difficult to see (see e.g. [FJ1]) that there exist functions $\varphi, \theta \in \mathcal{S}(\mathbb{R}^n)$ such that φ belongs to the class $\mathcal{A}$ defined in Section 2, supp $\theta \subseteq \{x : |x| < 1\}$,

$$\sum_{v \in \mathbb{Z}} \hat{\varphi}(2^{-v}\xi) \hat{\theta}(2^{-v}\xi) = 1 \text{ for } \xi \in \mathbb{R}^n \setminus \{0\}, \tag{3.1}$$

and $\int x^\gamma \theta(x)dx = 0$ for all $|\gamma| \leq N$, where N is any fixed positive integer. Formally, (3.1) yields the identity, for $f \in \mathcal{S}'(\mathbb{R}^n)$ (modulo polynomials):

$$f = \sum_{v \in \mathbb{Z}} \varphi_v * \theta_v * f, \qquad (3.2)$$

where $\theta_v(x) = 2^{vn}\theta(2^v x)$, and similarly for φ_v. Recalling that $\mathcal{Q}_v = \{Q \in \mathcal{Q} : \ell(Q) = 2^{-v}\}$, (3.2) gives

$$f = \sum_{v \in \mathbb{Z}} \sum_{Q \in \mathcal{Q}_v} \int_Q \theta_v(x - y)\varphi_v * f(y)dy.$$

For $Q \in \mathcal{Q}_v$, let $s_Q = |Q|^{1/2} \sup_{y \in Q} |\varphi_v * f(y)|$, and

$$a_Q(x) = \frac{1}{s_Q} \int_Q \theta_v(x - y)\varphi_v * f(y)dy.$$

Thus we obtain $f = \sum_{Q \in \mathcal{Q}} s_Q a_Q$. It is not difficult to check that each a_Q is a smooth N-atom for Q, by which we mean that supp $a_Q \subseteq 3Q$ (the cube concentric with Q having three times the side length), $\int x^\gamma a_Q(x)dx = 0$, if $|\gamma| \leq N$, and $|D^\gamma a_Q(x)| \leq c_\gamma \ell(Q)^{-|\gamma|-n/2}$ for all multi-indices γ. It is a simple consequence (see [FJ2]) of Peetre's maximal function characterization of the $\dot{F}_p^{\alpha q}$ spaces ([P1]) that the sequence $s = \{s_Q\}_{Q \in \mathcal{Q}}$ satisfies

$$\| s \|_{\dot{f}_p^{\alpha q}} \leq c \| f \|_{\dot{F}_p^{\alpha q}}. \qquad (3.3)$$

Moreover, there is a converse. For N sufficiently large ($N \geq$ the greatest integer in $n/\min(1, p, q) - n - \alpha$), if a_Q is a smooth N-atom for Q for each $Q \in \mathcal{Q}$, and $s = \{s_Q\}_{Q \in \mathcal{Q}}$, then

$$\| \sum_{Q \in \mathcal{Q}} s_Q a_Q \|_{\dot{F}_p^{\alpha q}} \leq c \| s \|_{\dot{f}_p^{\alpha q}}. \qquad (3.4)$$

Thus we have

$$\|f\|_{\dot{F}_p^{\alpha q}} \approx \inf\{\|s\|_{\dot{f}_p^{\alpha q}} : f = \sum_{Q \in \mathcal{Q}} s_Q a_Q, \text{ ea. } a_Q \text{ a smooth } N\text{-atom for } Q\}.$$

$$(3.5)$$

The analogue of (3.4) for the Besov spaces can be proved by elementary, direct computation (see [FJ1]); for the $\dot{F}$ spaces the vector-valued maximal inequality of Fefferman-Stein ([FS1]) is used. We refer to [FJ1-3] for these results. Note that (3.4) gives an example of (1.5): any collection of smooth N-atoms is a family of molecules for $X = \dot{F}_p^{\alpha q}$, where $Y = \dot{f}_p^{\alpha q}$.

The φ-transform identity gives a decomposition $f = \sum_Q s_Q \psi_Q$ with a family $\{\psi_Q\}_{Q \in \mathcal{Q}}$ fixed independently of f. For $\varphi, \psi \in \mathcal{A}$, and $v \in \mathbb{Z}$, let $\varphi_v(x) = 2^{vn}\varphi(2^v x)$ and $\psi_v(x) = 2^{vn}\psi(2^v x)$, as before. Also set $\tilde{\varphi}_v(x) =$

$\overline{\varphi}_v(-x)$; this is equivalent to $\hat{\tilde{\varphi}}_v = \overline{\hat{\varphi}}_v$. For $\varphi \in \mathcal{A}$ given, it is not difficult to construct $\psi \in \mathcal{A}$ such that

$$\sum_{v \in \mathbb{Z}} \hat{\tilde{\varphi}}_v(\xi)\hat{\psi}_v(\xi) = 1 \text{ for } \xi \in \mathbb{R}^n \setminus \{0\}.$$

This gives the identity

$$f = \sum_{v \in \mathbb{Z}} \tilde{\varphi}_v * \psi_v * f, \tag{3.6}$$

similarly to (3.2). Since $\tilde{\varphi}_v$ and ψ_v are of exponential type (in fact supp $\hat{\tilde{\varphi}}_v, \hat{\psi}_v \subset \{\xi : 2^{v-1} \leq |\xi| \leq 2^{v+1}\}$), we can use a technique similar to one used in the proof of the Shannon sampling theorem (see the discussion in [FJW]): expand $(\tilde{\varphi}_v * f)^\wedge$ in a Fourier series on $D_v = [-2^v\pi, 2^v\pi]^n$. This gives

$$(\tilde{\varphi}_v * f)^\wedge(\xi) = \sum_{k \in \mathbb{Z}^n} 2^{-vn}(2\pi)^{-n} \int (\tilde{\varphi}_v * f)^\wedge(y)e^{i2^{-v}k \cdot y}dy \, e^{-i2^{-v}k \cdot \xi},$$

where the integral is over D_v or equivalently $\mathbb{R}^n$, since supp $\hat{\tilde{\varphi}}_v \subseteq D_v$. By Fourier inversion then, for $\xi \in D_v$,

$$(\tilde{\varphi}_v * f)^\wedge(\xi) = \sum_{v \in \mathbb{Z}} 2^{-vn}(\tilde{\varphi}_v * f)(2^{-v}k)e^{-i2^{-v}k \cdot \xi}.$$

Since supp $\hat{\psi}_v \subseteq D_v$, we can substitute this into $\tilde{\varphi}_v * \psi_v * f = (\hat{\psi}_v(\tilde{\varphi}_v * f)^\wedge)^\vee$, and use $(g(\cdot - t))^\wedge(\xi) = e^{-it \cdot \xi}\hat{g}(\xi)$, to obtain

$$\tilde{\varphi}_v * \psi_v * f(x) = \sum_{k \in \mathbb{Z}^n} 2^{-vn/2}\tilde{\varphi}_v * f(2^{-v}k)2^{-vn/2}\psi_v(x - 2^{-v}k). \tag{3.7}$$

For $Q = Q_{vk} = \{x \in \mathbb{R}^n : 2^{-v}k_i \leq x_i \leq 2^{-v}(k_i + 1), \ i = 1, \ldots, n\} \in \mathcal{Q}$, let $\varphi_Q(x) = 2^{-vn/2}\varphi_v(x - 2^{-v}k)$, and similarly for ψ_Q. Then (3.6-7) and a trivial computation give

$$f = \sum_{Q \in \mathcal{Q}} \langle f, \varphi_Q \rangle \psi_Q. \tag{3.8}$$

Letting $s_Q = \langle f, \varphi_Q \rangle = |Q|^{1/2}\tilde{\varphi}_v(x - 2^{-v}k)$ if $Q = Q_{vk}$, the same considerations leading to (3.3) give

$$\| \{\langle f, \varphi_Q \rangle\}_{Q \in \mathcal{Q}} \|_{\dot{f}_p^{\alpha q}} \leq c \| f \|_{\dot{F}_p^{\alpha q}}. \tag{3.9}$$

Thus, in the terminology of Section 1, the family $\{\varphi_Q\}_{Q \in \mathcal{Q}}$ is a norming family for $\dot{F}_p^{\alpha q}$.

The converse estimate also holds: for any sequence $s = \{s_Q\}_{Q \in \mathcal{Q}}$,

$$\| \sum_{Q \in \mathcal{Q}} s_Q \psi_Q \|_{\dot{F}_p^{\alpha q}} \leq c \| s \|_{\dot{f}_p^{\alpha q}}. \tag{3.10}$$

Thus the conditions of Proposition 1.1, including (1.1-4), are satisfied if $X = \dot{F}_p^{\alpha q}$ and $Y = \dot{f}_p^{\alpha q}$.

Lemarié and Meyer constructed a function $\psi \in \mathcal{S}(\mathbb{R}^1)$ satisfying supp $\hat{\psi} \subseteq [-8\pi/3, -2\pi/3] \cup [2\pi/3, 8\pi/3]$ such that the family $\{\psi_Q\}_{Q \in \mathcal{Q}}$ forms an orthonormal basis for $L^2(\mathbb{R}^1)$, where as above $\psi_Q(x) = 2^{-vn/2}\psi_v(x - 2^{-v}k)$ for $Q = Q_{vk} \in \mathcal{Q}$. This gives (3.8) with $\varphi = \psi$ (it is possible to take $\varphi = \psi$ also in the case of the φ-transform, but we do not obtain orthonormality). In $\mathbb{R}^n$ there are a $2^n - 1$ functions $\psi^{(i)}$ such that $\{\psi_Q^{(i)} : Q \in \mathcal{Q}, i \in \{1, \ldots, n\}\}$ forms an orthonormal basis for $L^2(\mathbb{R}^n)$. An orthonormal basis consisting of such translates and dilates of $2^n - 1$ functions in $\mathbb{R}^n$ is called an orthonormal wavelet basis. We will not discuss their construction here; see e.g. [LM], [Me3], [Dau1], [Ma], or Section 7 of [FJW].

The wavelet decomposition characterizes the Littlewood-Paley spaces just as the φ-transform identity does. For the traditional spaces, e.g. L^p, H^p, BMO, etc., this is discussed in [Me3]; for the $\dot{B}$ and $\dot{F}$ spaces in general, see [FJW]. In particular, with a few notational modifications to take into account the $2^n - 1$ functions $\psi^{(i)}$ in $\mathbb{R}^n$, (3.8-10) hold for the Lemarié-Meyer wavelet transform. Thus Proposition 1.1 can be applied to this transform as well as to the φ-transform.

We conclude this section by discussing the duality (noted in Section 1) between families of molecules and norming families. For $s = \{s_Q\}_{Q \in \mathcal{Q}}$ and $t = \{t_Q\}_{Q \in \mathcal{Q}}$, let $\langle s, t \rangle = \sum_{Q \in \mathcal{Q}} s_Q \bar{t}_Q$. Suppose Y_0 and Y_1 are two sequences spaces which norm each other; i.e. $\| s \|_{Y_0} \approx \sup \{|\langle s, t \rangle| : \| t \|_{Y_1} \leq 1\}$, and vice versa. (This happens frequently even in the non-reflexive case; e.g., ℓ^1 and ℓ^∞.) In Proposition 1.1, note that for the $\dot{f}$ and $\dot{b}$ spaces, $B = \{\langle \psi_P, m_Q \rangle\}_{Q, P \in \mathcal{Q}}$ is bounded if and only if $A^* = \{\langle \varphi_P, m_Q \rangle\}_{Q, P \in \mathcal{Q}}$ is bounded. (This follows from Proposition 1.1, since (1.5-6) are independent of the choice of φ and ψ, and since the conditions on φ and ψ required to obtain (1.1-4) are satisfied if φ and ψ are interchanged.) Suppose (1.1-4) hold for each pair (X_0, Y_0) and (X_1, Y_1) of $\dot{F}$ and $\dot{f}$ spaces, and suppose Y_0 and Y_1 norm each other. Then, in the terminology of Proposition 1.1, $\{m_Q\}_{Q \in \mathcal{Q}}$ is a family of molecules for $X_0 \Leftrightarrow A$ is bounded on $Y_0 \Leftrightarrow A^*$ is bounded on $Y_1 \Leftrightarrow B$ is bounded on $Y_1 \Leftrightarrow \{m_Q\}_Q$ is a norming family for X_1. These assumptions are satisfied for the pairs $(\dot{F}_p^{\alpha q}, \dot{f}_p^{\alpha q})$ and $(\dot{F}_{p'}^{-\alpha q'}, \dot{f}_{p'}^{-\alpha q'})$ if $\alpha \in \mathbb{R}$, $1 \leq p, q \leq +\infty$, $p' = p/(p-1)$, and $q' = q/(q-1)$ (see [FJ3]); this is the reflexive case if $1 < p, q < +\infty$.

4.   MATRIX BOUNDEDNESS: THE SIMPLE CASES

We begin our discussion of bounded matrices on $\dot{f}_p^{\alpha q}$ with the few special cases for which a complete, explicit characterization is known. These cases are $\dot{f}_p^{\alpha p}$, $\alpha \in \mathbb{R}$, $0 < p \le 1$, or $p = +\infty$.

Lemma 4.1.   *Let* $A = \{a_{QP}\}_{Q,P \in \mathcal{Q}}$ *act on sequences* $s = \{s_Q\}_{Q \in \mathcal{Q}}$ *by* $As = \{(As)_Q\}_{Q \in \mathcal{Q}}$ *where* $(As)_Q = \sum_{P \in \mathcal{Q}} a_{QP} s_P$. *Then* A *is bounded on* $\dot{f}_p^{\alpha p}$, $p \in (0, 1]$, *if and only if*

$$\sup_{P \in \mathcal{Q}} |P|^{(\alpha/n + 1/2 - 1/p)p} \sum_{Q \in \mathcal{Q}} (|Q|^{-\alpha/n - 1/2 + 1/p} |a_{QP}|)^p < +\infty. \qquad (4.1)$$

A is bounded on $\dot{f}_\infty^{\alpha \infty}$, $\alpha \in \mathbb{R}$, *if and only if*

$$\sup_{Q \in \mathcal{Q}} |Q|^{-\alpha/n - 1/2} \sum_{P \in \mathcal{Q}} |a_{QP}| |P|^{\alpha/n + 1/2} < +\infty. \qquad (4.2)$$

Proof.   For $0 < p \le 1$, we have by definition

$$\| s \|_{\dot{f}_p^{\alpha p}}^p = \| s \|_{\dot{b}_p^{\alpha p}}^p = \sum_{Q \in \mathcal{Q}} (|Q|^{-\alpha/n - 1/2 + 1/p} |s_Q|)^p.$$

Hence, by the p-triangle inequality $|a + b|^p \le |a|^p + |b|^p$,

$$\| s + t \|_{\dot{f}_p^{\alpha p}}^p \le \| s \|_{\dot{f}_p^{\alpha p}}^p + \| t \|_{\dot{f}_p^{\alpha p}}^p. \qquad (4.3)$$

For $P \in \mathcal{Q}$, define a sequence e^P by $(e^P)_Q = 1$ if $Q = P$ and $(e^P)_Q = 0$ otherwise. Then $s = \sum_P s_P e^P$, and (4.3) shows that A is bounded if and only if

$$\| A e^P \|_{\dot{f}_p^{\alpha p}}^p \le c \| e^P \|_{\dot{f}_p^{\alpha p}}^p = c |P|^{(-\alpha/n - 1/2 + 1/p)p}.$$

Since $(A e^P)_Q = a_{QP}$, this last condition is equivalent to (4.1).

For $p = +\infty$, note that $\| s \|_{\dot{f}_\infty^{\alpha \infty}} = \sup_{Q \in \mathcal{Q}} |Q|^{-\alpha/n - 1/2} |s_Q|$. From this, (4.2) implies the boundedness of A easily. To prove the converse, let $s_Q = |Q|^{\alpha/n + 1/2} \operatorname{sgn}(a_{QP})$, for $P \in \mathcal{Q}$ fixed. Then $\| s \|_{\dot{f}_\infty^{\alpha \infty}} = 1$, and (4.2) is equivalent to $\| As \|_{\dot{f}_\infty^{\alpha \infty}} \le c$. $\square$

Proposition 1.1A can now be used to characterize families of molecules for $\dot{F}_p^{\alpha p}$, $0 < p \le 1$. In this case, however, there is a trivial direct characterization. From the definition, we have

$$\| f + g \|_{\dot{F}_p^{\alpha p}}^p \le \| f \|_{\dot{F}_p^{\alpha p}}^p + \| g \|_{\dot{F}_p^{\alpha p}}^p, \quad \text{if } 0 < p \le 1,$$

similarly to (4.3). It follows that $\{m_P\}_{P\in\mathcal{Q}}$ is a family of molecules for $\dot{F}_p^{\alpha p}$, $0 < p \leq 1$, if and only if $m_P \in \dot{F}_p^{\alpha p}$ for all $P \in \mathcal{Q}$, and

$$\| m_P \|_{\dot{F}_p^{\alpha p}}^p = \| \sum_{Q\in\mathcal{Q}} (e^P)_Q m_P \|_{\dot{f}_p^{\alpha p}}^p \leq c \| e^P \|_{\dot{f}_p^{\alpha p}}^p = c|P|^{(-\alpha/n-1/2+1/p)p}.$$

$$(4.4)$$

For curiosity's sake, we note that this is consistent with Proposition 1.1 and Lemma 4.1. These give that $\{m_P\}_{P\in\mathcal{Q}}$ is a family of molecules for $\dot{F}_p^{\alpha p}$, $0 < p \leq 1$, if and only if

$$\sum_{Q\in\mathcal{Q}} (|Q|^{-\alpha/n-1/2+1/p}|\langle m_P, \varphi_Q\rangle|)^p \leq c|P|^{(-\alpha/n-1/2+1/p)p}$$

for all $P \in \mathcal{Q}$. By (3.8-10), we have $m_P = \sum_{Q\in\mathcal{Q}}\langle m_P, \varphi_Q\rangle\psi_Q$ with

$$\| m_P \|_{\dot{F}_p^{\alpha p}}^p \approx \| \{\langle m_P, \varphi_Q\rangle\}_{Q\in\mathcal{Q}} \|_{\dot{f}_p^{\alpha p}}^p = \sum_{Q\in\mathcal{Q}} (|Q|^{-\alpha/n-1/2+1/p}|\langle m_Q, \varphi_P\rangle|)^p.$$

Hence we recover (4.4).

On the other hand, the characterization of norming families for $\dot{F}_p^{\alpha p}$, $0 < p \leq 1$, is not so explicit. By Proposition 1.1B and Lemma 4.1, $\{m_Q\}_{Q\in\mathcal{Q}}$ is a norming family for $\dot{F}_p^{\alpha p}$, $0 < p \leq 1$, if and only if

$$\sum_{Q\in\mathcal{Q}} (|\langle m_Q, \varphi_P\rangle||Q|^{(-\alpha/n-1/2+1/p)})^p \leq c|P|^{(-\alpha/n-1/2+1/p)p}$$

for all $P \in \mathcal{Q}$. Even if $\alpha = 0$ and $p = 1$, when this reduces to

$$\sum_{Q\in\mathcal{Q}} |\langle m_Q, \varphi_P\rangle||Q|^{1/2} \leq c|P|^{1/2}, \quad \text{for all } P \in \mathcal{Q}, \qquad (4.5)$$

we do not know a more explicit characterization of this class of families $\{m_Q\}_{Q\in\mathcal{Q}}$.

As we expect from the remark at the end of Section 3, the case of $\dot{F}_\infty^{\alpha\infty}$ is dual to that of $\dot{F}_1^{-\alpha 1}$. By Proposition 1.1 and Lemma 4.1, $\{m_Q\}_{Q\in\mathcal{Q}}$ is a family of molecules for $\dot{F}_\infty^{\alpha\infty}$ if and only if

$$\sum_{Q\in\mathcal{Q}} |\langle m_Q, \varphi_P\rangle||Q|^{\alpha/n+1/2} \leq c|P|^{\alpha/n+1/2},$$

and $\{m_\varphi\}_{Q\in\mathcal{Q}}$ is a norming family for $\dot{F}_\infty^{\alpha\infty}$ if and only if

$$\sum_{P\in\mathcal{Q}} |\langle m_Q, \varphi_P\rangle||P|^{\alpha/n+1/2} \leq c|Q|^{\alpha/n+1/2}. \qquad (4.6)$$

(We can express our conditions in terms of φ rather than ψ because of their interchangeability, as noted above.) As we should expect, the characterization (4.6) is trivial by other means: by (3.8-10), $m_Q = \sum_P\langle m_Q, \varphi_P\rangle\psi_P$

with

$$\| m_Q \|_{\dot{F}_1^{-\alpha 1}} \approx \| \{ \langle m_Q, \varphi_P \rangle \}_{P \in \mathcal{Q}} \|_{\dot{f}_1^{-\alpha 1}} = \sum_{P \in \mathcal{Q}} |P|^{\alpha/n+1/2} |\langle m_Q, \varphi_P \rangle|,$$

so the fact that $\| m_Q \|_{\dot{F}_1^{-\alpha 1}} \approx \sup\{ |\langle f, m_Q \rangle| : \| f \|_{\dot{F}_\infty^{-\alpha \infty}} \le 1 \}$ (i.e. that $\dot{F}_\infty^{\alpha \infty}$ norms $\dot{F}_1^{-\alpha 1}$) gives (4.6).

As discussed in Section 1, Proposition 4.1 also gives a characterization of the bounded linear operators on $\dot{F}_p^{\alpha p}$, $0 < p \le 1$ or $p = +\infty$. For $0 < p \le 1$, the condition for T to be bounded on $\dot{F}_p^{\alpha p}$ is that $\{T\psi_Q\}_{Q \in \mathcal{Q}}$ must be a family of molecules for $\dot{F}_p^{\alpha p}$. As we have just seen, this merely requires that T map ψ_Q to $\dot{F}_p^{\alpha p}$ boundedly:

$$\| T\psi_Q \|_{\dot{F}_p^{\alpha p}} \le c|Q|^{-\alpha/n-1/2+1/p} \approx \| \psi_Q \|_{\dot{F}_p^{\alpha p}}.$$

The boundedness of T on $\dot{F}_\infty^{\alpha \infty}$ is equivalent to

$$\sum_{P \in \mathcal{Q}} |\langle T\psi_P, \varphi_Q \rangle| |P|^{\alpha/n+1/2} \le c|Q|^{\alpha/n+1/2}, \tag{4.7}$$

which does not have an immediate, direct interpretation. If T is a translation invariant operator, i.e.

$$Tf(x) = \int K(x-y)f(y)dy,$$

then (4.7) is equivalent to the condition $K \in \dot{B}_1^{0\infty}$; this is Taibleson's criterion ([Ta]) for boundedness of Fourier multipliers on $\dot{\Lambda}_\alpha$.

5. ALMOST DIAGONAL MATRICES AND SMOOTH MOLECULES

We say that a matrix $A = \{a_{QP}\}_{Q,P \in \mathcal{Q}}$ is almost diagonal for $\dot{f}_p^{\alpha q}$ if there exists $\varepsilon > 0$ such that

$$\sup_{Q,P \in \mathcal{Q}} |a_{QP}|/\omega_{QP}(\varepsilon) < +\infty,$$

where

$$\omega_{QP}(\varepsilon) = \left(\frac{\ell(Q)}{\ell(P)} \right)^\alpha \left(1 + \frac{|x_Q - x_P|}{\max(\ell(P), \ell(Q))} \right)^{-J-\varepsilon}$$
$$\times \min \left\{ \left(\frac{\ell(Q)}{\ell(P)} \right)^{(n+\varepsilon)/2}, \left(\frac{\ell(P)}{\ell(Q)} \right)^{(n+\varepsilon)/2+J-n} \right\},$$

for $J = n/\min(1,p,q)$. Note that $\omega_{QP}(\varepsilon)$ depends implicitly on α, p, and q. The almost diagonality condition requires $|a_{QP}|$ to decay at a specific rate away from the diagonal, i.e. both as $\ell(Q)/\ell(P)$ goes to 0 or ∞, or as $|x_Q - x_P|$ goes to ∞.

Theorem 5.1. *An almost diagonal matrix for $\dot{f}_p^{\alpha q}$ is bounded on $\dot{f}_p^{\alpha q}$.*

The proof can be reduced to the case $\alpha = 0$, $1 < p, q < +\infty$, where the almost diagonality condition is simpler and more symmetric:

$$|a_{QP}| \le c \left(1 + \frac{|x_Q - x_P|}{\max(\ell(P), \ell(Q))}\right)^{-n-\varepsilon} \min\left(\frac{\ell(P)}{\ell(Q)}, \frac{\ell(Q)}{\ell(P)}\right)^{(n+\varepsilon)/2}. \quad (5.1)$$

For this reduction, we first note that we can take $\alpha = 0$ since it follows easily from the definitions that $\{a_{QP}\}_{Q,P \in \mathcal{Q}}$ is bounded on $\dot{f}_p^{\alpha q}$ if and only if $\{(|P|/|Q|)^{\alpha/n} a_{QP}\}_{Q,P \in \mathcal{Q}}$ is bounded on $\dot{f}_p^{0q}$. The next step of the reduction depends on the following lemma, whose proof is reminiscent of the classical technique of reducing to the case $p > 1$ via Blaschke products in the H^p theory of one variable.

Lemma 5.2. *For $A = \{a_{QP}\}_{Q,P \in \mathcal{Q}}$ and $0 < t \le 1$, let $\tilde{A} = \{\tilde{a}_{QP}\}_{Q,P \in \mathcal{Q}}$ be defined by $\tilde{a}_{QP} = |a_{QP}|^t (|Q|/|P|)^{1/2-t/2}$. If $\tilde{A}$ is bounded on $\dot{f}_{p/t}^{0q/t}$, then A is bounded on $\dot{f}_p^{0q}$.*

Proof. For $s = \{s_Q\}_{Q \in \mathcal{Q}} \in \dot{f}_p^{0q}$, define $r = \{r_Q\}_{Q \in \mathcal{Q}}$ by $r_Q = |Q|^{1/2-t/2} |s_Q|^t$. The definitions show that

$$\| s \|_{\dot{f}_p^{0q}} = \| r \|_{\dot{f}_{p/t}^{0q/t}}^{1/t}. \quad (5.2)$$

The t-triangle inequality gives

$$|(As)_Q|^t \le \sum_{P \in \mathcal{Q}} |a_{QP}|^t |s_P|^t = |Q|^{t/2-1/2} (\tilde{A}r)_Q.$$

Hence, similarly to (5.2), $\| As \|_{\dot{f}_p^{0q}} \le \| \tilde{A}r \|_{\dot{f}_{p/t}^{0q/t}}^{1/t}$, which with (5.2), yields the desired conclusion. $\square$

Now suppose A is almost diagonal for $\dot{f}_p^{0q}$ with $\min(p, q) \le 1$. Then there exists $t < \min(p, q)$, sufficiently close to $\min(p, q)$ that the assumed almost diagonality estimate for A holds with $J = n/\min(p, q)$ replaced wth n/t (for some smaller ε). A calculation shows that $\tilde{A}$, as in Lemma 5.2, satisfies the almost diagonality estimate (5.1) for $\dot{f}_{p/t}^{0q/t}$, since $p/t, q/t > 1$. Thus Theorem 5.1 for $\alpha = 0$ and $p, q > 1$ implies the general case.

The case $\alpha = 0$, $1 < p, q < +\infty$ is done by a simple maximal function estimate and an application of the Fefferman-Stein vector valued maximal inequality. The cases $p = +\infty$ and/or $q = +\infty$ are proved via duality. We refer to [FJ3] for the details. However, to shed some light on this result, we discuss in more detail the special case of $\dot{f}_2^{02} = \ell^2(\mathcal{Q})$, the sequence space associated with $\dot{F}_2^{02} \approx L^2$. The following result is known as Schur's lemma; it is, of course, classical.

Lemma 5.3. *Let* $A = \{a_{QP}\}_{Q,P \in \mathcal{Q}}$. *Suppose there exists a sequence* $\{\lambda_Q\}_{Q \in \mathcal{Q}}$ *such that* $\lambda_Q > 0$ *for all* Q,

$$\sum_{P \in \mathcal{Q}} |a_{QP}|\lambda_P \le c\lambda_Q \quad \text{for all } Q \in \mathcal{Q}, \tag{5.3}$$

and

$$\sum_{Q \in \mathcal{Q}} |a_{QP}|\lambda_Q \le c\lambda_P \quad \text{for all } P \in \mathcal{Q}. \tag{5.4}$$

Then A *is bounded on* $\ell^2(\mathcal{Q})$ *with norm* $\le c$.

Proof. By the Cauchy-Schwarz inequality and (5.3), we obtain

$$|(As)_Q| \le \sum_{P \in \mathcal{Q}} |a_{QP}|^{1/2}|s_P|\lambda_P^{-1/2}\lambda_P^{1/2}|a_{QP}|^{1/2}$$

$$\le (c\lambda_Q)^{1/2}(\sum_{P \in \mathcal{Q}} |a_{QP}||s_P|^2\lambda_P^{-1})^{1/2}.$$

Hence, by (5.4),

$$\| As \|_{\ell^2(\mathcal{Q})}^2 \le c \sum_{Q \in \mathcal{Q}} \sum_{P \in \mathcal{Q}} |s_P|^2\lambda_P^{-1}|a_{QP}|\lambda_Q \le c^2 \| s \|_{\ell^2(\mathcal{Q})}^2 . \qquad \square$$

A direct calculation shows that if A is almost diagonal for $\dot{f}_2^{02}$ (i.e. (5.1) holds), then

$$\sum_{Q \in \mathcal{Q}} |a_{QP}||Q|^{1/2} \le c|P|^{1/2}, \tag{5.5}$$

and

$$\sum_{P \in \mathcal{Q}} |a_{QP}||P|^{1/2} \le c|Q|^{1/2}. \tag{5.6}$$

Hence A satisfies the assumptions of Schur's lemma with $\lambda_P = |P|^{1/2}$. This gives a simple proof that A is bounded on ℓ^2.

Alternatively, note that by Lemma 4.1, (5.5) and (5.6) are the necessary and sufficient conditions for A to be bounded on $\dot{f}_1^{01}$ and $\dot{f}_\infty^{0\infty}$, respectively. By interpolation then, A is bounded on $\dot{f}_2^{02}$ (in fact, on $\dot{f}_p^{0p}$ for $1 \le p \le +\infty$).

We now consider families of molecules and norming families for $\dot{F}_p^{\alpha q}(\mathbb{R}^n)$, using Proposition 1.1 and Theorem 5.1. If $K, N \in \{-1, 0, 1, 2, 3, \ldots\}$, $\delta \in (0, 1]$, $\sigma > N + n$, and $Q \in \mathcal{Q}$, we say that a function m_Q is a smooth (N, K, δ, σ)-molecule for Q if

$$|m_Q(x)| \le |Q|^{-1/2} \left(1 + \frac{|x - x_Q|}{\ell(Q)} \right)^{-\sigma}, \quad \text{if } x \in \mathbb{R}^n, \tag{5.7}$$

$$\int x^\gamma m_Q(x) dx = 0 \quad \text{if } |\gamma| \le N, \tag{5.8}$$

$$|D^\gamma m_Q(x)| \le |Q|^{-1/2 - |\gamma|/n} \left(1 + \frac{|x - x_Q|}{\ell(Q)} \right)^{-\sigma}$$
$$\text{if } |\gamma| \le K, \quad \text{and } x \in \mathbb{R}^n, \tag{5.9}$$

and

$$|D^\gamma m_Q(x) - D^\gamma m_Q(y)|$$
$$\le |Q|^{-1/2 - |\gamma|/n - \delta/n} |x - y|^\delta \sup_{|z| \le |x-y|} \left(1 + \frac{|x - z - x_Q|}{\ell(Q)} \right)^{-\sigma} \tag{5.10}$$

$$\text{if } |\gamma| = K, \quad \text{and } x, y \in \mathbb{R}^n.$$

Here $x_Q = 2^{-v}k$ and $\ell(Q) = 2^{-v}$ if $Q = Q_{vk} \in \mathcal{Q}$; by convention, (5.8) is void if $N = -1$ and (5.9-5.10) are void if $K = -1$.

Theorem 5.4. *Suppose that for each $Q \in \mathcal{Q}$, m_Q is a smooth (N, K, δ, σ)-molecule for Q. Let $\alpha \in \mathbb{R}$, and $0 < p, q \le +\infty$ be fixed. Let $J = n/\min(1, p, q)$. For $t \in \mathbb{R}$, let $[t]$ be the greatest integer in t; i.e. $[t] \le t < [t] + 1$, $[t] \in \mathbb{Z}$.*

A.) *If $N \ge [J - n - \alpha]$, $K \ge [\alpha]$, $K + \delta > \alpha$, and $\sigma > \max(J, J - \alpha)$, then the matrix $A = \{a_{QP}\}_{Q,P \in \mathcal{Q}} = \{\langle m_P, \varphi_Q \rangle\}_{Q,P \in \mathcal{Q}}$ is almost diagonal for $\dot{f}_p^{\alpha q}$, and $\{m_Q\}_{Q \in \mathcal{Q}}$ is a family of molecules for $\dot{F}_p^{\alpha q}$.*

B.) *If $N \ge [\alpha]$, $K \ge [J - n - \alpha]$, $K + \delta > J - n - \alpha$, and $\sigma > \max(J, n + \alpha)$, then the matrix $B = \{b_{QP}\}_{Q,P \in \mathcal{Q}} = \{\langle \psi_P, m_Q \rangle\}_{Q,P \in \mathcal{Q}}$ is almost diagonal for $\dot{f}_p^{\alpha q}$, and $\{m_Q\}_{Q \in \mathcal{Q}}$ is a norming family for $\dot{F}_p^{\alpha q}$.*

For the proof, the required estimates are tedious, but elementary; see Appendix B in [FJ3]. The remaining assertions follow immediately from Proposition 1.1 and Theorem 5.1.

In the case $\alpha = 0$, there is a converse result to Theorem 5.4.

Lemma 5.5. *Let* $\{m_Q\}_{Q \in \mathcal{Q}}$ *be a given family of functions, and suppose* $0 < p, q \le +\infty$.

A.) *If* $A = \{a_{QP}\}_{Q, P \in \mathcal{Q}} = \{\langle m_P, \varphi_Q \rangle\}_{Q, P \in \mathcal{Q}}$ *is almost diagonal for* $\dot{f}_p^{0q}$, *then each* m_Q *is (up to a universal constant multiple) a smooth* (N, K, δ, σ)-*molecule for* Q, *for some sufficiently small* N, K, δ, *and* σ *(independent of* Q*) satisfying the conditions in Theorem 5.4A for* $\alpha = 0$.

B.) *If* $B = \{b_{QP}\}_{Q, P \in \mathcal{Q}} = \{\langle \psi_P, m_Q \rangle\}_{Q, P \in \mathcal{Q}}$ *is almost diagonal for* $\dot{f}_p^{0q}$, *then each* m_Q *is (up to a universal factor) a smooth* (N, K, δ, σ)-*molecule for* Q, *for some sufficiently small* N, K, δ, *and* σ *(independent of* Q*) satisfying the conditions in Theorem 5.4B for* $\alpha = 0$.

For the proof, we write $m_Q = \sum_{P \in \mathcal{Q}} \langle m_Q, \varphi_P \rangle \psi_P$, by (3.8), for A, and, using the interchangeability of φ and ψ in (3.8), $m_Q = \sum_{P \in \mathcal{Q}} \langle m_Q, \psi_P \rangle \varphi_P$ for B. The desired estimates follow from the estimates for ψ_P and φ_P, and the assumptions by direct but technical calculations; see Lemma 9.14 in [FJ3].

Thus for $\alpha = 0$ we have a complete characterization of the families of molecules and norming families that correspond to the class of almost diagonal matrices. For $\alpha \ne 0$, we do not obtain the analogue of Lemma 5.5 due to some technical problems involving convergence, but perhaps there is an alternate formulation that gives a complete result in that case as well.

Theorem 5.4 gives us an easy way to see that the $\dot{F}_p^{\alpha q}$ spaces are independent of the choice of φ belonging to the class $\mathcal{A}$ defined in Section 2. Suppose $\varphi, \varphi^* \in \mathcal{A}$, and define $\dot{F}_p^{\alpha q}(\varphi)$ and $\dot{F}_p^{\alpha q}(\varphi^*)$ respectively. Clearly φ_Q^* is (up to a constant factor) a smooth (N, K, δ, σ)-molecule for Q for $\delta = 1$ and arbitrarily large N, K, and σ. By Lemma 5.4B then, and using (3.8-10),

$$\| \{\langle f, \varphi_Q^* \rangle\}_{Q \in \mathcal{Q}} \|_{\dot{f}_p^{\alpha q}} \le c \, \| f \|_{\dot{F}_p^{\alpha q}(\varphi)} \approx \| \{\langle f, \varphi_Q \rangle\}_{Q \in \mathcal{Q}} \|_{\dot{f}_p^{\alpha q}} \, .$$

By symmetry, we obtain the equality of $\dot{F}_p^{\alpha q}(\varphi)$ and $\dot{F}_p^{\alpha q}(\varphi^*)$. A similar argument can be used to see that the characterization of the $\dot{F}_p^{\alpha q}$ spaces by the wavelet transform follows from their characterization by the φ-transform, and vice-versa (see [FJW]). Changing from one family $\{\psi_Q\}_{Q \in \mathcal{Q}}$ to another, say $\{\psi_Q^*\}_{Q \in \mathcal{Q}}$, is, roughly speaking, like a harmless change of basis, since

$$\psi_Q^* = \sum_{P \in \mathcal{Q}} \langle \psi_Q^*, \varphi_P \rangle \psi_P$$

and the matrix $\{\langle \psi_Q^*, \varphi_P \rangle\}_{Q,P \in \mathcal{Q}}$ is always almost diagonal, hence bounded.

The class of almost diagonal matrices has the useful property of being an algebra. For $A = \{a_{QP}\}_{Q,P \in \mathcal{Q}}$ and $B = \{b_{QP}\}_{Q,P \in \mathcal{Q}}$, we define $AB = C = \{c_{QP}\}_{Q,P \in \mathcal{Q}}$ in the usual way: $c_{QP} = \sum_{R \in \mathcal{Q}} a_{QR} b_{RP}$.

Theorem 5.6. *Suppose $\alpha \in \mathbb{R}$ and $0 < p, q \leq +\infty$. If A and B are almost diagonal for $\dot{f}_p^{\alpha q}$, then AB is also.*

See Appendix D in [FJ3] for the direct but technical proof.

Theorem 5.6 can be used in conjunction with the following result regarding the smooth N-atoms introduced in Section 3.

Lemma 5.7. *Suppose that for each $Q \in \mathcal{Q}$, m_Q is a smooth (N, K, δ, σ)-molecule for Q, where N, K, δ, and σ satisfy the conditions of Theorem 5.4A. Then there exists a matrix $t = \{t_{QP}\}_{Q,P \in \mathcal{Q}}$ which is almost diagonal for $\dot{f}_p^{\alpha q}$ and a family $\{a_R^Q\}_{Q,R \in \mathcal{Q}}$ such that for each $R, Q \in \mathcal{Q}$, a_R^Q is a smooth N-atom for R, such that*

$$m_Q = \sum_{R \in \mathcal{Q}} t_{RQ} a_R^Q \quad \text{for each } Q \in \mathcal{Q}.$$

If $m_Q = \psi_Q$, the proof of the smooth atomic decomposition result above (see (3.1-3)) easily gives the result. Using (3.8) with $f = m_Q$ along with Theorems 5.4A and 5.6 leads to the general case. (See Proposition 9.13 in [FJ3]).

Hence any smooth molecule, and in particular any ψ_Q, can be expressed in terms of smooth atoms in a nice way. Conversely, any smooth N-atom a_Q for Q (which is obviously a smooth (N, K, δ, σ)-molecule for $\delta = 1$ and arbitrarily large K and σ) can be written, by (3.8) as $a_Q = \sum_R \langle a_Q, \varphi_R \rangle \psi_R$, with the approximate estimate for $|\langle a_Q, \varphi_R \rangle|$ coming from Theorem 5.4A.

Thus the classes of smooth atoms, of smooth molecules, and the class $\{\psi_Q\}_{Q \in \mathcal{Q}}$, are closely inter-related by almost diagonal matrices. Because of this and Theorem 5.6, virtually any criterion related to almost diagonality that is expressed in terms of one of these classes can be expressed in terms of either of the other two. For example, recall from Sections 1-3 that a linear operator T on $\dot{F}_p^{\alpha q}$ is bounded if and only if the matrix $\{\langle T\psi_P, \varphi_Q\rangle\}_{Q,P\in\mathcal{Q}}$ is bounded on $\dot{f}_p^{\alpha q}$. In the case where $\{\langle T\psi_P, \varphi_Q\rangle\}_{Q,P\in\mathcal{Q}}$ is almost diagonal for $\dot{f}_p^{\alpha q}$, we say that T (which is bounded by Theorem 5.1) is an almost diagonal operator on $\dot{F}_p^{\alpha q}$. By the remarks above, such operators have a variety of equivalent characterizations.

Lemma 5.8. *Suppose $\alpha \in \mathbb{R}$, and $0 < p, q \leq +\infty$. Then the following are equivalent:*

a.) *$\{\langle T\psi_P, \varphi_Q\rangle\}_{Q,P\in\mathcal{Q}}$ is almost diagonal for $\dot{f}_p^{\alpha q}$.*

b.) *$\{\langle Tm_P, \varphi_Q\rangle\}_{Q,P\in\mathcal{Q}}$ is almost diagonal for $\dot{f}_p^{\alpha q}$ whenever each m_Q is a smooth (N, K, δ, σ)-molecule for Q, for N, K, δ, σ as in Theorem 5.4A, with almost diagonality estimates independent of the family $\{m_Q\}_{Q\in\mathcal{Q}}$.*

c.) *$\{\langle Ta_P, \varphi_Q\rangle\}_{Q,P\in\mathcal{Q}}$ is almost diagonal for $\dot{f}_p^{\alpha q}$ whenever each a_Q is a smooth N-atom for Q, for N as in Theorem 5.4A, with the matrix estimates independent of the family $\{a_Q\}_{Q\in\mathcal{Q}}$.*

In the next section we will see that there is a precise characterization of the almost diagonal operators on $\dot{F}_p^{0q}$ if $1 \leq p, q \leq +\infty$.

6. THE REDUCED DAVID-JOURNÉ CLASS

The classical theory of Calderón-Zygmund singular integral operators is concerned with convolution operators $Tf(x) = \int K(x-y)f(y)dy$ where the kernel K just fails to be integrable at the origin and at infinity: $|K(x)| \leq c/|x|^n$. (If $K \in L^1(\mathbb{R}^n)$, then simple methods, like Young's inequality, can be used to study T.) Calderón and Zygmund showed that under a certain cancellation condition and a weak smoothness assumption on K, the operator T, interpreted in the principal value sense, is bounded on L^p for $1 < p < +\infty$ (see e.g. [S]). This allowed the generalization to $\mathbb{R}^n$ of many results that had been first understood for $\mathbb{R}^1$ using complex variable methods.

More recently, attention has been focused on the case of a non-convolution kernel satisfying similar estimates. For some motivation for studying such operators, see e.g. [Cal3] or the discussion in [FJW]. We say that $K(x, y)$ is a (generalized) Calderón-Zygmund kernel if there exists $\varepsilon > 0$ such that

$$|K(x, y)| \leq c/|x - y|^n, \tag{6.1}$$

$$|K(x,y) - K(x',y)| \le c|x - x'|^\varepsilon / |x - y|^{n+\varepsilon} \text{ if } |x - x'| < |x - y|/2, \quad (6.2)$$

and

$$|K(x,y) - K(x,y')| \le c|y - y'|^\varepsilon / |x - y|^{n+\varepsilon} \text{ if } |y - y'| < |x - y|/2. \quad (6.3)$$

We say $T \in CZO(\varepsilon)$ if T is a continuous linear operator from $\mathcal{S}(\mathbb{R}^n)$ to $\mathcal{S}'(\mathbb{R}^n)$, with a kernel K satisfying (6.1-3) which represents T in the sense that if $\theta, \eta \in \mathcal{D}(\mathbb{R}^n)$ with supp $\theta \cap$ supp $\eta = \emptyset$, then

$$(T\theta, \eta) = \int \int K(x,y)\theta(y)\eta(x)dydx.$$

Here $(\cdot, \cdot)$ represents the bilinear pairing between $\mathcal{S}'$ and $\mathcal{S}$.

It turns out, although we refer to [DJ], [FHJW], or [Torr] for the details, that any $T \in CZO(\varepsilon)$ can be extended to act on $C^\infty \cap L^\infty(\mathbb{R}^n)$ in a natural way. In particular, we can define $T1$. Also, we need to introduce the notion of the "weak boundedness property". For $\theta \in \mathcal{D}(\mathbb{R}^n)$, $t > 0$, and $z \in \mathbb{R}^n$, let $\theta_t^z(x) = t^{-n}\theta((x - z)/t)$. If $T : \mathcal{D} \to \mathcal{D}'$ is linear and continuous, we say $T \in WBP$ if for every bounded subset $\mathcal{B}$ of $\mathcal{D}$, there exists $C = C(\mathcal{B})$ such that

$$|(T(\theta_t^z), \eta_t^z)| \le Ct^{-n}, \text{ for all } z \in \mathbb{R}^n \text{ and } t > 0.$$

It is easy to see that if T is bounded on L^2 (or on L^p for some $p \in [1, \infty]$) then $T \in WBP$. Note that $T \in WBP$ if and only if $T^* \in WBP$; here T^* is the adjoint of T. Note also that if $T \in CZO(\varepsilon)$ with kernel $K(x,y)$, then $T^* \in CZO(\varepsilon)$ with kernel $K^*(x,y) = K(y,x)$.

For a convolution operator T with kernel K, L^2 boundedness is usually easy to determine, since T is L^2 bounded if and only if $\hat{K} \in L^\infty$. For non-convolution operators the question is not so simple. Hence the following result of David and Journé [DJ] was a breakthrough.

Theorem 6.1. (David-Journé theorem). *Suppose $T \in CZO(\varepsilon)$ for some $\varepsilon > 0$. Then T is (or extends to be) bounded on L^2 if and only if $T1 \in BMO$, $T^*1 \in BMO$, and $T \in WBP$.*

One direction of the proof is easy. If T is L^2-bounded, then $T \in WBP$, as we have remarked, and classical techniques show that T maps L^∞ to BMO boundedly, so that $T1, T^*1 \in BMO$. For the other direction, the general case ($T1, T^*1 \in BMO$) can be reduced to the following special case.

Theorem 6.2. *Suppose $T \in CZO(\varepsilon)$, for some $\varepsilon > 0$, $T \in WBP$, and $T1 = 0 = T^*1$. Then T is bounded on L^2.*

We will discuss the reduction step in Section 9. We call the class of operators satisfying the assumptions of Theorem 6.2 the reduced David-Journé class; it is also known as the Lemarié algebra. This class has the following characterization.

Theorem 6.3.   *Assume $T : \mathcal{S} \to \mathcal{S}'$ is linear and continuous. Then T belongs to the reduced David-Journé class if and only if T is almost diagonal for L^2.*

Theorem 6.2 follows from the "only if" part of Theorem 6.3, along with Theorem 5.1, the characterization of L^2 by the φ or wavelet transforms, and the remarks in Section 1 about operators and their matrices.

For the proof of Theorem 6.3, suppose first that T belongs to the reduced David-Journé class. By Lemma 5.8, T is almost diagonal for L^2 if and only if $\{\langle Ta_P, \varphi_Q \rangle\}_{Q,P \in \mathcal{Q}}$ is almost diagonal for $\dot{f}_2^{02} \approx \ell^2(\mathcal{Q})$, whenever each $\{a_Q\}_{Q \in \mathcal{Q}}$ is a smooth 0-atom for Q (with matrix estimates independent of $\{a_Q\}_{Q \in \mathcal{Q}}$). By Lemmas 5.4-5, this is the case if and only if each Ta_Q is (up to a universal constant) a smooth $(0, 0, \delta, \sigma)$-molecule for Q for some $\delta > 0$ and $\sigma > n$. In fact, a more general result is true, which gives some insight into the role of ε in the definition of $CZO(\varepsilon)$.

Theorem 6.4.   *Suppose $0 < \varepsilon \le 1$, $T \in CZO(\varepsilon) \cap WBP$, $T1 = T^*1 = 0$, and a_Q is a smooth 0-atom for Q. Then there exists $C > 0$ (independent of a_Q) such that Ta_Q/C is a smooth $(0, 0, \varepsilon, n + \varepsilon)$-molecule for Q.*

We note that if T satisfies the assumptions of Theorem 6.4, then Theorems 5.1 and 5.4, and Lemma 5.8 show that T is bounded on $\dot{F}_p^{\alpha q}$ for $|\alpha| < \varepsilon$ and $1 \le p, q \le +\infty$. The proof of Theorem 6.4 depends on an important lemma of Y. Meyer [Me1] which guarantees that T maps each a_Q into L^∞ with an estimate $\| Ta_Q \|_{L^\infty} \le c|Q|^{-1/2}$. Assuming this, the rest of the proof amounts to making a series of elementary estimates on the relevant integrals, taking advantage of the cancellation properties of a_Q and T. The assumption $T^*1 = 0$ is used to obtain

$$\int Ta_Q = \langle Ta_Q, 1 \rangle = \langle a_Q, T^*1 \rangle = \langle a_Q, 0 \rangle = 0. \qquad (6.4)$$

We refer to [FHJW] or [FJW] for the details regarding the estimates.

For general α, p, and q, it can be shown that T takes smooth atoms into smooth molecules if certain additional smoothness and cancellation assumptions are satisfied. Hence in this case also, T is almost diagonal and thus bounded on $\dot{F}_p^{\alpha q}$. The exact conditions required are somewhat technical to state; we refer to [FTW] or [Torr]. The sharpness of these conditions is still open; that is, we do not know whether the almost diagonality of T on $\dot{F}_p^{\alpha q}$ implies these smoothness and cancellation conditions.

By Theorem 6.4 for $\alpha = 0$ and $p = q = 2$, we have one direction of Theorem 6.3. The other direction depends on the following result.

Lemma 6.5. *Suppose there exist $\delta > 0$ and $\sigma > n$ such that for each $Q \in \mathcal{Q}$, g_Q and m_Q satisfy (5.7) and (5.10) for $K = 0$. Let*

$$K(x, y) = \sum_{Q \in \mathcal{Q}} g_Q(x) m_Q(y).$$

Then K is a generalized Calderón-Zygmund kernel of order $\varepsilon = \min(\delta, \sigma - n)$.

The proof is a direct computation. Assuming this result, suppose T is almost diagonal for $L^2 \approx \dot{F}_2^{02}$. By Lemma 5.5, each $T\psi_Q$ is a smooth $(0, 0, \delta, \sigma)$-molecule for Q for some $\delta > 0$ and $\sigma > n$. The kernel K of T is

$$K(x, y) = \sum_{Q \in \mathcal{Q}} T\psi_Q(x) \varphi_Q(y).$$

(This is easy to see from (3.8), since then $Tf = \sum_{Q \in \mathcal{Q}} \langle f, \varphi_Q \rangle T\psi_Q$.) It follows from Lemma 6.5 that K is a generalized Calderón-Zygmund kernel. The other properties required of T in Theorem 6.3 are easier. Since T is almost diagonal, T is bounded on L^2, hence $T \in WBP$. Since each $T\psi_Q$ is a smooth $(0, 0, \delta, \sigma)$-molecule for Q, we have $0 = \langle T\psi_Q, 1 \rangle = \langle \psi_Q, T^*1 \rangle$. But this implies that $T^*1 = 0$ as an element of BMO since

$$\| T^*1 \|_{BMO} \approx \| \langle T^*1, \psi_Q \rangle \|_{\dot{f}_\infty^{02}}$$

by (3.8-10) with ψ in place of φ. Since T is almost diagonal, so is T^*; thus for the same reason as above, $T1 = 0$. These conclusions give that T belongs to the reduced David-Journé class, completing our discussion of Theorem 6.3.

The condition that T be almost diagonal for $\dot{F}_p^{0q}$ is the same for $1 \leq p, q \leq +\infty$. Thus Theorem 6.3 is a complete characterization of the bounded linear operators on $\dot{F}_p^{0q}$ corresponding to the class of almost diagonal matrices on $\dot{f}_p^{0q}$, $1 \leq p, q \leq +\infty$. It would be interesting to have such a characterization for $\min(p, q) < 1$ or for $\alpha \neq 0$.

7. THE NON-SMOOTH ATOMIC DECOMPOSITION

The theory of the real variable Hardy spaces $H^p(\mathbb{R}^n)$, $0 < p \leq 1$, was significantly clarified and simplied by Coifman's discovery of the "atomic decomposition" of $H^p(\mathbb{R}^1)$ ([Co]), and Latter's extension ([L]) of this result to $\mathbb{R}^n$, $n > 1$. A function $a(x)$ is an L^q-atom, $1 < q \leq +\infty$, for H^p if there exists a cube $Q \subseteq \mathbb{R}^n$ (not necessarily dyadic) such that supp $a \subseteq Q$, $\| a \|_{L^q} \leq |Q|^{1/q - 1/p}$, and $\int x^\gamma a(x) dx = 0$ if $|\gamma| \leq [n(1/p - 1)]$, where $[t]$ is the greatest integer in t. Coifman and Latter proved that if $f \in H^p$, $0 < p \leq 1$, and $1 < q \leq +\infty$, then there exist L^q-atoms $\{a_i\}_{i=1}^\infty$ and scalars $\{\lambda_i\}_{i=1}^\infty$ such that $f = \sum_{i=1}^\infty \lambda_i a_i$ and $(\sum_{i=1}^\infty |\lambda_i|^p)^{1/p} \leq c \| f \|_{H^p}$. The converse is easy; hence for $1 < q \leq +\infty$, $0 < p \leq 1$,

$$\| f \|_{H^p} \approx \inf \left\{ (\sum_{i=1}^\infty |\lambda_i|^p)^{1/p} : f = \sum_{i=1}^\infty \lambda_i a_i, \right.$$
$$\left. \text{and each } a_i \text{ is an } L^q \text{ atom for } H^p \right\}.$$

More recent proofs of this result involve considering the level sets of some Littlewood-Paley expression whose L^p-norm is equivalent to $\| f \|_{H^p}$ (see e.g. [Cal2], [Ch-F], [W], [Fo-S]). We will use this method to obtain an atomic decomposition of the sequence space $\dot{f}_p^{\alpha q}$, $\alpha \in \mathbb{R}$, $0 < p \leq 1$, $p \leq q \leq +\infty$. From the sequence space result and the smooth atomic decomposition described in Section 3, we easily obtain a corresponding "non-smooth" atomic decomposition of $\dot{F}_p^{\alpha q}$ for these indices. This gives, for example, a way of seeing the equivalence $\dot{F}_p^{02} \approx H^p$, $0 < p \leq 1$. We will use the sequence space decomposition result in the next chapter to further investigate matrix boundedness.

The restrictions noted above on the indices are exactly what is needed to obtain the "p-sublinearity" of the quasi-norms that we are considering.

Lemma 7.1. *Suppose* $\alpha \in \mathbb{R}$, $0 < p \leq 1$, $p \leq q \leq +\infty$, $s = \{s_Q\}_{Q \in \mathcal{Q}}$, $t = \{t_Q\}_{Q \in \mathcal{Q}}$, *and* $f, g \in \mathcal{S}'$. *Then*

$$\| s + t \|_{\dot{f}_p^{\alpha q}}^p \leq \| s \|_{\dot{f}_p^{\alpha q}}^p + \| t \|_{\dot{f}_p^{\alpha q}}^p, \tag{7.1}$$

and

$$\| f + g \|_{\dot{F}_p^{\alpha q}}^p \leq \| f \|_{\dot{F}_p^{\alpha q}}^p + \| g \|_{\dot{F}_p^{\alpha q}}^p. \tag{7.2}$$

Proof. We prove (7.1), (7.2) being similar. By the p-triangle inequality

$|a + b|^p \leq |a|^p + |b|^p$, we have

$$\| \, s + t \, \|_{\dot{f}_p^{\alpha q}}^p = \int (\sum_{Q \in \mathcal{Q}} (|Q|^{-\alpha/n} |s_Q + t_Q| \tilde{\chi}_Q)^{p \cdot q/p})^{p/q}$$

$$\leq \int (\sum_{Q \in \mathcal{Q}} [(|Q|^{-\alpha/n} |s_Q| \tilde{\chi}_Q)^p + (|Q|^{-\alpha/n} |t_Q| \tilde{\chi}_Q)^p]^{q/p})^{p/q}.$$

Using Minkowski's inequality for $q/p \geq 1$, we obtain the result. $\square$

For a sequence $s = \{s_Q\}_{Q \in \mathcal{Q}}$, we recall from Section 2 the notation

$$g^{\alpha q}(s) = (\sum_{Q \in \mathcal{Q}} (|Q|^{-\alpha/n} |s_Q| \tilde{\chi}_Q)^q)^{1/q},$$

so that $\| \, s \, \|_{\dot{f}_p^{\alpha q}} = \| \, g^{\alpha q}(s) \, \|_{L^p}$. We say that a sequence $r = \{r_Q\}_{Q \in \mathcal{Q}}$ is an ∞-atom for $\dot{f}_p^{\alpha q}$ if there exists Q' dyadic such that $r_Q = 0$ if $Q \nsubseteq Q'$, and

$$\| \, g^{\alpha q}(r) \, \|_{L^\infty} \leq |Q'|^{-1/p}.$$

Lemma 7.2.   *If r is an ∞-atom for $\dot{f}_p^{\alpha q}$, then $\| \, r \, \|_{\dot{f}_p^{\alpha q}} \leq 1$.*

Proof.   For Q' as in the definition of an ∞-atom, we have $\| \, r \, \|_{\dot{f}_p^{\alpha q}}^p = \int_{Q'} |g^{\alpha q}(r)|^p \leq |Q'|^{-1} |Q'| = 1$. $\square$

The main result in this chapter is the following theorem. It was proved in [FJ3] using real interpolation methods. We remarked there that a more direct proof, in the spirit of the recent proofs of the H^p atomic decomposition, could be given. For the sake of variety, we give this alternate proof now.

Theorem 7.3.   *Suppose $\alpha \in \mathbb{R}$, $0 < q \leq +\infty$, $0 < p < +\infty$, and $s = \{s_Q\}_{Q \in \mathcal{Q}} \in \dot{f}_p^{\alpha q}$. Then there exists a sequence $\{r_j\}_{j=1}^\infty$ of ∞-atoms for $\dot{f}_p^{\alpha q}$ and scalars $\{t_j\}_{j=1}^\infty$ such that $s = \sum_{j=1}^\infty t_j r_j$ and $(\sum_{j=1}^\infty |t_j|^p)^{1/p} \leq c \, \| \, r \, \|_{\dot{f}_p^{\alpha q}}$.*

Proof.   For $P \in \mathcal{Q}$, let

$$g_P^{\alpha q}(s) = (\sum_{Q \in \mathcal{Q} : P \subseteq Q} (|Q|^{-\alpha/n - 1/2} |s_Q|)^q)^{1/q}.$$

If $x \in P$, $g_P^{\alpha q}(s)$ coincides with $\sum_{Q \in \mathcal{Q} : P \subseteq Q} (|Q|^{-\alpha/n} |s_Q| \tilde{\chi}_Q(x))^q)^{1/q}$, since this expression is constant on P. Note that if $P_1, P_2 \in \mathcal{Q}$ and $P_1 \subseteq P_2$, then $g_{P_1}^{\alpha q}(s) \geq g_{P_2}^{\alpha q}(s)$. Note also that $\lim_{\ell(P)\to\infty,x\in P} g_P^{\alpha q}(s) = 0$, while $\lim_{\ell(P)\to 0,x\in P} g_P^{\alpha q}(s) = g^{\alpha q}(s)(x)$. For $k \in \mathbb{Z}$, let

$$A_k = \{P \in \mathcal{Q} : g_P^{\alpha q}(s) > 2^k\}.$$

It is easy to see that

$$\{x : g^{\alpha q}(s)(x) > 2^k\} = \bigcup_{A_k} P. \tag{7.3}$$

Also, note that

$$(\sum_{Q\in\mathcal{Q}\setminus A_k} (|Q|^{-\alpha/n}|s_Q|\tilde{\chi}_Q(x))^q)^{1/q} \leq 2^k \text{ for all } x. \tag{7.4}$$

If $g^{\alpha q}(s)(x) \leq 2^k$, this is clear. If not, then there exists a maximal $P \in A_k$ such that $x \in P$. Let $\tilde{P}$ be the unique dyadic cube satisfying $P \subseteq \tilde{P}$ and $\ell(\tilde{P}) = 2\ell(P)$. Then the left side of (7.4) is $g_{\tilde{P}}^{\alpha q}(s)$, which is $\leq 2^k$ since $\tilde{P} \notin A_k$.

Since $g^{\alpha q}(s) \in L^p$, by assumption, and $g^{\alpha q}(s)(x) > 2^k$ for $x \in Q$ if $Q \in A_k$, the cubes in A_k are bounded above in size. Let $\mathcal{B}_k$ be the collection of maximal dyadic cubes belonging to $A_k \setminus A_{k+1}$. If $J \in \mathcal{B}_k$, define a sequence $t_{k,J}$ by $(t_{k,J})_Q = s_Q$ if $Q \subseteq J$ and $Q \in A_k \setminus A_{k+1}$, and $(t_{k,J})_Q = 0$ otherwise. Then

$$s = \sum_{k\in\mathbb{Z}} \sum_{J\in\mathcal{B}_k} t_{k,J}.$$

(Note that if $P \notin \bigcup_{k\in\mathbb{Z}} A_k$, then $s_P = 0$.) By (7.4),

$$|g^{\alpha q}(t_{k,J})(x)| = (\sum_{Q\subseteq J,Q\in A_k\setminus A_{k+1}} (|Q|^{-\alpha/n}|s_Q|\tilde{\chi}_Q(x))^q)^{1/q}$$

$$\leq (\sum_{Q\in\mathcal{Q}\setminus A_{k+1}} (|Q|^{-\alpha/n}|s_Q|\tilde{\chi}_Q(x))^q)^{1/q} \leq 2^{k+1}.$$

Let $r_{k,J} = |J|^{-1/p}2^{-k-1}t_{k,J}$ and $\lambda_{k,J} = |J|^{1/p}2^{k+1}$. Then $s = \sum_{k\in\mathbb{Z}} \sum_{J\in\mathcal{B}_k} \lambda_{k,J}r_{k,J}$. Also each $r_{k,J}$ is an ∞-atom for $\dot{f}_p^{\alpha q}$: $(r_{k,J})_Q = 0$ if $Q \not\subseteq J$, and

$$\| g^{\alpha q}(r_{k,J}) \|_{L^\infty} \leq |J|^{-1/p},$$

by the estimate on $g^{\alpha q}(t_{k,J})$. By (7.2) we have

$$\sum_{k \in \mathbb{Z}} \sum_{J \in \mathcal{B}_k} |\lambda_{k,J}|^p = \sum_{k \in \mathbb{Z}} 2^{(k+1)p} \sum_{J \in \mathcal{B}_k} |J|$$

$$\leq 2^p \sum_{k \in \mathbb{Z}} 2^{kp} |\bigcup_{A_k} P| = 2^p \sum_{k \in \mathbb{Z}} 2^{kp} |\{x : g^{\alpha q}(s)(x) > 2^k\}|$$

$$\leq c_p \| g^{\alpha q}(s) \|_{L^p}^p = c_p \| s \|_{\dot{f}_p^{\alpha q}}^p .$$

Taking the pth root gives the conclusion. $\qquad \square$

This result and Lemma 7.1 immediately yield the following.

Corollary 7.4. *Suppose $\alpha \in \mathbb{R}$, $0 < p \leq 1$, and $p \leq q \leq +\infty$. Then*

$$\| s \|_{\dot{f}_p^{\alpha q}} \approx \inf \left\{ (\sum_{j=1}^{\infty} |\lambda_j|^p)^{1/p} : s = \sum_{j=1}^{\infty} \lambda_j r_j, \right.$$

$$\left. \text{and each } r_j \text{ is an } \infty\text{-atom for } \dot{f}_p^{\alpha q} \right\} .$$

Notice that Theorem 7.3 holds for all indices; however it only character-izes $\dot{f}_p^{\alpha q}$ for the range of indices in Corollary 7.4, since the converse requires (7.1).

Naturally, Corollary 7.4 has an analogue for the function spaces $\dot{F}_p^{\alpha q}$. We say that a function $A(x)$ is an ∞-atom for $\dot{F}_p^{\alpha q}$ if $A = \sum_{Q \in \mathcal{Q}} r_Q a_Q$, where $r = \{r_Q\}_{Q \in \mathcal{Q}}$ is an ∞-atom for $\dot{f}_p^{\alpha q}$, and each a_Q is a smooth N-atom for $\dot{F}_p^{\alpha q}$, for $N = [n(1/p - 1) - \alpha]$ (where $[x]$ denotes the greatest integer in x).

Theorem 7.5. *Suppose $\alpha \in \mathbb{R}$, $0 < p \leq 1$, and $p \leq q \leq +\infty$. Then $\|f\|_{\dot{F}_p^{\alpha q}} \approx \inf\{(\sum_{j=1}^{\infty} |\lambda_j|^p)^{1/p} : f = \sum_{j=1}^{\infty} \lambda_j A_j$ and each A_j is an ∞-atom for $\dot{F}_p^{\alpha q}\}$.*

Proof. By Theorem 7.3 and the smooth atomic decomposition described in Section 3, we can write $f \in \dot{F}_p^{\alpha q}$ as $f = \sum_{Q \in \mathcal{Q}} s_Q a_Q$, where each a_Q is a smooth N-atom for Q, $s = \sum_{j=1}^{\infty} \lambda_j r_j$, each r_j is an ∞-atom for $\dot{f}_p^{\alpha q}$, and

$$(\sum_{j=1}^{\infty} |\lambda_j|^p)^{1/p} \leq c \| s \|_{\dot{f}_p^{\alpha q}} \leq c \| f \|_{\dot{F}_p^{\alpha q}} .$$

Then

$$f = \sum_{Q \in \mathcal{Q}} \sum_{j=1}^{\infty} \lambda_j (r_j)_Q a_Q = \sum_{j=1}^{\infty} \lambda_j \sum_{Q \in \mathcal{Q}} (r_j)_Q a_Q = \sum_{j=1}^{\infty} \lambda_j A_j,$$

and each A_j is an ∞-atom for $\dot{F}_p^{\alpha q}$.

Conversely, by (3.4) and Lemma 7.2, an ∞-atom A for $\dot{F}_p^{\alpha q}$ satisfies $\| A \|_{\dot{F}_p^{\alpha q}} \leq c$. Hence (7.2) yields the converse. □

To compare to the Hardy space atomic decomposition, we make the following observation.

Lemma 7.6. *Suppose $0 < p \leq 1$ and A is an ∞-atom for $\dot{F}_p^{02}$. Then there exists $c > 0$ such that A/c is an L^2 atom for H^p.*

Proof. Suppose $A = \sum_{Q \in \mathcal{Q}} r_Q a_Q$ is as in the definition of an ∞-atom for $\dot{F}_p^{\alpha q}$, for some ∞-atom r for $\dot{f}_p^{02}$ associated to a cube Q'. Then supp $A \subseteq 3Q'$ since $r_Q = 0$ unless $Q \subseteq Q'$ and supp $a_Q \subseteq 3Q$. Note that $g^{02}(r)$ is supported in Q'. Hence

$$\| A \|_{L^2} \approx \| A \|_{\dot{F}_2^{02}} \leq c \| r \|_{\dot{f}_2^{02}} = c \| g^{02}(r) \|_{L^2}$$

$$= c(\int_{Q'} |g^{02}(r)|^2)^{1/2} \leq c \| g^{02}(r) \|_{L^{\infty}} |Q'|^{1/2} = c|Q'|^{1/2 - 1/p}.$$

Since $\sum_{Q \in \mathcal{Q}} r_Q a_Q$ converges to A in L^2 and is supported in $3Q'$, it converges in L^1 as well. Thus the cancellation required for A is inherited from the similar condition for the a_Q's. □

This gives a way to see the equivalence $\dot{F}_p^{02} \approx H^p$, $0 < p \leq 1$. The atomic characterization of H^p, Theorem 7.5, and Lemma 7.6 show that $\dot{F}_p^{02} \subseteq H^p$. Conversely, it suffices to show that an L^2 atom $a(x)$ for H^p satisfies $\| a \|_{\dot{F}_p^{02}} \leq c$. This is proved by direct estimates on the discrete g-function $(\sum_{v \in \mathbb{Z}} |\varphi_v * a|^2)^{1/2}$ of a. The details are very similar to the standard estimates for the continuous g-function, which can be found, for example, in [Torch].

8. A SHARPER CONDITION FOR MATRIX BOUNDEDNESS

From Lemma 4.1, we know that a matrix $A = \{a_{QP}\}_{Q,P \in \mathcal{Q}}$ is bounded on $\dot{f}_1^{01}$ if and only if

$$\sup_{P \in \mathcal{Q}} |P|^{-1/2} \sum_{Q \in \mathcal{Q}} |a_{QP}||Q|^{1/2} < +\infty, \tag{8.1}$$

and that A is bounded on $\dot{f}_\infty^{0\infty}$ if and only if

$$\sup_{Q \in \mathcal{Q}} |Q|^{-1/2} \sum_{P \in \mathcal{Q}} |a_{QP}||P|^{1/2} < +\infty. \tag{8.2}$$

(As we remarked in Section 5, we may as well assume $\alpha = 0$ in studying matrix boundedness on $\dot{f}_p^{\alpha q}$). To consider the range $1 \le p, q \le +\infty$, it is natural to look for conditions for boundedness in the other two extreme cases, $\dot{f}_1^{0\infty}$ and $\dot{f}_\infty^{01}$. We do not have sharp results for a general matrix A; however if we assume that A is positive ($a_{QP} \ge 0$ for all $Q, P \in \mathcal{Q}$) then the atomic decomposition from Section 7 can be used to give a sharp result for $\dot{f}_1^{0\infty}$, and duality yields a sharp result for $\dot{f}_\infty^{01}$.

Theorem 8.1. *Suppose* $A = \{a_{QP}\}_{Q,P \in \mathcal{Q}}$ *satisfies* $a_{QP} \ge 0$ *for all* $Q, P \in \mathcal{Q}$.

A.) A is bounded on $\dot{f}_1^{0\infty}$ if and only if

$$\sup_{P_0 \in \mathcal{Q}} \frac{1}{|P_0|} \| \{ \sum_{P \subseteq P_0, P \in \mathcal{Q}} a_{QP}|P|^{1/2} \}_{Q \in \mathcal{Q}} \|_{\dot{f}_1^{0\infty}} < +\infty. \tag{8.3}$$

B.) A is bounded on $\dot{f}_\infty^{01}$ if and only if

$$\sup_{Q_0 \in \mathcal{Q}} \frac{1}{|Q_0|} \| \{ \sum_{Q \subseteq Q_0, Q \in \mathcal{Q}} a_{QP}|Q|^{1/2} \}_{P \in \mathcal{Q}} \|_{\dot{f}_1^{0\infty}} < +\infty. \tag{8.4}$$

Proof. Recall from Section 7 that $r = \{r_Q\}_{Q \in \mathcal{Q}}$ is an ∞-atom for $\dot{f}_1^{0\infty}$ if there exists $Q' \in \mathcal{Q}$ such that $r_Q = 0$ if $Q \not\subseteq Q'$, and

$$\| g^{0\infty}(r) \|_{L^\infty} = \sup_{Q \in \mathcal{Q}} |Q|^{-1/2}|r_Q| \le |Q'|^{-1}.$$

We say that $r = \{r_Q\}_{Q \in \mathcal{Q}}$ is a maximal ∞-atom for $\dot{f}_1^{0\infty}$ if there exists $Q' \in \mathcal{Q}$ such that $r_Q = |Q|^{1/2}|Q'|^{-1}$ for $Q \subseteq Q'$, and $r_Q = 0$ otherwise. If r is a maximal ∞-atom for $\dot{f}_1^{0\infty}$, then

$$\| r \|_{\dot{f}_1^{0\infty}} = \| g^{0\infty}(r) \|_{L^1} = \| \sup_{Q : x \in Q} |Q|^{-1/2}|r_Q|\chi_Q \|_{L^1} = \int_{Q'} |Q'|^{-1} = 1.$$

Hence (8.3) is the condition that $\| Ar \|_{\dot{f}_1^{0\infty}} \leq c$ for every maximal ∞-atom r for $\dot{f}_1^{0\infty}$. Thus, (8.3) is necessary for the boundedness of A. Conversely, if $s = \{s_Q\}_{Q \in \mathcal{Q}}$ is an ∞-atom for $\dot{f}_1^{0\infty}$ associated to Q', then $|s_Q| \leq r_Q$ where r is the maximal ∞-atom for $\dot{f}_1^{0\infty}$ for Q'. Since A is positive, $|(As)_Q| \leq (Ar)_Q$ for all $Q \in \mathcal{Q}$, so $\| As \|_{\dot{f}_1^{0\infty}} \leq \| Ar \|_{\dot{f}_1^{0\infty}} \leq c$ if (8.3) holds. By Theorem 7.3, A is bounded on $\dot{f}_1^{0\infty}$.

Part B follows from A by a duality argument; see [FJ3]. $\square$

The sequence spaces $\dot{f}_p^{\alpha q}$ have the property that they are lattices: if $|s_Q| \leq |t_Q|$ for all $Q \in \mathcal{Q}$, then $\| \{s_Q\}_{Q \in \mathcal{Q}} \|_{\dot{f}_p^{\alpha q}} \leq \| \{t_Q\}_{Q \in \mathcal{Q}} \|_{\dot{f}_p^{\alpha q}}$. Also, note that if $A = \{a_{QP}\}_{Q,P \in \mathcal{Q}}$ is a general matrix and $|A|$ is the matrix with entries $|a_{QP}|$, then $|(As)_Q| \leq (|A||s|)_Q$, for all $Q \in \mathcal{Q}$, where $|s|_Q = |s_Q|$ for $Q \in \mathcal{Q}$. Hence the boundedness of $|A|$ on $\dot{f}_p^{\alpha q}$ implies the boundedness of A. Therefore, Theorem 8.1 gives sufficient conditions for boundedness of a non-positive matrix: if (8.3) (respectively (8.4)) holds with a_{QP} replaced by $|a_{QP}|$, then A is bounded on $\dot{f}_1^{0\infty}$ (respectively $\dot{f}_\infty^{01}$).

We let **b** denote the collection of all matrices A such that $|A|$ is bounded on $\dot{f}_p^{0q}$ for all $1 \leq p, q \leq +\infty$. By the previous remarks, any $A \in$ **b** is bounded on $\dot{f}_p^{0q}$.

Corollary 8.2. *A matrix A belongs to* **b** *if and only if A satisfies (8.1-2) and, with a_{QP} replaced by $|a_{QP}|$, (8.3-4).*

This corollary follows from the remarks we have made and an interpolation argument showing that boundedness on the "four corners" $\dot{f}_1^{01}$, $\dot{f}_\infty^{0\infty}$, $\dot{f}_1^{01}$, and $\dot{f}_1^{0\infty}$ of the $1/p, 1/q$ square implies boundedness on $\dot{f}_p^{0q}$ for all $1 \leq p, q \leq +\infty$ (see [FJ3]).

We norm **b** by the maximum of the four suprema required to be finite in Corollary 8.2. It is easy to see that **b** is a lattice: if $|a_{QP}| \leq |b_{QP}|$ for all $Q, P \in \mathcal{Q}$, then $\| \{a_{QP}\}_{Q,P \in \mathcal{Q}} \|_{\mathbf{b}} \leq \| \{b_{QP}\}_{Q,P \in \mathcal{Q}} \|_{\mathbf{b}}$. Since (8.1), (8.3) are the transposes of (8.2), (8.4), **b** is closed under taking adjoints. Also, since $|(AB)_Q| \leq (|A||B|)_Q$, **b** is an algebra. In particular, **b** is an algebra of bounded matrices on $\dot{f}_p^{0q}$, $1 \leq p, q \leq +\infty$, which by Section 5, contains the almost diagonal matrices.

Although the conditions in Corollary 8.2 are sharp for positive matrices, (8.3-4) are not sufficiently explicit to be easily checked in some cases. However, this result can be used to obtain explicit conditions which are sufficient, although not necessary. Let $\omega_{QP}(\varepsilon)$ be as in Section 5, for $\alpha = 0$ and $1 \leq p, q \leq +\infty$.

Theorem 8.3. *Suppose $A = \{a_{QP}\}_{Q,P \in \mathcal{Q}}$ satisfies*

$$\sup_{Q \in \mathcal{Q}} \sum_{P \in \mathcal{Q}} (|a_{QP}|/\omega_{QP}(\varepsilon))^s \omega_{QP}(\varepsilon)(|P|/|Q|)^{1/2} < +\infty$$

and

$$\sup_{P \in \mathcal{Q}} \sum_{Q \in \mathcal{Q}} (|a_{QP}|/\omega_{QP}(\varepsilon))^s \omega_{QP}(\varepsilon)(|Q|/|P|)^{1/2} < +\infty,$$

for some $\varepsilon > 0$ and $s > 1$. Then $A \in \mathbf{b}$.

The proof consists of verifying that the assumed conditions imply the necessary criterion of Corollary 8.2. We refer to [FJ3] for the details. Since $\sup_{Q \in \mathcal{Q}} \sum_{P \in \mathcal{Q}} \omega_{QP}(\varepsilon)(|P|/|Q|)^{1/2} < +\infty$ and $\sup_{P \in \mathcal{Q}} \sum_{Q \in \mathcal{Q}} \omega_{QP}(\varepsilon)$ $(|Q|/|P|)^{1/2} < +\infty$ (cf. (5.5-6)), every almost diagonal matrix satisfies the criteria of Theorem 8.3. In particular, we have obtained an alternate proof of the boundedness of almost diagonal matrices.

Using Lemma 5.2, we can obtain from Theorem 8.3 a criterion for the boundedness of A on $\dot{f}_p^{0q}$ for $\min(p, q) < 1$ also. We refer to [FJ3] for the exact statement. Also we can obtain results for $\alpha \neq 0$ trivially from the case $\alpha = 0$, as we have noted before.

The families of molecules and norming families for $\dot{F}_p^{0q}$, $1 \leq p, q \leq +\infty$, corresponding to $\mathbf{b}$ via Proposition 1, are not yet well-understood. These collections of such families obviously contain the families of smooth molecules (as in Lemma 5.4), since the almost diagonal matrices belong to $\mathbf{b}$. An exact characterization of the families corresponding to $\mathbf{b}$ is perhaps unlikely, but it would be interesting to obtain some sufficient conditions weaker than the conditions for smooth molecules.

Similarly, the operators T on $\dot{F}_p^{\alpha q}$ with associated matrices

$$\{\langle T\psi_P, \varphi_Q \rangle\}_{Q,P \in \mathcal{Q}}$$

belonging to $\mathbf{b}$ are not yet well-understood. By Sections 5-6, this collection is an algebra of operators bounded on $\dot{F}_p^{0q}$, $1 \leq p, q \leq +\infty$, which contains the reduced David-Journé class. In [FJ3] we considered the special case of Fourier multiplier operators. We showed that the class of multiplier operators satisfying Hörmander's criterion (or even some slightly weaker assumptions) have matrices belonging to **b** and hence are bounded. For these and some other associated results, we refer to [FJ3].

9.   PARAPRODUCTS AND THE FULL DAVID-JOURNÉ CLASS

We say that an operator $T \in CZO(\varepsilon)$ for some $\varepsilon > 0$ belongs to the full David-Journé class if $T \in WBP$, $T1 \in BMO$, and $T^*1 \in BMO$. The David-Journé theorem (Theorem 6.1) states that such operators are bounded on L^2. This result is obtained from the boundedness on L^2 of operators belonging to the reduced David-Journé class, consisting of those $T \in CZO(\varepsilon) \cap WBP$ satisfying $T1 = T^*1 = 0$ (Theorem 6.2). We now discuss the reduction step; i.e. the derivation of Theorem 6.1 from Theorem 6.2 (cf. [DJ] and [Me2]).

The reduction is accomplished using operators known as paraproducts. Let $\{\varphi_Q\}_{Q \in \mathcal{Q}}$ and $\{\psi_Q\}_{Q \in \mathcal{Q}}$ be as in the φ-transform identity (3.8). (Alternatively, the same argument can be carried out using the Lemarié-Meyer wavelets.) Let Φ belong to the class $\mathcal{B}$ defined in Section 2; in particular $\Phi \in \mathcal{S}$ and $\int \Phi = \hat{\Phi}(0) = 1$. For $v \in \mathbb{Z}$, let $\Phi_v(x) = 2^{vn}\Phi(2^v x)$, and, for $Q = Q_{vk} \in \mathcal{Q}$, let $\Phi_Q(x) = 2^{-vn/2}\Phi_v(x - 2^{-v}k)$, as usual. For $g \in BMO$, we define the paraproduct operator Π_g by

$$\Pi_g(f) = \sum_{Q \in \mathcal{Q}} \langle g, \varphi_Q \rangle |Q|^{-1/2} \langle f, \Phi_Q \rangle \psi_Q. \qquad (9.1)$$

A few remarks about Π_g are in order. Note that

$$|Q|^{-1/2}\langle f, \Phi_Q \rangle = f * \tilde{\Phi}_v(2^{-v}k) \text{ if } Q = Q_{vk}, \qquad (9.2)$$

where $\tilde{\Phi}_v(x) = \overline{\Phi_v(-x)}$. If $f * \tilde{\Phi}_v(2^{-v}k)$ is replaced in (9.1) by $f(x)$, the operator becomes the product operator $f \to f \cdot g$, by (3.8). Since $\tilde{\Phi}_v$ is an approximate identity, $f * \tilde{\Phi}_v(2^{-v}k)$ approximates $f(2^{-v}k)$, which is roughly equivalent to $f(x)$ for $x \in Q_{vk}$ (where $\psi_{Q_{vk}}$ is concentrated). This justifies the name "paraproduct." Of course the product operator $f \to f \cdot g$ is bounded on L^2 if and only if $g \in L^\infty$. By the next result, the paraproduct is L^2-bounded for $g \in BMO$.

Theorem 9.1.   *Suppose $g \in BMO$, and define Π_g by (9.1). Then*

$$\Pi_g \in CZO(1), \qquad (9.3)$$

$$\| \Pi_g f \|_{L^2} \leq c \| g \|_{BMO} \| f \|_{L^2}, \text{ for all } f \in L^2, \qquad (9.4)$$

$$\Pi_g 1 = g, \qquad (9.5)$$

and

$$\Pi_g^* 1 = 0. \qquad (9.6)$$

Assuming Theorem 9.1, the reduction step of the David-Journé theorem is easy. Suppose $T \in CZO(\varepsilon) \cap WBP$ for some $\varepsilon > 0$, $T1 = f \in BMO$, and $T^*1 = g \in BMO$. Let $S = T - \Pi_f - \Pi_g^*$. We have $S \in CZO(\varepsilon)$ since $T \in CZO(\varepsilon)$ and $\Pi_f, \Pi_g^* \in CZO(1) \subseteq CZO(\varepsilon)$. Also $\Pi_f, \Pi_g^* \in WBP$ since they are L^2-bounded; hence $S \in WBP$. We have

$$S1 = T1 - \Pi_f 1 - \Pi_g^* 1 = f - f = 0$$

and

$$S^*1 = T^*1 - \Pi_f^* 1 - \Pi_g 1 = g - g = 0.$$

Thus S belongs to the reduced David-Journé class. By Theorem 6.2, S is L^2-bounded; since Π_f and Π_g^* are L^2-bounded, so is T.

To prove Theorem 9.1 we require a few standard, important results in harmonic analysis.

Lemma 9.2. (Carleson's lemma, discrete version) *Let* $\{\lambda_Q\}_{Q \in \mathcal{Q}}$ *and* $\{s_Q\}_{Q \in \mathcal{Q}}$ *be complex sequences. Suppose*

$$\sup_{P \in \mathcal{Q}} \frac{1}{|P|} \sum_{Q \subseteq P, Q \in \mathcal{Q}} |\lambda_Q| = A < +\infty. \qquad (9.7)$$

Let $\omega(x) = \sup\{|s_Q| : x \in Q,\ Q \in \mathcal{Q}\}$. *Then*

$$\sum_{Q \in \mathcal{Q}} |\lambda_Q||s_Q| \le A \int_{\mathbb{R}^n} \omega(x)dx. \qquad (9.8)$$

Proof. (As in [Me2].) For $t > 0$, let

$$\Omega_t = \bigcup \{Q \in \mathcal{Q} : |s_Q| > t\} = \{x : \omega(x) > t\}.$$

Assuming $\omega \in L^1$, we have $|\Omega_t| < +\infty$. Let $\{J_t^{(k)}\}_{k=1}^\infty$ be the collection of maximal cubes $Q \in \mathcal{Q}$ such that $|s_Q| > t$. Then $|\Omega_t| = \sum_{k=1}^\infty |J_t^{(k)}|$.

For $Q \in \mathcal{Q}$ and $t > 0$, let $\chi(Q, t) = 1$ if $0 < t < |s_Q|$, and $\chi(Q, t) = 0$

otherwise. If $\chi(Q, t) = 1$, then $Q \subseteq \Omega_t$. Hence by (9.7),

$$\sum_{Q \in \mathcal{Q}} |\lambda_Q||s_Q| = \int_0^\infty \sum_{Q \in \mathcal{Q}} |\lambda_Q|\chi(Q, t)dt$$

$$\leq \int_0^\infty \sum_{\substack{Q \subseteq \Omega_t \\ Q \in \mathcal{Q}}} |\lambda_Q|dt = \int_0^\infty \sum_{k=1}^\infty \sum_{Q \subseteq J_t^{(k)}, Q \in \mathcal{Q}} |\lambda_Q|dt$$

$$\leq A \int_0^\infty \sum_{k=1}^\infty |J_t^{(k)}|dt = A \int_0^\infty |\Omega_t|dt = A \int_{\mathbb{R}^n} \omega(x)dx. \quad \square$$

Lemma 9.3. (L^2 **boundedness of the non-tangential maximal function**) *Suppose* $\Phi \in \mathcal{S}(\mathbb{R}^n)$ *and* $\int \Phi = 1$. *For* $x \in \mathbb{R}^n$, $t > 0$, *and* $f \in L^2(\mathbb{R}^n)$, *let*

$$u(x, t) = \int f(x, y)t^{-n}\Phi(y/t)dy.$$

For $\gamma \in (0, \infty)$, *let* $u_\gamma^*(x) = \sup\{|u(y, t)| : |x - y| < \gamma t\}$. *Then*

$$\| u_\gamma^* \|_{L^2(\mathbb{R}^n)} \leq c_{\gamma, \Phi} \| f \|_{L^2(\mathbb{R}^n)}. \tag{9.9}$$

This result is standard: u_γ^* is dominated by the Hardy-Littlewood maximal function of f. See, for example, [S].

Lemma 9.4. *Let* $f \in BMO$, *and let* $\{\varphi_Q\}_{Q \in \mathcal{Q}}$ *be as in (3.8). Then*

$$\sup_{P \in \mathcal{Q}} \frac{1}{|P|} \sum_{Q \subseteq P, Q \in \mathcal{Q}} |\langle f, \varphi_Q \rangle|^2 \leq c \| f \|_{BMO}^2. \tag{9.10}$$

In fact, the two sides of (9.10) are equivalent, by (3.9-10) for $\dot{F}_\infty^{02} \approx BMO$. Alternatively, a direct proof of (9.10) can be given by following the proof of one direction of the Carleson measure characterization of BMO in [FS2].

Assuming Lemmas 9.2-4, we turn to the proof of Theorem 9.1. It is easy to see that the kernel of Π_g is

$$K(x, y) = \sum_{Q \in \mathcal{Q}} \langle g, \varphi_Q \rangle |Q|^{-1/2}\overline{\Phi_Q(y)}\psi_Q(x).$$

By (9.10), taking $Q = P$, we have $|\langle g, \varphi_Q \rangle| \leq c|Q|^{1/2} \| g \|_{BMO}$. Since $\psi, \Phi \in \mathcal{S}$, ψ_Q and Φ_Q satisfy (5.7) and (5.10) with $K = 0$, $\delta = 1$, and $\sigma = n + 1$. Therefore Lemma 6.5 gives (9.3).

By (3.10) for $\dot{F}_2^{02} \approx L^2$,

$$\| \Pi_g f \|_{L^2}^2 \leq c \sum_{Q \in \mathcal{Q}} |\langle g, \varphi_Q \rangle|^2 |Q|^{-1} |\langle f, \Phi_Q \rangle|^2.$$

With $\lambda_Q = |\langle g, \varphi_Q \rangle|^2$, Lemma 9.4 shows that (9.7) holds with $A = c \| g \|_{BMO}^2$. Using (9.2) and Lemma 9.2, we have

$$\| \Pi_g f \|_{L^2}^2 \leq c \| g \|_{BMO}^2 \int_{\mathbb{R}^n} \sup \{ |f * \tilde{\Phi}_v(2^{-v}k)|^2 : x \in Q_{vk}, \, Q_{vk} \in \mathcal{Q} \} dx.$$

It is easy to see that the sup in the last integral is dominated by u_γ^* for $\tilde{\Phi}$ and some finite γ. Hence, by Lemma 9.3, (9.4) holds.

By the definition of Φ_Q, we have $\langle 1, \Phi_Q \rangle = \int \overline{\Phi}_Q = |Q|^{1/2} \int \overline{\Phi} = |Q|^{1/2}$. Hence, by (3.8),

$$\Pi_g 1 = \sum_{Q \in \mathcal{Q}} \langle g, \varphi_Q \rangle \psi_Q = g,$$

so (9.5) holds. It is easy to see that the adjoint of Π_g is defined by

$$\Pi_g^* f = \sum_{Q \in \mathcal{Q}} \overline{\langle g, \varphi_Q \rangle} |Q|^{-1/2} \langle f, \psi_Q \rangle \Phi_Q.$$

Since $\langle 1, \psi_Q \rangle = \int \overline{\psi}_Q = \overline{\hat{\psi}_Q(0)} = 0$, (9.6) holds. This completes the proof of Theorem 9.1.

The David-Journé theorem, the proof of the reduction step, and Theorem 6.3 show that an operator $T \in CZO(\varepsilon)$ for some $\varepsilon > 0$ is L^2-bounded if and only if $T = S + \Pi_f + \Pi_g^*$, where S is almost diagonal for L^2, and Π_f and Πg are paraproducts with $f, g \in BMO$.

Based on our discussion so far, it is natural to consider the class of paraproducts, or of all L^2-bounded CZO's, in terms of their matrices. Using orthonormal wavelets, a certain degree of analysis, which we will not go into here, is easy. However, some interesting questions remain open. For instance, we know ([Me2]) that the composition of CZO's is not necessarily a CZO; what is the algebra of matrices generated by the L^2-bounded CZO's? What natural, larger algebras of L^2-bounded operators exist, and how are they characterized in terms of their matrices? Returning to the theme of Section 1, what are the families of molecules and norming families corresponding to the paraproducts or to the algebra generated by the L^2-bounded CZO's?

REFERENCES

[BGS]   Burkholder, D.L., Gundy, R.F., and Silverstein, M.L., A maximal function characterization of the class H^p, Trans. Amer. Math. Soc. 157 (1971), 137-153.

[Cal1]   Calderón, A.P., Intermediate spaces and interpolation, the complex method, Studia Math. 24 (1964), 113-190.

[Cal2]   Calderón, A.P., An atomic decomposition of distributions in parabolic H^p spaces, Advances in Math. 25 (1977), 216-225.

[Cal3]   Calderón, A.P., Commutators, singular integrals on Lipschitz curves and applications, Proc. of Int. Congress of Math., Helsinki (1978), 85-96.

[Ch-F]   Chang, S.-Y. A., and Fefferman, R., A continuous version of duality of H^1 and BMO on the bidisc, Ann. of Math. 112 (1980), 179-201.

[Co]   Coifman, R.R., A real variable characterization of H^p, Studia Math. 51 (1974), 269-274.

[CMS]   Coifman, R.R., Meyer, Y., and Stein, E.M., Some new function spaces and their applications to harmonic analysis, J. Func. Anal. 62 (1985), 304-335.

[Dau1]   Daubechies, I., Orthonormal bases of compactly supported wavelets, Comm. Pure Appl. Math. 41 (1988), 909-996.

[Dau2]   Daubechies, I., The wavelet transform, time-frequency localization and signal analysis, IEEE Trans. Inf. Th. 36 (1990), 961-1005.

[Dau3]   Daubechies, I., Ten Lectures on Wavelets, NSF-CBMS Series in Applied Mathematics, SIAM (in press, 1991).

[DGM]   Daubechies, I., Grossman, A., and Meyer, Y., Painless nonorthogonal expansions, J. Math. Phys. 27 (1986), 1271-1283.

[DJ]   David, G., and Journé, J.-L., A boundedness criterion for generalized Calderón-Zygmund operators, Ann. of Math. 120 (1984), 371-397.

[FHJW]   Frazier, M., Han, Y.-S., Jawerth, B., and Weiss, G., The $T1$ theorem for Triebel-Lizorkin spaces, in Harmonic Analysis and

Partial Differential Equations, J. Garcia-Cuerva, et al., eds., Springer Lecture Notes in Math. 1384 (1989), 168-181.

[FJ1]   Frazier, M., and Jawerth, B., Decomposition of Besov spaces, Indiana Univ. Math. J. 34 (1985), 777-799.

[FJ2]   Frazier, M., and Jawerth, B., The φ-transform and applications to distribution spaces, in Function Spaces and Applications, M. Cwikel et al., eds., Springer Lecture Notes in Math. 1302 (1988), 223-246.

[FJ3]   Frazier, M., and Jawerth, B., A discrete transform and decompositions of distribution spaces, J. Func. Anal. 93 (1990), 34-170.

[FJW]   Frazier, M., Jawerth, B., and Weiss, G., Littlewood-Paley Theory and the Study of Function Spaces, CBMS Regional Conference Series 79, AMS, Providence, RI, 1991.

[FTW]   Frazier, M., Torres, R., and Weiss, G., The boundedness of Calderón-Zygmund operators on the spaces $\dot{F}_p^{\alpha q}$, Rev. Mat. Iberoamericana 4 (1988), 41-72.

[Fo-S] Folland, G.B., and Stein, E.M., Hardy Spaces on Homogeneous Groups, Princeton University Press, Princeton, NJ, 1982.

[FS1]   Fefferman, C., and Stein, E.M., Some maximal inequalities, Amer. J. Math. 93 (1971), 107-115.

[FS2]   Fefferman, C., and Stein, E.M., H^p spaces of several variables, Acta. Math. 129 (1972), 137-193.

[GM1]   Grossman, A., and Morlet, J., Decomposition of Hardy functions into square integrable wavelets of constant shape, SIAM J. Math. Anal. 15 (1984), 723-736.

[GM2]   Grossman, A., and Morlet, J., Decomposition of functions into wavelets of constant shape, and related transforms, in "Mathematics and Physics, Lectures on Recent Results," ed. L. Streit, World Scientific Publishing (Singapore), 1985.

[GMP]   Grossman, A., Morlet, J., and Paul, T., Transforms associated to square integrable group representations I. General results, J. Math. Phys. 27 (1985), 2473-2479; II. Examples, Ann. Inst. Henri Poincaré 45 (1986) 293-309.

[HW]   Heil, C., and Walnut, D., Continuous and discrete wavelet transforms, SIAM Review 31 (1989), 628-666.

[J]   Jawerth, B., On Besov spaces, Lund technical report, 1977:1.

[L]   Latter, R., A decomposition of $H^p(\mathbb{R}^n)$ in terms of atoms, Studia Math. 62 (1978), 92-101.

[LM]   Lemarié, P., and Meyer, Y., Ondelettes et bases hilbertiennes, Rev. Mat. Iberoamericana 2 (1986), 1-18.

[LP]   Littlewood, J.E., and Paley, R.E.A.C., Theorems on Fourier series and power series I, II, J. London Math. Soc. 6 (1931), 230-233, Proc. London Math. Soc. 42 (1936), 52-89.

[Ma]   Mallat, S., Multiresolution representation and wavelets, Ph.D. Dissertation, Elect. Eng. Dept., Univ. of Pennsylvania (1988).

[Me1]   Meyer, Y., Les nouveaux opérateurs de Calderón-Zygmund, Colloque en l'honneur de L. Schwartz, I, Astérisque 131 (1985), 237-254.

[Me2]   Meyer, Y., Wavelets and operators, in Analysis at Urbana, E. Berkson, et al., eds., London Math. Soc. Lecture Note Series 137, Cambridge Univ. Press 1989, 256-364.

[Me3]   Meyer, Y., Ondelettes et Opérateurs I, II, III, Hermann ed., Paris 1990.

[P1]   Peetre, J., On spaces of Triebel-Lizorkin type, Ark. Mat. 13 (1975), 123-130.

[P2]   Peetre, J., New Thoughts on Besov Spaces, Duke Univ. Math. Series, Durham, N.C. 1976.

[S]   Stein, E.M., Singular Integrals and Differentiability Properties of Functions, Princeton Univ. Press, Princeton, NJ, 1970.

[SW]   Stein, E.M., and Weiss, G., On the theory of harmonic functions of several variables, I: the theory of H^p spaces, Acta Math. 103 (1960), 25-62.

[Str]   Strömberg, J.-O., A modified Franklin system and higher order spline systems on $\mathbb{R}^n$ as unconditional bases for Hardy spaces, Conference on Harmonic Analysis in Honor of Antoni Zygmund, Vol. II, Beckner et al., eds., Univ. of Chicago Press, Chicago (1981), 475-494.

[Ta]   Taibleson, M., On the theory of Lipschitz spaces of distributions on Euclidean n-space I, II, III, J. Math. and Mech. 13 (1964), 407-480, 14 (1965), 821-840, 15 (1966), 973-981.

[Torch] Torchinsky, A., Real Variable Methods in Harmonic Analysis, Academic Press, Orlando, Florida, 1986.

[Torr] Torres, R., Boundedness results for operators with singular kernels on distribution spaces, Mem. Amer. Math. Soc. 442, 1991.

[Tr] Triebel, H., Theory of Function Spaces, Monographs in Mathematics Vol. 78, Birkhauser Verlag, Basel, 1983.

[U] Uchiyama, A., A constructive proof of the Fefferman-Stein decomposition of $BMO(\mathbb{R}^n)$, Acta. Math. 148 (1982), 215-241.

[W] Wilson, J.M., On the atomic decomposition for Hardy spaces, Pacific J. Math. 116 (1985), 201-207.

[Z] Zygmund, A., Trigonometric Series, Cambridge Univ. Press, Cambridge, 1959.

ON CARDINAL SPLINE-WAVELETS

CHARLES K. CHUI*

Center for Approximation Theory
Texas A&M University
College Station, Texas

1. INTRODUCTION

A distinct feature of the short-time Fourier transform (STFT, also known as the windowed Fourier transform) introduced by D. Gabor [26] is that for time-frequency localizations the time-frequency window has constant size at all frequencies. This inflexibility of the STFT limits its range of applications in the study of non-stationary signals which vary between very low and very high frequencies, since high-frequency signals require narrow time-windows for accuracy and low-frequency signals require wide time-windows for studying complete cycles. Hence, a collection of time-frequency windows whose widths are frequency dependent adjusts by itself is needed. The integral wavelet transform (IWT):

$$(W_\psi f)(b, a) := \frac{1}{\sqrt{a}} \int_{-\infty}^{\infty} f(t) \overline{\psi\left(\frac{t-b}{a}\right)} dt, \qquad (1.1)$$

*Supported by NSF Grant DMS 89-0-01345 and SDIO/IST managed by the U.S. Army Research Office under Contract No. DAAL 03-90-G-0091.

419

introduced by Grossmann and Morlet [29], indeed has this so-called "zoom-in" and "zoom-out" capability, if the "scale" a is used to represent a constant multiple of the reciprocal of the frequency, or if $\log a$ is used as the frequency itself (cf. [28]). For more details, the reader is referred to the CBMS Lecture Notes by Daubechies [22].

For effective time-frequency localizations the "mother wavelet" ψ in (1.1) together with its Fourier transform $\hat{\psi}$ must have rapid decay; and in order to recover the "signal" f from its IWT $(W_\psi f)(b, a)$, ψ must also satisfy:

$$\int_{-\infty}^{\infty} \frac{|\hat{\psi}(\omega)|^2}{|\omega|} d\omega < \infty, \qquad (1.2)$$

so that under certain very mild conditions, we have

$$\int_{-\infty}^{\infty} \psi(t) dt = \hat{\psi}(0) = 0.$$

This zero-mean condition and the decay property of ψ suggests the name of "ondelette" or "wavelet" for ψ (cf. [35]).

If ψ is "suitably" selected, then the values of the IWT, $(W_\psi f)(b, a)$ at the dyadic values $b = \frac{j}{2^k}$ and on the binary scale-levels $a = 2^{-k}$, where j and k are integers, can be obtained by very efficient algorithms; and conversely, the "signal" f can be recovered from the dyadic-binary samples of $(W_\psi f)(b, a)$ via equally efficient algorithms. This is the main theme of our paper. Hence, in the following, we will only consider "wavelets" ψ that admit such algorithms. To be more specific, in order to consider such a ψ, we must also consider its "dual" $\tilde{\psi}$, where both ψ and $\tilde{\psi}$ (along with their Fourier transforms) must have sufficiently fast decay at $\pm\infty$, and they both satisfy (1.2). Our requirements on ψ and $\tilde{\psi}$ are as follows:

(i) Both doubly indexed sequences $\{2^{k/2}\psi(2^k x - j)\}$ and $\{2^{k/2}\tilde{\psi}(2^k x - j)\}$, $j, k \in \mathbb{Z}$, are unconditional bases of $L^2 := L^2(-\infty, \infty)$; and

(ii) these two bases are biorthogonal, in the sense of

$$\int_{-\infty}^{\infty} \psi(2^k x - j)\overline{\tilde{\psi}(2^\ell x - m)} dx = 2^{-k}\delta_{k,\ell}\delta_{j,m}. \qquad (1.3)$$

The main reason for the above requirements is that with ψ and its dual $\tilde{\psi}$, we have a reproducing kernel $K(x, y)$ of L^2, namely:

$$K(x, y) = \sum_{k,j \in \mathbb{Z}} 2^k \overline{\tilde{\psi}(2^k y - j)}\psi(2^k x - j), \qquad (1.4)$$

so that any $f \in L^2$ has the wavelet-series representation:

$$f(x) = \sum_{k,j \in \mathbb{Z}} c_{k,j} 2^{k/2} \psi(2^k x - j).$$ (1.5)

Here, the coefficient sequence $\{c_{k,j}\}$ yields the IWT of f, with the mother wavelet $\tilde{\psi}$, evaluated at the dyadic-binary samples, namely:

$$c_{k,j} = (W_{\tilde{\psi}} f) \left(\frac{j}{2^k}, \frac{1}{2^k} \right).$$ (1.6)

Hence, algorithms to compute these coefficients in the wavelets series (1.5) and to recover f from these coefficients are now plausible.

Let us consider the following situations:

Case I.   Suppose that

$$\{2^{k/2} \psi(2^k x - j)\}, \qquad j, k \in \mathbb{Z},$$

is an orthonormal basis of L^2. Then obviously ψ is self-dual; that is, $\tilde{\psi} \equiv \psi$.

Case II.   Suppose that for any fixed k and ℓ in $\mathbb{Z}$, with $k \neq \ell$, we have:

$$\{\psi(2^k x - j)\}_{j \in \mathbb{Z}} \perp \{\psi(2^\ell x - j)\}_{j \in \mathbb{Z}}.$$

Then we can find the dual $\tilde{\psi}$ of ψ by

$$\hat{\tilde{\psi}}(\omega) = \frac{\hat{\psi}(\omega)}{\left[\sum_{j \in \mathbb{Z}} |\hat{\psi}(\omega + 2\pi j)|^2 \right]^{1/2}}.$$ (1.7)

Here, because of the assumption (i) on ψ, the denominator in (1.7) never vanishes.

Case III.   No orthogonality is assumed.

It is obvious that Case II is more general than Case I, and Case III, which is obviously the most general, is itself a special case of the yet more general concept of frames. As discussed elsewhere (see the introduction and the chapter by Feichtinger and Grochenig), frames satisfy a reconstruction formula which is formally identical to (1.5); however, the two sets of dual functions are biorthogonal if and only if each set is linearly independent, i.e. forms a basis.

By using the structure of "multiresolution analysis" of L^2 due to Meyer [35] and Mallat [34] (which we will briefly discuss in Section 2), Daubechies [21] was able to construct a large class of compactly supported "orthonormal" wavelets ψ. Other classes of examples in Case I are the Meyer wavelets (cf. [35]), and the globally supported spline-wavelets constructed independently by Battle [2] and Lemarié [32]. From both the theoretical and applied points of view, the compactly supported wavelets of Daubechies are certainly most attractive. However, with the exception of the Haar function, it was also shown in [21] that real-valued compactly supported orthonormal wavelets are not symmetric nor antisymmetric. Since a wavelet is used as a bandpass filter in signal analysis (cf. e.g. [17, 22]), and symmetry and antisymmetry are equivalent to linear-phase and generalized linear phase, respectively (cf. [14]), other wavelets are needed in order to avoid any distortion. So, in order to retain the minimum support of ψ, one must give up a certain structure in ψ. This brings us to Case II. In Chui and Wang [13], compactly supported spline-wavelets of arbitrary orders are constructed as examples in this case. These spline-wavelets have minimum supports and are symmetric or antisymmetric depending on the even or odd order of the spline spaces. Furthermore, explicit formulas are also given in [13]. However, their duals $\tilde{\psi}$ are not compactly supported, although they have exponential decay. The formulations and properties of these wavelets will be discussed in Section 2. Generalizations to cover all the situations in Case 2 were done independently by Micchelli [36] and Chui and Wang [14]. In order to obtain compactly supported ψ and $\tilde{\psi}$, while requiring symmetry or antisymmetry, we have to give up orthogonality. Examples for this general setting were constructed in Cohen [17], Feauveau [25], and Cohen, Daubechies, Feauveau [18]. This approach was also considered independently by Vetterli and Herley [40] in the design of filter banks. Wavelets with much less structure have also been used very successfully in many applications, such as data compression in DeVore, Jawerth, and Lucier [24].

2.   CARDINAL SPLINE-WAVELETS

It is well-known that every cardinal B-spline generates a multiresolution analysis of $L^2 = L^2(-\infty, \infty)$ (cf. [34,35]). The nth order cardinal B-spline is nothing but the probability density of the random sum $\sum_{j=1}^{n} t_i$ where $t_1, \ldots, t_n$ are random variables uniformly distributed on $[0,1]$. From a deterministic point of view, N_1 is the characteristic function of the interval $[0,1)$, and the cardinal B-spline of order n is obtained recursively by

$$N_n(x) = \int_0^1 N_{n-1}(x - t)dt,$$

$n = 2, 3, \ldots$. Explicit formulas of N_n in terms of piecewise Bernstein polynomials can be easily computed (cf. [5]) and very efficient recursive formulas
of N_n and its derivative N_n' in terms of N_{n-1} are available (cf. [3, 5, 39]).
For an arbitrary but fixed positive integer n, the space of cardinal splines
of order n with knot sequence $\mathbb{Z}$, the set of all integers, is the collection $\mathcal{S}_n$
of functions f in $C^{n-2}(\mathbb{R})$ such that the restrictions of f on $[j, j+1]$, $j \in \mathbb{Z}$,
are polynomials of degree at most $n-1$. It is well-known that every $f \in \mathcal{S}_n$
can be expressed as a cardinal B-spline series, namely:

$$f(x) = \sum_{j \in \mathbb{Z}} a_j N_n(x - j), \qquad x \in R,$$

where, since the support of N_n, given by

$$\text{supp } N_n = [0, n],$$

is finite, pointwise convergence of the B-spline series is valid for all sequences $\{a_j\}$ (cf. [3]). To describe the multiresolution analysis generated
by N_n, let V_0 denote the L^2-closure of $\mathcal{S}_n$; that is,

$$V_0 = \mathcal{S}_n \cap L^2,$$

and for each $k \in \mathbb{Z}$, set

$$V_k = \{f(x): \ f(2^{-k}x) \in V_0\}.$$

Then it follows that

(i) $\cdots \subset V_{-2} \subset V_{-1} \subset V_0 \subset V_1 \subset \cdots$;

(ii) $f(x) \in V_k \Leftrightarrow f(2x) \in V_{k+1}$, $k \in \mathbb{Z}$; and it can also be shown that

(iii) $\text{clos}_{L^2} \left(\bigcup_{k \in \mathbb{Z}} V_k \right) = L^2$;

(iv) $\bigcap_{k \in \mathbb{Z}} V_k = \{0\}$; and

(v) for each $k \in \mathbb{Z}$,

$$2^{-k} A \|\{c_n\}\|_{\ell^2}^2 \leq \left\| \sum_{n \in \mathbb{Z}} c_n N_n(2^k x - n) \right\|_{L^2}^2 \leq 2^{-k} B \|\{c_n\}\|_{\ell^2}^2,$$

for all ℓ^2-sequences $\{c_n\}$, where A and B are the infimum and supremum, respectively, of the spectrum of the positive definite symmetric
banded Toeplitz matrix $[a_{i-j}]$ with

$$a_j = \int_{-\infty}^{\infty} N_n(x)N_n(x-j)dx = N_{2n}(n+j).$$

Property (v) says that $\{N_n(2^k x - j)\}$, $j \in \mathbb{Z}$, is an unconditional basis
of V_k. With the properties (i)–(v), the nth order cardinal B-spline N_n is
said to generate a multiresolution analysis of L^2 (cf. [34, 35]).

Following Mallat [34] and Meyer [35], we consider for each $k \in \mathbb{Z}$, the
orthogonal complementary subspace W_k of V_{k+1} relative to V_k; that is,

$$\begin{cases} W_k \text{ is a subspace of } V_{k+1} \text{ ;} \\ V_{k+1} = V_k + W_k \text{ ; and} \\ V_k \bot W_k, \end{cases}$$

and the notation

$$V_{k+1} = V_k \oplus W_k$$

will be used. Hence, it follows from (i), (iii), and (iv) that $W_k \bot W_\ell$ for
all $k \neq \ell$, and

$$L^2 = \bigoplus_{k \in \mathbb{Z}} W_k;$$

and consequently, every function $f \in L^2$ has a (mutually) orthogonal decomposition:

$$f = \cdots \oplus g_{-2} \oplus g_{-1} \oplus g_0 \oplus g_1 \oplus \cdots , \qquad (2.1)$$

where $g_k \in W_k$, $k \in \mathbb{Z}$. We will call the spaces W_k "wavelet spaces"
and (2.1) the complete orthogonal wavelet decomposition of $f \in L^2$. In
the same manner as the cardinal B-spline N_n generates the spline spaces
V_k, $k \in \mathbb{Z}$, the compactly supported spline function

$$(2.2)\psi_n(x) = 2^{-n+1} \sum_{j=0}^{2n-2} (-1)^j N_{2n}(j+1)N_{2n}^{(n)}(2x-j)$$

$$= 2^{-n+1} \sum_{j=0}^{3n-2} (-1)^j \left\{ \sum_{\ell=0}^{n} \binom{n}{\ell} N_{2n}(j-\ell+1) \right\} N_n(2x-j),$$

introduced in [13], generates the wavelet spaces W_k, $k \in \mathbb{Z}$; that is, for
each $k \in \mathbb{Z}$,

$$W_k = \text{clos}_{L^2}(\text{span}\{\psi_n(2^k x - j): j \in \mathbb{Z}\}).$$

(We remark that a Fourier transform consideration was discussed in [1].) In addition, analogous to N_n having minimum support among all nontrivial functions in V_0, ψ_n also has minimum support among all nontrivial functions in W_0 (cf. [13, 14]). For this reason, we call ψ_n a *cardinal B-spline-wavelet (or* simply, *B-wavelet) of order* n.

We remark that although $\psi_n(2^k x - j)$ is orthogonal to $\psi_n(2^m x - \ell)$ for all j and $\ell \in \mathbb{Z}$, provided $k \neq m$, the unconditional basis $\{\psi_n(2^k x - j): j \in \mathbb{Z}\}$ of W_k is not an orthogonal basis, unless $n = 1$. This is Case II in Section 1. That is, we do not have a compactly supported "orthogonal wavelet" as those constructed by Daubechies [21]. Hence, as discussed in Section 1 we also need the dual $\widetilde{\psi}_n$ of ψ_n, defined by the biothogonality condition (1.3) (cf. [13,14,18]). In [13], it was shown that this dual $\widetilde{\psi}_n$ is given by

$$\widetilde{\psi}_n(x) = (-1)^{n+1} 2^{-n+1} \sum_{j \in \mathbb{Z}} \alpha_j \eta_n(x - n + 1 - j), \qquad (2.3)$$

where

$$\eta_n(x) = L_{2n}^{(n)}(2x - 1) \qquad (2.4)$$

and $\{\alpha_j\}$ is the coefficient sequence in

$$L_{2n}(x) = \sum_{j \in \mathbb{Z}} \alpha_j N_{2n}(x + n - j). \qquad (2.5)$$

Here, L_{2n} is the $(2n)^{\text{th}}$ order fundamental cardinal spline in $\mathcal{S}_{2n}$, determined uniquely by the interpolatory condition:

$$L_{2n}(j) = \delta_{j,0}, \quad j \in \mathbb{Z}.$$

For this reason, we call η_n the nth order *interpolatory wavelet*. This wavelet η_n in (2.4) was introduced in [12] and the properties of $\{\alpha_j\}$ are well documented in the cardinal spline literature (cf. [38]). We remark that both

$$\{\eta_n(x - j): j \in \mathbb{Z}\}$$

and

$$\{\widetilde{\psi}_n(x - j): j \in \mathbb{Z}\}$$

are also unconditional bases of W_0. (For a more general framework of such wavelets, see [14 and 36].) Now, with both the cardinal B-wavelet ψ_n and its dual $\widetilde{\psi}_n$, we have a reproducing kernel

$$K_n(x, y) = \sum_{k \in \mathbb{Z}} \sum_{j \in \mathbb{Z}} 2^k \widetilde{\psi}_n(2^k y - j) \psi_n(2^k x - j) \qquad (2.6)$$

of L^2 as in (1.4) (cf. [13]). That is,

$$\int_{-\infty}^{\infty} f(y)K_n(x,y)dy = f(x)$$

for all $f \in L^2$. Consequently, every $f \in L^2$ can be expressed as

$$f(x) = \sum_{k \in \mathbb{Z}} \sum_{j \in \mathbb{Z}} 2^{\frac{k}{2}} \left(W_{\widetilde{\psi}_n} f \right) \left(\frac{j}{2^k}, \frac{1}{2^k} \right) \psi_n(2^k x - j), \qquad (2.7)$$

where $(W_{\widetilde{\psi}_n} f)(b, a)$ is the IWT of f with $\widetilde{\psi}_n$ as the mother wavelet, defined in (1.1) (cf. [28]). By the uniqueness of the complete orthogonal wavelet decomposition of f in (2.1), we note from (2.7) that the kth wavelet component g_k of f is given by

$$g_k(x) = \sum_{j \in \mathbb{Z}} 2^{\frac{k}{2}} \left(W_{\widetilde{\psi}_n} f \right) \left(\frac{j}{2^k}, \frac{1}{2^k} \right) \psi_n(2^k x - j),$$

where the jth coefficient is the $2^{\frac{k}{2}}$ multiple of the integral wavelet transform of f relative to $\widetilde{\psi}_n$ at the dyadic point $\frac{j}{2^k}$ of the kth binary scale.

In applications, every $f \in L^2$ must first be projected onto some appropriate spline space V_N. For instance, in numerical solutions of differential or integral equations, standard spline methods combined with the Galerkin and collocation methods, can be applied; and in other applications, where more explicit information of the data function f can be extracted, quasi-interpolation and even local interpolation methods by splines (cf. [7, 8,9,10]) are particularly useful. (In a somewhat different approach, periodic spline wavelets have been used [33] to solve non-linear partial differential equations; this work is summarized in the chapter by Liandrat and Tchamatchian. By contrast, Lawton and collaborators at Aware [27] have used the Daubechies wavelets in the standard Galerkin method to solve non-linear PDE.) Let $f_N \in V_N$ be such a spline approximant of $f \in L^2$. We then have

$$f_N = g_{N-1} \oplus \cdots \oplus g_{N-M} \oplus f_{N-M}, \qquad (2.8)$$

where

$$\begin{cases} g_k(x) = \sum_{j \in \mathbb{Z}} d_j^k \psi_n(2^k x - j) \in W_k; \\ f_k(x) = \sum_{j \in \mathbb{Z}} c_j^k N_n(2^k x - j) \in V_k. \end{cases} \qquad (2.9)$$

The (mutually) orthogonal decomposition of f_N in (2.8) may be considered as a finite orthogonal wavelet decomposition of f, provided that both

$$\|f - f_N\|_{L^2} \quad \text{and} \quad \|f_{N-M}\|_{L^2}$$

are sufficiently small. This can always be achieved by taking sufficiently large values of M and N, in view of Properties (iii) and (iv) of the multiresolution analysis.

A pyramid decomposition algorithm can be used to compute

$$\begin{cases} \mathbf{d}^k := \{d_j^k\} \\ \mathbf{c}^m := \{c_j^m\} \end{cases} \tag{2.10}$$

for $k = N-1, \ldots, N-M$ and $m = N-M$ from $\mathbf{c}^N$; and a pyramid reconstruction algorithm can be applied to recover $\mathbf{c}^N$ from $\mathbf{c}^{N-M}, \mathbf{d}^{N-M}, \ldots,$ $\mathbf{d}^{N-1}$ (cf. [12]). The sequences that are needed for the "moving average" reconstruction algorithm are finite sequences, while those which give the decomposition algorithm have exponential decay.

The objective of this paper is to highlight some of the important results on cardinal-spline wavelets as introduced above, including an error estimate induced by "truncation" of the decomposition sequences, symmetric (or anti-symmetric) properties of ψ_n and $\widetilde{\psi}_n$ required for linear-phase filtering, oscillatory behavior of any wavelet series $g_k \in W_k$, and computational aspects of the cardinal B-wavelet ψ_n. In addition we mention, very briefly, the possibility of constructing (non-periodic) spline-wavelets on a bounded interval and compactly supported (non-tensor-product) box-spline wavelets in two and more variables.

3. DECOMPOSITION AND RECONSTRUCTION SEQUENCES

That V_k is a subspace of V_{k+1} for all $k \in \mathbb{Z}$ is governed by the so-called "two-scale" identity:

$$N_n(2^k x) = \sum_{j \in \mathbb{Z}} p_j N_n(2^{k+1} x - j), \tag{3.1}$$

where $\{p_j\}$ is a finite sequence with support $[0, n]$ and

$$p_j = \frac{1}{2^{n-1}} \binom{n}{j}, \qquad 0 \le j \le n. \tag{3.2}$$

From (2.2), we note that the relation $W_k \subset V_{k+1}$, $k \in \mathbb{Z}$, is described by the other "two-scale" identity:

$$\psi_n(2^k x) = \sum_{j \in \mathbb{Z}} q_j N(2^{k+1} x - j), \tag{3.3}$$

where $\{q_j\}$ is also a finite sequence with support $[0, 3n - 2]$ and

$$q_j = \frac{(-1)^j}{2^{n-1}} \sum_{\ell=0}^{n} \binom{n}{\ell} N_{2n}(j - \ell + 1), \qquad 0 \le j \le 3n - 2. \tag{3.4}$$

The two finite sequences $\{p_j\}$ and $\{q_j\}$ yield a reconstruction algorithm:

$$c_j^{k+1} = \sum_{m \in \mathbb{Z}} \{p_{j-2m} c_m^k + q_{j-2m} d_m^k\}, \tag{3.5}$$

where $\mathbf{c}^k = \{c_j^k\}$ and $\mathbf{d}^k = \{d_j^k\}$ uniquely determine the functions f_k and g_k as in (2.9) and (2.10), and a schematic diagram of the "pyramid" algorithm is shown below

$$\mathbf{d}^{N-M} \qquad \mathbf{d}^{N-M+1} \qquad\qquad\qquad \mathbf{d}^{N-1}$$
$$\searrow \qquad\quad \searrow \qquad\quad \searrow \qquad\quad \searrow$$
$$\mathbf{c}^{N-M} \ \longrightarrow\ \mathbf{c}^{N-M+1} \ \longrightarrow\ \cdots \ \longrightarrow\ \mathbf{c}^{N-1} \ \longrightarrow\ \mathbf{c}^N \ .$$

To describe the decomposition algorithm, we apply the coefficient sequence $\{\alpha_j\}$ in (2.5) and the finite sequences in (3.2) and (3.3) to form the infinite sequences

$$(3.6) \qquad\qquad a_j = \sum_{\ell \in \mathbb{Z}} q_{j+2n-1-2\ell} \alpha_\ell;$$

$$(3.7) \qquad\qquad b_j = -\sum_{\ell \in \mathbb{Z}} p_{j+2n-1-2\ell} \alpha_\ell.$$

Note that since $\{\alpha_j\}$ is of exponential decay (cf. [38]), so are $\{a_j\}$ and $\{b_j\}$. It was established in [13] that the relation $V_{k+1} = V_k + W_k$, $k \in \mathbb{Z}$, is described by the identity

$$N_n(2^{k+1} x - \ell) = \sum_{j \in \mathbb{Z}} [a_{\ell-2j} N_n(2^k x - j) + b_{\ell-2j} \psi_n(2^k x - j)]; \tag{3.8}$$

which gives a decomposition algorithm:

$$\begin{cases} c_j^k = \displaystyle\sum_{m \in \mathbb{Z}} a_{m-2j} c_m^{k+1}; \\[2mm] d_j^k = \displaystyle\sum_{m \in \mathbb{Z}} b_{m-2j} c_m^{k+1} \end{cases} \tag{3.9}$$

A schematic diagram of this "pyramid" algorithm is shown below:

$$
\begin{array}{ccccccccc}
\mathbf{d}^{N-1} & & \mathbf{d}^{N-2} & & & & \mathbf{d}^{N-M} \\
& \nearrow & & \nearrow & & \nearrow & & \nearrow \\
\mathbf{c}^N & \longrightarrow & \mathbf{c}^{N-1} & \longrightarrow & \mathbf{c}^{N-2} & \longrightarrow & \cdots & \longrightarrow & \mathbf{c}^{N-M}
\end{array}
$$

Observe, however, that although $\{a_j\}$ and $\{b_j\}$ are of exponential decay, they must still be truncated in order to apply the algorithm (3.9). In [16], instead of truncating these sequences directly, we truncate the sequence $\{\alpha_j\}$, so that a fairly sharp estimate can be attained. In other words, we consider

$$
\alpha_j^P := \begin{cases} \alpha_j & \text{for} \quad |j| \le P \\[2mm] 0 & \text{for} \quad |j| > P, \end{cases} \tag{3.10}
$$

and define, as in (3.6) and (3.7),

$$
\begin{cases} a_j^P := \displaystyle\sum_{\ell \in \mathbb{Z}} q_{j+2n-1-2\ell}\alpha_\ell^P \\[3mm] b_j^P := -\displaystyle\sum_{\ell \in \mathbb{Z}} p_{j+2n-1-2\ell}\alpha_\ell^P, \end{cases} \tag{3.11}
$$

where, of course, the finite sequences $\{p_j\}$ and $\{q_j\}$ are given in (3.2) and (3.3), respectively. Observe that

$$
\begin{cases} a_j^P = 0 & \text{for} \quad j > 2P + n - 1 \text{ or } j < -2P - 2n + 1 \\[2mm] b_j^P = 0 & \text{for} \quad j > 2P - n + 1 \text{ or } j < -2P - 2n + 1. \end{cases}
$$

Let us now use $\{a_j^P\}$ and $\{b_j^P\}$ in the decomposition algorithm (3.9), namely:

$$
\begin{cases} \tilde{c}_j^k = \displaystyle\sum_{m \in \mathbb{Z}} a_{m-2j}^P c_m^{k+1} \\[3mm] \tilde{d}_j^k = \displaystyle\sum_{m \in \mathbb{Z}} b_{m-2j}^P c_m^{k+1}, \end{cases} \tag{3.12}
$$

and apply the (exact) reconstruction algorithm (3.5) to compare $\tilde{\mathbf{c}}^k$ with $\mathbf{c}^k$, namely:

$$
\begin{aligned}
\tilde{c}_j^k &= \sum_{m \in \mathbb{Z}} \{ p_{j-2m}\tilde{c}_m^{k-1} + q_{j-2m}\tilde{d}_m^{k-1} \} \\
&= \sum_{m \in \mathbb{Z}} \left\{ p_{j-2m} \sum_{\ell \in \mathbb{Z}} a_{\ell-2j}^P c_\ell^k + q_{j-2m} \sum_{\ell \in \mathbb{Z}} b_{\ell-2j}^P c_\ell^k \right\} \\
&= \sum_{\ell \in \mathbb{Z}} \left\{ \sum_{m \in \mathbb{Z}} [p_{j-2m}a_{\ell-2j}^P + q_{j-2m}b_{\ell-2j}^P] \right\} c_\ell^k.
\end{aligned}
$$

This allows us to study the ℓ^2-error:

$$E_{n,P}(\mathbf{c}^k) := \left\{ \sum_{j \in \mathbb{Z}} (\tilde{c}_j^k - c_j^k)^2 \right\}^{1/2}, \tag{3.13}$$

where the subscript n indicates that the nth order cardinal splines and wavelets are used and the subscript P defines the truncation in (3.10). In [16], we obtained a fairly general result on this estimate, which yields the following:

$$E_{2,P}(\mathbf{c}^k) \leq 2.73205080757 \varepsilon_2^{P+1} \|\mathbf{c}^k\|_{\ell^2};$$
$$E_{3,P}(\mathbf{c}^k) \leq 4.3269115534 \varepsilon_3^{P+1} \|\mathbf{c}^k\|_{\ell^2};$$
$$E_{4,P}(\mathbf{c}^k) \leq 7.83737467851 \varepsilon_4^{P+1} \|\mathbf{c}^k\|_{\ell^2},$$

where $\varepsilon_2 = 0.267949192431$, $\varepsilon_3 = 0.4305753471$, and $\varepsilon_4 = 0.535280430796$. Details and derivations can be found in [16].

4. PROPERTIES OF CARDINAL SPLINE-WAVELETS

It is clear that all the cardinal B-splines N_n are symmetric with respect to $n/2$. Hence, it follows from the symmetry or anti-symmetry of the sequence $\{q_j\}$, that the cardinal B-wavelet ψ_n is symmetric for even n relative to the center of its support and anti-symmetric for odd n. Also, since the fundamental cardinal spline L_{2n} is an even function and the sequence $\{\alpha_j\}$ is symmetric with respect to $j = 0$, the interpolatory wavelets η_n and dual wavelets $\widetilde{\psi}_n$ are also symmetric for even n and anti-symmetric for odd n. The importance of this property is two-fold: first, storage and computation are both reduced; and second, when wavelets are used for time-frequency analysis, linear-phase filtering is always achieved. For details, the reader is referred to [14].

Perhaps the most exciting property of cardinal spline-wavelets as observed in [15] is its contrast to the so-called "total positivity" of B-splines (cf. [31]). As is well-known, total positivity is the most important ingredient in B-splines that governs sign changes, zero counts, and shape characteristics of spline functions. For instance, if we use the standard notation S^- and S^+ for the counts of strong and weak sign changes, respectively (in the interior of the supports) (cf. [3,31,39]), then it is well known that

$$S^- \left(\sum_{j=0}^{N} c_j N_n(\cdot - j) \right) \leq S^-(\{c_j\}) \tag{4.1}$$

for any finite sequence $\{c_j\}$. For the cardinal B-wavelet series

$$g(x) = \sum_{j=0}^{N} c_j \psi_n(x - j),$$

where we may assume, without loss of generality,

$$c_0 c_N \neq 0, \quad |c_\ell| + \cdots + |c_{\ell+2n-2}| > 0, \quad \ell = 0, \ldots, N - 2n + 2 \quad (4.2)$$

(in order to ensure that supp g is an interval), the following result was obtained in [15].

Theorem 4.1. *Under the assumption (4.2),*

$$S^- \left(\sum_{j=0}^{N} c_j \psi_n(\cdot - j) \right) \geq N + 3n - 2. \quad (4.3)$$

In particular, ψ_n has exactly $3n - 2$ simple zeros in $(0, 2n - 1)$, which is the interior of supp ψ_n.

Since it is clear that the above lower bound is bounded below by

$$S^+(\{c_j\}) + 3n - 2,$$

the "oscillation" property (4.3) of wavelets is a contrast to the property of total positivity of splines. In fact, we believe that even the following statement holds.

Conjecture. *Under the assumption (4.2) and certain restriction of the change of magnitudes of c_j,*

$$2N + 3n - 2 - S^-(\{c_j\}) - s_n \leq S^+ \left(\sum_{j=0}^{N} c_j \psi_n(\cdot - j) \right) \leq 2N + 3n - 2 - s_n \quad (4.4)$$

and in particular, if $S^-(\{c_j\}) = 0$, then

$$S^+ \left(\sum_{j=0}^{N} c_j \psi_n(\cdot - j) \right) = 2N + 3n - 2 - s_n, \quad (4.5)$$

where s_n is some non-negative integer depending only on the order n.

To support this conjecture, we established the following result in [15].

Theorem 4.2.   *The above conjecture (4.4), and hence (4.5), holds for* $n = 2$ *with* $s_2 = 0$.

In other words, in contrast to the property of "total positivity" of the cardinal B-spline N_n, we believe that the cardinal B-wavelet ψ_n possesses the property of "**complete oscillation**". An accurate statement of this concept has yet to be formulated. Graphs of the wavelets $\psi_n, \tilde{\psi}_n,$ and η_n for $n = 2, 3, 4$ that illustrate this oscillatory behavior are shown in Figures 1-3.

5.   COMPUTATION OF CARDINAL SPLINE-WAVELETS

In order to work only with integers, we consider

$$\Psi_n(x) = (2n - 1)!\psi_n(x) \tag{5.1}$$

and set

$$\Psi_n(x) = 2^{-n+1} \sum_{j=0}^{3n-2} (-1)^j \gamma_j^n N_n(2x - j). \tag{5.2}$$

In view of (2.2), we then have

$$\begin{cases} \gamma_j^n = \displaystyle\sum_{\ell=0}^n b_\ell^n e^{2n-1}(j - \ell), \text{ where} \\[2mm] e^{2n-1}(j) = (2n - 1)!N_{2n}(j + 1), \text{ and} \\[2mm] b_\ell^n = \dbinom{n}{\ell} = \dfrac{n!}{\ell!(n - \ell)!}. \end{cases} \tag{5.3}$$

Since computational procedures for the cardinal B-splines are well documented in the literature (cf. [3,4,38,39]), we only give an algorithm to compute γ_j^n as given in [16], where a more direct computational scheme is also discussed. In the modified "Pascal triangle" to be displayed at the end of this section, we use the configurations

to mean that c and $\tilde{c}$ are obtained from a and b by using the linear combinations:

$$c = ra + sb \quad \text{and} \quad \tilde{c} = a + b$$

By carefully studying this procedure, the following result is derived in [16].

Theorem 5.1. *For every positive integer* n, *the sequence* $\{\gamma_j^n\}$, $j = 0, \ldots, 3n - 2$, *is a Pólya frequency sequence; that is, the upper triangular infinite Toeplitz matrix with first row given by* $(\gamma_0^n, \ldots, \gamma_{3n-2}^n, 0, \ldots)$ *is totally positive.*

Perhaps the property of total positivity of $\{\gamma_j^n\}$ and the alternation of signs in the formulation Ψ_n in (5.2) are the reasons for the property of "complete oscillation" of the cardinal wavelet series as discussed in Section 4.

6. EXTENSIONS

In the above discussions, we only considered cardinal splines; that is, splines with equally-spaced knots. One of the main advantages of our spline approach to wavelets is that splines with arbitrary knot-distribution can be considered in an analogous manner with convolutions replaced by matrix multiplications. In particular, in constructing (non-periodic) wavelets on a bounded interval, we may consider B-splines with stacked knots at both boundary points. This is being done in a joint work with E. Quak, and some special cases such as linear wavelets have been settled. In this respect we would like to mention that wavelets on bounded domains have also been considered by Jaffard and Meyer [30]. On the other hand, since (irreducible) box splines (cf. [5]) are natural generalizations of cardinal B-splines, we believe that compactly supported wavelets based on any box spline that generates a multiresolution analysis in $L^2(\mathbb{R}^s)$, $s \geq 2$, can be obtained. In fact, this has been done in [11] for box-splines with any unimodulo direction set for $s = 2$ and 3. However, since we rely on a technique, valid only for $s = 2$ and 3, from the work of Riemenschneider and Shen [37] in constructing (non-compactly supported) orthonormal wavelets, the problem of constructing compactly supported box-spline wavelets in $\mathbb{R}^s$, for $s \geq 4$, remains open.

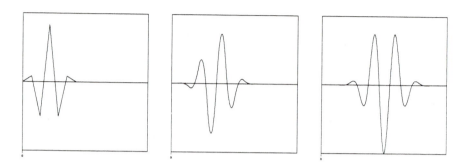

FIGURE 1.   Cardinal B-wavelets: ψ_2, ψ_3, ψ_3

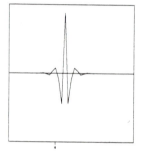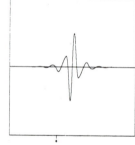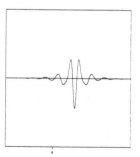

FIGURE 2.   Dual cardinal B-wavelets: $\tilde{\psi}_2$, $\tilde{\psi}_3$, $\tilde{\psi}_3$

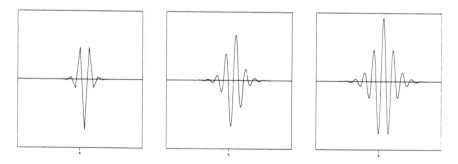

FIGURE 3.   Cardinal interpolatory B-wavelets: η_2, η_3, η_3

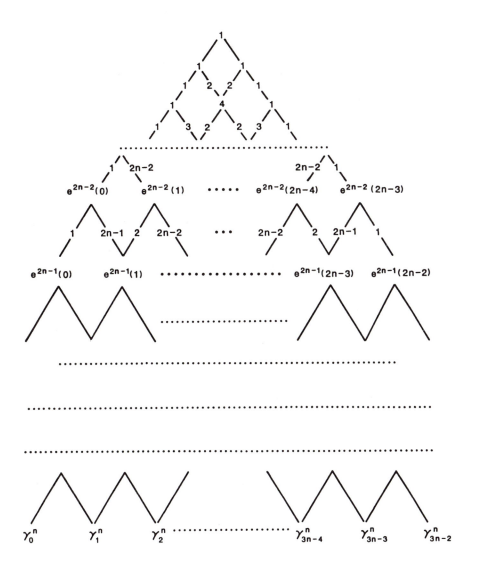

FIGURE 4. Computation of $\gamma_0^n, \ldots, \gamma_{3n-2}^n$

REFERENCES

[1] Auscher, P., Ondelettes fractales et applications, Doctoral Thesis, Univ. Paris-Dauphine, 1989.

[2] Battle, G., A block spin construction of ondelettes, Part I: Lemarié functions, Comm. Math. Phys. **110** (1987), 601-615.

[3] de Boor, C., *A Practical Guide to Splines*, Springer-Verlag, N.Y., 1978.

[4] Chan, A.K. and C.K. Chui, Real-time signal analysis with quasi-interpolatory splines and wavelets, in *Curves and Surfaces*, P.J. Laurent, A. Le Méhaute, and L.L. Schumaker (eds.), Academic Press, N.Y., 1991.

[5] Chui, C.K., *Multivariate Splines*, CBMS-NSF Series in Applied Math. #54, SIAM Publ., Philadelphia, 1988.

[6] Chui, C.K., Curve design and analysis using splines and wavelets, Trans. Eight Army Conf. on Appl. Math. and Computing, 1990.

[7] Chui, C.K., Vertex splines and their applications to interpolation of discrete data, in *Computation of Curves and Surfaces*, W. Dahman, M. Gasca, and C.A. Micchelli (eds.) Kluwer Academic Publ., 1990, pp. 137-181.

[8] Chui, C.K., Construction and applications of interpolation formulas, in *Multivariate Approximation and Interpolation*, W. Haussmann and K. Jetter (eds.), ISNM Series Math., Birkhaüser Verlag, Basel, 1990, pp. 11-23.

[9] Chui, C.K. and H. Diamond, A characterization of multivariate quasi-interpolation formulas and applications, Number. Math. **57** (1990), 105-121.

[10] Chui, C.K. and H. Diamond, A general framework for local interpolation, Numer. Math. **51** (1991), 569-581.

[11] Chui, C.K., J. Stöckler and J.D. Ward, Compactly supported box-spline wavelets, CAT Report #230, Texas A&M University, 1990.

[12] Chui, C.K. and J.Z. Wang, A cardinal spline approach to wavelets, Proc. Amer. Math. Soc., to appear.

[13] Chui, C.K. and J.Z. Wang, On compactly supported spline wavelets and a duality principle, Trans Amer. Math. Soc., to appear.

[14] Chui, C.K. and J.Z. Wang, A general framework of compactly supported splines and wavelets, CAT Report #219, Texas A&M University, 1990.

[15] Chui, C.K. and J.Z. Wang, An analysis of cardinal spline-wavelets, CAT Report #231, Texas A&M University, 1990.

[16] Chui, C.K. and J.Z.Wang, Computational and algorithmic aspects of cardinal spline-wavelets, CAT Report #235, Texas A&M University, 1990.

[17] Cohen, A., Doctoral Thesis, Univ. Paris-Dauphine, 1990.

[18] Cohen, A., I. Daubechies, and J.C. Feauveau, Biorthogonal bases of compactly supporeted wavelets, Comm. Pure and Appl. Math., to appear.

[19] Coifman, R.R., Wavelet analysis and signal processing in *Signal Processing I: Signal Processing Theory*, L. Auslander, T. Kailath, and S. Mitter (eds.), Springer-Verlag, 1990.

[20] Dahmen, W. and C.A. Micchelli, On stationary subdivision and the construction of compactly supported orthonormal wavelets, in *Multivariate Approximation and Interpolation*, W. Haussmann and K. Jetter (eds.), Birkhäuser Verlag, Basel, 1990, 69-89.

[21] 21. Daubechies, I., Orthonormal bases of compactly supported wavelets, Comm. Pure and Appl. Math. **41** (1988), 909-996.

[22] Daubechies, I., *Wavelets*, CBMS-NSF Series in Applied Math., SIAM Publ., Philadelphia, to appear.

[23] Daubechies, I., S. Jaffard, and J.-L. Journé, A simple Wilson orthonormal basis with exponential decay, preprint.

[24] DeVore, R., B. Jawerth, and B. Lucier, Surface compression, preprint.

[25] Feauveau, J.C., Analyse multirésolution par ondelettes non orthogonales et bancs de filtres numériques, Doctoral Thesis, Univ. Paris-Sud, 1990.

[26] Gabor, D., Theory of communication, J. IEE (London) **93** (1946), 429-457.

[27] Glowinski, R., W. Lawton, M. Ravachol, and E. Tenenbaum, Wavelet solution of linear and nonlinear elliptic, parabolic, and hyperbolic problems in one space dimension in *Proc. Ninth International*

Conf. on Computing Methods in Appl. Sciences and Engineering, R. Glowinski and A. Lichnewsky (eds.), SIAM Publ., Phil., 1990.

[28] Grossmann, A., R. Kronland-Martinet, and J. Morlet, Reading and understanding continuous wavelet transforms, in *Wavelets: Time-Frequency Methods and Phase Space*, J.M. Combes, A. Grossmann, and Ph. Tchamitchian (eds.), Springer-Verlag, N.Y., 1989.

[29] Grossmann, A. and J. Morlet, Decomposition of Hardy functions into square integrable wavelets of constant shape, SIAM J. Math. Anal. **15** (1984), 723-736.

[30] Jaffard, S. and Y. Meyer, Bases d' ondelettes dans des ouverts de $\mathbb{R}^n$, J. Math. Pures et Appl. **68** (1989), 95-108.

[31] Karlin, S., *Total Positivity*, Vol. 1, Stanford Univ. Press, 1968.

[32] Lemarié, P.G., Ondelettes à localisation exponentielle, J. de Math. Pures et Appl. **67** (1988), 227-236.

[33] Liandrat, J. and Ph. Tchamitchian, Resolution of the 1D regularized Burgers equation using a spatial wavelet approximation, ICASE Report # 90-83, NASA, 1990.

[34] Mallat, S.G., Multiresolution approximations and wavelet orthonormal bases of $L^2(\mathbb{R})$, Trans. Amer. Math. Soc., **315** (1989), 69-87.

[35] Meyer, Y., Ondelettes et functions splines, Seminaire Equations aux Derivees Partielles, Ecolé Polytechnique, Paris, France, Dec. 1986.

[36] Micchelli, C.A., Using the refinement equation for the construction of pre-wavelets, Numerical Algorithms, to appear.

[37] Riemenschneider, S.D. and Z.W. Shen, Box splines, cardinal series and wavelets, in *Approximation Theory and Functional Analysis*, C.K. Chui (ed.), Academic Press, N.Y. 1991, pp. 133-150.

[38] Schoenberg, I.J., *Cardinal Spline Interpolation*, CBMS-NSF Series in Appl. Math. #12, SIAM Publ., Philadelphia, 1973.

[39] Schumaker, L.L., *Spline Function: Basic Theory*, John Wiley & Sons, N.Y., 1981.

[40] Vetterli, M. and C. Herley, Wavelets and filter banks: Theory and design, CU/CTR/TR Report #206/90/36, Columbia University, 1990.

Wavelet Bases for $L^2(\mathbf{R})$ with Rational Dilation Factor

P. AUSCHER

Washington University
Department of Mathematics
*St. Louis, Missouri**

INTRODUCTION

In this chapter, we present some special constructions of wavelet bases for $L^2(\mathbf{R})$. It is now well-known that one can exhibit many "smooth" functions ψ such that the functions obtained by translation and dilation, $2^{j/2}\,\psi(2^j x - k)$, j, $k \in \mathbf{Z}$, form an orthonormal basis for $L^2(\mathbf{R})$ (see [3]). The dilations are the unitary transformations $f(x) \mapsto 2^{j/2} f(2^j x)$ and one may wonder what role plays the number 2; what if we replace 2 by an arbitrary number $M > 1$? In other words, we consider the following problem:

(P) *Given any real number $M > 1$, does there exist a finite set, ψ_1, ψ_2, ..., ψ_ℓ, of functions in $L^2(\mathbf{R})$ such that the family $M^{j/2}\,\psi_i(M^j x - k)$, j, $k \in \mathbf{Z}$, $1 \le i \le \ell$, is an orthonormal basis for $L^2(\mathbf{R})$?*

We call such a basis a wavelet basis with *dilation factor M*. Of course, we look for functions with the three basic properties of a wavelet : reg-

*new address: Université de Rennes I, Département de Mathématiques, 35042 Rennes Cedex, France.

ularity, decay at ∞ and cancellation (this explains the word "smooth"). By cancellation we mean that the moments $\int_{\mathbf{R}} x^m \psi_i(x)\,dx$ of the wavelets vanish when $m = 0, 1, \ldots n$, for some integer n. Linked to problem (**P**) is also finding a constructive algorithm to obtain these wavelets.

This problem is interesting for two reasons. First, in the case $M = 2$ the construction of a wavelet basis following [6] relies on the geometric property that the lattices $2^{-j}\mathbf{Z}$ are increasing for the inclusion. This property is preserved only when M is an integer. Therefore, it is not clear whether problem (**P**) has a solution when M is not an integer. Second, in signal analysis the number $1/M$ is the sampling ratio of an M-channel subband coder using a quadrature mirror filter. When M is an integer, the M channels in the frequency domain are of comparable size (see [3], [10] and [11]). When M is a rational number, say $3/2$ for example, a wavelet basis would give a 2-channel subband coding, the size of the high frequency channel being essentially half as big as the low frequency one. This may be useful for speech analysis.

At the time we undertook this study, in 1986 (this work is part of the author's doctoral dissertation [1] and still unpublished), only a few examples of smooth wavelet bases with dilation factor 2 were known: Meyer's basis [7]; Battle-Lemarié's spline basis [2], [5], and a construction, unnoticed until 1988, of Strömberg [9]. Moreover, it was noticed in [7] that a modification of the construction of the wavelet yields a wavelet basis with dilation factor $1 + 1/q$ for any non-negative integer q. We were also able to produce a spline wavelet basis with factor 3 in the spirit of Lemarié's; in this case we had to construct two wavelets.

Given these examples, problem (**P**) seems to have reasonable solutions, the number of wavelets solution in (**P**) depending on M. We show in this chapter that this problem has a positive answer when M is any rational number and give some examples of our solutions. An appropriate mathematical framework to do so is an extension of Mallat's notion of multiresolution analysis [6] and this is presented shortly in section 1. We then discuss examples in more details. We conclude by some comments and a short account on the relation of this problem to signal analysis.

1. EXTENDED THEORY OF MULTIRESOLUTION ANALYSIS FOR $L^2(\mathbf{R})$

Fix $M \in \mathbf{R}$, $M > 1$. A multiresolution analysis for $L^2(\mathbf{R})$ with dilation factor M (which we denote by $MRA(M)$) consists of a collection V_j, $j \in \mathbf{Z}$, of subspaces of $L^2(\mathbf{R})$ such that

(i) $V_j \subset V_{j+1}$, $j \in \mathbf{Z}$;

(ii) $\bigcap_{j \in \mathbf{Z}} V_j = \{0\}$ and $\bigcup_{j \in \mathbf{Z}} V_j$ is dense in $L^2(\mathbf{R})$;

(iii) $f(x) \in V_j$ if and only if $f(Mx) \in V_{j+1}$;

(iv) there exists a function $\varphi(x) \in V_0$ such that $\varphi(x - k)$, $k \in \mathbf{Z}$, forms an orthonormal basis for V_0.

When $M = 2$, this is the classical definition of an MRA in the sense of Mallat ([6], see also [8]). The choice of the translations by integers acting on V_0 is independent of M; by a change of variable, one can always reduce to this situation. Let us describe a few examples.

1) Let V_j be the subspace of functions in L^2 that are piecewise constant on the intervals $[kM^{-j}, (k+1)M^{-j}]$, $k \in \mathbf{Z}$. If M is an integer, then $\{V_j\}$ is easily seen to be an $MRA(M)$; the function $\varphi(x)$ in (iv) is given by $\chi_{[0,1[}(x)$, the characteristic function of $[0, 1[$.

2) One can also consider linear (or more general) splines, that is V_j is the space of piecewise linear continuous functions in $L^2(\mathbf{R})$ with nodes at kM^{-j}, $k \in \mathbf{Z}$. The function $\varphi(x)$ in (iv) is independent of M and we refer to [5] for its construction. One obtains an $MRA(M)$ only when M is an integer. Observe that the inclusions (i) are geometrically obvious; it suffices to draw a picture to be convinced of this restriction on M. The next example does not share the same feature.

3) Let $\widehat{f}(\xi) = \int_{-\infty}^{+\infty} f(x)e^{-ix\xi}\,dx$ denote the Fourier transform of $f(x)$ and V_j be the collection of functions that have Fourier transforms supported in $[-\pi M^j, \pi M^j]$. Let $\varphi(x) = (sin\pi x)/\pi x$, then $\widehat{\varphi}(\xi) = \chi_{[-\pi,\pi]}(\xi)$ and the well-known sampling theorem of Shannon shows that (iv) holds. This makes $\{V_j\}$ an $MRA(M)$ for any $M > 1$. Note that, although obvious considering the Fourier variable ξ, the inclusions (i) are not easily seen geometrically in the space variable x.

4) This last example is the key to the construction of our family of bases in section 2, and is a variation of the construction of Meyer [7]. Fix a non-negative integer q and $0 < \epsilon \leq (2q+1)^{-1}$. Define $\varphi(x)$ by

$$\widehat{\varphi}(\xi) = \cos \omega(\xi), \quad \xi \in \mathbf{R}, \tag{1}$$

where $\omega(\xi)$ is an even C^∞-function that satisfies

$$\omega(\xi) = 0, \quad 0 \leq \xi \leq \pi(1 - \epsilon);$$
$$\omega(\xi) + \omega(2\pi - \xi) = \pi/2, \quad \pi(1 - \epsilon) \leq \xi \leq \pi(1 + \epsilon); \tag{2}$$
$$\omega(\xi) = \pi/2, \quad \xi \geq \pi(1 + \epsilon).$$

Let $M = 1 + 1/q$; we claim that $V_j = \{\sum_k a_k \varphi(M^j x - k); \sum_k |a_k|^2 < \infty\}$ is an $MRA(M)$. The proof of (iv) is independent of the choice of M and reduces to the following identity (see [8]):

$$\sum_{k \in \mathbf{Z}} |\widehat{\varphi}(\xi + 2k\pi)|^2 = 1 \qquad \text{for all} \quad \xi \in \mathbf{R}, \tag{3}$$

whose verification is immediate. The only delicate point is (i), which we check for $j = 0$. Let $f(x)$ be in V_0, then $f(x) = \sum_k a_k \varphi(x - k)$ for some sequence $\{a_k\}$. Its Fourier transform, $\widehat{f}(\xi) = (\sum_k a_k e^{-ik\xi})\widehat{\varphi}(\xi)$, has support contained in $\text{Supp}\widehat{\varphi} = [-\pi(1 + \epsilon), \pi(1 + \epsilon)]$. But the relation between ϵ and M implies that $\widehat{\varphi}(\xi/M) = 1$ when $\xi \in \text{Supp}\widehat{\varphi}$. We may write therefore,

$$\widehat{f}(\xi) = \widehat{f}(\xi)\widehat{\varphi}(\xi/M) = \left(\sum_{\ell \in \mathbf{Z}} \widehat{f}(\xi + 2\ell\pi M)\right)\widehat{\varphi}(\xi/M),$$

the last inequality coming from the fact that for $\xi/M \in \text{Supp}\widehat{\varphi}$, $\widehat{f}(\xi + 2\ell\pi M) \neq 0$ only if $\ell = 0$. The latter sum is then expanded in a Fourier series with period $2\pi M$, and this makes $\widehat{f}$ the Fourier transform of a function in V_1. This proves $V_0 \subset V_1$.

The next two general results show what we can expect from such a notion.

Theorem 1. *Suppose that* $\{V_j\}$ *is an* $MRA(M)$ *for* $L^2(\mathbf{R})$ *and that* $M > 1$.
(i) *If* M *is an irrational number, then* $\varphi(x)$ *in* (iv) *has a Fourier transform whose modulus is the characteristic function of a measurable set,* E, *of Lebesgue measure* 2π. *Moreover, if* $x \in E$ *then* $x/M \in E$.
(ii) *If* $M = p/q$, p *and* q *being relatively prime integers with* $p > q > 1$, *then* φ *has neither compact support nor exponential decay at* ∞.

Let us quickly comment on this theorem. The first part tells us that the case M irrational does not lead to interesting examples in the sense that $\varphi(x)$ decays no faster than $1/|x|$ at ∞ (this is in contrast with higher dimensions where "smooth" MRA with irrational factors do exist, see [8]). Example 3 is therefore representative of all such MRA. The second assertion is restrictive as well since it excludes compactly supported functions $\varphi(x)$ whereas such functions exist when $M = 2$. Nevertheless we have seen a construction for which $\varphi(x)$ is C^∞ with a compactly supported Fourier transform when M is a rational number. As we mentioned in the introduction, the geometric inclusions of the lattices $M^{-j}\mathbf{Z}$ associated with the spaces V_j fail in this case and this makes the inclusions (i) not immediate. However we do obtain these inclusions in example 4 because of the support condition of $\widehat{\varphi}(\xi)$. Therefore unlike the case where $M = 2$, there is no evidence that one can derive a systematic way of producing $MRA(M)$ with non-integer rational factors.

Assertion (i) can be proved as follows. Since $\varphi_1(x) = \varphi(x - 1) \in V_1$, we have $\widehat{\varphi}_1(\xi) = m_1(\xi)\widehat{\varphi}(\xi/M)$ where $m_1(\xi)$ is $2\pi M$-periodic, and $\widehat{\varphi}_1(\xi) = \widehat{\varphi}(\xi)e^{-i\xi} = m_0(\xi)e^{-i\xi}\widehat{\varphi}(\xi/M)$. Hence $m_1(\xi) = m_0(\xi)e^{-i\xi}$ a. e. on $\{\widehat{\varphi}(\xi/M) \neq 0\} = F$. Since M is irrational, a measure theoretic argument using the different periods of the above functions implies that F has Lebesgue measure, $|F|$, not exceeding $2\pi M$. We deduce from this that $|\mathrm{Supp}\widehat{\varphi}| \leq 2\pi$. But (1) and (3) imply (since $|\widehat{\varphi}(\xi)| \leq 1$)

$$|\mathrm{Supp}\widehat{\varphi}| \geq \int_{\mathrm{Supp}\widehat{\varphi}} |\widehat{\varphi}(\xi)|^2 \, d\xi = \int_{\mathbf{R}} |\widehat{\varphi}(\xi)|^2 \, d\xi$$

$$= \int_{-\pi}^{\pi} \sum |\widehat{\varphi}(\xi + 2k\pi)|^2 \, d\xi = 2\pi.$$

Therefore $|\mathrm{Supp}\widehat{\varphi}| = 2\pi$, and the same computation gives that $\widehat{\varphi}$ is unimodular a. e. on its support.

The proof of (ii) begins in a similar way. Suppose that φ has exponential decay, then $\widehat{\varphi}$ is analytic and vanishes on a set that is at most numerable. Thus $m_1(\xi) = m_0(\xi)e^{-i\xi} \neq 0$ except if ξ belongs to a numerable set, say A. Then there is a real $\xi \notin A$ such that $\xi + 2\pi M \notin A$ as well and, since m_0 and m_1 are $2\pi M$-periodic, we see from the above formula that M should be an integer .

Let us now construct a wavelet basis for an arbitrary rational M. We assume from now on that $M = p/q$ as above (M could be an integer) and that $\{V_j\}$ is an $MRA(M)$. Following the idea of Mallat [6], let W_j be the orthogonal complement of V_j in V_{j+1}. This yields an orthogonal decomposition of $L^2(\mathbf{R})$;

$$L^2(\mathbf{R}) = \bigoplus_{j \in \mathbf{Z}} W_j.$$

Moreover $f(x) \in W_j$ if and only if $f(Mx) \in W_{j+1}$; we can, therefore, limit ourselves to study the space W_0. A last remark is that $\psi(x) \in W_0$ if and only if $\psi(x - kq) \in W_0$ for any $k \in \mathbf{Z}$; that is, W_0 is invariant under the translations by $kq, k \in \mathbf{Z}$ (note the difference with V_0).

Theorem 2. *With the above notations, there exists a set of $p - q$ wavelet functions $\psi_1, \ldots, \psi_{p-q}$ in W_0 such that the collection of functions $(p/q)^{j/2}$ $\psi_i\big((p/q)^j x - kq\big)$, $k, j \in \mathbf{Z}$, $1 \leq i \leq p - q$, forms an orthonormal wavelet* *basis for $L^2(\mathbf{R})$. Furthermore the number of wavelets in W_0 does not depend on the choice of the ψ_i's.*

Remarks.   1) If we set $\widetilde{\psi}_i(x) = \sqrt{1/q}\,\psi_i(x/q)$, then the functions $\widetilde{\psi}_1$, ..., $\widetilde{\psi}_{p-q}$ abstractly solve problem (**P**).

2) Consider for a moment an $MRA(M)$ with M irrational. One can define W_0 analogously and characterize this space in terms of the Fourier transform using Theorem 1. It is then easy to prove that there is no finite set of functions in W_0 from which we obtain a basis in the sense of problem (**P**). In fact, we do not know any solution to this problem when M is irrational. If any exists though, we have shown that it cannot be derived from an $MRA(M)$.

When $M = 2$ this theorem is nothing but the now classical result of Mallat [6]. The proof of the general case is an adaptation of this and can be understood from the following geometric argument. One starts by looking at the sum $V_0 \oplus W_0 = V_1$. Both V_0 and V_1 have an orthonormal basis which we rewrite in a form corresponding to the translation invariance structure of W_0. To do so, let $f_i(x) = \varphi(x - i)$ and $g_n(x) = \sqrt{p/q}\,\varphi((p/q)x - n)$; then $f_i(x - kq)$, $0 \le i \le q - 1$, $k \in \mathbf{Z}$, is a basis for V_0. It suffices therefore to complete this basis to a basis for V_1 with the same structure. Since $g_n(x - kq)$, $0 \le n \le p - 1$, $k \in \mathbf{Z}$, is already such a basis, we see, at least formally, that the "number" of missing functions to select in W_0 is $p - q$. One can achieve this completion; the choice of the ψ_i's is however not unique. In certain cases (M integer or $M = 1 + 1/q$) this allows us to choose these functions so that they are regular and decaying at ∞, provided φ is such. Their cancellation property is obtained as a direct consequence. The full proof provides a constructive algorithm to obtain these functions in the case where $M = 1 + 1/q$. We refer to [1] for further details.

2.   EXAMPLES OF WAVELET BASES WITH RATIONAL DILATION FACTOR

Here we deal only with rational non-integer dilation factors although our constructions work in the integer case. In the two following examples, we first exhibit an $MRA(M)$ and then produce an associated wavelet basis.

Example 1. $M = 1 + 1/q$, q integer greater than 2.

Consider example 4) of section 1 ; we already proved that the function φ defined in (1) induces an $MRA(M)$. Before proceeding further and to make clearer the rest of the construction, we adopt some notations. We refer the reader to the Figures for graphical support.

Let r be a non-negative integer and fix $0 < \epsilon < 1$. Define $S_r = [r\pi(1-\epsilon), r\pi(1+\epsilon)]$, and $-S_r = \{-x, x \in S_r\}$; notice that $-S_r = S_r - 2\pi r$. Define also T_r to be the interval adjacent to S_r and S_{r+1} (ϵ will always be taken so small that this definition makes sense).

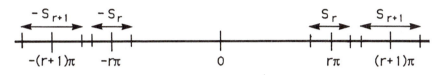

FIGURE 1. Position of the intervals S_r and S_{r+1}

Let $\theta_r(\xi) = \omega(\xi/r)$; by (2), θ_r is even, $\theta_r(\xi) = 0$ when $0 \le \xi \le r\pi(1-\epsilon)$, $\pi/2$ when $\xi \ge r\pi(1+\epsilon)$ and $\theta_r(\xi) + \theta_r(2\pi r - \xi) = \pi/2$ when $\xi \in S_r$. Let $\widehat{\varphi}_r(\xi) = \widehat{\varphi}(\xi/r) = \cos\theta_r(\xi)$ where φ is defined in (1). Observe that $\mathrm{Supp}\widehat{\varphi}_r \subset \{\xi; |\xi| \le r\pi(1+\epsilon)\}$. We also have the recurrence relations

$$\theta_r\left(\frac{r\xi}{r+1}\right) = \theta_{r+1}(\xi) \qquad \text{and} \qquad \widehat{\varphi}_r\left(\frac{r\xi}{r+1}\right) = \widehat{\varphi}_{r+1}(\xi). \qquad (4)$$

Finally introduce ψ_r by

$$\widehat{\psi}_r(\xi) = (\mathrm{sgn}\,\xi)^{r+1} e^{-i\xi/2} \sin\theta_r(\xi) \cos\theta_{r+1}(\xi), \qquad (5)$$

where $\mathrm{sgn}\,\xi = 1$ if $\xi > 0$, -1 if $\xi < 0$.

As formulas (4) and (5) show, the cosine factor of the function in (5) and $\widehat{\varphi}_{r+1}(\xi)$ are the same. This will ensure that $\widehat{\psi}_r(\xi)$ belongs to the space "V_1". Also the sine and phase factors will induce orthogonality to "V_0". This needs some explanations.

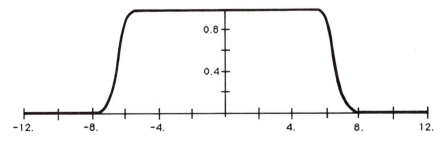

FIGURE 2. Graph of $\widehat{\varphi}_r$, $r = 2$

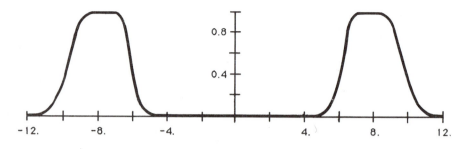

FIGURE 3.   Graph of the modulus of $\widehat{\psi}_r$, $r = 2$

Take $r = q$ and $0 < \epsilon < (2q+1)^{-1}$. The change of the space variable $x \mapsto qx$ implies that $\sqrt{1/q}\,\varphi_q(x - k/q)$, $k \in \mathbf{Z}$, is an orthonormal basis for $V_0^q = \{\, f(qx);\, f \in V_0 \,\}$. Also define W_0^q and V_1^q by the same change of variable from W_0 and V_1 respectively (note that this preserves inclusion, orthogonality, ...). By the general theory of section 1, we are looking for a single function $\psi \in W_0^q$ such that the collection $\psi(x - k)$, $k \in \mathbf{Z}$, is an orthonormal basis for W_0^q. As a consequence of Theorem 2, it suffices to take $\psi \in V_1^q$, orthogonal to V_0^q and such that the integer translates of ψ form an orthonormal family of $L^2(\mathbf{R})$; we claim that ψ_q defined by (5) does the job.

First the orthonormality follows from the relation $\sum_k |\widehat{\psi}_q(\xi + 2k\pi)|^2 = 1$ as in (3). To see this we take ξ within an interval of length 2π, $[q\pi(1 - \epsilon)$, $q\pi(1 - \epsilon) + 2\pi]$, and distinguish 4 cases:

If $\xi \in S_q$, the sum reduces to the $k = 0$ and $k = -q$ terms. Using the functional properties of the θ_r's, the $k = -q$ term is $\cos^2 \theta_q(\xi)$, the $k = 0$ term is $\sin^2 \theta_q(\xi)$ and these terms add up to 1.

If $\xi \in T_q$, the only non-zero term is the $k = 0$ term, whose value is 1.

If $\xi \in S_{q+1}$, we have two non-zero terms: the $k = 0$ term is $\cos^2 \theta_{q+1}(\xi)$ while the $k = -q - 1$ term is $\sin^2 \theta_{q+1}(\xi)$, and their sum is 1.

In the remaining case, we see that $\xi - 2(q + 1)\pi \in -T_q$ and reach the same conclusion as in the second case.

Second by the Parseval formula, the orthogonality of ψ to V_0^q is equivalent to

$$I_\ell = \int_{\mathbf{R}} \widehat{\psi}_q(\xi)\overline{\widehat{\varphi}_q(\xi)}\, e^{-i(\ell/q)\xi}\, d\xi = 0 \qquad \text{for all integers } \ell.$$

Fix ℓ and observe that the support of this integral is $S_q \cup (-S_q)$. Using $-S_q = S_q - 2q\pi$, we perform the change of variable $\xi \mapsto \xi + 2q\pi$ on $-S_q$

and obtain

$$I_\ell = \int_{S_q} \left(\widehat{\psi}_q(\xi)\overline{\widehat{\varphi}_q(\xi)} + \widehat{\psi}_q(\xi + 2q\pi)\overline{\widehat{\varphi}_q(\xi + 2q\pi)} \right) e^{-i(\ell/q)\xi} \, d\xi.$$

A straightforward calculation shows that the integrand is 0 ; this is where the factor $(\operatorname{sgn}\xi)^{q+1} e^{-i\xi/2}$ plays its major role. This proves that $I_\ell = 0$.

We finally prove that $\psi_q \in V_1^q$. This is equivalent to the relation

$$\widehat{\psi}_q(\xi) = P(\xi)\widehat{\varphi}_q \left(\frac{q\xi}{q+1} \right) = P(\xi)\widehat{\varphi}_{q+1}(\xi),$$

where $P(\xi)$ is $2\pi(q+1)$-periodic. This amounts to showing that

$$m(\xi) \overset{\text{def}}{=} (\operatorname{sgn}\xi)^{q+1} e^{-i\xi/2} \sin\theta_q(\xi)$$

is $2\pi(q+1)$-periodic on $\operatorname{Supp}\widehat{\varphi}_{q+1}$, which reduces to verifying the equality

$$m(\xi) = m(\xi - 2(q+1)\pi) \qquad \text{when } \xi \in S_{q+1}.$$

The latter follows from the observation that $\theta_q(\xi) = \theta_q(\xi - 2(q+1)\pi) = \pi/2$ for such ξ as $|\xi|$ and $|\xi - 2(q+1)\pi|$ exceed $q\pi(1+\epsilon)$ (recall that $\epsilon \le (2q+1)^{-1}$).

We have therefore obtained a wavelet ψ_q, defined by (5), such that the collection of functions $(1 + 1/q)^{j/2}\, \psi_q \left((1+1/q)^j\, x - k \right)$, $k, j \in \mathbf{Z}$, is an orthonormal basis for $L^2(\mathbf{R})$.

Remarks.   1) If we delete the term $(\operatorname{sgn}\xi)^{q+1}$ in (5) we obtain a wavelet basis as well. It is different though if q is even (this is the basis mentioned at the end of [7]). In such a case, one can show that there is a different $MRA(M)$ from which this basis can be derived.

2) This comment prepares the set up for the next construction. Let $V(r)$ be the space generated by the functions $\varphi_r(x - k/r)$, $k \in \mathbf{Z}$. In the course of the previous argument, we have seen that $V(q) = V_0^q$. Moreover $V(q+1) = V_1^q$, since $\varphi_{q+1}(x - k(q+1)^{-1}) = (q+1)\varphi\left((q+1)x - k\right)$. Set $W(q) = W_0^q$, we have just obtained that

$$V(q) \oplus W(q) = V(q+1) \qquad \text{provided} \qquad 0 \le \epsilon \le \frac{1}{2q+1}, \qquad (6)$$

and that $\psi_q(x - k)$, $k \in \mathbf{Z}$ is a basis for $W(q)$.

Example 2. $M = p/q$, p and q relatively prime with $p > q > 1$.

First of all we construct the $MRA(M)$. Start from φ defined in (1) and let $\mathbf{V}_j$ be the space generated by the functions $\varphi\left((p/q)^j x - k\right)$, $k \in \mathbf{Z}$. The conditions (ii), (iii) and (iv) are easily fulfilled (note that $\mathbf{V}_0$ coincides with V_0 of the previous example, so there is nothing to prove for (iv)). The inclusions (i) follow from the relation $\widehat{\varphi}(q\xi/p) = 1$ when $\xi \in \mathrm{Supp}\widehat{\varphi}$; this is the case when $0 \le \epsilon \le (p-q)/(p+q)$. In fact, we assume a somewhat more restrictive condition on ϵ, namely, $0 < \epsilon \le 1/(2p-1)$.

With the obvious notations, the change of variable $x \mapsto qx$ takes the spaces $\mathbf{V}_0$, $\mathbf{W}_0$ and $\mathbf{V}_1$ to $\mathbf{V}_0^q$, $\mathbf{W}_0^q$ and $\mathbf{V}_1^q$ respectively. We have already observed that $\mathbf{V}_0^q = V_0^q = V(q)$, and analoguous calculations involving (4) show that $\mathbf{V}_1^q = V(p)$. An induction on formula (6) gives

$$V(p) = V(q) \oplus W(q) \oplus W(q+1) \oplus \ldots \oplus W(p-1)$$

since $\epsilon \le 1/(2p-1) \le 1/(2r+1)$ for $q \le r \le p-1$. Hence

$$\mathbf{W}_0^q = W(q) \oplus W(q+1) \oplus \ldots \oplus W(p-1).$$

From this we obtain the orthonormal basis $\psi_r(x-k)$, $q \le r \le p-1$, $k \in \mathbf{Z}$ for $\mathbf{W}_0^q$. Taking all dilations of these functions with powers of p/q yields an orthonormal basis for $L^2(\mathbf{R})$.

Remark. Either function φ_r or ψ_r considered here is in the Schwartz class. Each depends on a parameter ϵ that goes to 0 with $1/r$, and for r large its Fourier transform is close to being a characteristic function of an interval (or the union of two intervals). Therefore its graph looks very much alike that of $(\sin x)/x$ which is very oscillatory with a long tail. The Figures 4 a–d) show a graph of ψ_r defined in (5) where r is 1, 2, 3 and 4 respectively.

CONCLUSION

We have shown how to construct some wavelet bases for $L^2(\mathbf{R})$ with a rational dilation factor. To this end, we have extended Mallat's definition of a multiresolution analysis ; we have shown as well the limitations of this approach.

In signal analysis, subband coding using uniform M-channel quadrature mirror filters has been studied for several years (see, e. g., [10] and [11]). From the work of Mallat and Daubechies, this theory is formally equivalent to that of $AMR(M)$ when $M = 2$ and this formal equivalence still holds when M is an integer (see, e. g., [1]); this is no longer true if M is only a rational number. We have amplified the functional aspect by looking at wavelet bases for $L^2(\mathbf{R})$. On the other hand in a recent paper [4],

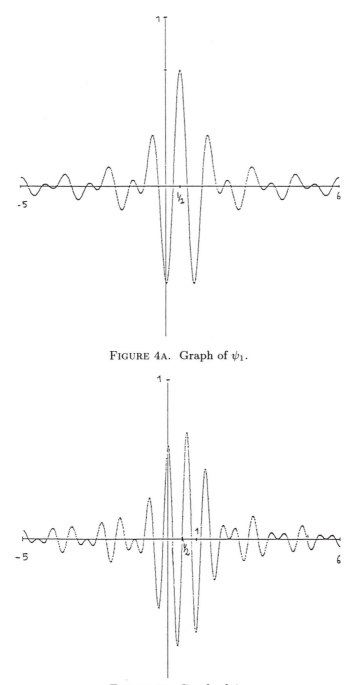

FIGURE 4A. Graph of ψ_1.

FIGURE 4B. Graph of ψ_2.

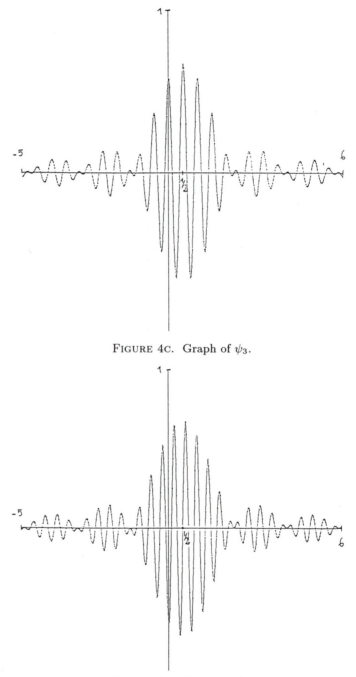

FIGURE 4C. Graph of ψ_3.

FIGURE 4D. Graph of ψ_4.

Kovačević and Vetterli have discussed a different approach that also uses rational dilation in order to obtain non-uniform subband coding of digital signals.

REFERENCES

[1]   Auscher, P., Ondelettes fractales et applications, PhD thesis, Université de Paris-Dauphine, 1989.

[2]   Battle, G., *A block spin construction of ondelettes, Part I : Lemarié functions*, Comm. Math. Phys. (1987).

[3]   Daubechies, I., Ten lectures on wavelets, NSF series in Applied Mathematics, S.I.A.M. (in press 1991).

[4]   Kovačević, J., and Vetterli, M., *Perfect reconstruction filter banks with rational sampling rates in one and two dimensions*, Proc. SPIE Conf. on Visual Communications and Image Processing, Philadelphia 1989, 1258–1265.

[5]   Lemarié, P. G., *Ondelettes à localisation exponentielle*, J. de Math. pures et appl. **67** (1989), 227–236.

[6]   Mallat, S., *Multiresolution approximations and orthonormal bases of wavelets for L²(**R**)*, Trans. Amer. Math. Soc. **315** (1989), 69–87.

[7]   Meyer, Y., *Principe d'incertitude, bases hilbertiennes et algèbres d'opérateurs*, Sem. Bourbaki, vol. 662 (1986).

[8]   Meyer, Y., Ondelettes et opérateurs, Vol. I, Hermann 1990.

[9]   Strömberg, J. O., *A modified Franklin system and higher order spline systems on* **Rⁿ** *as unconditional bases for Hardy spaces*, Conf. in honor of A. Zygmund, Vol. 2, W. Beckner et al. eds., Wadsworth math. series, Wadsworth, Belmont Calif. (1983) 475–495.

[10] Vaidyanathan, P. P., *Theory and design of M-channel maximally decimated quadrature mirror filters with arbitrary M, having the perfect-reconstruction property*, IEEE Trans. Acoust., Speech, Signal Processing ASSP-**35** (1987), 476–492.

[11] Vetterli, M., *A theory of multirate filter banks*, IEEE Trans. Acoust., Speech, Signal Processing ASSP-**35** (1987), 356–372.

SIZE PROPERTIES OF WAVELET-PACKETS

R. R. COIFMAN

Y. MEYER

V. WICKERHAUSER

1. INTRODUCTION

Wavelets are the building blocks of wavelet analysis in the same way as the functions $\cos nx$ are the building blocks of the ordinary Fourier analysis. But in contrast with sines and cosines, wavelets have a finite duration which can be arbitrarily small. This is the reason why the challenge of the construction of wavelets is to keep the best frequency localization which is allowed by Heisenberg's uncertainty principle.

The wavelet orthonormal basis with the best frequency localization was constructed in [4]. It is defined as the collection

$$\psi_{j,k}(x) \ = \ 2^{j/2} \, \psi(2^j x - k) \quad , \quad j \in \mathbb{Z} \, , \ k \in \mathbb{Z} \, , \tag{1.1}$$

where ψ has the following properties

$$\psi(x) \quad \text{belongs to the Schwartz class} \quad \mathcal{S}(\mathbb{R}) \tag{1.2}$$

$$\text{the Fourier transform} \quad \hat{\psi}(\xi) \quad \text{is supported by} \tag{1.3}$$
$$\frac{2\pi}{3} \ \leq \ |\xi| \ \leq \ \frac{8\pi}{3}$$

$$\hat{\psi}(\xi) \;=\; e^{-i\xi/2}\,\theta(\xi) \tag{1.4}$$

where $0 \le \theta(\xi) \le 1$ and $\theta(-\xi) = \theta(\xi)$

$$\theta^2(2\pi - \xi) + \theta^2(\xi) \;=\; 1 \qquad \text{if} \quad \frac{2\pi}{3} \le \xi \le \frac{4\pi}{3} \tag{1.5}$$

$$\theta(2\xi) \;=\; \theta(2\pi - \xi) \qquad \text{if} \quad \frac{2\pi}{3} \le \xi \le \frac{4\pi}{3}\,. \tag{1.6}$$

The frequency localization of ψ is given by (1.3) while (1.4), (1.5) and (1.6) are convenient to provide an orthonormal sequence. The fact that this collection $\psi_{j,k}$, $j \in \mathbb{Z}$, $k \in \mathbb{Z}$, is complete in $L^2(\mathbb{R}; dx)$ is, as often, related to some operator theory which will be described in section 2.

The Fourier transform $\hat{\psi}_{j,k}$ is supported by the "dyadic annulus" $\frac{2\pi}{3}2^j \le |\xi| \le \frac{8\pi}{3}2^j$ and this frequency localization is poor when j is large. Even if it means minor modification in the construction of ψ, one can achieve a slightly better frequency localization and replace $2\pi/3$ by $\pi - \delta$, $8\pi/3$ by $2\pi + 2\delta$. Then ψ_δ still belongs to the Schwartz class when $\delta > 0$ but the limiting case $\delta = 0$ gives the "Shannon wavelets" $\psi_0(x) = \frac{\sin 2\pi x}{2\pi x} - \frac{\sin \pi x}{\pi x}$. The relation with cardinal sines will be explained in section 4.

In some applications such as speech signal processing one would like to be able to switch from a wavelet expansion to some orthonormal expansions offering a better frequency localization. This flexibility should not be ruined by the computational cost. In other words, most computations leading to wavelet coefficients should also provide the new coefficients. *Basic wavelet-packets* will be defined in section 4 and *general wavelet-packets* in section 8. They provide these new and efficient expansions (theorem 6). This remarkable efficiency is verified in numerical experiments on speech signal processing.

We would like to understand why wavelet-packets work so well and to investigate their frequency localization. It will be proved (see theorem 3) that wavelet-packets do not enjoy the sharp frequency localization which has been announced in [2]. By S. Bernstein's inequalities, a sharp frequency localization would imply a uniform bound on L^∞-norms of the basic wavelet-packets $w_n(x)$. But theorem 3 shows that the average growth of $\|w_n\|_\infty$ is n^γ for some positive γ.

The fact that γ is rather small plays a key role in the construction of a large library of wavelet-packet orthonormal bases. Even if the problem of describing the full collection of such bases is still unsolved, the already known bases offer enough flexibility for the applications to speech signal processing.

2. THE SCALING FUNCTION φ

In order to prove that the collection $\psi_{j,k}$, $j \in \mathbb{Z}$, $k \in \mathbb{Z}$, is complete in $L^2(\mathbb{R})$, one tries to construct an approximation to the identity which is related to our wavelets.

This approximation to the identity will follow naturally from the following scheme.

Definition 1. *A multiresolution analysis of $L^2(\mathbb{R})$ is an increasing sequence V_j, $j \in \mathbb{Z}$, of closed subspaces of $L^2(\mathbb{R})$ with the following properties*

$$\bigcap_{-\infty}^{\infty} V_j = \{0\} \quad , \quad \bigcup_{-\infty}^{\infty} V_j \quad \text{is dense in } L^2(\mathbb{R}) \tag{2.1}$$

$$\forall f \in L^2(\mathbb{R}) \, , \, \forall j \in \mathbb{Z} \, , \qquad f(x) \in V_j \iff f(2x) \in V_{j+1} \tag{2.2}$$

$$\text{there exists a function} \quad \varphi \in \mathcal{S}(\mathbb{R}) \quad \text{such that} \tag{2.3}$$
$$\varphi(x-k) \, , \, k \in \mathbb{Z} \, , \quad \text{is an orthonormal basis of } V_0.$$

If we are given a multiresolution analysis, (2.2) implies

$$\frac{1}{2}\varphi\left(\frac{x}{2}\right) = \sum_{-\infty}^{\infty} \gamma_k \varphi(x+k) \tag{2.4}$$

where

$$\gamma_k = \int_{-\infty}^{\infty} \varphi\left(\frac{x}{2}\right) \overline{\varphi(x+k)} \, dx = 0(|k|^{-m})$$

for any $m \geq 1$.

Passing to the Fourier transform, one obtains

$$\hat{\varphi}(2\xi) = m_0(\xi)\hat{\varphi}(\xi) \quad , \quad m_0(\xi) = \sum_{-\infty}^{\infty} \gamma_k e^{ik\xi} \, . \tag{2.5}$$

We then define

$$m_1(\xi) = e^{-i\xi} \overline{m_0(\xi + \pi)} \tag{2.6}$$

and $\psi \in \mathcal{S}(\mathbb{R})$ by

$$\hat{\psi}(2\xi) = m_1(\xi) \, \hat{\varphi}(\xi) \, . \tag{2.7}$$

Denoting by W_j the orthogonal complement of V_j in V_{j+1} it is easy to check that

$$\psi(x-k) \, , \, k \in \mathbb{Z} \, , \quad \text{is an orthonormal basis of } W_0. \tag{2.8}$$

An obvious rescaling shows that $2^{j/2}\psi(2^j x - k)$, $k \in \mathbb{Z}$, is an orthonormal basis of W_j. Since $\bigcup_{-\infty}^{\infty} V_j$ is dense in $L^2(\mathbb{R})$, the full collection $\psi_{j,k}$, $j \in \mathbb{Z}$, $k \in \mathbb{Z}$, is an orthonormal basis of $L^2(\mathbb{R})$.

It remains to be shown that the explicit ψ which is defined by (1.4) can also be obtained by (2.7). To prove this assertion, we define $\varphi \in \mathcal{S}(\mathbb{R})$ by the following conditions : $\varphi(-x) = \varphi(x)$, the Fourier transform $\hat{\varphi}(\xi)$ of $\varphi(x)$ is non-negative, $\hat{\varphi}(\xi) = 1$ on $[-2\pi/3,\, 2\pi/3]$ and

$$\sum_{-\infty}^{\infty} |\hat{\varphi}(\xi + 2k\pi)|^2 \;=\; 1 . \tag{2.9}$$

Condition (2.9) alone implies that $\varphi(x - k)$, $k \in \mathbb{Z}$, is an orthonormal sequence. This sequence spans a closed subspace denoted V_0. The other V_j's are defined by (2.2). It is easy to verify that all the other conditions in definition 1 are satisfied and that this algorithm leads to the function ψ as defined by (1.4).

3.   QUADRATURE MIRROR FILTERS

S. Mallat working on image processing made a fundamental discovery. He pointed out that some discrete algorithms named quadrature mirror filters (QMF's) were intimately related to multiresolution analysis (the latter concept was created by S. Mallat and one of the authors).

Quadrature mirror filters belong to a larger group of algorithms called *subband coding* which are used in speech processing as well as in image processing. The reader is referred to [3] or [4].

In our approach, a pair of quadrature mirror filters provides a dichotomy for every infinitely dimensional separable Hilbert space H, equipped with an orthonormal basis e_k, $k \in \mathbb{Z}$. A trivial dichotomy would be given by $H = H_0 \oplus H_1$ where H_0 is generated by (e_{2k}) and H_1 by (e_{2k+1}), $k \in \mathbb{Z}$. In a second example, H_0 is the closed linear span of the orthonormal sequence $\frac{e_{2k}+e_{2k+1}}{\sqrt{2}}$, $k \in \mathbb{Z}$, while H_1 is similarly spanned by $\frac{e_{2k}-e_{2k+1}}{\sqrt{2}}$, $k \in \mathbb{Z}$.

We now pass to the general case. Let (u_k) and (v_k) be two sequences in $l^2(\mathbb{Z})$. We consider the sequence (f_k) of vectors of H defined by

$$\begin{cases} f_{2k} = \displaystyle\sum_{-\infty}^{\infty} u_{2k-l}\, e_l \\[2ex] f_{2k+1} = \displaystyle\sum_{-\infty}^{\infty} v_{2k-l}\, e_l . \end{cases} \tag{3.1}$$

We would like to know whether (f_k) is still an orthonormal basis of H.

If so, $H = H_0 \oplus H_1$ where the sum is direct and orthonormal, H_0 being spanned by (f_{2k}) and H_1 by (f_{2k+1}).

We consider the following symbols

$$\begin{cases} m_0(\theta) = \dfrac{1}{\sqrt{2}} \sum_{-\infty}^{\infty} u_k \, e^{ik\theta} \\[2mm] m_1(\theta) = \dfrac{1}{\sqrt{2}} \sum_{-\infty}^{\infty} v_k \, e^{ik\theta} \,, \end{cases} \qquad (3.2)$$

and we have

Proposition 1.   *The three following properties are equivalent*

$$(f_k)_{k \in \mathbb{Z}} \quad \text{is an orthonormal sequence in } H \qquad (3.3)$$

$$(f_k)_{k \in \mathbb{Z}} \quad \text{is an orthonormal basis of } H \qquad (3.4)$$

for every $\theta \in [0, 2\pi)$, the matrix

$$S(\theta) = \begin{pmatrix} m_0(\theta) & m_1(\theta) \\ m_0(\theta + \pi) & m_1(\theta + \pi) \end{pmatrix} \quad \text{is unitary.} \qquad (3.5)$$

The first example corresponds to $m_0(\theta) = \frac{1}{\sqrt{2}}$ and $m_1(\theta) = \frac{1}{\sqrt{2}} e^{-i\theta}$. The second example to $m_0(\theta) = \frac{1}{2}(1 + e^{i\theta})$, $m_1(\theta) = \frac{1}{2}(1 - e^{i\theta})$.

We now consider the mapping $F = (F_0, F_1)$ which transforms the "old coordinates" (α_k) into the "new coordinates" β_{2k} and γ_{2k} as defined by the relation

$$\sum_{-\infty}^{\infty} \alpha_k e_k = \sum_{-\infty}^{\infty} \beta_{2k} f_{2k} + \sum_{-\infty}^{\infty} \gamma_{2k} f_{2k+1} \,. \qquad (3.6)$$

We have $(\beta_{2k}) = F_0[(\alpha_k)]$ and $(\gamma_{2k}) = F_1[(\alpha_k)]$. The mapping F is a unitary isomorphism between $l^2(\mathbb{Z})$ and $l^2(2\mathbb{Z}) \times l^2(2\mathbb{Z})$.

These two operators F_0 and F_1 will be called *quadrature mirror filters*.

4.   WAVELETS AND QUADRATURE MIRROR FILTERS

Let us return to the multiresolution framework as defined in section 2. We have at our disposal two orthonormal bases for V_j, j being kept fixed. The first one is simply $e_k = 2^{j/2} \varphi(2^j \cdot -k)$ while the second one is f_k where

$$\begin{cases} f_{2k} = 2^{(j-1)/2} \, \varphi(2^{j-1} \cdot -k) \\[2mm] f_{2k+1} = 2^{(j-1)/2} \, \psi(2^{j-1} \cdot -k) \,. \end{cases} \qquad (4.1)$$

These two bases are connected by (3.1) if $m_0(\theta) = \frac{1}{\sqrt{2}} \sum_{-\infty}^{\infty} u_k e^{ik\theta}$ and $m_1(\theta) = \frac{1}{\sqrt{2}} \sum_{-\infty}^{\infty} v_k e^{ik\theta}$ are defined by

$$\hat{\varphi}(2\xi) = m_0(\xi)\,\hat{\varphi}(\xi) \tag{4.2}$$

and

$$\hat{\psi}(2\xi) = m_1(\xi)\,\hat{\varphi}(\xi) . \tag{4.3}$$

In other words, $m_0(\xi)$ is 2π -periodic, even, C^∞, non-negative, $m_0(\xi) = 1$ on $[-\pi/3, \pi/3]$ and

$$m_0^2(\xi) + m_0^2(\xi + \pi) = 1 . \tag{4.4}$$

If $\left[-\frac{\pi}{3}, \frac{\pi}{3}\right]$ is replaced by $\left[-\frac{\pi}{2} + \delta, \frac{\pi}{2} - \delta\right]$ $\delta > 0$, the other properties of m_0 can be kept and this new $m_0(\theta)$ is closer to the ideal filter.

If $m_0(\xi) = 1$ on $\left[-\frac{\pi}{2}, \frac{\pi}{2}\right)$, $m_0(\xi) = 0$ on $[-\pi, \pi/2)$ and $[\pi/2, \pi)$, then $\hat{\varphi}(\xi) = 1$ on $[-\pi, \pi)$ and $\hat{\varphi}(\xi) = 0$ outside which gives $\varphi(x) = \frac{\sin \pi x}{\pi x}$. In that case V_j is the subspace of $L^2(\mathbb{R})$ defined by the condition that the Fourier transform of $f \in V_J$ is supported by $[-2^j, 2^j)$ and, in the same way, W_j is defined by the condition that $\hat{f}$ is supported by $2^j \leq |\xi| < 2^{j+1}$. The price to be paid for this sharp frequency localization is the corresponding lack of localization of $\varphi(x)$ and $\psi(x)$ with respect to the x variable.

It should be noticed that (4.2) and $\hat{\varphi}(0) = 1$ imply

$$\hat{\varphi}(\xi) = m_0(\xi/2)\,m_0(\xi/4)\,m_0(\xi/8)\ldots \tag{4.5}$$

Similarly we have

$$\hat{\psi}(\xi) = m_1(\xi/2)\,m_0(\xi/4)\,m_0(\xi/8)\ldots \tag{4.6}$$

That leads us to define $w_\varepsilon \in L^2(\mathbb{R})$ by

$$\hat{w}_\varepsilon(\xi) = m_{\varepsilon_1}(\xi/2)\,m_{\varepsilon_2}(\xi/4)\ldots m_{\varepsilon_j}(\xi/2^j)\ldots \tag{4.7}$$

when $\varepsilon = (\varepsilon_1, \varepsilon_2, \ldots)$, $\varepsilon_j \in \{0, 1\}$ and $\varepsilon_j = 0$ when j is large enough. These functions $w_\varepsilon(x)$ will be our *basic wavelet-packets* and our goal is to investigate their properties.

Another approach to basic wavelet-packets will be proposed in section 5 and the L^p-norms of these basic wavelet-packets will be estimated in section 6 and 7 when p is large.

5. DEFINITION OF WAVELET-PACKETS

We consider two sequences (u_k) and (v_k) satisfying one of the equivalent conditions in proposition 1. It will be assumed that there exists a multires-olution analysis (V_j) of $L^2(\mathbb{R})$ which is connected to this pair of quadrature mirror filters by (2.5), (2.7) and (3.2). But we do not need more specific information on the construction of φ and ψ.

The *basic wavelet-packets* $w_n(x)$, $n = 0, 1, 2, \ldots$ are defined by the following recursion

$$w_{2n}(x) = \sqrt{2} \sum_{-\infty}^{\infty} u_k w_n(2x + k) \tag{5.1}$$

coupled with

$$w_{2n+1}(x) = \sqrt{2} \sum_{-\infty}^{\infty} v_k w_n(2x + k) , \tag{5.2}$$

the function $w_0(x)$ belonging to $L^1(\mathbb{R})$ and being normalized by

$$\int_{-\infty}^{\infty} w_0(x) \, dx = 1 . \tag{5.3}$$

Let us start with $w_0(x)$. By (5.1), we have

$$w_0(x) = \sqrt{2} \sum_{-\infty}^{\infty} u_k w_0(2x + k) \tag{5.4}$$

and therefore

$$\hat{w}_0(2\xi) = m_0(\xi) \, \hat{w}_0(\xi) \tag{5.5}$$

$$\hat{w}_0(0) = 1 . \tag{5.6}$$

But the unique continuous function satisfying (5.5) and (5.6) is $\hat{\varphi}(\xi)$ and therefore $w_0(x) = \varphi(x)$.

We now turn to (5.2) with $m = 0$. We obtain $w_1(x) = \psi(x)$. We can proceed and (5.1) gives $w_2(x)$. Then (5.2) gives $w_3(x)$ and so on...

Let us modify the labelling of the basic wavelet-packets. They will be labelled by the denumerable set E of all sequences $\varepsilon = (\varepsilon_1, \varepsilon_2, \ldots)$ where $\varepsilon_j \in \{0, 1\}$ and $\varepsilon_j = 0$ eventually. Let $E_j \subset E$ be defined by $0 = \varepsilon_{j+1} = \varepsilon_{j+2} = \ldots$. Then $E_j \uparrow E$. Finally the new labelling is given by $w_n(x) = w_\varepsilon(x)$ when $n = \varepsilon_1 + 2\varepsilon_2 + \ldots + 2^{j-1}\varepsilon_j$.

Then the Fourier transform $\hat{w}_\varepsilon(\xi)$ of $w_\varepsilon(x)$ is given by

$$\hat{w}_\varepsilon(\xi) = m_{\varepsilon_1}(\xi/2) \, m_{\varepsilon_2}(\xi/4) \ldots m_{\varepsilon_j}(\xi/2^j) \, \hat{\varphi}(\xi/2^j). \tag{5.7}$$

Since $\hat{\varphi}(\xi) = m_0(\xi/2) \, m_0(\xi/4) \ldots$ and since $\varepsilon \in E_j$, (5.7) can be rewritten $\hat{w}_\varepsilon(\xi) = m_{\varepsilon_1}(\xi/2) \, m_{\varepsilon_2}(\xi/4) \ldots$ and the fact that ε belongs to E_j can be ignored.

Wavelet-packets provide new orthonormal bases as theorem 1 shows.

Theorem 1.   *For each j, $j = 0, 1, 2 \ldots$ the collection $w_\varepsilon(x - k)$, $\varepsilon \in E_j$, $k \in \mathbb{Z}$, is an orthonormal basis of V_j.*

Roughly speaking, theorem 1 means that the space V_j has been decoupled into 2^j orthogonal channels $W^{(\varepsilon)}$, $\varepsilon \in E_j$. Since the band width of W_j, as defined by (1.3), is of the order of magnitude of 2^j, it was natural to expect the bandwidth of each $w_\varepsilon(x)$, $\varepsilon \in E_j$, to be $0(1)$. One of the goals of this work is to disprove this conjecture.

For proving theorem 1, we return to the labelling $n = 0, 1, \ldots$ We want to prove that the collection

$$w_n(x - k) \quad , \quad 0 \le n < 2^j \quad , \quad k \in \mathbb{Z} \tag{5.8}$$

is an orthonormal basis of V_j.

When $j = 0$, we have $n = 0$, $w_0(x) = \varphi(x)$ and we know that $\varphi(x - k)$, $k \in \mathbb{Z}$, is an orthonormal basis of V_0. Let us assume that $w_n(x - k)$, $0 \le n < 2^{j-1}$, $k \in \mathbb{Z}$, is an orthonormal basis of V_{j-1}. Then (2.2) implies that $\sqrt{2}\, w_n(2x - k)$, $0 \le n < 2^{j-1}$, $k \in \mathbb{Z}$, is an orthonormal basis of V_j. But (5.1) and (5.2) can be rewritten

$$w_{2n}(x - k) = \sqrt{2} \sum_{-\infty}^{\infty} u_{2k-l}\, w_n(2x - l) \tag{5.9}$$

and

$$w_{2n+1}(x - k) = \sqrt{2} \sum_{-\infty}^{\infty} v_{2k-l}\, w_n(2x - l) \,. \tag{5.10}$$

This transformation is orthogonal since it has the same form as the one defined in (3.1). Therefore $w_n(x - k)$, $0 \le n < 2^j$, $k \in \mathbb{Z}$, is an orthonormal basis of V_j and theorem 1 is proved by induction on j.

Corollary.   *The collection $w_n(x - k)$, $n = 0, 1, 2, \ldots$, $k \in \mathbb{Z}$, is an orthonormal basis of $L^2(\mathbb{R})$.*

6. L^∞-NORMS OF WAVELET-PACKETS

Our goal is to study the frequency localization of the basic wavelet-packets $w_n(x)$. A convenient way for estimating this frequency localization is to compute

$$\sigma_n = \inf_{\xi_0 \in \mathbb{R}} \int_{-\infty}^{\infty} |\xi - \xi_0|^2\, |\hat{w}_n(\xi)|^2\, \frac{d\xi}{2\pi} \tag{6.1}$$

Since $\int_{-\infty}^{\infty} |\hat{w}_n(\xi)|^2 \, d\xi = 2\pi$, we have

$$\int_{-\infty}^{\infty} |\hat{w}_n(\xi)| \, d\xi \ \leq \ \pi \sqrt{2 + \sigma_n} \ . \tag{6.2}$$

But $\hat{w}_\varepsilon(\xi) = m_{\varepsilon_1}(\xi/2) \, m_{\varepsilon_2}(\xi/4) \ldots = e^{-i\lambda(\varepsilon)\xi} \, m_0(\xi/2 + \varepsilon_1\pi) \, m_0(\xi/4 + \varepsilon_2\pi) \ldots$ where $\lambda(\varepsilon) = \frac{\varepsilon_1}{2} + \frac{\varepsilon_2}{4} + \cdots$.

Since $0 \leq m_0(\xi) \leq 1$, we obtain

$$\|w_\varepsilon\|_\infty \ = \ w_\varepsilon(\lambda(\varepsilon)) \ = \ \frac{1}{2\pi} \int_{-\infty}^{\infty} |\hat{w}_\varepsilon(\xi)| \, d\xi \tag{6.3}$$

and therefore

$$\|w_n\|_\infty \ = \ \frac{1}{2} \sqrt{2 + \sigma_n} \ . \tag{6.4}$$

That means that the growth of $\|w_n\|_\infty$ as n tends to infinity gives a lower bound of the frequency localization. In a still unpublished work, E. Séré assumed that $m_0(\xi)$ is strictly increasing on $\left[-\frac{2\pi}{3}, -\frac{\pi}{3}\right]$ and satisfies the following condition

$$\sup_{\left\{-\frac{2\pi}{3} \leq \xi \leq -\frac{\pi}{3}\right\}} \left(m_0(\xi) + \left(\frac{\pi}{3} + \xi\right)m_0'(\xi)\right) \ = \ r \ < \ 2 \ . \tag{6.5}$$

One can construct examples of 2π-periodic C^∞ functions $m_0(\xi)$ satisfying these two conditions and the ones mentioned above : (4.4), $m_0(\xi) = 1$ on $[-\pi/3, \pi/3]$ and $0 \leq m_0(\xi) \leq 1$.

Defining $\operatorname{var}(\varepsilon)$ as $\sum_1^\infty |\varepsilon_{j+1} - \varepsilon_j|$, E. Séré proved the existence of two constants $\beta > \alpha > 1$, depending on $m_0(\xi)$, such that, for every $\varepsilon \in E$,

$$C_1 \alpha^{\operatorname{var}(\varepsilon)} \ \leq \ \|w_\varepsilon\|_\infty \ \leq \ c_2 \beta^{\operatorname{var}(\varepsilon)} \tag{6.6}$$

where $c_2 > c_1 > 0$ are two other constants.

Dropping (6.5), we want to prove a more general estimate.

Theorem 2. *Let us assume that* $m_0(\xi) = 1$ *on* $\left[-\frac{\pi}{3}, \frac{\pi}{3}\right]$, $m_0(-\xi) = m_0(\xi)$, $0 \leq m_0(\xi) \leq 1$, $m_0^2(\xi) + m_0^2(\xi + \pi) = 1$ *and lastly*

$$m_0(\xi) \quad \text{is decreasing on} \quad [0, \pi] \ . \tag{6.7}$$

Then we have (for $n \geq 1$)

$$\|w_n\|_\infty \ \leq \ Cn^{1/4} \ . \tag{6.8}$$

Moreover if $m_0(\xi) = 1$ *on* $\left[-\frac{\pi}{2} + \delta, \frac{\pi}{2} - \delta\right]$, $0 < \delta < \pi/2$, *we obtain*

$$\|w_n\|_\infty \ \leq \ Cn^{\gamma(\delta)} \tag{6.9}$$

where $\gamma(\delta)$ tends to 0 as δ tends to 0.

The two proofs are similar and we begin with (6.8).

We already know that, if $\varepsilon \in E_j$, we have

$$
\begin{aligned}
\|w_\varepsilon\|_\infty &= \frac{1}{2\pi} \int_{-\infty}^{\infty} m_0\left(\frac{\xi}{2} + \varepsilon_1\pi\right) \ldots m_0\left(\frac{\xi}{2^j} + \varepsilon_j\pi\right) \hat{\varphi}\left(\frac{\xi}{2^j}\right) d\xi \\
&= \frac{2^j}{2\pi} \int_{-\infty}^{\infty} m_0(\xi + \varepsilon_j\pi) \ldots m_0(2^{j-1}\xi + \varepsilon_1\pi) \, \hat{\varphi}(\xi) \, d\xi \\
&\leq \frac{\sqrt{2}}{2\pi} 2^j \int_{-\pi}^{\pi} m_0(\xi + \varepsilon_j\pi) \ldots m_0(2^{j-1}\xi + \varepsilon_1\pi) \, d\xi \\
&= \frac{\sqrt{2}}{2\pi} 2^j J(\varepsilon) .
\end{aligned}
$$

where $J(\varepsilon)$ denotes the integral above. This estimate follows from the fact that $\hat{\varphi}$ is compactly supported and m_0 is 2π-periodic.

Since $\hat{\varphi}(\xi) \geq \frac{1}{\sqrt{2}}$ on $[-\pi, \pi]$, we obtain the following two-sided estimate for $\varepsilon \in E_j$

$$
\frac{2^j}{2\pi\sqrt{2}} J(\varepsilon) \leq \|w_\varepsilon\|_\infty \leq \frac{\sqrt{2}}{2\pi} 2^j J(\varepsilon) . \tag{6.10}
$$

For estimating $J(\varepsilon)$, we apply the following observations (lemma 1).

Lemma 1. *If both $P(\xi)$ and $Q(\xi)$ are 2π-periodic and continuous functions of the real variable ξ, then, for any integer $q \geq 1$,*

$$
\begin{aligned}
I = \int_{-\pi}^{\pi} P(\xi) \, Q(2^q\xi) \, d\xi = \\
2^{-q} \int_{-\pi}^{\pi} \left[P(2^{-q}\xi) + \ldots + P(2^{-q}\xi + 2^{-q}(2^q - 1)2\pi) \right] Q(\xi) \, d\xi
\end{aligned}
$$

and therefore

$$
|I| \leq M\|Q\|_1 \tag{6.12}
$$

where $\|Q\|_1 = \int_{-\pi}^{\pi} |Q(\xi)| \, d\xi$ and

$$
M = 2^{-q} \sup_{0 \leq \xi \leq 2\pi} \left(|P(\xi)| + \ldots + |P(\xi + 2^{-q}(2^q - 1)2\pi)| \right) .
$$

This observation will be applied to $P(\xi) = m_0(\xi + \varepsilon_j\pi) \, m_0(2\xi + \varepsilon_{j-1}\pi)$, $Q(\xi) = m_0(\xi + \varepsilon_{j-2}\pi) \ldots m_0(2^{j-3}\xi + \varepsilon_1\pi)$ and $q = 2$. We check by brute force that $M = \frac{\sqrt{2}}{4}$. Therefore

$$\int_{-\pi}^{\pi} m_0(\xi + \varepsilon_j \pi) \, m_0(2\xi + \varepsilon_{j-1}\pi) \ldots m_0(2^{j-1}\xi + \varepsilon_1 \pi) \, d\xi$$

$$\leq \frac{\sqrt{2}}{4} \int_{-\pi}^{\pi} m_0(\xi + \varepsilon_{j-2}\pi) \ldots m_0(2^{j-2}\xi + \varepsilon_1 \pi) \, d\xi$$

and an obvious induction gives (6.8).

To prove (6.9) we first consider the limiting case where $m_0(\xi)$ is replaced by the characteristic function $\chi_0(\xi)$ of $[-\pi/2, \pi/2)$. Then $\chi_0(\xi + \varepsilon_0\pi) \ldots \chi_0(2^{q-1} + \varepsilon_{q-1}\pi)$, once restricted to $[-\pi, \pi)$, is either the characteristic function of $[-\pi 2^{-q}, \pi 2^{-q})$ or the characteristic function of the union $U(\varepsilon)$ of two intervals of length $\pi 2^{-q}$. If $0 < \delta < \pi 2^{-q-1}$ then the product $m_0(\xi + \varepsilon_0\pi) \ldots m_0(2^{q-1}\xi + \varepsilon_{q-1}\pi)$ defines a bump function which is supported by $U(\varepsilon) + [-2\delta, 2\delta]$. It follows that the mean values of our product on $\xi_0 + 2k\pi 2^{-q}$, $0 \leq k < 2^q$, do not exceed $5 \cdot 2^{-q}$.

Returning to our problem of estimating $J(\varepsilon)$, we write $mq \leq j < (m+1)q$ where q will be frozen and m tends to infinity. Then lemma 1 implies $J(\varepsilon) = J_m(\varepsilon) \leq 5 \cdot 2^{-q} J_{m-1}(\varepsilon)$ and an obvious iteration gives $J(\varepsilon) \leq C5^m \cdot 2^{-j}$. Finally, $\|w_\varepsilon\|_\infty \leq C5^m$.

An optimal choice of q is to pick the largest integer such that $\delta < \pi 2^{-q-1}$. Therefore $\gamma(\delta) = 0(\log 1/\delta)^{-1}$.

Theorem 2 is now completely proved. The estimate given by (6.8) is sharp since, in a way, the reverse inequality is true, as theorem 3 shows.

Theorem 3.   *The assumptions on $m_0(\xi)$ being the same as in theorem 2, there exists a constant $r > 1$ such that, for $j \geq 0$,*

$$2^{-j} \sum_{\varepsilon \in E_j} \|w_\varepsilon\|_\infty \geq \frac{1}{\sqrt{2}} r^j \, . \tag{6.13}$$

To prove (6.12), we return to (6.10) and are led to estimating

$$S_j = \sum_{\varepsilon \in E_j} \int_{-\pi}^{\pi} m_0(\xi + \varepsilon_1\pi) \ldots m_0(2^{j-1}\xi + \varepsilon_j \pi) \, d\xi$$

$$= \int_{-\pi}^{\pi} \sigma(\xi)\sigma(2\xi) \ldots \sigma(2^{j-1}\xi) \, d\xi$$

where

$$\sigma(\xi) = m_0(\xi) + m_0(\xi + \pi) \geq 1 \, .$$

But

$$\log \left\{ \frac{1}{2\pi} \int_{-\pi}^{\pi} \sigma(t)\sigma(2t)\ldots\sigma(2^{j-1}t)\, dt \right\}$$

$$\geq \frac{1}{2\pi} \int_{-\pi}^{\pi} \log \left\{ \sigma(t)\ldots\sigma(2^{j-1}t) \right\} dt$$

$$= \frac{j}{2\pi} \int_{-\pi}^{\pi} \log \sigma(t)\, dt = \beta j \qquad \text{where} \quad \beta > 0.$$

If $m_0(\xi) = 1$ on $\left[-\frac{\pi}{2} + \delta, \frac{\pi}{2} - \delta\right]$, then $\sigma(t) \geq 1$ on $\left[\frac{\pi}{2} - \delta, \frac{\pi}{2} + \delta\right]$ and in general β will be of the order of magnitude of $C\delta$. Finally the average lower bound of $\|w_n\|_\infty$ is $n^{\beta(\delta)}$ where $\beta(\delta)$ tends to 0 as δ tends to 0.

7. L^p-NORMS OF WAVELET-PACKETS

Theorem 4. *Let us keep the notations and assumptions of theorem 2 and theorem 3. Then there exists a $p_0 \geq 2$ such that, for each $p > p_0$, one can find a positive $\gamma = \gamma(p)$ with the property that*

$$\overline{\lim_{n \to +\infty}} \; n^{-\gamma} \left(\|w_1\|_p + \ldots + \|w_n\|_p \right) > 0. \tag{7.1}$$

In other words, the average growth of $\|w_n\|_p$ is $n^{\gamma(p)}$ where $\gamma(p) > 0$ when p is large. It means that $w_n(x)$ cannot be a product $u_n(x)v_n(x)$ between some highly oscillating bounded factor $u_n(x)$ and some envelope $v_n(x)$ which would keep a given shape with bounded sizes.

Theorem 4 easily follows from S. Bernstein's inequalities. When $\varepsilon \in E_j$, the Fourier transform $\hat{w}_\varepsilon(\xi)$ of $w_\varepsilon(x)$ is supported by the interval $|\xi| \leq \frac{8\pi}{3} 2^j$. Bernstein's inequality gives

$$\|w_\varepsilon\|_\infty \leq C 2^{j/p} \|w_\varepsilon\|_p \tag{7.2}$$

and since the average value of $\|w_\varepsilon\|_\infty$ is large, so is $\|w_\varepsilon\|_p$ as long as $2^{1/p} < r$ (and therefore p_0 tends to ∞ as δ tends to 0).

8. OTHER ORTHONORMAL BASES

Let us begin with the description of a rather general splitting scheme indexed by a dyadic tree. Let us fix two sequences (u_k) and (v_k) defining a pair of quadrature mirror filters, as in section 3. We start with a Hilbert space H equipped with a given orthonormal basis $(e_k)_{k \in \mathbb{Z}}$ and we split H accordingly to (3.1). Let us write $e_k^{(0)} = f_{2k}$ and $e_k^{(1)} = f_{2k+1}$. We now consider H_0 equipped with $e_k^{(0)}$, $k \in \mathbb{Z}$, and we go on splitting H_0 with the same sequences u_k and v_k. We obtain two new subspaces $H_{0,0}$ and $H_{0,1}$ equipped with the corresponding orthonormal bases $e_k^{(0,0)}$ and $e_k^{(0,1)}$ as defined by (3.1). Similarly H_1 is split into $H_{1,0} \oplus H_{1,1}$.

At the j-th step we have obtained 2^j subspaces H_α, $\alpha \in \{0,1\}^j$, of H. A very convenient notation will be to write $H_\alpha = H_I$ where I is the dyadic interval $\left[\frac{\alpha_1}{2} + \ldots + \frac{\alpha_j}{2^j} , \frac{\alpha_1}{2} + \ldots + \frac{\alpha_j}{2^j} + \frac{1}{2^j}\right)$ when $\alpha = (\alpha_1, \ldots, \alpha_j)$. This labelling has the following advantage. If I is a dyadic interval (contained in $[0,1)$) and if $I = I_1 \cup I_2 \cup \ldots \cup I_m$ is a partition of I by dyadic intervals, then

$$H_I = H_{I_1} \oplus \ldots \oplus H_{I_m} \tag{8.1}$$

the sum being direct and orthogonal.

Does this identity still hold when an infinite sequence I_m, $m = 1, 2, \ldots$ of dyadic intervals forms a partition of I? If this is true we can raise the more difficult problem where I, except for a null set, is covered by a sequence I_m of disjoint dyadic intervals. By null set we mean either a null set with respect to the Lebesgue measure or to some other measure adapted to the given quadrature mirror filters.

A first answer is given by the following theorem.

Theorem 5. *Let us assume that, except for a denumerable set, a dyadic interval I is covered by the union $\bigcup_1^\infty I_m$ of disjoint dyadic intervals I_m, $m = 1, 2, \ldots$.*

Let us also assume that $m_0(\theta) = \frac{1}{\sqrt{2}} \sum_{-\infty}^\infty u_k e^{ik\theta}$ satisfies the same hypothesis as in theorem 2. Then

$$H_I = H_{I_1} \oplus H_{I_2} \oplus \ldots \oplus H_{I_m} \oplus \ldots$$

where the sum is direct and orthogonal.

Before proving theorem 5, let us give an application. The basic wavelet-packets are $w_n(x-k)$, $n = 0, 1, 2, \ldots$, $k \in \mathbb{Z}$, and the *general* wavelet-packets will be defined as

$$2^{q/2} w_n(2^q x - k) \quad , \quad n \in \mathbb{N} , q \in \mathbb{Z} , k \in \mathbb{Z} . \tag{8.2}$$

This full collection clearly is an over-complete system in $L^2(\mathbb{R})$ and our goal will be to construct orthonormal bases of $L^2(\mathbb{R})$ with sub-collections of the form

$$2^{q/2} w_n(2^q x - k) \quad , \quad k \in \mathbb{Z} , (n,q) \in E .$$

An obvious solution is given by $q = 0$, $n = 0, 1, \ldots$ and an other one by $n = 1$ and $q \in \mathbb{Z}$.

To describe some other possibilities, let us associate the dyadic interval $I(n, q) = [2^q n, 2^q(n + 1))$ to each of the wavelet-packets $2^{q/2} w_n(2^q x - k)$, $k \in \mathbb{Z}$.

We then have

Theorem 6. *If a subset $E \subset \mathbb{N} \times \mathbb{Z}$ has the property that, except for a denumerable set, $[0, \infty)$ is covered by the disjoint union of the dyadic intervals $I(n, q)$, $(n, q) \in E$, then the corresponding wavelet-packets*

$$2^{q/2} w_n(2^q x - k) \quad , \quad k \in \mathbb{Z}, \ (n, q) \in E, \tag{8.3}$$

form an orthonormal basis of $L^2(\mathbb{R})$.

Theorem 6 can be deduced from theorem 5 if the following simple remark which has already been used is kept in mind. We first identify the abstract Hilbert space H equipped with an orthonormal basis $(e_k)_{k \in \mathbb{Z}}$ to the space V_N equipped with the orthonormal basis $2^{N/2} \varphi(2^N x - k)$, $k \in \mathbb{Z}$. We denote by E_N the subset of E defined by $2^{-N} I(n, q) \subset I = [0, 1)$. If the intervals I_m appearing in theorem 5 are precisely these $2^{-N} I(n, q)$, $(n, q) \in E_N$, then theorem 5 states that the collection

$$2^{q/2} w_n(2^q x - k) \quad , \quad k \in \mathbb{Z}, \ (n, q) \in E_N \tag{8.4}$$

is an orthonormal basis of V_N.

It suffices to let N tend to infinity to obtain theorem 6.

The proof of theorem 5. To prove theorem 5 as stated, it suffices to consider the case where $I = [0, 1)$ and $H_I = H$. We denote by $\pi_m : H \to H_{I_m}$ the orthogonal projector and we want to show that, for each $x \in H$, we have

$$\|w\|^2 = \sum_1^{\infty} \|\pi_m(x)\|^2 . \tag{8.5}$$

This relation will ensure that H is the closed linear span of the orthogonal subspaces H_{I_m}.

To prove (8.5), we consider a given $x \in H$ and without losing generality we can assume $\|x\| = 1$. If I is any dyadic subinterval of $[0, 1)$, we write $\omega(I) = \|\pi_I(x)\|^2$ where $\pi_I(x)$ is the orthogonal projection of x on H_I. This functional ω is finitely additive and we want to extend this property and prove

$$\omega(I) = \omega(I_1) + \omega(I_2) + \ldots + \omega(I_m) + \ldots \tag{8.6}$$

when, except for a denumerable set, I is the union $\bigcup_1^\infty I_m$ of the pairwise disjoint dyadic intervals I_m.

The following lemma 2 will immediately imply (8.5).

Lemma 2. *For $x \in H$, $\|x\| = 1$, there exists a continuous measure μ on $[0,1)$ such that, for every dyadic interval $I \subset [0,1)$,*

$$\|\pi_I(x)\|^2 = \int_I d\mu(t) . \tag{8.7}$$

To prove the existence of μ, it suffices to check the following continuity property of the additive set functional $\omega(I)$:

$$\begin{array}{l} \text{if} \quad I^{(1)} \supset I^{(2)} \supset \ldots \supset I^{(j)} \supset \ldots \quad \text{where the length} \\ \text{of} \quad I^{(j)} \quad \text{is} \quad 2^{-j} \;, \text{ then} \quad \omega(I^{(j)}) \quad \text{tends to} \quad 0 \;. \end{array} \tag{8.8}$$

We will show a more precise estimate.

Lemma 3. *For $x = \sum_{-N}^N \xi_k e_k$, $N = 1, 2, \ldots$ and $\|x\| = 1$, we have*

$$\|\pi_I(x)\| \leq CN^2 |I|^{1/4} \tag{8.9}$$

where C is an absolute constant and $|I|$ denotes the length of I.

If we admit lemma 3, (8.7) follows easily. Assuming $\|x\| = 1$, we write $x = \sum_{-\infty}^\infty \xi_k e_k$ and denote by x_N the finite sum $\sum_{-N}^N \xi_k e_k$. We then have $\|\pi_I(x)\| = \|\pi_I(x - x_n)\| + \|\pi_I(x_n)\| \leq \|x - x_N\| + CN^2 |I|^{1/4} \leq \varepsilon$ whenever $\|x - x_N\| \leq \varepsilon/2$ (which fixes N) and then $CN^2 |I|^{1/4} \leq \varepsilon/2$ (which gives $|I| \leq \eta(\varepsilon)$).

To prove lemma 3, we will use a specific realization of H and of the corresponding subspaces H_I. There are several (isometrically equivalent) such realizations and our choice will be dictated by convenience.

We consider the realization where H is $L^2\big([0, 2\pi], \frac{d\theta}{2\pi}\big)$ and where e_k becomes $e^{-ik\theta}$, $k \in \mathbb{Z}$. Then the vectors f_{2k} will be $\sqrt{2}\, m_0(\theta)\, e^{-2ik\theta}$ and f_{2k+1} will be $\sqrt{2}\, m_1(\theta)\, e^{-2ik\theta}$. We can proceed further and finally the orthonormal basis of H_I will be

$$2^{j/2} m_{\varepsilon_0}(\theta) \ldots m_{\varepsilon_{j-1}}(2^{j-1}\theta)\, e^{-i2^j k\theta}$$

when

$$I = \left[\frac{\varepsilon_0}{2} + \ldots + \frac{\varepsilon_{j-1}}{2^j} \;,\; \frac{\varepsilon_0}{2} + \ldots + \frac{\varepsilon_{j-1}}{2^j} + \frac{1}{2^j} \right) .$$

Finally our vector $x = \sum_{-N}^{N} \xi_k e_k$ will be a trigonometric polynomial f and

$$
\begin{aligned}
\|\pi_I(f)\|^2 &= 2^j \sum_k \left| \int_0^{2\pi} f(\theta) m_{\varepsilon_0}(\theta) \ldots m_{\varepsilon_{j-1}}(2^{j-1}\theta)\, e^{i2^j k\theta}\, \frac{d\theta}{2\pi} \right|^2 \\
&= 2^j \sum_k |\lambda(j,k)|^2
\end{aligned}
\tag{8.10}
$$

We need exactly the same estimate as the one in theorem 2:

$$
\int_0^{2\pi} |m_{\varepsilon_1}(\theta) \ldots m_{\varepsilon_j}(2^{j-1}\theta)|\, d\theta \;\leq\; C 2^{-j} 2^{j/4}\,.
\tag{8.11}
$$

As we know this estimate is not optimal and the factor $2^{j/4}$ can be replaced by $2^{\gamma(\delta)j}$ depending on the properties of $m_0(\theta)$.

If $k = 0$, $|\lambda(j,0)| \leq C\|f\|_\infty\, 2^{-j}\, 2^{j/4}$ which is the required bound.

When $k \neq 0$, we integrate by parts and rewrite

$$
\begin{aligned}
\lambda(j,k) =\ & \\
& -i(k2^j)^{-1} \int_0^{2\pi} \frac{d}{d\theta} \left\{ f(\theta) m_{\varepsilon_0}(\theta) \ldots m_{\varepsilon_{j-1}}(2^{j-1}\theta) \right\} e^{i2^j k\theta}\, \frac{d\theta}{2\pi}\,.
\end{aligned}
$$

The term where $f(\theta)$ is differentiated is treated as above. The term where $m_{\varepsilon_q}(2^q\theta)$ is differentiated ($0 \leq q < j$) is bounded by $C|k|^{-1} 2^{q-j} \lambda(q,j,k)$ where

$$
\begin{aligned}
\lambda(q,j,k) =\ & \\
& \int_0^{2\pi} \left| m_{\varepsilon_0}(\theta) \ldots m_{\varepsilon_{q-1}}(2^{q-1}\theta) \right| \left| m_{\varepsilon_{q+1}}(2^{q+1}\theta) \ldots m_{\varepsilon_{j-1}}(2^{j-1}\theta) \right|\, d\theta\,.
\end{aligned}
$$

To estimate this integral, we return to the argument used in theorem 2. We group the first q terms (or $q-1$ terms if q is odd) by pairs and apply lemma 1 inductively. We gain a factor $(\sqrt{2}/4)^{(q-1)/2}$. We then repeat this treatment on the second half of the integrand, starting from $m_{\varepsilon_{q+1}}(2^{q+1}\theta)$. We obtain a factor $(\sqrt{2}/4)^{(j-q-1)/2}$. All together, we have obtained $C(\sqrt{2}/4)^{j/2}$ and the sum over q gives $\frac{C}{|k|}(\sqrt{2}/4)^{j/2}$. Finally the l^2 norm of this sequence is $C'(\sqrt{2}/4)^{j/2}$ as announced.

Theorem 5 is completely proved.

It would be interesting to know whether all the measures $\mu = \mu_x$, $x \in H$, are absolutely continuous with respect to the Lebesgue measure on $[0,1)$. In that case, theorem 5 could be extended to the situation where except for a null set, $[0,1)$ is covered by $\bigcup_1^\infty I_m$.

The study of these measures μ_x can be simplified by the following remark.

Lemma 4.

$$\|\mu_x - \mu_y\| \leq \|x - y\|(\|x\| + \|y\|) .$$

For proving this estimate, it suffices to write

$$\|\mu_x - \mu_y\| = \lim_{m \uparrow +\infty} \sum_{|I|=2^{-m}} |\mu_x(I) - \mu_x(I)|$$

where the sum runs over all dyadic intervals $I \subset [0,1)$ with length 2^{-m}. But

$$|\mu_x(I) - \mu_y(I)| = \left| \|\pi_I(x)\|^2 - \|\pi_I(y)\|^2 \right|$$
$$= \left| \|\pi_I(x)\| - \|\pi_I(y)\| \right| \left(\|\pi_I(x)\| - \|\pi_I(y)\| \right)$$
$$\leq \|\pi_I(x)\| \, \|\pi_I(x-y)\| + \|\pi_I(y)\| \, \|\pi_I(x-y)\| .$$

The two terms are similar and the Cauchy-Schwarz inequality applied to

$$\sum_{|I|=2^{-m}} \|\pi_I(x)\| \, \|\pi_I(x-y)\|$$

gives

$$\left(\sum_{|I|=2^{-m}} \|\pi_I(x)\|^2 \right)^{1/2} \left(\sum_{|I|=2^{-m}} \|\pi_I(x-y)\|^2 \right)^{1/2} = \|x\| \, \|x-y\| .$$

9. CONCLUSION

To conclude, we describe an example showing that theorem 5 or theorem 6 do not give the final answer to the problem which has been raised. More precisely we show that some Cantor sets may play the role of the exceptional denumerable set. We denote by $K \subset [0,1]$ a symmetric Cantor set with dissection ratio $1/4$. It means that K can be covered by 2^j intervals of length 4^{-j} and this will be the only property we shall use. Consider the open intervals (a_m, b_m) which are the components of $[0,1] \setminus K$ and assume that $I_m = [a_m, b_m]$ in theorem 5 or 6. Returning to theorem 2, let us assume that $\|w_n\|_\infty \leq Cn^\gamma$ with $0 < \gamma < 1/4$. We know that this can be achieved. Then the reasoning which was used for theorem 5 gives $\mu(I) \leq C2^{-(1-2\gamma)j}$ for each dyadic interval I contained in $[0,1)$ with length 2^{-j}. Finally $\mu(K) = 0$ and the conclusions of theorem 5 or 6 are valid.

This example shows that there exist some wavelet-packet orthonormal bases far beyond the ones described in theorem 6. It also shows that this fact is related to the slow growth of $\|w_n\|_\infty$ is announced in the introduction.

REFERENCES

[1] R. Coifman, *Adaptive multiresolution analysis, computation, signal processing and operator theory,* ICM 90 (Kyoto).

[2] R. Coifman, Y. Meyer, S. Quake and V. Wickerhauser, *Signal Processing and Compression with wavelet packets,* Numerical Algorithms Research Group, Yale University (1990).

[3] Y. Meyer, *Wavelets and applications,* ICM 90 (Kyoto).

[4] Y. Meyer, *Ondelettes,* Hermann (1990).

INDEX